Mathematics for Computer Scientists

Peter Hartmann

Mathematics for Computer Scientists

A Practice-Oriented Approach

Peter Hartmann
Informatik, Hochschule Landshut
Landshut, Bayern, Germany

ISBN 978-3-658-40422-2   ISBN 978-3-658-40423-9   (eBook)
https://doi.org/10.1007/978-3-658-40423-9

This Springer imprint is published by the registered company Springer Fachmedien Wiesbaden GmbH, part of Springer Nature.
The registered company address is: Abraham-Lincoln-Str. 46, 65189 Wiesbaden, Germany

Preface

Mathematics is an essential root of computer science: In all its areas, mathematical methods are used again and again. Mathematical thinking is typical work for computer scientists. Up until now, I believe too little attention has been given to the close interlocking of the two disciplines.

This textbook contains in one volume the essential areas of mathematics needed to understand computer science. In addition, I constantly present concrete applications of mathematical techniques for computer science. For example, logic is used to test programs, methods of linear algebra are used in robotics and in graphical data processing. With eigenvectors, the importance of nodes in networks can be assessed. The theory of algebraic structures turns out to be useful for hashing, in cryptography and for data security. Differential calculus is used to compute interpolation curves, the Fourier transform plays an important role in data compression. Queueing theory and principal component analysis are important applications of statistics. Throughout the book connections between mathematics and computer science are presented.

In the textbook presentation, it is not only about the results, but also the practice of mathematical thinking in solving problems. Computer scientists require the same analytical approach to tasks as mathematicians.

The textbook is primarily intended to supplement the mathematics lectures of computer science students. The presentation is a practice-oriented approach and each lesson explains how you can apply what you have learned by giving you many real world examples, and by constantly cross-referencing math and computer science. I put a lot of emphasis on the motivation of the results, the derivations are detailed, and the book is therefore well suited for self-study. Practitioners who are looking for the mathematics underlying their applications can use it for reference. However, no matter how you use the textbook, keep in mind just as you cannot learn a programming language by reading the syntax, it is impossible to understand mathematics without working on problems with paper and pencil.

The three parts of the book cover discrete mathematics and linear algebra, analysis including numerical methods, and finally the basics of probability theory and statistics. Within each chapter, the definitions and theorems are consecutively numbered, a second consecutive numbering in round brackets is given to formulas to which reference is later made.

At the beginning of each chapter I summarize the learning objectives. At the end of each chapter you will find comprehension questions and exercises. The comprehension questions serve to quickly check understanding. After working through a chapter you should be able to answer them without much calculation. The exercises are intended to deepen and practice the presented material. Most of them can be solved on paper, some are intended for programming or need a computer algebra system. For this I use the open source tool SageMath (www.sagemath.org). The exercises are mostly not too difficult; if you have completed the corresponding chapter, you should be able to solve most of the associated problems. In each chapter, you will find a weblink to the answers for the comprehension questions and to suggested solutions to the end of chapter exercises.

This textbook grew out of lectures given at Landshut University of Applied Sciences in the Computer Science Department over many years. I have repeatedly given sections of the book to my colleagues at Landshut University for critical review. I would like to thank them, especially Ludwig Griebl, who read large parts carefully and helped me with many tips. I would also like to thank my family very much: Renate and Felix have created the freedom for me that I needed to be able to carry out this project.

Finally, I would like to thank all the students who, through their work, their questions and discussions in many courses over the past years, have significantly influenced the content of this book.

Learning works best when it's fun. This book should help you see mathematics in your studies as more than a necessary evil. I hope you will find mathematics is useful to you and can even sometimes be fun.

This English edition is the translation of the seventh German edition. I have received many helpful comments on all editions. Many thanks! With all this feedback, I was able to revise and expand the book in each edition.

Please continue to help me and all readers to further improve the quality of the book through your feedback. The manuscript was translated using artificial intelligence and then revised by me. Please also write to me if you find parts that may have been translated incorrectly. My email is peter.hartmann@haw-landshut.de — I will respond! In addition I would like to know which computer science references are missing for you, and also what you consider superfluous. Above all, let me know if you have found any errors. Thank you!

Landshut   Peter Hartmann
May 2023

Contents

Part I
Discrete Mathematics and Linear Algebra

Sets and Mappings

1

Abstract

Sets, operations on sets and map pings between sets are part of the language of mathematics. Therefore, we begin the book with them. If you have worked through this first chapter, you will know

- the concept of a set and set operations,
- relations, in particular equivalence relations and order relations,
- important properties of mappings,
- the concept of the power set of finite and infinite sets, and
- you have learned how to carry out simple proofs.

1.1 Set Theory

What is a set actually? The following definition of a set is from Georg Cantor (1845–1918), the founder of modern set theory:

▶ **Definition 1.1** A set is a collection of definite well-differentiated objects of our perception or our thinking into a whole.

Do you already feel quite uncomfortable with this first definition? After all, mathematics claims exactness and precision; but this definition does not sound like precise mathematics at all: What is a "collection", what are "well-differentiated objects"? As we will see, all

Supplementary Information The online version contains supplementary material available at https://doi.org/10.1007/978-3-658-40423-9_1.

areas of mathematics need solid foundations on which to build. Finding these foundations is usually very difficult (just as it is difficult to formulate the requirements in a software project precisely). Cantor tried to put the set theory, which had been used more or less intuitively until then, on solid mathematical feet and formulated the above definition for this purpose. But this soon led to unexpected difficulties. Perhaps you know the story of the barber in a village who says of himself that he shaves all the men in the village who do not shave themselves. Does he shave himself or not? A similar problem arises when one considers the set of all sets that do not contain themselves as an element. Does the set contain itself or not? The guild of mathematicians has since extricated itself from this swamp of paradoxes, but the theory has not become any simpler as a result. Fortunately, as computer scientists, we do not have to deal with such problems; we can simply use the concept of a set and make a wide detour around things like "the set of all sets".

If an object x belongs to the set M, we write $x \in M$ (x is an element of M), if an object y does not belong to M, we write $y \notin M$. Two sets M, N are called equal if they contain the same elements. We then write: $M = N$. If M and N are not equal, we write $N \neq M$.

How do you describe sets?

On the one hand, you can list all the elements in curly brackets: For the set containing the numbers 1, 4, 8, 9, we write, for example $\{1, 4, 8, 9\}$ or $\{8, 4, 9, 1\}$. The order of the elements does not matter here. Even if an element occurs several times in the brackets, we are not bothered by it: it only counts once.

On the other hand, one can define a set by specifying characteristic properties of its elements:

$$\mathbb{N} := \{x \mid x \text{ is a natural number}\} = \{1, 2, 3, 4, \ldots\}$$

":=" is read as "is defined by" and thus differs from the "=" sign. We often use the ":=" to make clear that a new expression is on the left, which has not been defined before, and which can now be used as a synonym for the expression on the right. In this respect, the ":=" is comparable to the assignment operator "=" in C++ and Java, while our "=" is more like the comparison operator "==". I will sometimes omit the ":" when the meaning is clear. (You can't afford that when programming!)

The vertical bar is read as "with the property".

The second way to write the set $\mathbb{N}$ ($\mathbb{N} = \{1, 2, 3, 4, \ldots\}$) shows I will sometimes be lax if it is clear what the points mean. So in this book the natural numbers start with 1. In the literature, the 0 is sometimes also counted among the natural numbers, but the representation is not uniform. There is agreement, however, in the terms: $\mathbb{N}$ for $\{1, 2, 3, \ldots\}$ and $\mathbb{N}_0$ for $\{0, 1, 2, \ldots\}$.

A few more examples of sets with which we will often have to do are the integers $\mathbb{Z}$, the rational numbers $\mathbb{Q}$ and the real numbers $\mathbb{R}$:

$$\mathbb{Z} := \{x \mid x \text{ is an integer}\} = \{0, +1, -1, +2, -2, \ldots\}$$
$$= \{y \mid y = 0 \text{ or } y \in \mathbb{N} \text{ or } -y \in \mathbb{N}\}.$$

$$\mathbb{Q} := \{x \mid x \text{ is a fraction}\} := \left\{x \;\middle|\; x = \frac{p}{q}, p \in \mathbb{Z}, q \in \mathbb{N}\right\}.$$

Here you have to take a moment to see you catch all the fractions with the definition on the right. Is $\frac{7}{-4}$ in the set? Yes, if you know the rules of fractions, because $\frac{7}{-4} = \frac{-7}{4}$. Here elements also occur several times: According to the rules of fractions, for example, $\frac{4}{5} = \frac{8}{10}$.

And why don't we just write $p \in \mathbb{Z}, q \in \mathbb{Z}$ in the definition? Every time a mathematician sees a fraction, an alarm goes off and he checks whether the denominator is not equal to 0. $q \in \mathbb{Z}$ would also include the case $q = 0$. You can't divide by 0! In Chap. 5 on algebraic structures, we will see why this does not work.

Without knowing exactly what a decimal number is, we call $\mathbb{R} := \{x \mid x \text{ is a decimal number}\}$, the real numbers and calculate with these numbers as you have learned in school. The more precise characterization of $\mathbb{R}$ we postpone to the second part of the book (compare Chap. 12). In addition to the elements of $\mathbb{Q}$, the set $\mathbb{R}$ also contains, for example, all roots and numbers like e and π.

The set $M = \{x \mid x \in \mathbb{N} \text{ und } x < 0\}$ does not contain any element. M is called the empty set and is denoted by $\emptyset$.

Sets are again objects, that is, sets can be combined to sets again, and sets can contain sets as elements. Here are two strange sets:

$$M := \{\mathbb{N}, \mathbb{Z}, \mathbb{Q}, \mathbb{R}\}, \quad N := \{\emptyset, 1, \{1\}\}.$$

How many elements do these sets have? M does not have infinitely many, but exactly four elements! N has three elements: the empty set, "1" and the set which contains the 1.

If a set consists of a single element ($M = \{m\}$), one must carefully distinguish between m and $\{m\}$. These are two different objects!

Relationships between Sets

A set M is called a *subset* of the set N ($M \subset N$, $N \supset M$), if every element of M is also an element of N. We also say: M is contained in N or N is a superset of M. Relationships between sets are often represented graphically by bubbles (the *Venn diagrams*, Fig. 1.1).

If M is not a subset of N, then $M \not\subset N$ is written. The possibilities are shown in Fig. 1.2.

Caution: $M \not\subset N$ does not necessarily imply $N \subset M$!

If $M \subset N$ and $M \neq N$, then M is called a proper subset of N ($M \subsetneq N$).

If $M \subset N$ and $N \subset M$, then $M = N$ applies. This property is often used to prove the equality of sets. Here, the task to show $M = N$ is decomposed into the two simpler subtasks $M \subset N$ and $N \subset M$.

For all sets M, $M \subset M$ and also $\emptyset \subset M$ apply, because each element of the empty set is contained in M.

You don't believe that? Tell me an element of the empty set that is not in M! See, there is none!

Fig. 1.1 $M \subset N$

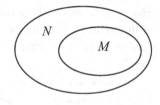

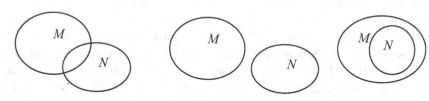

Fig. 1.2 $M \not\subset N$

▶ **Definition 1.2** If M is a set, then the set of all subsets of M is called the *power set* of M. It is denoted by $P(M)$.

Examples

$$M = \{1, 2\}, \qquad P(M) = \{\emptyset, \{1\}, \{2\}, \{1, 2\}\},$$
$$M = \{a, b, c\}, \quad P(M) = \{\emptyset, \{a\}, \{b\}, \{c\}, \{a, b\}, \{a, c\}, \{b, c\}, \{a, b, c\}\}.$$

$P(\mathbb{R})$ is the set of all subsets of $\mathbb{R}$. For example, since $\mathbb{N} \subset \mathbb{R}$, $\mathbb{N} \in P(\mathbb{R})$ is true. ◀

Operations with Sets

The *intersection* of two sets M and N is the set of elements that are contained in both M and N, see Fig. 1.3.

$$M \cap N := \{x \mid x \in M \text{ and } x \in N\} \tag{1.1}$$

M and N are called *disjoint*, if $M \cap N = \emptyset$.

Fig. 1.3 $M \cap N$

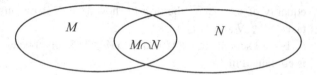

$$M = \{1,3,5\}, \qquad N = \{2,3,5\}, \qquad S = \{5,7,8\}.$$
$$M \cap N = \{3,5\}, \qquad N \cap S = \{5\}, \qquad M \cap N \cap S = \{5\}. \quad \blacktriangleleft$$

Wait! Does something strike you about the last expression? What is $M \cap N \cap S$? We haven't defined that yet! Maybe we are implicitly assuming that we are forming $(M \cap N) \cap S$. That's possible, it is a double application of the definition in (1.1). Or are we maybe forming $M \cap (N \cap S)$? Is it the same thing? As long as we don't know that, we really ought to put parentheses around it.

The *union* of two sets M and N is the set of elements contained in M or N, see Fig. 1.4. It is also allowed for an element to be contained in both sets. Mathematicians and computer scientists don't use the word "or" as exclusive or—that would be "either or".

$$M \cup N := \{x \mid x \in M \text{ or } x \in N\}$$

In the example from above, $M \cup N = \{1,2,3,5\}$.

The *difference set* $M \setminus N$ or $M - N$ of two sets M, N is the set of elements contained in M, but not in N, see Fig. 1.5.

$$M \setminus N := \{x \mid x \in M \text{ and } x \notin N\}$$

If $N \subset M$, then $M \setminus N$ is also called the *complement* of N in M. The complement is denoted by $\overline{N}^M$.

Often one calculates with subsets of a certain fixed set, the universe (for example, with subsets of $\mathbb{R}$). If the universe is clear from the context, the complement is always formed with respect to it, without mentioning it explicitly. Then we simply write $\overline{N}$ for the complement. (Occasionally one also reads $\complement N$.)

Fig. 1.4 $M \cup N$

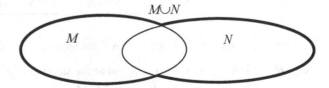

Fig. 1.5 $M \setminus N$

Calculation Rules for Set Operations

▶   **Theorem 1.3** Let M, N, S be sets. Then:

$$M \cup N = N \cup M$$
$$M \cap N = N \cap M$$
$\left.\right\}$ commutative laws

$$(M \cup N) \cup S = M \cup (N \cup S)$$
$$(M \cap N) \cap S = M \cap (N \cap S)$$
$\left.\right\}$ associative laws

$$M \cap (N \cup S) = (M \cap N) \cup (M \cap S)$$
$$M \cup (N \cap S) = (M \cup N) \cap (M \cup S)$$
$\left.\right\}$ distributive laws

For every set M it holds: $M \cup \emptyset = M, M \cap \emptyset = \emptyset, M \setminus \emptyset = M$.

Because of the associative law, we can now define:

$$M \cup N \cup S := (M \cup N) \cup S,$$

and, in the future, we are allowed to omit the brackets; the union of three sets is independent of the order in which it is carried out. The same holds for the intersection and also for the intersection and union of more than three sets.

> Try to write down, for four sets, how much work is saved due to the fact that we are allowed to omit brackets.

From school you know the distributive law for real numbers: It is

$$a(b + c) = ab + ac.$$

Surprisingly, the corresponding rule for sets also applies if the signs of arithmetic are interchanged—with real numbers this is false!

You shouldn't believe everything just because it's in a book; now it's time for our first proof. We will prove the first distributive law in detail:

show: $\underbrace{S \cap (M \cup N)}_{A} = \underbrace{(S \cap M) \cup (S \cap N)}_{B}$.

We prove $A = B$ by $A \subset B$ and $B \subset A$. This splits the problem into two sub-steps.

"$A \subset B$":   Let $x \in A$. Then $x \in S$ and at the same time $x \in M$ or $x \in N$.
If $x \in M$ is true, then $x \in S \cap M$ and thus $x \in B$.
If $x \in N$ is true, then $x \in S \cap N$ and thus $x \in B$,
so always $x \in B$.

"$B \subset A$":   Let $x \in B$. Then $x \in S \cap M$ or $x \in S \cap N$.
If $x \in S \cap M$ is true, then $x \in S$ and $x \in M$, thus also $x \in M \cup N$ and this
implies $x \in S \cap (M \cup N)$.

If $x \in S \cap N$ is true, then $x \in S$ and $x \in N$, thus also $x \in M \cup N$ and therefore $x \in S \cap (M \cup N)$,

so in any case $x \in A$. □

With the building block □ I will always mark the end of a proof: This way we have created something we can build on.

> Prove the second distributive law yourself. Make a similar case distinction and don't be surprised if it takes you some time. It is always harder to come up with something than to take something in. But trying it out for yourself is immensely important for true understanding!

Here are a few more rules for complement formation:

▶ **Theorem 1.4** For calculations in a universe G the following rules apply:

$$M \setminus N = M \cap \overline{N}$$

$$\overline{M \cup N} = \overline{M} \cap \overline{N}$$

$$\overline{M \cap N} = \overline{M} \cup \overline{N}$$

$$\overline{\overline{M}} = M$$

$$M \subset N \text{ implies } \overline{N} \subset \overline{M}.$$

So building the complement reverses the operation sign. You should at least partially check these rules by yourself.

Occasionally, we will also need infinite intersections and infinite unions:

▶ **Definition 1.5** Let M be a set of indices (for example $M = \mathbb{N}$). For all $n \in M$ let a set A_n be given. Then:

$$\bigcup_{n \in M} A_n := \{x \mid x \in A_n \text{ for at least one } n \in M\},$$

$$\bigcap_{n \in M} A_n := \{x \mid x \in A_n \text{ for all } n \in M\}.$$

If M is a finite set, for example $M = \{1, 2, 3, \ldots, k\}$, then these definitions agree with our previous definition of union and intersection:

$$\bigcup_{n \in M} A_n = A_1 \cup A_2 \cup \ldots \cup A_k,$$

$$\bigcap_{n \in M} A_n = A_1 \cap A_2 \cap \ldots \cap A_k.$$

In this case, we write $\bigcup_{i=1}^{k} A_i$ for the union and $\bigcap_{i=1}^{k} A_i$ for the intersection.

Example

$M = \mathbb{N}$, $A_n := \{x \in \mathbb{R} \mid 0 < x < \frac{1}{n}\}$. Then $\bigcap_{n \in \mathbb{N}} A_n = \emptyset$. ◀

This is strange: Every finite intersection contains elements: $\bigcap_{n=1}^{k} A_n = A_k$, but one cannot find a real number that is contained in all A_n, so the infinite intersection is empty.

The Cartesian Product of Sets

▶   **Definition 1.6** Let M, N be sets. Then the set

$$M \times N := \{(x, y) \mid x \in M, y \in N\}$$

of all ordered pairs (x, y) with $x \in M$ and $y \in N$ is called the *Cartesian product* of M and N.

Ordered means that, for example, $(5, 3)$ and $(3, 5)$ are different elements (in contrast to $\{5, 3\} = \{3, 5\}$!).

Examples

1. $M = \{1, 2\}$, $N = \{a, b, c\}$.

$$M \times N = \{(1, a), (1, b), (1, c), (2, a), (2, b), (2, c)\},$$
$$N \times M = \{(a, 1), (a, 2), (b, 1), (b, 2), (c, 1), (c, 2)\}.$$

   So, in general, $M \times N \neq N \times M$.
2. $M \times \emptyset = \emptyset$ for all sets M.
3. $\mathbb{R}^2 := \mathbb{R} \times \mathbb{R} = \{(x, y) \mid x, y \in \mathbb{R}\}$.
   These are the Cartesian coordinates that you are probably familiar with. The points of the $\mathbb{R}^2$ can be drawn in the Cartesian coordinate system by drawing two perpendicular axes, each representing the real number line (Fig. 1.6).
   Analogously, we can define the n-fold Cartesian product of sets $M_1, M_2, \ldots, M_n$ (the set of ordered n-tuples):

$$M_1 \times M_2 \times M_3 \times \ldots \times M_n = \{(m_1, m_2, m_3, \ldots, m_n) \mid m_i \in M_i\}.$$

   If all sets are equal, we abbreviate: $M^n := \underbrace{M \times M \times M \times \ldots \times M}_{n\text{-times}}$.

In the Cartesian product, we can again consider (as with all sets) subsets, intersections, unions, and so on:

4. $P := \{(x, y) \in \mathbb{R}^2 \mid y = x^2\} = \{(x, x^2) \mid x \in \mathbb{R}\}$
   This set forms a parabola in the $\mathbb{R}^2$.
5. $R_{\leq} := \{(x, y) \in \mathbb{R}^2 \mid x \leq y\}$
   This is the half-plane shown in Fig. 1.7. ◀

Fig. 1.6 Cartesian coordinates

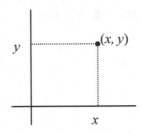

Fig. 1.7 The relation $R_\le$

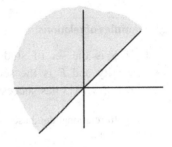

1.2 Relations

▶ **Definition 1.7** Let M and N be sets, $R \subset M \times N$. Then R is called a *relation* on $M \times N$. If $M = N$, we simply say R is a relation on M.

A relation on $M \times N$ describes a relationship between elements of the sets M and N, such as $x > y$, $x \ne y$. If such a relationship can only take on the values "true" or "false", it can be completely described by a subset of $M \times N$: This subset contains exactly the elements (x, y) for which the relationship is true.

We just saw an example of such a relation (No. 5): The half-plane $R_\le$ precisely describes the relationship "$\le$", it represents a relation on $\mathbb{R}$.

Instead of $(x, y) \in R$, one usually writes xRy. In Example 5 this means:

$$x \le y \text{ if and only if } xR_\le y.$$

Let's go one step further and simply designate the set $R_\le$ with "$\le$" (why should we always designate sets with letters?), then from now on we can read $x \le y$ as "(x, y) is an element of $\le$", that is "x is less than or equal to y". So apparently nothing new. Why are we doing all this at all? On the one hand, the definition should make it clear any two-way relationship between elements determines a set of tuples (pairs) for which this relationship is fulfilled. On the other hand, our new concept of relations is much more general than what we could express so far, for example, with $\le$. The set shown in Fig. 1.8 also represents a relation.

Abb. 1.8 A relation

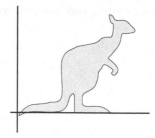

Examples of relations

1. If S is the set of students of the Department of Computer Science of a university and F is the set of courses offered in the degree program, then the set $B = \{(s,f) \mid s$ has successfully attended module $f\}$ is a relation on $S \times F$.

   In relational databases, data sets are characterized by the relations that exist between them.

2. $R := \{(x,y) \in \mathbb{R}^2 \mid (x,y)$ lies below the main diagonal$\}$ just describes the relation ">" (Fig. 1.9). If the main diagonal belongs to R, then R determines the relation "$\geq$".
3. All comparison operators $<, >, \leq, \geq, =, \neq$ form relations on $\mathbb{R}$.
4. "$\subset$" is a relation between subsets of $\mathbb{R}$. Thus, the relation $R_\subset$ is a subset of the cartesian product of all subsets of $\mathbb{R}$:

$$R_\subset = \{(A, B) \in P(\mathbb{R}) \times P(\mathbb{R}) \mid A \subset B\}.$$

Take a moment and try to really understand this line. It already contains a considerable degree of abstraction. But there is nothing mysterious about it, it just uses the known definitions quite mechanically, step by step. It is very important to learn how to work out results step by step. In the end, there is a result that you know is correct, even if one can no longer grasp the correctness at a glance. ◀

We now focus on the special case $M = N$, i.e. relations of a set M. Here are some important properties of relations:

Fig. 1.9 The relation $R_>$

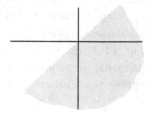

▶   **Definition 1.8** R is a relation on the set M. R is called

reflexive,   if for all $x \in M$ it holds: xRx.
symmetric, if for all $x, y \in M$ with xRy it holds: yRx.
transitive, if for all $x, y, z \in M$ with xRy and yRz it holds: xRz.

Examples thereof on the set $\mathbb{R}$ of real numbers:

Examples

$<$ is not reflexive, since $x < x$ does not hold.
is not symmetric, since from $x < y$ it does not follow that $y < x$.
is transitive, since from $x < y$ and $y < z$ it follows that $x < z$.

$\leq$ is reflexive, since $x \leq x$ holds.
is not symmetric, but
transitive (just like "$<$").

$>, \geq$ have the same properties as $<, \leq$.

$\neq$ is not reflexive, since $x \neq x$ does not hold.
is symmetric, since from $x \neq y$ it follows that $y \neq x$.
is not transitive, since from $x \neq y$ and $y \neq z$ it does not follow that $x \neq z$ (it
could be $x = z$).

$=$ is reflexive, since $x = x$,
symmetric, since from $x = y$ it follows that $y = x$,
transitive, since from $x = y$ and $y = z$ it follows that $x = z$. ◀

Equivalence Relations

▶   **Definition 1.9** A relation is called an *equivalence relation*, if it is reflexive, symmetric and transitive.

First of all an

Example

$$R_5 := \{(m, n) \in \mathbb{Z} \times \mathbb{Z} \mid m - n \text{ is divisible by 5}\} \subset \mathbb{Z} \times \mathbb{Z}. \qquad (1.2)$$

Here $n \in \mathbb{Z}$ is divisible by 5, means there is a $k \in \mathbb{Z}$ with $5k = n$.

R_5 is an equivalence relation:

nR_5n, because $n - n = 0$ is divisible by 5. So R_5 is reflexive.

If nR_5m, then $n - m = 5k$ for a $k \in \mathbb{Z}$. Then $m - n = 5(-k)$, thus m - n is also divisible by 5. So R_5 is symmetric.

Let nR_5m and mR_5s, thus $n - m = 5k$, $m - s = 5l$. Then $(n - m) + (m - s) = n - s = 5k + 5l = 5(k + l)$, thus $n - s$ is divisible by 5 and therefore nR_5s.

We will deal intensively with this equivalence relation (and others with different divisors than 5) later. ◄

A special property of equivalence relations is they decompose the underlying set into disjoint subsets, the so-called equivalence classes:

▶   **Definition 1.10** Let R be an equivalence relation on M and $a \in M$. Then the set

$$[a] := \{x \in M \mid xRa\}$$

is called *equivalence class* of a. The elements of $[a]$ are called the elements which are equivalent to a.

These are all elements in relation to a.

Examples

1. The hairdresser is interested in the equivalence relation: "has the same hair color as".
2. Let R be "=" on $\mathbb{R}$. Then $[a] = \{x \in \mathbb{R} \mid x = a\} = \{a\}$.
3. Let $R_5 \subset \mathbb{Z} \times \mathbb{Z}$ be defined as in (1.2). Then, for example:
   $[1] = \{n \in \mathbb{Z} \mid n - 1 \text{ is divisible by } 5\} = \{1, 6, 11, 16, \ldots, -4, -9, -14, \ldots\}$,
   $[0] = \{0, 5, 10, 15, \ldots, -5, -10, -15, \ldots\}$,
   $[2] = \{2, 7, 12, 17, \ldots, -3, -8, -13, \ldots\}$.
   It is $[5] = \{n \in \mathbb{Z} \mid n - 5 \text{ is divisible by } 5\} = [0]$ and also: $[6] = [1]$, $[7] = [2]$ and so on. ◄

▶   **Theorem 1.11** If R is an equivalence relation on M, then:

a) Different equivalence classes are disjoint.
b) The union of all equivalence classes gives the whole set M.

In Example 1 this is immediately clear. For the relation R_5 from (1.2) this means: $\mathbb{Z} = [0] \cup [1] \cup [2] \cup [3] \cup [4]$. There are no more equivalence classes than these five and these classes are disjoint.

Pairwise disjoint subsets M_i, $i = 1, \ldots, n$ of a set M, whose union is M, we call *partition* of M. The equivalence classes of an equivalence relation therefore form a partition of the underlying set.

To prove the theorem:

a) We show: If A, B are equivalence classes and $A \cap B \neq \emptyset$, then $A = B$. Conversely, it follows: If $A \neq B$, then A, B are disjoint. (This is our first proof by contradiction! The arrow "$\Rightarrow$" stands for "it follows".)
Let $A = [a]$, $B = [b]$, $a, b \in M$, and $y \in A \cap B$.
$A \subset B$:

$$
\left.
\begin{array}{r}
\left.
\begin{array}{r}
x \in A \Rightarrow xRa \\
yRa \Rightarrow aRy
\end{array}
\right\} \Rightarrow xRy \\
y \in B \Rightarrow yRb
\end{array}
\right\} \Rightarrow xRb \Rightarrow x \in B
$$

$B \subset A$: goes just the same way.
b) For each $a \in M$, $a \in [a]$ and thus: $\bigcup_{a \in M} [a] = M$.   □

> In this union, the equivalence classes are probably repeated several times. But this does not matter at all. It is only important that each element of M is element of an equivalence class and that we thus capture all elements of M in the union.

In particular, Theorem 1.11a) implies $b \in [a]$ if and only if $[a] = [b]$.

Order Relations

The natural, integer and real numbers carry an order that we have already studied in detail as examples of relations. You can also define orders on other sets. The properties of such an order relation are the same as for the known order on the numbers:

▶ **Definition 1.12** Let M be a set and $\preccurlyeq$ a relation on M, which is transitive and reflexive and which has the property: For all $a, b \in M$ with $a \preccurlyeq b$ and $b \preccurlyeq a$, $a = b$ holds. Then the set M is called *partially ordered* and $\preccurlyeq$ is called a *partial order* on the set M. If further for all $a, b \in M$ it holds $a \preccurlyeq b$ or $b \preccurlyeq a$ then M is called *ordered set* and $\preccurlyeq$ *order on M*.

The property of a relation "from $a \preccurlyeq b$ and $b \preccurlyeq a$ it follows $a = b$" is also called *antisymmetric*.

In ordered sets, each two elements are comparable with respect to $\preccurlyeq$. Such orders are often called *total* or *linear orders* to distinguish them from partial orders.

From an order relation, another relation can be obtained by specifying

$$a \prec b \Leftrightarrow a \preccurlyeq b \text{ and } a \neq b$$

which is often used on ordered sets.

1. $\mathbb{N}, \mathbb{Z}, \mathbb{Q}, \mathbb{R}$ are ordered sets with the known relation $\leq$.
2. A family tree is a partially ordered set with the relation $A \preccurlyeq B$, which is true if A is a descendant of B or if $A = B$. The family tree is not totally ordered, there are also uncomparable people, for example all descendants of a parent couple. $A \prec B$ is here the relationship: A is a descendant of B.
3. The elements of $\mathbb{R}^2$ can be ordered by the following definition:

$$(a,b) \preccurlyeq (c,d): \Leftrightarrow \begin{cases} a < c & \text{or} \\ a = c \text{ and } b < d & \text{or} \\ a = c \text{ and } b = d \end{cases}$$

In this way, all Cartesian products of linearly ordered sets can be ordered. This order is called the *lexicographic order*. The dictionary entries in a dictionary are ordered in this way. ◄

1.3   Mappings

▶   **Definition 1.13** Let M, N be sets and let each $x \in M$ be associated with exactly one $y \in N$. Through M, N and this assignment, a mapping (or map) from M to N is defined.

Especially when M and N are subsets of the real numbers, mappings are also called functions.

Mappings are often denoted by small Latin letters, such as f or g. The element associated with x (the image of x) is then denoted by $f(x)$ or $g(x)$.

The notation that has become established is: $f: M \rightarrow N$, $x \mapsto f(x)$. The arrows are to be distinguished:

"$\rightarrow$"   the arrow between the sets, denotes the mapping of M to N,
"$\mapsto$"   the arrow between the elements, denotes the assignment of individual elements.

A mapping can be specified by listing all function values individually, for example in a table, or by giving an explicit assignment rule by means of which $f(x)$ can be determined for all $x \in M$. The former is of course only possible for finite sets.

Examples

$$f: \{a,b,c,\ldots,z\} \to \mathbb{N}, \quad \begin{array}{c|cccc|c} x & a & b & c & \ldots & z \\ \hline f(x) & 1 & 2 & 3 & & 26 \end{array}$$

$$g: \mathbb{R} \to \mathbb{R}, \qquad x \mapsto x^2$$

$$h: \mathbb{R} \to \mathbb{R}, \qquad x \mapsto \sqrt{|x|}$$

◀

Attention: In the last example, $x \mapsto \sqrt{x}$ would not be a reasonable assignment. Why not?

I would like to put together a few terms about maps:

▶ **Definition 1.14** If $f: M \to N$ is a mapping, then it is called

$D(f) := M$	the *domain* of f,
$x \in M$	*argument* of f,
N	the *codomain* or *set of destination* of f,
$f(M) :=$	$\{y \in N \mid$ there is a $x \in M$ with $y = f(x)\} = \{f(x) \mid x \in M\}$ the *image*

of f. If $x \in M$, $y \in N$ and $y = f(x)$, then y is called the *image* of x and x is called the *preimage* or *inverse image* of y.

If $U \subset M$, then the set of images of the $x \in U$ is called the *image* of U. The image of U is denoted by $f(U) := \{f(x) \mid x \in U\}$.

If $V \subset N$, then the set of preimages of the $y \in V$ is called the *preimage* or *inverse image* of V. It is denoted by $f^{-1}(V) := \{x \in M \mid f(x) \in V\}$.

If $U \subset M$, then the mapping $f|_U : U \to N, x \mapsto f(x)$ is called the *restriction* of f to U.

You have to be careful with the names for the image set and the preimage set: In the brackets behind f or f^{-1} there are sets, not elements. Also $f(U)$, $f^{-1}(V)$ are again sets, not elements!

Examples of mappings

1. See Fig. 1.10.
   Image $W = \{1,3\} \subsetneq N$. 1 is image of a and b, 2 is not an image, a,b are preimages of 1.
   For $U = \{a,b\}$ is $f(U) = \{1\}$,
   for $V = \{2\}$ is $f^{-1}(V) = \emptyset$, for $V = \{2,3\}$ is $f^{-1}(V) = \{c,d\}$.
2. See Fig. 1.11.
   This is not a mapping! (c has no image, d has 2 images.)

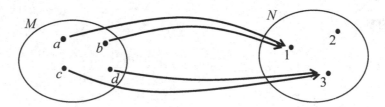

Fig. 1.10 A mapping

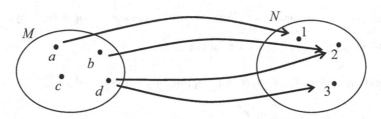

Fig. 1.11 Not a mapping

3. If M is a set, then a mapping $a \colon \mathbb{N} \to M$, $n \mapsto a(n) =: a_n$ called *sequence*. Each natural number n is assigned an element $a_n \in M$.

    Sequences are often referred to simply as $(a_n)_{n \in \mathbb{N}}$. If $M = \mathbb{R}$ ($\mathbb{Q}$), then one also speaks of *of real (rational) sequences*.

4. Let p be a prime number:

    $$h \colon \mathbb{N}_0 \to \{0, 1, 2, \ldots, p - 1\}$$

    $$n \mapsto n \bmod p := \text{remainder of division } n \text{ by } p$$

    for $p = 5$ for example we get $h(4) = 4$, $h(7) = 2$.

    Remember the relation R_5 from (1.2)? What does h do with the elements of an equivalence class? This mapping is an example of a hash function. Hash functions play an important role in computer science, for example in the efficient determination of the memory address of a data record. Here p is usually a large prime number. More about this in Sect. 4.4.

5.

    $$f \colon \{A, B, C, \ldots, Z\} \to \{0, 1, 2, \ldots, 127\}$$

    $$A \mapsto 65,$$

    $$B \mapsto 66,$$

    $$\vdots$$

    $$Z \mapsto 90.$$

    This is part of the ASCII code. Letters (characters) are represented by numbers in every computer, numbers in turn by 0-1-sequences. In total, the ASCII code can represent 128 characters, because the target set contains exactly $128 = 2^7$ elements. This means that 7 bits are needed to include all image elements. 8 bits = 1 byte is exactly the size of the data type `char` in C++, which can thus encode all ASCII characters and in the 8th bit also extensions of the ASCII code, such as country-specific special characters. In Java, the type `char` has the size 2 bytes. This makes it possible to encode $2^{16} = 65\,536$ symbols.

6. A simple but important example: $f\colon M \to M$, $x \mapsto x$ is the *identical mapping* of the set M. It is called id_M (or just id) and exists on any set.

7. In the section on logic, we will deal with predicates (propositional functions). These are mappings into the target set $\{true, false\} = \{t, f\}$. The statement "$x < y$" is either true or false for two real numbers x, y. This fact is described by the predicate:

$$P\colon \mathbb{R}^2 \to \{t, f\}, \ (x, y) \mapsto P(x, y) = \begin{cases} t & \text{if } x < y \\ f & \text{if not } x < y \end{cases}$$

8. When programming, you constantly have to deal with mappings, they are usually called functions or methods there. You put something in as an input parameter (send a message) and get an output (a response). A function in C++ that determines the greatest common divisor of two integers (`int gcd(int m, int n)`) is a mapping:

$$\gcd\colon M \times M \to M, \ (m, n) \mapsto \gcd(m, n).$$

Here M is the domain of an integer, for example $M = [-2^{31}, 2^{31} - 1]$. With this example we also see a mapping of "several variables" is in fact nothing other than an mapping of *one* set into another. The domain is then just the cartesian product of the domains of the individual variables. ◀

Do you know methods that are not mappings in our sense?

▶ **Definition 1.15: Composition of mappings** Let $f\colon M \to N$, $x \mapsto f(x)$ and $g\colon N \to S$, $y \mapsto g(y)$ be mappings. Then $h\colon M \to S$, $x \mapsto g(f(x))$ is also a mapping. It is denoted by $h = g \circ f$. Since we read from left to right, we say "g after f".

Arabs would say "first f then g".

In this definition, not only the sign "$\circ$" is defined, but also an assertion is made, namely that $g \circ f$ is again a mapping. We have to think about this before we can write the definition with a good conscience. Since we have for all $x \in M$ the unique assignment $(g \circ f)(x) := g(f(x))$ here, the assumption is correct.

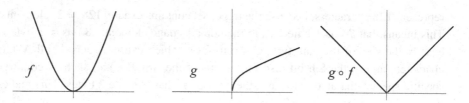

Fig. 1.12 Composition of mappings

This happens often: When defining a term, you always have to make sure it is reasonable and does not contain any implicit contradictions. The mathematicians have coined the beautiful term "well-defined" for this purpose. Only then can you relax comfortably in your armchair.

To be able to form $g(f(x))$, it is enough if the domain N of the mapping g includes the value set of f:$D(g) \supset f(M)$. The definition can therefore be extended to this case.

Example of composition of mappings

$$f: \mathbb{R} \to \mathbb{R}, \ x \mapsto x^2, \quad g: \mathbb{R}_0^+ \to \mathbb{R}, \ x \mapsto \sqrt{x}.$$

Here is $\mathbb{R}_0^+ := \{x \mid x \geq 0\}$. $W(f) \subset \mathbb{R}_0^+$, thus the composition is possible (Fig. 1.12):

$$g \circ f: \mathbb{R} \to \mathbb{R}, \quad x \mapsto \sqrt{x^2} = |x|. \ \blacktriangleleft$$

The following three terms are important properties of mappings, which are needed again and again throughout the book. You should memorize them:

▶ **Definition 1.16** Let $f: M \to N$ be a mapping. Then f is called:

injective,   if for all $x_1, x_2 \in M$ with $x_1 \neq x_2$ it holds that $f(x_1) \neq f(x_2)$.
surjective,   if for all $y \in N$ there is a $x \in M$ with $f(x) = y$.
bijective,   if f is surjective and injective.

In an injective mapping, different arguments are always mapped to different elements of the codomain, in a surjective mapping, each element of the codomain has a preimage. The codomain is then equal to the image.

We examine the examples from before:

Examples

1. (compare Fig. 1.10) The mapping is not surjective, because 2 has no preimage, and not injective, because $f(a) = f(b)$.
2. In Fig. 1.11 no mapping was shown, so the terms injective, surjective, bijective are meaningless here.
3. Sequences can be injective and surjective. The question of whether there are surjective rational or real number sequences is not trivial! More about this after Definition 1.21.
4. The hash function $h(n) = n \bmod p$ is surjective, but not injective, since for example $h(0) = h(p)$.

   In computer science this means that collisions can occur when determining the hash addresses of different data records, which require special treatment.
5. The ASCII code is surjective, 128 characters are encoded, but not all of them are printable. The part shown in the example is not surjective (0 has no image, for example). The ASCII code is also injective; it must be, otherwise you could not reverse the encoding.

   A *code* is always an injective mapping of a source alphabet into a code alphabet. The injectivity ensures the reversibility of the mapping, that is, the decodability. The target alphabet has very specific properties the source alphabet does not have, for example the processing ability for a computer (ASCII code, Unicode), or the readability for the blind (Braille), or the easy transferability by radio (Morse alphabet).
6. The identical mapping is surjective and injective, that is, bijective.
7. The predicate

$$P \colon \mathbb{R}^2 \to \{t, f\}, (x, y) \mapsto P(x, y) = \begin{cases} t & \text{if } x < y \\ f & \text{if not } x < y \end{cases}$$

   is surjective, $(P(2, 3) = t, \quad P(3, 2) = f)$, but obviously not injective: $P(1, 2) = P(2, 3)$.
8. The greatest common divisor gcd is surjective, because the pair (n, n) has the gcd n, but not injective, because for example $(3, 6)$ and $(9, 12)$ have the same gcd.
9. $f \colon \mathbb{R} \to \mathbb{R}, x \mapsto x^2, g \colon \mathbb{R}_0^+ \to \mathbb{R}, x \mapsto \sqrt{x}, g \circ f \colon \mathbb{R} \to \mathbb{R}, x \mapsto \sqrt{x^2} = |x|$: f and $g \circ f$ are nothing, g is injective, but not surjective. ◀

Perhaps you have noticed in these examples it is usually easier to disprove the property of being injective or surjective than to prove it. That is because it is enough to find a single counterexample to prove the property of "not injective":

„It exists x, y with $x \neq y$ and $f(x) = f(y)$."

To prove the property of "injective", much more has to be done:

$$\text{„For all } x, y \text{ with } x \neq y \text{ it holds } f(x) \neq f(y)\text{.“}$$

Similarly for the properties of "not surjective":

$$\text{„It exists } y, \text{ which doesn't have a preimage.“}$$

and "surjective":

$$\text{„Every } y \text{ has a preimage.“}$$

Remember: It is much easier to be destructive than constructive. Sometimes you can use this to your advantage when you cleverly reformulate the question you are investigating.

▶ **Theorem 1.17** Let $f: M \to N$ and $g: N \to S$ be mappings. Then the following results hold:

a) If f, g are injective or surjective, then so is $g \circ f$.
b) If $g \circ f$ is bijective, then g is surjective and f is injective.

As an example, I prove part b), the rest is a simple exercise.

Let $g \circ f$ be bijective. Show: f is injective.

Assume f is not injective. Then there is $x \neq y$ with $f(x) = f(y)$. Then $(g \circ f)(x) = g(f(x)) = g(f(y)) = (g \circ f)(y)$ is also not injective, in contradiction to the assumption.   □

▶ **Definition 1.18** If $f: M \to N$, $x \mapsto f(x)$ are bijective, then a mapping is defined by $g: N \to M$, $y \mapsto x$ with $y = f(x).g$ is called the *inverse mapping* of f. The inverse mapping is denoted by f^{-1}. We also say: g is inverse to f.

Again, one must realize that g really gives a unique assignment rule. For the inverse mapping f^{-1}, both $f^{-1} \circ f = id_M$ and $f \circ f^{-1} = id_N$ hold.

▶ **Theorem 1.19** Let $f: M \to N$ and $g: N \to M$ be mappings with $f \circ g = id_N$ and $g \circ f = id_M$. Then f and g are bijective, g is the inverse mapping of f and f is the inverse maping of g.

The proof is a direct consequence of the second part of the last theorem: Since the identity map is bijective, f and g must also be bijective.   □

For bijectivity and thus for the existence of inverses, the statements $f \circ g = id_N$ and $g \circ f = id_M$ are both necessary. Only one of the statements is not sufficient for this, as the following example shows:

$$f\colon \mathbb{R} \to \mathbb{R}^2 \qquad g\colon \mathbb{R}^2 \to \mathbb{R}$$

$$x \mapsto (x,x) \qquad (x,y) \mapsto \frac{1}{2}(x+y)$$

It is indeed $g \circ f = id_{\mathbb{R}}$, but f is not surjective, g is not injective and $f \circ g \neq id_{\mathbb{R}^2}$!

The Cardinality of Sets

We come back to the concept of sets; we want to deal with the cardinality of sets, that is, with the number of elements of a set M. We denote this by $|M|$. What is particularly interesting is the case where a set has infinitely many elements. For two finite sets, the following statement is obvious:

If $|M| = |N|$, then there is a bijective map between the two sets.

You just have to take one element from each set in turn and map them to each other. This property is the basis for the following definition, which is to apply to both finite and infinite sets:

▶ **Definition 1.20** Two sets M, N are called *equipotent* ($|M| - |N|$) if there is a bijective map between M and N.

A typical approach in mathematics is to simply transfer a characteristic property for a set of objects (here for the finite sets) to a larger class of objects (the infinite sets) and then see what comes out. In this case it has been shown the extension of the concept of cardinality was very fruitful for mathematics. "equipotent" is, by the way, an equivalence relation on any set of sets. (We must not talk about the set of all sets, it does not exist.)

In this way one can introduce the *cardinal numbers*. The cardinality of M is less than or equal to the cardinality of N, if M is equipotent to a subset of N. So cardinalities can be compared.

▶ **Definition 1.21** The set M is called *finite with the cardinality* $k \in \mathbb{N}$, if $|M| = |\{1, 2, 3, \ldots, k\}|$.

Of course, for example, $|\{0, 1, 2\}| = |\{a, b, c\}| = |\{0, \{1\}, \mathbb{R}\}|$ applies; these sets have the same cardinality "3". The cardinalities of the finite sets correspond exactly to the natural numbers.

The sets $\mathbb{N}, \mathbb{Z}, \mathbb{Q}, \mathbb{R}$ have no finite cardinality, they are infinitely large. But are the sets equipotent? It looks as if $\mathbb{Z}$ has more elements than $\mathbb{N}$. But look at the following mapping ($\mathbb{N}_0 := \mathbb{N} \cup \{0\}$):

$$f\colon \mathbb{N}_0 \to \mathbb{Z}, \ 0 \mapsto 0, 1 \mapsto 1, 2 \mapsto -1, 3 \mapsto 2, 4 \mapsto -2, \ldots$$

f is obviously bijective, so $|\mathbb{N}_0| = |\mathbb{Z}|$! It will now not be a problem for you to write down a bijective mapping between $\mathbb{N}$ and $\mathbb{N}_0$, and thus we have $|\mathbb{N}| = |\mathbb{Z}|$. Cantor has shown that one can even find a bijective mapping between $\mathbb{Z}$ and $\mathbb{Q}$ (the procedure is constructive and well understandable, but it does exceed the scope of this book). So there are just as many rational numbers as natural numbers. Actually logical, or not? It should be infinity equal to infinity. But now something amazing happens: Cantor was also able to prove that there is no bijective mapping between $\mathbb{Q}$ and $\mathbb{R}$. So there are more real numbers than rational, infinity is not equal to infinity! In fact, one can find many, many more different infinite cardinalities.

With this we now also know that there are surjective rational number sequences, but no surjective real number sequences.

Sets that are finite or have the same cardinality as $\mathbb{N}$ are called *countable sets*, sets of greater cardinality, such as $\mathbb{R}$ are called *uncountable*.

For our purposes, we do not need to further deal with cardinalities. But I still want to tell you a little story that has driven many mathematicians almost to despair for decades: The cardinality of $\mathbb{N}$ is called $\aleph_0$ (the Hebrew letter Alef), the cardinality of $\mathbb{R}$ is called C (for continuum). It can be calculated that the cardinality of the set of all subsets of $\mathbb{N}$, that is, of $P(\mathbb{N})$, is also equal to C. Cantor now wondered whether there might be no other cardinality between $\aleph_0$ and C at all. Or can you still stuff a set between $\mathbb{Q}$ and $\mathbb{R}$? Around 1900, Cantor made the conjecture that this is not possible (the continuum hypothesis). Through the results of the mathematicians Kurt Gödel (1938) and Paul Cohen (1963), this problem found a very surprising solution: Neither the continuum hypothesis nor its opposite can be proven. A mathematical sensation one might have to put on a level with the theory of relativity in physics in the world of mathematics: *You can prove there are things you can neither prove nor disprove*. This solves the problem, but many mathematicians had to get used to this solution.

The research in connection with the decidability or undecidability of mathematical assertions also had significant effects on computer science: In the course of his work on this topic, Alan Turing invented his famous Turing machine in the 1930s. In a thought experiment, he wanted to solve every problem solvable in a logical way with the universal Turing machine. Although this did not succeed, he incidentally laid the theoretical foundations of computer architecture.

1.4   Comprehension Questions and Exercises

Comprehension Questions

1. What is the difference between $=$ and $:=$?
2. What is the power set of the empty set $P(\emptyset)$?
3. Are there sets A, B with $A \subset B$, $A \neq B$, but $|A| = |B|$?
4. What is $\bigcup_{n \in \mathbb{N}} \{x \in \mathbb{R} \mid -n < x < n\}$?
5. Can an order relation be an equivalence relation at the same time?
6. An equivalence relation divides a set M into disjoint equivalence classes. Conversely, let a partition of M into disjoint subsets be given, whose union is just M. Is there an equivalence relation whose equivalence classes are exactly these subsets?

7. What is the difference between a partial order and a linear order?
8. Let $f: M \rightarrow N$ and $g: N \rightarrow M$ be mappings with the property that $g \circ f: M \rightarrow M$ is the identical mapping on M. Is then also $f \circ g: N \rightarrow N$ the identical mapping on N ?
9. Is there an injective mapping from $\mathbb{R}$ to $\mathbb{N}$?

Exercises

1. Prove the formula $S \cup (M \cap N) = (S \cup M) \cap (S \cup N)$.
2. Prove the formula $\overline{M \cup N} = \overline{M} \cap \overline{N}$.
3. Examine the relations $\subset$ and $\not\subset$ for reflexivity, symmetry and transitivity.
4. Let $R = \{(m, n) \in \mathbb{Z} \times \mathbb{Z} \mid m - n$ is divisible by 5$\}$ and $r \in \{0, 1, 2, 3, 4\}$.   We already know R is an equivalence relation. Show that $[r]$, the equivalence class of r is exactly the set of numbers that leave remainder r when divided by 5.

   m has remainder r when divided by q if and only if it exists a $k \in \mathbb{Z}$ with $m = qk + r$.

5. Let the following sets of natural numbers be given:

$$M = \{x \mid 4 \text{ divides } x\}$$
$$N = \{x \mid 100 \text{ divides } x\}$$
$$T = \{x \mid 400 \text{ divides } x\}$$
$$S = \{x \mid x \text{ is a leap year}\}$$

Express the set S using the set operations $\cup, \cap$ and $\setminus$ from the sets M, N, T. Eliminate the $\setminus$ sign in this representation by using the complement.

   Leap years are the years divisible by 4, except for the years divisible by 100, but not by 400. (2000 and 2400 are leap years, 1900 and 2100 are not leap years!)

6. Check whether the following mappings are surjective or injective:
   a) $f: \mathbb{R}^3 \rightarrow \mathbb{R}^2, (x, y, z) \mapsto (x + y, y + z)$
   b) $f: \mathbb{R}^2 \rightarrow \mathbb{R}^3, (x, y) \mapsto (x, x + y, y)$
7. If $f: M \rightarrow N$ is a mapping, then two mappings are generated between the power sets:

$$F: P(M) \rightarrow P(N), \quad G: P(N) \rightarrow P(M),$$
$$U \mapsto f(U) \qquad\qquad V \mapsto f^{-1}(V).$$

Are these really mappings? (Check the definition.) Using a simple example for f, show F and G are not inverses of each other.
8. If f and g are mappings, and if the composition $f \circ g$ is possible, then show: If f and g are surjective or injective, then $f \circ g$ is surjective or injective.

Logic

2

Abstract

As a computer scientist you are constantly dealing with tasks from logic. In this chapter you will learn

- what a proposition is,
- the difference between syntax and semantics,
- how you can evaluate propositions and propositional formulas,
- the predicate logic,
- important principles of proof in mathematics,
- the use of preconditions and postconditions in programming.

Why is logic so important in computer science? It starts with the fact that the control flow of a program works with logical (Boolean) variables that have the value true or false. It continues when testing a program, where certain preconditions must be met by the input variables and then the results must meet specific output conditions.

Even when defining a system, it is important to make sure that the requirements are not contradictory. When creating a system architecture up to maintenance, logical thinking is always required. Especially in safety-critical systems, more and more work is being done with formal specification and partly also verification methods that require a high level of knowledge of logical methods.

Supplementary Information The online version contains supplementary material available at https://doi.org/10.1007/978-3-658-40423-9_2.

27

Therefore, logic takes up a higher position in a mathematics textbook for computer scientists than in books for other disciplines.

2.1   Propositions and Propositional Variables

We will start with the concept of proposition:

Facts of reality are captured in the form of statements. In mathematical logic we call them propositions. We limit ourselves to propositions for which it makes sense to ask whether they are true or false. The principle of the excluded third applies.

You have probably already heard of fuzzy logic: It takes into account that there is also warm between cold and hot; there are truth functions in which something can only be "a little true". My washing machine usually works very successfully with fuzzy logic. We will not deal with it here, but rather investigate classical logic, the knowledge of which of course also forms the basis for fuzzy logic.

Examples

1. 5 is smaller than 3.
2. Paris is the capital of France.
3. The study of computer science is very difficult.
4. Brush your teeth after eating!
5. It's colder at night than it is outside. ◄

1. and 2. are propositions in our sense. 3. is very subjective, the answer is certainly different from person to person; but you can ask about it! 4. is not a fact, it makes no sense to talk about truth here. 5. is not even a sentence that has a reasonable meaning.

For your consolation: We will not venture into the borderline areas of theory in which we stand helpless before a sentence and wonder whether it is a proposition or not. Let others do that.

The words "true" and "false" are called *truth values*. Every proposition has one of these two truth values. However, the presence of a proposition does not mean that one can immediately say which truth value a proposition has; perhaps it is still unknown.

Two more

Examples

6. The World Climate Conference will succeed in stopping the climate catastrophe.
7. Every even number greater than 2 is the sum of two prime numbers (the *Goldbach conjecture*). ◄

Proposition 6 will turn out at some point, 7 is unknown, perhaps one will never find out.

> The Goldbach conjecture is about 300 years old. So far, the "Goldbach test" has worked well for every even number: $6 = 3 + 3$, $50 = 31 + 19$, $98 = 19 + 79$, and so on. Even in the age of high-performance computers, no one has found a single counterexample. But in 300 years no mathematician has been able to prove that the statement is really true for all even numbers. Maybe you have a chance to become famous here: Finding a single number is enough!

In the future, the content of a proposition is no longer interesting to us, but only the property of the proposition being either true or false. Therefore, from now on we will also refer to propositions with symbols, mostly with large Latin letters, $A, B, C, D, \ldots$, which can stand for any proposition and can take on the truth values true or false. We call these symbols *propositional variables* and we will use them as variables just like, for example, x and y in the expression $3x^2 + 7y$. Values can be assigned here to the variables x and y. The truth values "true" and "false" will from now on be abbreviated with t resp. f.

Compound Propositions

In everyday language, propositions are combined with words like "and", "or", "not", "if—then", "except" and others. We now want to reproduce such combinations of propositions for our propositional variables and determine the truth value of the compound propositions depending on the contained sub-propositions. For this purpose, we will set up tables.

In the example of numerical variables, such a table looks like this:

x	y	$3x^2 + 7y$
1	1	10
1	2	17
2	1	19
$\vdots$	$\vdots$	$\vdots$

We will proceed in the same way for propositional variables. The big advantage we have is that we only have finitely many assignment values for a variable, namely exactly two. We can therefore, in contrast to the above example, capture all possible values in a finite table, we call this truth table. How many such assignment possibilities are there for the combination of two propositional variables? For A and B we have four possibilities whose combination can result in true or false:

A	B	$A * B$
t	t	t/f
f	t	t/f
t	f	t/f
f	f	t/f

Let's denote the connection of A and B with $*$, then you can write down 16 combinations in total in the column $A * B$, from four times *true* to four times *false*. Some of these are particularly interesting, which I will introduce below.

1. **"and"**, notation: $\wedge$ (the *conjunction*).

$$
\begin{aligned}
&\text{Examples:} \quad A: \ \ 2 \text{ is even.} &&\text{(true)}\\
&\qquad\qquad\quad B: \ \ 5 < 3. &&\text{(false)}\\
&\qquad\qquad\quad C: \ \ \text{for all real numbers } x \text{ is } x^2 \geq 0. &&\text{(true)}
\end{aligned}
$$

The proposition "2 is even and $5 < 3$" is false, the proposition "2 is even and for all real numbers x is $x^2 \geq 0$" is true. The complete truth table looks like this:

A	B	$A \wedge B$
t	t	t
f	t	f
t	f	f
f	f	f

The compound proposition is therefore only true if each individual proposition is true.

2. **"or"**, notation: $\vee$ (the *disjunction*).

$$
\begin{aligned}
&\text{Examples:} \quad A: \ \ \text{Christmas is on December 24th} &&\text{(true)}\\
&\qquad\qquad\quad B: \ \ \text{The moon is made of green cheese} &&\text{(false)}\\
&\qquad\qquad\quad C: \ \ \text{Today is Monday} &&\text{(true or false)}
\end{aligned}
$$

The proposition "Christmas is on December 24th or the moon is made of green cheese" is true, the proposition "Christmas is on December 24th or today is Monday" is always true. "Today is Monday or the moon is made of green cheese" is only true on Monday. This gives us the truth table:

A	B	$A \vee B$
w	w	w
f	w	w
w	f	w
f	f	f

The compound proposition is only false if both individual propositions are false. Note again here the logical or does not describe the exclusive or, both propositions may also be true: "$5 > 3$ or 2 is even" is a true proposition. In everyday language, these two "or" are not always clearly distinguished.

3. **"if—then"**, notation: $\rightarrow$ (the *implication*).

Examples: A: Christmas is on December 24th   (true)

B: I'll eat my hat   (definitely false!)

C: Christmas falls on Easter   (false)

D: Goldbach's conjecture is true   (?)

The proposition "If Christmas is on December 24th, then I'll eat my hat" is false, while you can believe me the proposition "If Christmas falls on Easter, then I'll eat my hat" is true! What about: "If Goldbach's conjecture is true, then Christmas is on December 24th"? I think you will accept it as true, regardless of whether Goldbach was right or not. With these examples we can fill in the table completely:

$$
\begin{array}{cc|c}
A & B & A \rightarrow B \\
\hline
t & t & t \\
f & t & t \\
t & f & f \\
f & f & t
\end{array}
\tag{2.1}
$$

The initially surprising thing about this table is line 2, which says you can also infer something true from something false. But you probably know such situations from everyday life: if you start from false assumptions, you can be lucky or unlucky with a conclusion. Maybe something reasonable comes out by chance, maybe not.

Sometimes it is difficult to verify this table with propositions from everyday life. Try it yourself! A student gave me this example in a lecture: "If today is Tuesday, then tomorrow is Thursday." Isn't this proposition always false, no matter whether I announce it on Monday, Tuesday or Wednesday? What do you think about it?

If this table gives you a stomach ache, I can calm you down from the side of mathematics: we simply take the table as a definition for the connection "$\rightarrow$" and don't care at all whether it has anything to do with reality or not. Mathematicians are allowed to do that! That's our ivory tower. The downside is that we eventually want to start something with our mathematics again. This gets us out of the ivory tower. That only works well if our original assumptions were reasonable. In the case of logic, it turned out our truth tables are really meaningful exactly in the form presented.

Easier to understand is the *if and only if* relationship:

4. **"if and only if"**, notation: $\leftrightarrow$ (the *equivalence*).

$$
\begin{array}{cc|c}
A & B & A \leftrightarrow B \\
\hline
t & t & t \\
f & t & f \\
t & f & f \\
f & f & t
\end{array}
$$

The compound proposition is true here only when both individual propositions have the same truth value.

Finally, an operation that only applies to one logical variable, the negation:

5. **"not"**, notation: ¬ (the *negation*).

This is simple: The "not" just reverses the truth value of a proposition:

$$
\begin{array}{c|c}
A & \neg A \\
\hline
t & f \\
f & t
\end{array}
$$

What do we have from our connections now? We can use them to build up increasingly complex propositions from given propositions by putting them together. A few

---

Examples

$$A \to (B \lor C)$$

$$\neg(A \leftrightarrow (B \lor (\neg C)))$$

$$(A \lor B) \land (\neg(B \land (A \lor C))) \blacktriangleleft$$

With *and*, *or*, *if—then*, and *if and only if* I have defined the four of the 16 possible connections of two propositional variables with which we usually work in logic. What about the other twelve? Some are boring: for example "everything true" or "everything false". You might know the *exclusive or*, *nand* and *nor* from digital technology. Do we need them too? In fact, all 16 possible connections can be generated as a combination of the four basic operations together with the *not*. Yes, it is even possible to represent all 16 operations and the *not* as a combination of two variables with only one operation, the *nand* operator ↑. Here $A \uparrow B := \neg(A \land B)$. As a result, in digital technology, every logical connection can be implemented solely by *nand* gates.

When programming, you also work with formulas built up from propositions: The conditions behind if, for, while contain propositional variables, in Java of type boolean, and connections of propositional variables, which in a specific program run are replaced by the truth values "*true*", "*false*" and evaluated.

Just as algebraic formulas are usually built up from $a, b, c, d, \ldots, +, \cdot, -, :, \ldots$ we get formulas of *propositional algebra* in this way. And just as one can evaluate algebraic formulas by substituting numbers, one can evaluate propositional formulas by substituting truth values, that is, one can determine whether the entire formula is true or false.

What we have done here with the construction of propositional formulas and their evaluation from elementary propositions is a first example of a *formal language*. Formal languages have an extraordinary importance in computer science. A formal language consists, like natural languages, of three elements:

- A set of symbols from which the sentences of the language can be constructed, the *alphabet*.

- Rules to specify how correct sentences can be formed from the symbols of the alphabet, the *syntax*.
- An assignment of meaning to syntactically correctly formed sentences, the *semantics*.

As the first formal language, the computer scientist usually has to do with a programming language: The alphabet consists of the keywords and the allowed characters of the language, the syntax defines how correct programs can be formed from these characters. The correctness of the syntax can be checked by the compiler. But as you surely know from bitter experience, not every syntactically correct program performs meaningful actions. The task of the programmer is to ensure the semantics of the program corresponds to the specification.

The formulas of propositional logic also build a formal language.

- The alphabet consists of the identifiers for propositional variables, the logical symbols $\wedge, \vee, \rightarrow, \leftrightarrow, \neg$ as well as the brackets (and), which serve to group together parts of the formula.
- The rules of syntax state:
  - The identifiers for propositional variables are formulas,
  - if F_1 and F_2 are formulas, then so are $(F_1 \wedge F_2)$, $(F_1 \vee F_2)$, $(\neg F_1)$, $(F_1 \rightarrow F_2)$, $(F_1 \leftrightarrow F_2)$.
- The semantics is finally determined by the truth tables for the elementary logical connections. As a result, every syntactically correct formula can be assigned a truth value, provided the identifiers are assigned concrete truth values.

In order to improve the readability of formulas, some precedence rules for compound propositions are set, similar to those in arithmetic, which allow the omission of many brackets. These rules are:

$\neg$ has the strongest binding,
$\wedge$ binds stronger than $\vee$,
$\vee$ binds stronger than $\rightarrow$,
$\rightarrow$ binds stronger than $\leftrightarrow$.

For example, $((\neg A) \wedge B) \rightarrow (\neg(C \vee D))$ is equivalent to $\neg A \wedge B \rightarrow \neg(C \vee D)$.

> But beware! The argument with the better readability is dangerous if you do not have all the precedents in your head. Do you know exactly what happens in Java or C++ with the instruction `if(!a&&b||c==5)`? In such cases, it is better to write a few more brackets than necessary to avoid unnecessary errors.

The evaluation of a composite formula can be traced back to the elementary truth tables using connection tables again. First, the contained propositional variables are listed in the columns with all possible combinations of truth values, then the evaluation is carried out step by step, "from the inside out". When evaluating $(A \wedge B) \rightarrow C$, we get something like:

A	B	C	$A \wedge B$	$(A \wedge B) \to C$
t	t	t	t	t
t	t	f	t	f
t	f	t	f	t
t	f	f	f	t
$\vdots$	$\vdots$	$\vdots$	$\vdots$	$\vdots$

There are still four cases missing in the table. You can see with more than two propositional variables in the formula, this evaluation method quickly becomes very extensive. Fortunately, there is another tool that can help us determine the truth value of a formula. Parts of a formula can be replaced by other equivalent and possibly simpler formula parts.

What are logically equivalent formulas? Let's evaluate $B \vee \neg A$ once:

$$
\begin{array}{cc|c|c}
A & B & \neg A & B \vee \neg A \\
\hline
t & t & f & t \\
f & t & t & t \\
t & f & f & f \\
f & f & t & t \\
\end{array}
\tag{2.2}
$$

Compare this table with the Table (2.1) for $A \to B$. You can see that $A \to B$ is true (false) if and only if $B \vee \neg A$ is true (false). The two formulas are logically equivalent:

▶ **Definition 2.1** Two propositional formulas are called *logically equivalent* if they have the same truth value for all possible assignments of truth values.

For the logician, those formulas provide calculation rules: Part of a formula can be replaced by a logically equivalent formula without changing the truth value of the overall formula.

Example

In most programming languages there are logical operators for "and" (in C++ and Java, for example, `&&`), "or" (`||`) and "not" (`!`), but no operator for "if then". But since $A \to B$ and $B \vee \neg A$ are equivalent, you can define an arrow operator in C++ (or similarly in Java):

```
bool arrow(bool A, bool B){return (B || !A);}
```

`arrow(A,B)` then means A→B. ◀

We introduce a new symbol to identify logically equivalent formulas:

▶   **Definition 2.2** If the formulas F_1 and F_2 are logically equivalent, we write for this

$$F_1 \Leftrightarrow F_2.$$

Please note: The sign $\Leftrightarrow$ is not a symbol of the language of propositional logic, it says something about the truth value of formulas, that is, about the semantics.

Unfortunately, no uniform notation is used for logical equivalence and equivalence in the literature. You will occasionally also find the symbol $\equiv$ for logical equivalence, the double arrow $\Leftrightarrow$ is sometimes also used to designate the equivalence of propositions.

I will put together a few important logically equivalent formulas:

▶   **Theorem 2.3:** Calculation rules for propositional formulas

$$\left. \begin{array}{l} A \vee B \Leftrightarrow B \vee A \\ A \wedge B \Leftrightarrow B \wedge A \end{array} \right\} \text{commutative laws} \qquad (2.3)$$

$$\left. \begin{array}{l} (A \vee B) \vee C \Leftrightarrow A \vee (B \vee C) \quad (=: A \vee B \vee C) \\ (A \wedge B) \wedge C \Leftrightarrow A \wedge (B \wedge C) \quad (=: A \wedge B \wedge C) \end{array} \right\} \text{associative laws}$$

$$\left. \begin{array}{l} A \wedge (B \vee C) \Leftrightarrow (A \wedge B) \vee (A \wedge C) \\ A \vee (B \wedge C) \Leftrightarrow (A \vee B) \wedge (A \vee C) \end{array} \right\} \text{distributive laws}$$

$$\neg(A \wedge B) \Leftrightarrow \neg A \vee \neg B$$
$$\neg(A \vee B) \Leftrightarrow \neg A \wedge \neg B \qquad (2.4)$$
$$\neg\neg A \Leftrightarrow A$$

Do you notice anything here? Compare these formulas with Theorem 1.3 and 1.4. There you will find the same rules for calculating with sets. $\cup$ and $\vee$ correspond to each other, $\cap$ and $\wedge$ as well as "complement" and $\neg$.

The just discovered analogy between set operations and logical operations leads to the theory of Boolean algebras, Theorem 2.7.

However, the proof is much simpler here, if also much more boring: We only have to insert a finite number of values in the truth tables. However, for the distributive and associative laws, we have three initial propositions here and thus $2^3 = 8$ different combinations of truth values to check. You write for a while. Believe me, the rules are correct!

What is the result of evaluating the formula $B \vee \neg A \leftrightarrow (A \rightarrow B)$?

If you have read the last page carefully, you will already know what the result is. But let's write the table again:

A	B	$B \vee \neg A$	$(A \rightarrow B)$	$B \vee \neg A \leftrightarrow (A \rightarrow B)$
t	t	t	t	t
f	t	t	t	t
t	f	f	f	t
f	f	t	t	t

You can see this formula is always true, no matter what values A and B take. This is because the formulas to the right and left of the bidirectional arrow are logically equivalent:

▶ **Theorem 2.4** Two formulas F_1, F_2 are logically equivalent, that is $F_1 \leftrightarrow F_2$, if $F_1 \leftrightarrow F_2$ is true for all possible assignments of truth values.

The correctness of this theorem can be read directly from the corresponding truth table:

F_1	F_2	$F_1 \leftrightarrow F_2$
t	t	t
f	f	t
t	f	f
f	t	f

If you assign truth values to F_1 and F_2, these are logically equivalent in the first two rows, and exactly in these rows is also $F_1 \leftrightarrow F_2$ true.   □

▶ **Definition 2.5** A formula is called *valid* or *tautology*, if it is true for all possible assignments of truth values.

$B \vee \neg A \leftrightarrow (A \rightarrow B)$ is therefore a tautology and all calculation rules from Theorem 2.3 become tautologies if you replace the sign $\Leftrightarrow$ by $\leftrightarrow$.

> By using tautologies in everyday life, one can pretend competence that is actually not present: "If the rooster crows from dungeons top, the weather will change or it will not" (an old German folk wisdom). Written as a formula: $A \rightarrow B \vee \neg B$. Check for yourself that this is a tautology. By the way, this statement is even true if the rooster is just cold!

Here are some more tautologies, most of which will appear again in the next section:

$$A \vee \neg A$$

$$\neg(A \wedge \neg A) \tag{2.5}$$

$$A \wedge (A \to B) \to B \tag{2.6}$$

$$(A \to B) \wedge (B \to A) \leftrightarrow (A \leftrightarrow B) \tag{2.7}$$

$$(A \to B) \leftrightarrow (\neg B \to \neg A) \tag{2.8}$$

Maybe you check one or the other of these tautologies yourself. Substitute some statement-propositions for A and B. Can you interpret the formulas?

Another symbol of semantics we will use below is the double arrow $\Rightarrow$:

▶ **Definition 2.6** If F_1 and F_2 are formulas and if, for every assignment of truth values for which F_1 is true, F_2 is also true, we write

$$F_1 \Rightarrow F_2.$$

We also say: F_2 follows from F_1.

▶ **Theorem 2.7** $F_1 \Rightarrow F_2$ holds if and only if $F_1 \to F_2$ is a tautology, that is, it is true for every assignment of truth values.

To see this, we just have to take another look at the truth table of the implication:

F_1	F_2	$F_1 \to F_2$
t	t	t
$(t$	f	$f)$
f	t	t
f	f	t

I have put the second line in parentheses because it cannot occur: If $F_1 \Rightarrow F_2$ holds, then the combination t, f for F_1, F_2 is not possible, and thus $F_1 \to F_2$ is always true. And if $F_1 \to F_2$ is always true, then if F_1 is true then also F_2 is true. □

Boolean Algebras

Different concrete structures often have very similar properties. Mathematicians are then happy. They try to work out the core of these similarities and to define an abstract structure that is characterized by exactly these common properties. Then they can study these structure. All results obtained for it of course also apply to the concrete models from which they originally started.

Something similar happens when you define a common superclass for different classes of a program: properties of the superclass are automatically inherited by all derived classes.

This is also the case with set operations and logical operators: the underlying common structure is called Boolean algebra. The operations are often denoted as in the case of sets with ∪, ∩ and ⁻:

▶   **Definition 2.8** Let a set B be given, which contains at least two different elements 0 and 1 and on which two operations ∪ and ∩ between elements of the set as well as an operation ⁻ on the elements are defined. $(B, ∪, ∩, ⁻)$ is called *Boolean algebra*, if for all elements $x, y, z \in B$ the following properties apply:

$$x \cup y = y \cup x, \qquad\qquad x \cap y = y \cap x$$
$$x \cup (y \cup z) = (x \cup y) \cup z \qquad x \cap (y \cap z) = (x \cap y) \cap z$$
$$x \cup (y \cap z) = (x \cup y) \cap (x \cup z) \quad x \cap (y \cup z) = (x \cap y) \cup (x \cap z)$$
$$x \cup \bar{x} = 1 \qquad\qquad\qquad x \cap \bar{x} = 0$$
$$x \cup 0 = x \qquad\qquad\qquad x \cap 1 = x$$

The power set $P(M)$ of a set M with *union, intersection* and *complement* forms a Boolean algebra. The empty set is the 0 and the set M itself is the 1-element.

A minimal Boolean algebra is obtained by taking the set $B = \{0, 1\}$ as the basic set. This is the *two-element algebra*, which is used for the design of technical circuits. Try it yourself to write down truth tables for this algebra!

The set of all propositional formulas that can be formed with the operations ∧, ∨, ¬ from n propositional variables (the n-ary propositional formulas) with the operations ∧, ∨, ¬ is also a Boolean algebra, an *algebra of formulas*. 1 is the always true proposition, 0 is the always false proposition.

Evaluation of Propositional Formulas in a Program

As an example, let's look at the calculation of leap years:

Leap years are the years divisible by 4, except for the years divisible by 100, but not by 400. Let x be a natural number (a year). Let's look at the following propositions:

A: x is divisible by 4.
B: x is divisible by 100.
C: x is divisible by 400.
S: x is a leap year.

The connections "except" and "but not" can both be represented by the symbols "∧¬". This gives us for

D: x is divisible by 100 but not by 400:

$$D \leftrightarrow B \wedge \neg C$$

and finally for S:

$$S \leftrightarrow A \wedge \neg D \leftrightarrow A \wedge \neg (B \wedge \neg C). \tag{2.9}$$

This allows us to formulate a query in Java, for example. The operator $\%$ gives the remainder in an integer division, and so we write:

```
if((x%4 == 0)&&!((x%100 == 0)&&(x%400 != 0)))
    // February has 29 days
else
    // February has 28 days
```
(2.10)

We can also convert the formula (2.9) using our rules: From (2.4) and (2.3) we get, for example:

$$A \wedge \neg (B \wedge \neg C) \Leftrightarrow (C \vee \neg B) \wedge A$$

and from that the Java instruction:

```
if(((x%400 == 0)||(x%100 != 0)) && (x%4 == 0)).
```
(2.11)

Is there a difference between (2.10) and (2.11)? Of course not in the result of the evaluation, otherwise our logic would be wrong. But there can be differences in the runtime of the program: The program evaluates a logical expression from left to right and stops immediately if it can decide whether a statement is true or false. If now a number x is inserted, then in 75 % of the cases (2.10) the evaluation can be ended after the evaluation of A ($x\%4 == 0$), because if A is false, then the whole proposition is also false. In the case (2.11) first C ($x\%400 == 0$) must be checked, because it is usually false, also $\neg B$ ($x\%100 != 0$), and because this is usually true, then A must be checked at the end. You can see this is much more time consuming. Therefore, in time-critical applications, it makes sense to check and possibly reformulate control conditions carefully!

2.2   Proof Principles

I am often asked whether proofs are actually necessary in computer science. After all, there is the profession of mathematician, who lives to prove theorems. The users of mathematics should be relieved by this. It is enough to provide them with the tools, that is, the right formulas into which the problems to be solved are fed, and which spit out the result.

There may be application areas in which this is partly true, but computer science is certainly not one of them. Computer scientists are constantly "proving" even though they do not call it that: They think about whether a system architecture is feasible, they analyze whether a protocol can do what it is supposed to do, they ponder whether an algorithm works correctly in all special cases ("what happens if, …"), they check whether the switch statement in the program also overlooks nothing, they look for test cases that ensure the highest possible path coverage of the program, and much more. All of this is nothing other than "proving".

Do not be afraid, you can learn to prove as well! There are tricks used over and over again. In this section, I would like to introduce you to some of these tricks and try them out with concrete examples. The contents of the theorems are rather secondary, even though I take this opportunity to present you two famous pearls of mathematics in Theorems 2.10 and 2.11. It is about the techniques behind the proofs.

But often the contents of a mathematical assertion only become clear through the proof, perhaps because one deals intensively with the assertion in the proof.

Throughout the book, I will prove again and again, not for the sake of proving, because I hope you have enough confidence in me that I do not introduce any false concepts to you (at least not intentionally). We will prove when we can learn something from it or when it serves to understand the material.

In proofs, the validity of certain propositions, the assertions, is always concluded from the validity of other propositions, the assumptions. We speak here of the semantics, that is, of true and false propositions. The mapping of propositions to logical formulas and the rules of calculation for logical formulas often help to determine the truth value. For this purpose, the conditions and the assertions must be precisely identified and formulated cleanly.

The Direct Proof

The simplest form of proof is the direct proof: From an assumption A an assertion B is derived. An example:

▶   **Theorem 2.9** If $\underbrace{n \in \mathbb{Z} \text{ is odd}}_{A}$, then also $\underbrace{n^2 \text{ is odd}}_{B}$.

Proof: Assume A is true. Then there is an integer m with $n = 2m + 1$. Then $n^2 = (2m + 1)^2 = 4m^2 + 4m + 1$, that is, n^2 is also odd.   □

In the direct proof $A \Rightarrow B$ is shown, that is, $A \rightarrow B$ is always true. Note one can do this without even knowing whether the condition A is fulfilled. Also, from the Goldbach conjecture, one can derive many things without knowing whether it is actually true. But if at some point A is recognized as true and we already know that $A \rightarrow B$ is always true, then B is also always true. (To this belongs the tautology $A \wedge (A \rightarrow B) \rightarrow B$, compare (2.6).)

The Proof of Equivalence

Here an assertion A is proven to be true (or false) if and only if another assertion B is true (or false), that is $A \Leftrightarrow B$.

An equivalence proof is nothing more than the consecutive execution of two direct proofs. One shows $A \Rightarrow B$ and $B \Rightarrow A$. The task is thereby divided into two easier subtasks. This can also be expressed by a tautology: If $A \to B$ and $B \to A$ are always true, then $A \leftrightarrow B$ and vice versa: $(A \to B) \wedge (B \to A) \leftrightarrow (A \leftrightarrow B)$, compare (2.7).

The last part of Theorem 1.4 is such an equivalence statement: $M \subset N \Leftrightarrow \overline{N} \subset \overline{M}$. Try to proof this theorem by breaking it down into two direct proofs yourself.

A variant of the equivalence proof is the *circular reasoning*. To show three or more assertions are equivalent, for example the assertions A, B, C, D, it is enough to show: $A \Rightarrow B, B \Rightarrow C, C \Rightarrow D, D \Rightarrow A$.

The Proof by Contradiction

Again, actually $A \Rightarrow B$ should be shown. Often it is easier to carry out the conclusion $\neg B \Rightarrow \neg A$ instead. Since we already know that $(A \to B) \leftrightarrow (\neg B \to \neg A)$ is a tautology (compare (2.8)), this is equivalent to $A \Rightarrow B$. In Theorem 1.11 we have carried out such a proof by contradiction. Take a look at it again!

Proofs by contradiction are a powerful and often used tool. One reason for this is that, in addition to the assumption A, one can also assume $\neg B$; one therefore has more in hand to work with. Strictly speaking, one could also describe the proof by contradiction with the tautology: $(A \to B) \leftrightarrow (A \wedge \neg B \to \neg A)$

I would like to introduce you to a famous example. You probably know $\sqrt{2}$ is not a rational number, that is, it cannot be represented as a fraction of two integers. Why is that so? The following proof shows especially well that it depends on the precise formulation of the assumption and the assertion.

▶ **Theorem 2.10** $\sqrt{2}$ is not a rational number.

In the proof we show: $\underbrace{\text{Let } m, n \in \mathbb{N}, \text{ and } m, n \text{ reduced}}_{A}$, then $\underbrace{2 \neq (\frac{m}{n})^2}_{B}$ applies.

The brilliant idea lies in the additional assumption that m and n are reduced. That doesn't matter, because if m, n still have a common divisor, then we just divide by it. But the proof is essentially based on this.

Assume there is m, n with $2 = (\frac{m}{n})^2$. (Proposition $\neg B$)
Then $m^2 = 2n^2$ and thus m^2 is an even number.

But then m is an even number too, because according to Theorem 2.9 the square of an odd number is always odd.

Here we have packed a small proof of contradiction into the proof of contradiction.

So m has the form $m = 2l$ and $(2l)^2 = 2n^2$ applies. Dividing by 2 we get $2l^2 = n^2$. Just like before, this means n^2 and thus n is an even number.

m and n are therefore divisible by 2 and thus not reduced. (Proposition $\neg A$)   □

Another form of proof by contradiction is that one derives the assertion itself from the opposite of the assertion ($\neg B \Rightarrow B$). But $\neg B$ and B cannot be true both, because $\neg(B \wedge \neg B)$ is a tautology (see (2.5)). Then the assumption $\neg B$ must be false, that is, B itself must be true. One last example:

▶   **Theorem 2.11** There is no largest prime number (assertion B).

Proof:

Assume: a largest prime p exists.   Assumption: $\neg B$

Let $p_1, p_2, ..., p_n$ all prime numbers and let $q = p_1 p_2 \cdots p_n + 1$.   (2.12)

    Assertion: Then q is prime.   Assertion: A
    Assume q is not prime   Assumption: $\neg A$
    Then q has a prime divisor $p_i \in \{ p_1, p_2, ..., p_n\}$ and
    because of (2.12) is $q = p_i \cdot \alpha = \underbrace{p_1 p_2 \cdots p_n}_{p_i \beta} + 1 = p_i \cdot \beta + 1$.

    $\Rightarrow \boxed{p_i(\alpha - \beta) = 1}$, where $\alpha - \beta$ is an integer.   Wrong proposition.
    (1 is never product oft two integers $\neq 1$).
    So: q is prime.   Thus: A

As q is product of all primes + 1 ist, it follows $q > p$.   Thus: B

                                                                                 □

Here again, two proofs by contradiction are nested. In the "inner" proof, the assumption $\neg A$ has been used to derive a false assertion, so A must be true. This is yet another version of the proof by contradiction.

> This proof is due to Euclid and is thus about 2300 years old. Can you imagine some mathematicians are happy about a beautiful proof, just as many others are about a good piece of music? What makes a "beautiful" proof? It is usually short and concise, nevertheless understandable—at least for a mathematician—and often contains surprising conclusions and twists not so easy to come by when looking at the assertion to be proven. I think the last two proofs definitely fall into this category. The Hungarian mathematician Paul Erdős (1913–1996) believed God keeps a book in which he records the perfect proofs for mathematical

theorems. At least parts of this book have reached Earth by unknown means and have been compiled in the earthly book "Proofs from the BOOK". Euclid's proof for the infinite number of prime numbers is the first one in it. It really deserves it.

Later we will learn another proof principle, mathematical induction. But first we have to work out a few more logical concepts:

2.3   Predicate Logic (First-Order Logic)

Look again at the example for calculating leap years in Sect. 2.1 after Definition 2.8: $S =$ "x is a leap year". Is the proposition true or false? Does it even make sense to ask if it is true or false? Obviously, the truth of the statement depends on x; that is, it only becomes a proposition when the value of x is known. However, only a certain set of values is allowed for x, here the natural numbers.

Propositions that only become true or false by inserting certain values occur frequently. They are called predicates or propositional functions. Predicates are nothing more than mappings. For the predicate A that depends on x, we write $A(x)$. It is also possible to plug in several variables.

> **Examples**
>
> | $A(x,y)$: | x ist larger than y | $(x, y \in \mathbb{R})$ |
> |---|---|---|
> | $S(x)$: | x is a leap year | $(x \in \mathbb{N})$ ◀ |

▷ **Definition 2.12** Let M be a set and M^n the n-fold cartesian product of M. A n-ary *predicate P* is a mapping that assigns to each element from M^n a truth value t or f. M is called the *domain of individuals* of the predicate P.

Often we only speak of predicates without mentioning the number n of variables.

$A(x, y)$ is a mapping from $\mathbb{R}^2$ to $\{t, f\}$ with domain of individuals $\mathbb{R}$.

$S(x)$ is a mapping from $\mathbb{N}$ to $\{t, f\}$ with domain of individuals $\mathbb{N}$.

$A(x, y)$, $S(x)$ are "propositions" that are true or false depending on x, y. For each x, y we thus obtain propositions with which we can again do propositional logic, that is, we can again connect them with $\wedge, \vee, \neg, \rightarrow, \leftrightarrow$. This gives rise to formulas of predicate logic, which also is called first-order logic:

> **Example**
>
> $(S(x) \vee S(y)) \wedge \neg A(x, y)$, $x, y \in \mathbb{N}$, is a formula of predicate logic. ◀

Predicates can be used just like propositions. All rules for building formulas carry over. Each formula thus formed is again a predicate. We assume now that combined predicates have the same domain of individuals. This is not a significant restriction and saves us writing.

Connections of predicates are of course again predicates.

So far we have not gained anything essentially new. But there is another method, in addition to the insertion of values, to obtain new, interesting formulas from predicates.

If $A(x)$ is a predicate, the following alternatives for the truth of $A(x)$ are possible:

1. $A(x)$ is always true.
2. $A(x)$ is always false.
3. $A(x)$ is true for some x.
4. $A(x)$ is false for some x.

Examples

with the domain of individuals $\mathbb{R}$:

1. $A(x)$: $x^2 > -1$ is true for all $x \in \mathbb{R}$.
2. $A(x)$: $x^2 < -1$ is always false.
3. $A(x)$: $x^2 > 2$ is true for all $x > \sqrt{2}$ and $x < -\sqrt{2}$.
4. $A(x)$: $x^2 > 2$ is false for all $-\sqrt{2} \leq x \leq \sqrt{2}$. ◀

For cases 1 and 3, logicians have introduced abbreviations:

$\forall x A(x)$ ($\forall$: the universal quantifier), read: "for all x, $A(x)$ is true."

$\exists x A(x)$ ($\exists$: the existential quantifier), read "there is an x with $A(x)$."

$\exists x A(x)$ states there is at least one such x, there may be more, perhaps $A(x)$ is also true for all x. Cases 2 and 4 can be reduced to 1 and 3: $\forall x \neg A(x)$, respectively $\exists x \neg A(x)$, so we don't need new symbols for this.

Now it is interesting: If $A(x)$ is a predicate, then $\forall x A(x)$ and $\exists x A(x)$ are again propositions that are true or false.

Example

x is even $\Leftarrow$ is a predicate

for all $x \in \mathbb{N}$ x is even. $\Leftarrow$ is a false proposition

it exists $x \in \mathbb{N}$ which is even $\Leftarrow$ is a true proposition ◀

n-ary predicates can also be quantified.

Example

$$A(x, y): \qquad x^2 > y \qquad\qquad (x, y \in \mathbb{R}) \quad \Leftarrow \text{ is a binary predicate}$$

$$\forall x A(x, y): \quad \text{for all } x \text{ is } x^2 > y \quad (y \in \mathbb{R}) \quad \Leftarrow \text{ is a unary predicate}$$

$$\exists x A(x, y): \quad \text{it exists } x \text{ with } x^2 > y \quad (y \in \mathbb{R}) \quad \Leftarrow \text{ is a unary predicate} \blacktriangleleft$$

The binary predicate $A(x, y)$ is thus quantified by $\forall x A(x, y)$ or $\exists x A(x, y)$ to a unary predicate. Only y is still variable, x is bound by the quantifier. x is called in this case *bound variable*, y is called *free variable*.

Free variables are still available for further quantification:

$\exists y \forall x A(x, y)$: it exists y, such that for all x holds $x^2 > y$ $\Leftarrow$ is a proposition (true).

Attention: $\exists x \forall x A(x, y)$ is not allowed! x is already used for quantification, just "bound".

New propositions or predicates are thus obtained from existing predicates by linking or by quantifying:

▶   **Definition 2.13: Formulas of predicate logic**
   a) A predicate (according to Definition 2.12) is a formula of predicate logic.
   b) If P, Q are formulas of predicate logic with the same domain of individuals, then $(P \wedge Q)$, $(P \vee Q)$, $(\neg P)$, $(P \to Q)$, $(P \leftrightarrow Q)$ are formulas of predicate logic.
   c) If P is a formula of predicate logic with at least one free variable x, then $\forall x P$ and $\exists x P$ are formulas of predicate logic.

We need to look at this definition a little more closely. It is the first example of a so-called *recursive definition*. These play an important role in computer science. Why can't we just define it roughly like this: "Formulas of predicate logic are all possible combinations and quantifications of predicates"? For us humans, this is more readable and perhaps initially more understandable. However, Definition 2.13 provides more: It simultaneously provides a precise recipe for how to build formulas, and—almost even more importantly—a precise instruction for how to check a given expression to see if it is a valid formula or not. This works recursively, backwards, by repeatedly applying the definition until you arrive at elementary formulas.

Example

We investigate whether $P := \forall y \forall x (A(x, y) \wedge B(x))$ is a formula of predicate logic:
P is a formula according to c), if $Q := \forall x (A(x, y) \wedge B(x))$ is a formula.
Q is a formula according to c), if $R := (A(x, y) \wedge B(x))$ is a formula.
R is a formula according to b), if $A(x, y)$ and $B(x)$ are formulas.
$A(x, y)$ and $B(x)$ are formulas according to a).
So P is a formula of predicate logic. $\blacktriangleleft$

The brackets in Definition 2.13b) are necessary to exclude that, for example, $P \wedge Q \vee R$ would pass as a formula. However, for reasons of readability, brackets are often omitted in logical expressions. The following rules apply: $\neg$ is the most binding operator, $\wedge$ and $\vee$ bind stronger than $\rightarrow$ and $\leftrightarrow$.

The verification using such a recursive definition follows precise rules, so it can be carried out mechanically. Language definitions for programming languages are usually built recursively (for example, using the Backus Naur Form, BNF). The first thing a compiler does is *parse*, analyze whether the syntax of the program is correct.

As we have seen, quantification turns a n-ary predicate (for $n > 1$) into a $(n - 1)$-ary predicate and a 1-ary predicate into a proposition. Propositions are therefore also called 0-ary predicates and, according to Definition 2.13, are formulas of predicate logic.

All formulas of predicate logic, including propositions, will be referred to briefly as predicates. Sometimes I will not be quite as precise and simply say "proposition" again for a predicate, as I have already done in the example with the leap years in Sect. 2.1 after Definition 2.8.

Negation of Quantified Predicates

The negation of the predicate:

> *There is a real number that is not rational.*

is:

> *All real numbers are rational.*

The negation of the predicate:

> *For all even numbers $z > 2$ it holds that z is the sum of two prime numbers.*

is:

> *There is an even number $z > 2$ that is not the sum of two prime numbers.*

For all predicates it holds:

▶   **Theorem 2.14** If $A(x)$ is a predicate, then:

$$\neg(\exists x A(x)) \text{ is equivalent to } \forall x(\neg A(x)),$$
$$\neg(\forall x A(x)) \text{ is equivalent to } \exists x(\neg A(x)).$$

The negation therefore reverses the quantifier and the statement inside. Note this is not only true for one quantifier, but for any number of quantifiers. Let's calculate the negation of $\exists y \forall x A(x, y)$, for example:

$$\neg(\exists y \forall x A(x, y)) \quad \Leftrightarrow \quad \forall y \neg (\forall x A(x, y)) \quad \Leftrightarrow \quad \forall y \exists x (\neg A(x, y)).$$

All quantifiers are therefore reversed and the statement inside is negated. This also works with 25 quantifiers. If you carefully formulate the statements contained in a contradiction proof, the application of this trick is often a great relief!

2.4   Logic and Testing of Programs

For the sake of simplicity, we will assume for the moment a program runs without further user interaction and comes to an end after it starts. At least parts of a program (methods or parts of methods) meet this requirement. Such a program is nothing more than a set of rules for transforming variable values. Depending on the values of the variables at the beginning of a program run, a variable assignment results at the end of the program.

Now when is a program correct? This decision can only be made if the possible input states and the resulting output states have been specified beforehand. The program is then correct if every allowed input state generates the specified output state.

Before and after the program runs, the variables must therefore fulfill certain conditions. These conditions are nothing other than predicates: The variables are placeholders for the individuals. During a program run, individuals are inserted and thus a proposition results. Allowed states result in true propositions, forbidden states usually result in false propositions (Fig. 2.1).

The program is correct if, for each variable assignment, from V, after execution of A, the condition N follows. We write for this: $V \xrightarrow{A} N$.

After program implementation, the program is tested. For this purpose, as many allowed variables as possible are used and the result is checked. If this is done carefully, one can be confident that the program is correct. However, all allowed input states cannot usually be checked. Is it still possible to prove $V \xrightarrow{A} N$?

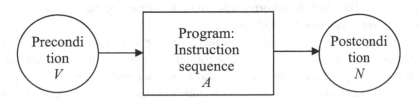

Fig. 2.1 Precondition and postcondition

Example

The specification of A is: "Exchange the content of the two integer variables a and b".
The preconditions and postconditions are:

V: Variable a has the value x, variable b has the value y. (x, y Integer)
N: Variable a has the value y, variable b has the value x. (x, y Integer) ◀

You can see here these are predicates, not propositions. We cannot say anything about the truth or falsity of V and N, but we will prove that for the following sequence of instructions A in (2.13) $V \underset{A}{\rightarrow} N$ is always true.

$$A: \quad a = a+b; b = a-b; a = a-b; \tag{2.13}$$

The predicate V is transformed into another predicate line by line when the instructions are executed:

$V \xrightarrow[a=a+b]{} V_1$ V_1: a has the value $x + y$, b has the value y.

$V_1 \xrightarrow[b=a-b]{} V_2$ V_2: b has the value $x + y - y = x$, a has the value $x + y$.

$V_2 \xrightarrow[a=a-b]{} V_3$ V_3: a has the value $x + y - x = y$, b has the value $x.V_3$ is the same as N.
This proves the validity of the program.

If you test this program cleverly, you will find another error: I have not considered over-flows. The precondition must restrict the range of individuals to avoid this. Can you do that?

For most of you, this will have been the first and the last program proof of your life. You can see that it is very, very time-consuming even for small programs. But in theory you can really prove the correctness of programs.

Practically, this has only been possible for small programs so far. But the importance of proving is increasing more and more. In particular, with safety-critical software, you cannot afford any errors. Smartcards are well suited for this: The software on it is manageable, and after all a cashcard is—at least from the perspective of the banks—always in potential enemy hands, freedom from errors is very important.

Since proving programs is a formal process, it can also be automated: You can write proof programs. But then their correctness must be proven by hand. This is a current topic in software development technology.

One drawback is a proof can only show that a specification is fulfilled. For this, on the one hand, the specification must be strictly formalized, that is, it must contain complete preconditions and postconditions. This is also a problem with larger software projects. On the other hand, errors in the specification cannot be detected by a program proof.

The formulation of preconditions and postconditions specifies the operation of a program. Even without proving programs, preconditions and postconditions help us a great deal in our daily programming work, and above all in program testing.

Example

Your project manager assigns you the task of writing a module that calculates the quotient of two integers. (That is his specification.) You implement the module and deliver the "program":

$$q = x/y;$$

The module is integrated into another module of your colleague and is used. If the Airbus pilot reads the message "floating point exception" on his display, there is trouble. Who is to blame? You are of the opinion your colleague should have checked the denominator for 0 before. Your colleague thinks the same of you.

> If you work in larger software projects once, you will notice that the correct and unambiguous agreement of system interfaces always belongs to the most difficult tasks of the whole process.

You are out of the woods when you deliver the corresponding preconditions and postconditions together with your program. To the program $q = x/y;$ belongs the precondition $V =$ "Variables x, y are from $\mathbb{Z}$ and $y \neq 0$" and the postcondition $N =$ "q has the value: integer part of x/y".

You could have specified differently: $V =$ "Variables x, y are from $\mathbb{Z}$", $N =$ "q has the value: integer part of x/y if $y \neq 0$ and is otherwise undefined" Then your implementation would have to look like this:

$$if(y != 0) \quad q = x/y; \blacktriangleleft$$

Get used to specifying program modules with preconditions and postconditions and implementing these conditions. If a condition is violated, an exception can be thrown. Even if we don't think about program proofs, this makes testing much easier and thus increases the quality of the program.

2.5   Comprehension Questions and Exercises

Comprehension Questions

1. Can you always decide from a proposition whether it is true or false?
2. Explain the difference between ↔ and ⇔.
3. And what is the difference between → and ⇒?
4. You can make a proposition from a unary predicate using two methods. What are they?

5. What does a formal language consist of?
6. Are there propositions that are not predicates?
7. Is there a difference between a tautology (which is always true) and the truth value "true"?

1. Check whether the following sentences are propositions and whether they are true or false:
   a) Either $5 < 3$ or from $2 + 3 = 5$ it follows that $3 \cdot 4 = 12$.
   b) When I am big, I am small.
   c) This sentence is not a proposition.
2. For the conjunctions "neither ... nor", "either ... or" and "namely ..., but not" set up truth tables and try to express these connectives with $\land, \lor, \neg$.
3. Form the negation of
   a) The triangle is right-angled and isosceles.
   b) Boris can speak Russian or German.
4. Prove a distributive and an associative law for the logical operations $\land, \lor$.
5. Show that $(\neg B \to B) \to B$ is a tautology. (If one can infer the assumption from the opposite of an assumption, then the assumption is true.) Is also $(\neg B \to B) \leftrightarrow B$ a tautology?
6. For each natural number n let a_n be a real number. In the second part of the book we will see the sequence a_n converges to zero if and only if the following logical proposition is true:

$$\forall t \exists m \forall n \left(n > m \to |a_n| < \frac{1}{t} \right), \quad m, n, t \in \mathbb{N}.$$

Form the negation of this proposition.
7. From Theorem 2.11 it does not follow that the product of the first n prime numbers $+ 1$ is a prime number. Find the smallest n with the property that $p_1 \cdot p_2 \cdot \ldots \cdot p_n + 1$ is not prime.

Natural Numbers, Mathematical Induction, Recursion

3

Abstract

At the end of this chapter

- you will know what an axiomatic system is, and know the axioms of natural numbers,
- you can represent natural numbers in different numeral systems,
- you will master the proof principle of mathematical induction and have proved some important theorems with mathematical induction,
- you will know what recursive functions are and know the relationship between recursion and induction,
- you will have carried out runtime calculations for some recursive algorithms.

3.1   The Axioms of Natural Numbers

In mathematics, new assertions (theorems) are obtained from given assertions by means of logical conclusions. This whole process has to start somewhere. At the beginning there must be a set of facts that are assumed to be true without being proven themselves. These unproven basic facts of a theory are called *axioms*.

In applied mathematics, one tries to imitate situations from the real world. The axioms should describe basic facts as simply and plausibly as possible. When this is done, the mathematician forgets the real world behind the axioms and starts to calculate. If he

Supplementary Information The online version contains supplementary material available at https://doi.org/10.1007/978-3-658-40423-9_3.

51

has created a good model, then the results he obtains can be applied to real world problems again.

The formulation of axiom systems is a lengthy and difficult task, often involving entire generations of mathematicians. The axiom system of a theory should meet the following requirements:

1. As few and simple (plausible) axioms should be formulated as possible, but of course also enough to be able to completely describe the theory.
2. The axioms must be independent of each other, that is, one axiom should not follow from the others.
3. The axioms must be free of contradiction.

The natural numbers and their basic properties are known to every human being from childhood. We become familiar with them, for example, by counting gummi bears (here the natural numbers denote the cardinality of a set) or by counting off (here the sequence of numbers is more interesting). Intuitively we can deal with natural numbers. But what are the characteristic properties we can rely on when we want to do mathematics? At the end of the 19th century, Giuseppe Peano formulated the following axiom system, which today is the generally accepted basis of the theory of natural numbers:

▶   **Definition 3.1: The Peano axioms**   The set $\mathbb{N}$ of natural numbers is characterized by the following properties:

(P1)   1 is a natural number.
(P2)   Each natural number has exactly one successor that is different from 1 and is a natural number.
(P3)   Different natural numbers have different successors.
(P4)   For each subset $M \subset \mathbb{N}$ with the properties
        a) $1 \in M$,
        b) if $n \in M$, then the successor is an element of M,

        it follows that: $M = \mathbb{N}$.

The first three axioms are self-evident, the fourth is plausible by "trying it out": $(1 \in M \Rightarrow 2 \in M \Rightarrow 3 \in M \Rightarrow \cdots)$.

The successor of the natural number n is denoted by $n + 1$.

Incredibly, one can deduce everything from these few axioms that is known about natural numbers. For example, one can define the addition $n + m$ as "form m-times the successor of n". It follows that addition is associative and commutative. Similarly, one obtains a description of multiplication. Further, from $\mathbb{N}$, with the help of the axioms, one can construct the integers $\mathbb{Z}$ and the rational numbers $\mathbb{Q}$ with all their rules. The whole number theory with its theorems can be deduced from it. We will not do this here, we simply believe from now on all the calculation rules that we know from school are correct.

3.2   The Mathematical Induction

We investigate the axiom (P4) a little more closely and deduce our first theorem from it:

▶   **Theorem 3.2: The proof principle of mathematical induction** Let $A(n)$ be an assertion about the natural number n. If it holds:

a) $A(1)$ is true,
b) for all $n \in \mathbb{N}$: if $A(n)$ is true, then $A(n+1)$ is also true.

Then $A(n)$ is true for all $n \in \mathbb{N}$.

Before the proof, a simple example to illustrate what is behind this principle:
Let $A(n)$ be the assertion: $1 + 2 + 3 + \cdots + n = \frac{n(n+1)}{2}$.
By trying it out, one finds that $A(1)$ is true: $1 = \frac{1 \cdot 2}{2}$, $A(2)$ is true: $1 + 2 = \frac{2 \cdot 3}{2}$, $A(3)$ is true: $1 + 2 + 3 = \frac{3 \cdot 4}{2}$ and so on. But what about $A(1000)$ or $A(10^{13})$?
Since there are infinitely many natural numbers, one would actually need an infinite amount of time to check all cases. The induction principle says: If we can generally infer from one number to its successor, we do not have to check everything individually. Then $A(4), A(5), \ldots, A(999), A(1000)$ is true and also the assertion for all further numbers. This way we can prove it in finite time. Therefore, mathematical induction is a very important and widely used proof method in mathematics.
We only show: $A(n) \rightarrow A(n+1)$:
Assume $A(n)$ is true, that is $1 + 2 + 3 + \cdots + n = \frac{n(n+1)}{2}$. Then it holds:

$$1 + 2 + 3 + \cdots + n + (n+1) = \frac{n(n+1)}{2} + (n+1) \quad \text{and}$$

$$\frac{n(n+1)}{2} + (n+1) = \frac{n(n+1) + 2(n+1)}{2} = \frac{n^2 + 3n + 2}{2} = \frac{(n+1)(n+2)}{2},$$

so $A(n+1)$ is also true.   □

The proof of the induction principle is quite simple: Let the conditions a), b) from Theorem 3.2 be satisfied for $A(n)$. Now let $M = \{n \in \mathbb{N} \mid A(n) \text{ ist true for } n\}$. Then $1 \in M$, and if $n \in M$, so is $n+1 \in \mathbb{N}$, and by axiom (P4) it follows: $M = \mathbb{N}$.   □

You should always carry out a mathematical induction according to the following scheme:

Base case: Show: $A(1)$ is correct.

Induction hypothesis: We assume: $A(n)$ is correct.
Induction step: Show: then $A(n + 1)$ is also correct.

If you formulate and write down the base case and induction hypothesis precisely, often a large part of the work is already done for simpler tasks. Another example:

▶   **Theorem 3.3** Let $q \in \mathbb{R}$. Then: $(1 - q)(1 + q + q^2 + q^3 + \cdots + q^n) = 1 - q^{n+1}$.

Base case: $A(1)$: $(1 - q)(1 + q^1) = 1 - q^2$ is correct.
Induction hypothesis: It holds $(1 - q)(1 + q + q^2 + q^3 + \cdots + q^n) = 1 - q^{n+1}$.
Induction step:

$$(1 - q)(1 + q + q^2 + q^3 + \cdots + q^n + q^{n+1})$$
$$= (1 - q)(1 + q + q^2 + q^3 + \cdots + q^n) + (1 - q)(q^{n+1})$$
$$\underset{\substack{\uparrow \\ \text{Induction} \\ \text{hypothesis}}}{=} \quad (1 - q^{n+1}) + (1 - q)(q^{n+1}) = (1 - q^{n+1}) + (q^{n+1} - q^{n+2}) = 1 - q^{n+2}.$$

In the language of logic from Sect. 2.3 $A(n)$ is a predicate with domain of individuals $\mathbb{N}$. As a proof principle, Theorem 3.2 states the formula

$$\underbrace{A(1)}_{\text{Base case}} \wedge \underbrace{\forall n(A(n) \to A(n + 1))}_{\text{Induction step}} \to \forall n A(n)$$

is always true.

You should always keep the following hints in mind when carrying out an induction:

1. Never forget the base case! If that is missing, the whole rest may be in vain.
2. The induction hypothesis "for $n \in \mathbb{N}$ it is assumed $A(n)$ is true" may not be confused with the assertion "$A(n)$ is true for *all* $n \in \mathbb{N}$".
3. If you did not use the induction hypothesiswhen carrying out the induction step, you did something wrong.

Always try to incorporate the hypothesis into the proof. If you do not use the hypothesis, your calculation may still be correct, but it is no longer an induction.

For practice, we carry out two slightly more complex induction proofs. At the same time, these are theorems we will need later.

▶   **Definition 3.4** For $n \in \mathbb{N}$ is $n! := 1 \cdot 2 \cdot 3 \cdot \ldots \cdot n$ (read: n factorial). It is $0! := 1$.

"$0! = 1$" looks a bit odd. But it has been shown in all formulas with $n!$ that it is clever to define it this way, and so we just do it. But we will have to pay special attention to this special case most of the time.

▶   **Theorem 3.5** Let M and N be finite sets of equal cardinality, let $|M| = |N| = n$. Then there are $n!$ bijective maps from M to N.

What is the meaning of this theorem? Let's look at the example $M = N = \{1, 2, 3\}$. Here we can still write down all the bijections in a table:

	φ_1	φ_2	φ_3	φ_4	φ_5	φ_6
1	1	1	2	2	3	3
2	2	3	1	3	1	2
3	3	2	3	1	2	1

In this case (and always when $M = N$ is), $n!$ is exactly the number of different arrangements of the elements of the set M. So a soccer coach has $11! = 39\,916\,800$ ways to position his 11 players in different positions.

To prove Theorem 3.5:

Base case: $n = 1$: There is exactly one bijection, the identity.

Induction hypothesis: For all sets of size n, the assertion is true.

Induction step: Let $M = \{x_1, x_2, x_3, \ldots, x_{n+1}\}$ and $N = \{y_1, y_2, y_3, \ldots, y_{n+1}\}$ sets of size $n + 1$. Compute the numbers of bijections between N and M.

The number of bijections φ with $\varphi(x_{n+1}) = y_1$ is equal to the number of bijections from $M \setminus \{x_{n+1}\}$ to $N \setminus \{y_1\}$. That is, according to the induction hypothesis $n!$.

Likewise,   the number of bijections with $\varphi(x_{n+1}) = y_2$ is equal to $n!$,
                                    the number of bijections with $\varphi(x_{n+1}) = y_3$ is equal to $n!$,
                                    the number of bijections with $\varphi(x_{n+1}) = y_{n+1}$ is equal to $n!$.

Each bijection from M to N appears in this list, so we have a total of $(n + 1)n! = (n + 1)!$ bijections.   □

The natural numbers fall more or less from the sky, but not the way we write them down: that has to do with the fact that we have 10 fingers. Computers work with 0 and 1, for them the 2-finger system, the base 2 system or *binary system*, is suitable.

In a byte you can place exactly $2^8 = 256$ numbers. The base 256 system is difficult for humans to handle; but we often use the base 16 system, the *hexadecimal system*, because in this one a byte is represented by two digits: the first 4 bit by the first digit, the second 4 bit by the second digit.

For some computational purposes, the precision of a standard integer in the computer is no longer sufficient. The Integer data type usually contains 32 bit and covers a range of 2^{32}, that is, about 4.3 billion numbers. If you have to work with larger numbers, you need large number arithmetic. A common implementation is to work in the base 256 system. One byte represents a digit, a number is represented by an array of bytes.

The following theorem gives a recipe for how we can obtain the representation of a number in different numeral systems:

▶   **Theorem 3.6: Representation of natural numbers in different base systems** Let $b > 1$ be a fixed natural number (the base). Then every natural number n can be uniquely represented in the form

$$n = a_m b^m + a_{m-1} b^{m-1} + \cdots + a_1 b^1 + a_0 b^0. \tag{3.1}$$

Here $m \in \mathbb{N}_0$, $a_i \in \{0, 1, \ldots, b-1\}$ and $a_m \neq 0$.

The possible coefficients $0, 1, \ldots, b-1$ we call the *digits* of the numeral system with base b. For all natural numbers b it is defined $b^0 := 1$, so the last summand in (3.1) is equal to a_0.

Proof by induction on n:

Base case: For $n = 1$ it is $a_0 = 1$ and $m = 0$.

Induction hypothesis: The number n has the unique representation

$$n = a_m b^m + a_{m-1} b^{m-1} + \cdots + a_1 b^1 + a_0 b^0.$$

Induction step: It is

$$n + 1 = (a_m b^m + a_{m-1} b^{m-1} + \cdots + a_1 b^1 + a_0 b^0) + 1 b^0$$
$$= a_m b^m + a_{m-1} b^{m-1} + \cdots + a_1 b^1 + (a_0 + 1) b^0.$$

If $a_0 + 1 < b$, we have the desired representation. If $a_0 + 1 = b$, then we have:

$$n + 1 = a_m b^m + a_{m-1} b^{m-1} + \cdots + (a_1 + 1) b^1 + 0 b^0.$$

If $a_1 + 1 < b$, we now have the desired representation. If $a_1 + 1 = b$, then we have:

$$n + 1 = a_m b^m + a_{m-1} b^{m-1} + \cdots + (a_2 + 1) b^2 + 0 b^1 + 0 b^0$$

and so on. If for one of the a_i holds $a_i > b - 1$, the process ends at this i. If always $a_i = b - 1$, we finally get the representation

$$n + 1 = 1 b^{m+1} + 0 b^m + 0 b^{m-1} + \cdots + 0 b^1 + 0 b^0.$$

I would like to forego the proof of the uniqueness of this representation.   □

Examples

$$b = 10: \quad 256 = \qquad\qquad 2 \cdot 10^2 + 5 \cdot 10^1 + 6$$
$$b = 16: \quad 256 = \qquad\qquad 1 \cdot 16^2 + 0 \cdot 16^1 + 0$$
$$255 = \qquad\qquad\qquad 15 \cdot 16^1 + 15$$
$$b = 2: \quad 256 = 1 \cdot 2^8 + 0 \cdot 2^7 + 0 \cdot 2^6 + \cdots + 0 \cdot 2^1 + 0$$
$$255 = \qquad 1 \cdot 2^7 + 1 \cdot 2^6 + \cdots + 1 \cdot 2^1 + 1$$

◀

We introduce a shortened notation. For

$$n = a_m b^m + a_{m-1} b^{m-1} + \cdots + a_1 b^1 + a_0 b^0$$

we write $(a_m a_{m-1} a_{m-2} \cdots a_0)_b$.

So it is $256 = (256)_{10} = (100)_{16} = (1\,0000\,0000)_2$.

$255 = (15\,15)_{16}$ is easily misunderstood. If $b > 10$, one therefore uses new abbreviations for the digits: A for 10, B for 11 and so on. So $255 = (FF)_{16}$.

But how does one find the coefficients of the number representation for a given basis? If we divide n by b, we get

$$(a_m b^m + \cdots + a_1 b^1 + a_0 b^0): b = a_m b^{m-1} + \cdots + a_1 b^0 \text{ remainder } a_0.$$

If we divide the quotient by b again, we get:

$$(a_m b^{m-1} + \cdots + a_2 b^1 + a_1 b^0): b = a_m b^{m-2} + \cdots + a_2 b^0 \text{ remainder } a_1.$$

We continue like this: We divide the quotients by b over and over again until finally the quotient is 0:

$$a_m b^0: b = 0 \text{ remainder } a_m.$$

Thus, when we continue to divide by b, the remainders are exactly the digits of the number representation in reverse order.

Example: 456 in the base 3 system

$$456 : 3 = 152 \quad \text{remainder } 0$$
$$152 : 3 = 50 \quad \text{remainder } 2$$
$$50 : 3 = 16 \quad \text{remainder } 2$$
$$16 : 3 = 5 \quad \text{remainder } 1$$
$$5 : 3 = 1 \quad \text{remainder } 2$$
$$1 : 3 = 0 \quad \text{remainder } 1$$

Then $456 = (121220)_3$. ◀

Now two reformulations of the induction principle are sometimes very useful for calculation. They can be derived from theorem 3.2. We omit the proofs here:

▶   **Theorem 3.7: The induction principle II (shifting the base case)** Let $A(n)$ be an assertion about the integer n and let $n_0 \in \mathbb{Z}$ be given. If it holds:

a) $A(n_0)$ is true,
b) for all $n \geq n_0$: If $A(n)$ is true, then $A(n+1)$ is true as well.

Then $A(n)$ is true for all $n \geq n_0$.

▶   **Theorem 3.8: The induction principle III (generalized induction hypothesis)** Let $A(n)$ be an assertion about the integer n and let $n_0 \in \mathbb{Z}$ be given. If it holds:

a) $A(n_0)$ is true,
b) for all $n \geq n_0$: From the validity of $A(k)$ for all $n_0 \leq k \leq n$ follows $A(n+1)$.

Then $A(n)$ is true for all $n \geq n_0$.

3.3   Recursive Functions

Often functions with the domain $\mathbb{N}$ are not given explicitly, but only a rule for how to calculate later function values from earlier ones. For example, one can define $n!$ by the rule:

$$
n! = \begin{cases} 1 & \text{for } n = 0 \\ n(n-1)! & \text{for } n > 0 \end{cases}. \tag{3.2}
$$

This provides a calculation rule for all $n \in \mathbb{N}$.

When is such a backward-stepping calculation possible? Let us recall a function with domain $\mathbb{N}$ is nothing but a sequence $(a_n)_{n \in \mathbb{N}}$. (See Example 3 in Sect. 1.3 after Definition 1.14.) For sequences, I formulate the recursion theorem:

▶   **Theorem 3.9** A sequence $(a_n)_{n \in \mathbb{N}}$ is completely determined if it is given:

a) a_1,
b) for all $n \in \mathbb{N}$, $n > 1$ a rule F_n, with which a_n can be computed from a_{n-1}:
   $a_n = F_n(a_{n-1})$.

The proof is carried out—as it should be—with mathematical induction:

Base case: a_1 is computable.
Induction hypothesis: Let a_n be computable.
Induction step: Then $a_{n+1} = F_{n+1}(a_n)$ is computable.   □

Just like the induction principle, the recursion theorem can also be easily generalized: one does not have to start with 1, and a_n can be calculated from $a_1, a_2, \ldots, a_{n-1}$ or from a part of these values. The latter are the recursions of higher order, see Definition 3.10.

The recursion principle allows for exact definitions of fuzzy formulations, for example, those that use dots. We have already seen this with the example $n!$ The rule for the sequence $a_n = n!$ is:

$$0! = 1, \ n! = F_n((n-1)!) := n \cdot (n-1)!.$$

Although this is not as readable for us humans as the one from Definition 3.4:

$$n! := 1 \cdot 2 \cdot 3 \cdot \ldots \cdot n,$$

a computer can not do anything with the dots from the second definition, but it is able to interpret the recursive definition. Another

---

Example

A precise definition of the summation sign is:

$$\sum_{i=1}^{n} a_i := \begin{cases} 0 & \text{if } n = 0 \\ \sum_{i=1}^{n-1} a_i + a_n & \text{else} \end{cases} \qquad \text{instead of } \sum_{i=1}^{n} a_i = a_1 + a_2 + \cdots + a_n,$$

or in the notation of Theorem 3.9:

$$\sum_{i=1}^{0} a_i = 0, \quad \sum_{i=1}^{n} a_i = F_n\left(\sum_{i=1}^{n-1} a_i\right) := \sum_{i=1}^{n-1} a_i + a_n.$$

◀

Most higher programming languages support recursive ways of writing. In C++, for example, a program for the recursive calculation of $n!$ looks like this:

```
long nfac(int n){
    if(n==0)
        return 1;
    else
        return n*nfac(n-1);
}
```

This is a literal translation of (3.2) into C++ syntax.

Recursion is a very elegant and often used method in computer science. The syntax of programming languages is usually defined recursively. There are many algorithms in which the principle of recursion leads not only to a better clarity, but also to much higher efficiency than traditional iterative algorithms. Look at the two following recursive definitions of the power function:

$$x^n = \begin{cases} 1 & \text{if } n = 0 \\ x^{n-1}x & \text{if } n > 0 \end{cases} \tag{3.3}$$

or

$$x^n = \begin{cases} 1 & \text{if } n = 0 \\ x^{\frac{n}{2}}x^{\frac{n}{2}} & \text{if } n \text{ is even} \\ x^{n-1}x & \text{if } n \text{ is odd} \end{cases} \tag{3.4}$$

Check for a few numbers n (for example $n = 15, 16, 17$), how many multiplications you need to calculate x^n. (3.3) corresponds in runtime to the iterative calculation, (3.4) is a huge improvement.

> Such algorithms are called *Divide-and-Conquer-Algorithms*: The problem is solved by reducing it to several subproblems of the same type, but with less complexity.

With the following small checklist you should check every recursive program to avoid unpleasant surprises:

1. Is there a termination condition?
2. Does the recursive function call have a smaller argument than the function call itself?
3. Is the termination condition reached in any case at some point?

Recursions of Higher Order

If the function value of n depends not only on the immediately preceding function value, but on further function values, one speaks of a recursion of higher order. A well-known example of this is the Fibonacci sequence:

$$a_n = a_{n-1} + a_{n-2},$$
$$a_0 = 0, \ a_1 = 1.$$

The first function values are: 0, 1, 1, 2, 3, 5, 8, 13, 21, 34.

▶ **Definition 3.10** A *recursion of k-th order* is an equation

$$a_n = F_n(a_{n-1}, a_{n-2}, \ldots, a_{n-k}), \quad n \in \mathbb{N},$$

with which the value of the sequence at the position n can be calculated from the k previous sequence values. A sequence $(a_n)_{n \in \mathbb{N}}$, which for all n satisfies the recursion equation, is called *solution of the recursion*.

Linear recursions play an important role in the theory of recursive functions:

▶ **Definition 3.11** A recursion of k-th order of the form

$$a_n = \lambda_1 a_{n-1} + \lambda_2 a_{n-2} + \ldots + \lambda_k a_{n-k} + g(n)$$

with $\lambda_1, \lambda_2, \ldots \lambda_k \in \mathbb{R}$, $g(n) \in \mathbb{R}$ is called a *linear recursion with constant coefficients*. If $g(n) = 0$ for all n, the recursion is called *homogeneous*, otherwise *inhomogeneous*.

Recursive functions, especially those of higher order, are also called *difference equations*. To uniquely determine such a function, more than just one initial value is necessary. In general it holds

▶ **Theorem 3.12** A recursion of k-th order has a unique solution if the k consecutive values $a_1, a_2, \ldots, a_k$ are given.

Is it possible to give a closed form for this solution in addition to the recursive description, that is, a calculation rule that directly calculates a_n from the initial values $a_1, a_2, \ldots, a_k$? Sometimes that works. For example, the recursive function of first order

$$a_n = x \cdot a_{n-1}$$
$$a_1 = x$$

has the solution $a_n = x^n$, we have already seen this in the last section. In general, however, this is not possible.

> A famous recursive function that looks very simple but for which there is no explicit solution is the logistic equation. It is $a_n = \alpha a_{n-1}(1 - a_{n-1})$. This equation plays an important role in chaos theory. Depending on the value α, the function shows chaotic behavior, it jumps wildly back and forth. Calculate the function values for a few values α between 2 and 4 in a small program!

However, an explicit solution can always be calculated for linear recursions with constant coefficients. I would like to do this for constant recursions of first order with a constant inhomogeneous term.

▶ **Theorem 3.13** The first-order linear recursion

$$a_n = \lambda \cdot a_{n-1} + c, \quad \lambda, c, a_1 \in \mathbb{R},$$

has the solution

$$a_n = \begin{cases} a_1 \cdot \lambda^{n-1} + \frac{1-\lambda^{n-1}}{1-\lambda} c & \text{if } \lambda \neq 1 \\ a_1 + (n-1)c & \text{if } \lambda = 1 \end{cases}.$$

We solve the problem by successively substituting the function values:

$$
\begin{aligned}
a_n &= \lambda a_{n-1} + c \\
&= \lambda(\lambda a_{n-2} + c) + c \\
&= \lambda^2 a_{n-2} + \lambda c + c \\
&= \lambda^2(\lambda a_{n-3} + c) + \lambda c + c = \cdots \\
&= \lambda^{n-1} a_1 + (\lambda^{n-2} + \lambda^{n-3} + \cdots + 1)c.
\end{aligned}
\tag{3.5}
$$

Now we have to distinguish whether $\lambda = 1$ or $\lambda \neq 1$. In the first case, the sum is $\lambda^{n-2} + \lambda^{n-3} + \cdots + 1 = n - 1$ and thus $a_n = a_1 + (n-1)c$, in the second case, according to Theorem 3.3, $\lambda^{n-2} + \lambda^{n-3} + \cdots + 1 = (1 - \lambda^{n-1})/(1 - \lambda)$ and thus

$$
a_n = a_1 \cdot \lambda^{n-1} + \frac{1 - \lambda^{n-1}}{1 - \lambda} c.
$$

In this calculation, "..." appears again; I said that with the help of recursion, such inaccuracies can be avoided. If that bothers you, then "guess" first, as in (3.5), how the sequence a_n looks, and then you can use induction to prove that exactly this sequence satisfies the conditions of the theorem.

Example

On a savings account you have an initial balance of K_a and save S Euros per month. The interest rate is Z. How much is your balance after n months? Let k_n be the balance at the beginning of the n-th month. Then:

$$
k_n = (1 + Z)k_{n-1} + S, \quad k_1 = K_a.
$$

So this is a recursion to which we can apply Theorem 3.13. Let $q = 1 + Z$. At the end of month n, the balance is $K_e = k_{n+1}$:

$$
K_e = K_a q^n + \frac{1 - q^n}{1 - q} S.
$$

◀

For higher-order linear recursions, a solution can always be given, but the calculation is much more complex and shows an amazing similarity to the theory of differential equations, which we will deal with in Chap. 17. Difference equations can be seen as the discrete brother of differential equations. I don't go into that any further.

The closed form of the Fibonacci sequence, for example, is

$$
a_n = \frac{1}{\sqrt{5}}\left(\left(\frac{1}{2} + \frac{\sqrt{5}}{2}\right)^n + \left(\frac{1}{2} - \frac{\sqrt{5}}{2}\right)^n\right)
$$

Try it out!

Runtime Calculations for Recursive Algorithms

Not every time the recursive programming of algorithms leads to the goal. You have to be careful there and estimate the computational effort and the memory requirements. Often, recursive functions for runtime calculation can be easily derived from recursive algorithms. With what we have just learned, we will calculate some such functions explicitly.

The runtime of an algorithm is essentially dependent on the performance of the machine and on the size n of input variables of the algorithm. The machine provides a constant factor. The more powerful the machine is the smaller the factor is. The size n of input values (numbers/words/length of numbers/...) makes itself felt in a function $f(n)$ for the runtime. This function $f(n)$ we want to determine more closely at some examples. First for a non-recursive algorithm:

1. Calculation of $n!$

runtime:

```
fac = 1;                          ←α = const
for(int i = 1; i <= n; ++i)
    fac *= i;                     ←β = const
```
$$\left. \right\} n \cdot \beta$$

Here you get $f(n) = n\beta + \alpha$. The runtime increases linearly with n.

2. Sorting with *Selection Sort*. The recursive algorithm is:

runtime:

sort list with n entries	$←f(n)$
find smallest element in list	$←\alpha n$
put it on front of the list	$←\beta$
sort list with $n - 1$ entries	$←f(n - 1)$

Here you get for the runtime the recursive function specification

$$f(n) = f(n - 1) + \alpha n + \beta$$
$$f(1) = \gamma$$

with machine-dependent constants α, β and γ.

This is a linear recursion of first order, unfortunately with a non-constant inhomogeneous part, so that Theorem 3.13 cannot be applied. Please calculate yourself, similar to the proof of this theorem, that a linear homogeneous recursion of first order of the form $a_n = a_{n-1} + g(n)$ has the solution:

$$a_n = a_1 + \sum_{i=2}^{n} g(i).$$

This gives for $g(n) = \alpha n + \beta$ using the summation formula, which I have derived after Theorem 3.2:

$$\sum_{i=2}^{n} g(i) = \sum_{i=2}^{n}(\alpha i + \beta) = \alpha \sum_{i=2}^{n} i + \beta(n-1) = \alpha\left(\frac{n(n+1)}{2} - 1\right) + \beta(n-1),$$

and from that:

$$f(n) = \gamma + \alpha\left(\frac{n(n+1)}{2} - 1\right) + \beta(n-1)$$

$$= \frac{\alpha}{2}n^2 + \underbrace{\frac{\alpha}{2}n - \alpha + \beta(n-1) + \gamma}_{\text{no longer plays a role with a large } n}$$

The runtime increases quadratically, one also says $f(n)$ *is of order* n^2. The orders of magnitude of runtimes we will look at a little more closely in Part 2 of the book, see the Big O-notation in Sect. 13.1.

3. You are certainly familiar with the Towers of Hanoi: n wooden discs with decreasing diameters are stacked on top of each other. You can move one disc at a time and the task is to rebuild the pyramid in the same order at another location. For this purpose, a third storage space is available, but a larger disc may never be placed on a smaller one.

   If you don't know the solution, try it out (for example, with different sized books). You will see it always works and the algorithm can be described recursively as follows:

	runtime:
move the tower of height n:	$\leftarrow f(n)$
move the tower of height $n-1$ to auxiliary position	$\leftarrow f(n-1)$
move bottom disc to target position	$\leftarrow \gamma$
move the tower of height $n-1$ onto this disc	$\leftarrow f(n-1)$

   Amazing: There is no mention here of how a tower can be moved. Nevertheless, this is a precisely programmable prescription!

Now we have $f(n) = 2f(n-1) + \gamma$, $f(1) = \gamma$ and Theorem 3.13 is applicable, with $\lambda = 2$ and $c = \gamma$. We get:

$$f(n) = f(1) \cdot 2^{n-1} + \frac{1 - 2^{n-1}}{1 - 2}\gamma = \gamma \cdot 2^{n-1} + (2^{n-1} - 1)\gamma = 2^n\gamma - \gamma.$$

$f(n)$ is of order 2^n; the runtime grows exponentially. Exponential algorithms are considered unusable for computation by machines.

According to legend, a monastic order in Hanoi believed that the world would end if the problem was solved for 64 discs. Do the monks have a point?

4. Sorting with *Merge Sort*. The recursive algorithm is:

	runtime:
sort a list with n entries	$\leftarrow f(n)$
sort first half of the list with $n/2$ entries	$\leftarrow f(n/2)$
sort second half of the list with $n/2$ entries	$\leftarrow f(n/2)$
merge both lists	$\leftarrow \gamma n$

The recursive function definition for the runtime is:

$$f(n) = 2f(n/2) + \gamma n$$
$$f(1) = \alpha.$$

This problem has two hooks: On the one hand, $n/2$ is not always an integer (which is also annoying when implementing), on the other hand, the inhomogeneous term γn is not constant, so we cannot apply Theorem 3.13. We will calculate the function f by hand. First, let's take the simplest case where $n = 2^m$. Then we get:

$$f(2^m) = 2f(2^{m-1}) + \gamma 2^m.$$

Now let's set $g(m) := f(2^m)$, so we at least get a linear recursion of first order for g:

$$f(2^m) = g(m) = 2g(m-1) + \gamma 2^m$$
$$= 2(2g(m-2) + \gamma 2^{m-1}) + \gamma 2^m = 2^2 g(m-2) + \gamma 2^m + \gamma 2^m$$
$$= 2^2(2g(m-3) + \gamma 2^{m-2}) + 2\gamma 2^m = 2^3 g(m-3) + 3\gamma 2^m = \ldots$$
$$= 2^m g(0) + m\gamma 2^m = 2^m f(1) + m\gamma 2^m.$$

Now we replace 2^m with n and m with $\log_2 n$. This gives us

$$f(n) = \underbrace{n\alpha}_{\text{does not play a role}} + (\log_2 n) \cdot \gamma \cdot n.$$

For intermediate values between the powers of two, this formula also applies approximately, and so we get the statement that Merge Sort is of order $n \log_2 n$.

We will deal with the logarithm in the second part of the book. Just so much: The inverse function to the function $x \mapsto a^x$ is called the logarithm to the base a, that is $x \mapsto a^x = y \Rightarrow x = \log_a y$ and here: $2^m = n \Rightarrow m = \log_2 n$. The logarithm to the base 2 plays a particularly important role in computer science.

3.4   Comprehension Questions and Exercises

Comprehension Questions

1. Can an axiom be proven?
2. Can a proof of a theorem be correct if the induction hypothesisn is not used in the induction step?
3. Why is the hexadecimal system so important for computer scientists?
4. When representing a natural number in the base b system, can a negative base be used?
5. In the calculation of the base b representation of a number, a continued division with remainder is performed. Can you really be sure that this process always ends?
6. An integer in common programming languages has a defined size, for example 4 bytes. How would you proceed if you wanted to implement a class that contains arbitrarily large integers?
7. What do you have to pay attention to when implementing recursive programs?

Exercises

1. Show that for $n \geq 3$ it holds: $2n + 1 \leq 2^n$.
2. For which natural numbers does $n^2 \leq 2^n$ hold?
3. Show by mathematical induction:
   $n \in \mathbb{N}, x \in \mathbb{R}, x \geq -1 \Rightarrow (1 + x)^n \geq 1 + nx$.
4. Show by mathematical induction:
   $\sum_{k=1}^{n} (2k - 1) = n^2$.
5. Show by mathematical induction:
   $\sum_{i=0}^{n} 2^i = 2^{n+1} - 1$.

   This is due to the fact that when representing numbers in binary form, it holds: $\underbrace{111\ldots111}_{n} + 1 = 1\underbrace{000\ldots000}_{n}$. Formulate corresponding statements for other bases.

6. Show by mathematical induction: Every non-empty subset of the natural numbers has a smallest element.

This property plays a big role in number theory. It is also called: The natural numbers are "well-ordered". The rational numbers or the real numbers do not have this property!

7. Check the correctness of the following induction proof:

Assertion:	All horses have the same color.
Base case:	A horse obviously has the same color.
Induction hypothesis:	In every set of n horses all have the same color.
Induction step:	Let $n + 1$ horses be given. If you take out one horse, then the remaining n horses have the same color according to the induction hypothesis. If you add the $(n + 1)$th

horse again and remove another horse, then the remaining horses n have the same color again. Since at least one horse is contained in both n-element horse sets, all horses of the n-element sets have the same color and thus all $n + 1$ horses.

8. Representation of numbers in different numeral systems:
   a) Represent the number $(10000)_{10}$ in the base 4, 5 and 8 system.
   b) Represent $(FB97B7FE)_{16}$ in the binary system.
   c) Represent $(3614)_7$ in the base 11 system.
9. Examine the following recursive algorithm:

$$x^n = \begin{cases} 1 & \text{if } n = 0 \\ x^{\frac{n}{2}} \cdot x^{\frac{n}{2}} & \text{if } n \text{ is even} \\ x^{n-1} \cdot x & \text{else} \end{cases}$$

Calculate the runtime of this algorithm depending on n in the special cases $n = 2^m$ and $n = 2^m - 1$. Check that this is the best resp. the worst case (that is, the real runtime is in between).

Implement the algorithm and check your calculation by counting the number of recursive calls made during execution.

Some Number Theory

4

Abstract

The central learning content of this chapter is

- the binomial coefficient and its most important properties,
- the use of the mathematical summation symbol $\sum$,
- rules for divisibility of natural numbers and the Euclidean algorithm,
- calculating with residue classes,
- the application of modular arithmetic for hashing.

4.1   Combinatorics

We begin this chapter with the compilation of some theorems on combinatorics. Combinatorics describes ways to select, combine, or permute elements from given sets. The first result we have already formulated in Theorem 3.5 as an example for mathematical induction. Here it is again for the sake of remembrance:

▶   **Theorem 4.1** The elements of a set with n elements can be arranged in exactly $n!$ different ways.

▶   **Theorem 4.2** Let M be a finite set with $n > 0$ elements. Then there are 2^n different mappings from M to $\{0, 1\}$.

Supplementary Information The online version contains supplementary material available at https://doi.org/10.1007/978-3-658-40423-9_4.

Example

Let $M = \{a, b, c, d\}$:

	φ_1	φ_2	φ_3	$\cdots$	φ_{15}	φ_{16}
a	0	0	0	$\cdots$	1	1
b	0	0	0	$\cdots$	1	1
c	0	0	1	$\cdots$	1	1
d	0	1	0	$\cdots$	0	1

◀

At first sight, it is not at all clear what this result should mean, but it has quite practical applications:

Interpretation 1: Through one data element of n bit length, exactly 2^n symbols can be represented (encoded).

You can see this in the example: Each mapping determines a unique combination of 0 and 1, so each mapping can be associated with a symbol. In a byte (8 bits), you can encode 256 symbols, in 2 bytes 65 536, in 4 bytes 2^{32}, that is about 4.3 billion. 4 bytes is the size of an integer, for example, in Java.

Interpretation 2: A set with n elements has exactly 2^n subsets.

Each subset can be uniquely identified with a mapping. This subset is the set of elements that are mapped to 1. Precisely formulated:

$$\Phi : P(M) \rightarrow \{\text{mappings from } M \text{ to } \{0, 1\}\}$$

$$A \mapsto \varphi_A \text{ mit } \varphi_A(x) = \begin{cases} 1 & \text{if } x \in A \\ 0 & \text{if } x \notin A \end{cases}$$

is a bijective mapping, so the two sets are equipotent (compare Definition 1.20). In the example above, φ_1 and $\emptyset$ correspond to each other, φ_2 and $\{d\}$, ..., φ_{15} and $\{a, b, c\}$, φ_{16} and M.

It is typical for mathematicians that they formulate theorems as generally and abstractly as possible. But by simple specializations one is often able to achieve different concrete results.

The proof of Theorem 4.2 is by induction:

Base case: $M = \{m\}$: $\varphi_1 : m \mapsto 0$, $\varphi_2 : m \mapsto 1$. There are no more map-
                                pings.

Induction hypothesis: Let the assertion be true for any set with n elements.

Induction step: Let $M = \{x_1, x_2, \ldots, x_n, x_{n+1}\}$ be a set with $n + 1$ elements. Then there are 2^n different mappings with $\varphi(x_{n+1}) = 0$. These correspond exactly to the 2^n mappings to $\{0, 1\}$ on the set $\{x_1, x_2, \ldots, x_n\}$.

Likewise, there are 2^n different mappings with $\varphi(x_{n+1}) = 1$. There are no other possibilities, and thus M has exactly $2^n + 2^n = 2 \cdot 2^n = 2^{n+1}$ mappings to $\{0, 1\}$.   □

▶ **Definition 4.3** Let $n, k \in \mathbb{N}_0, n \geq k$. Then

$$\binom{n}{k} := \text{number of subsets of cardinality } k \text{ of a set with } n \text{ elements}$$

is called the *binomial coefficient n choose k*.

Obviously, for all $n \in \mathbb{N}$: $\binom{n}{0} = 1 = \binom{n}{n}$. Since there are n 1-element subsets, it follows $\binom{n}{1} = n$. The set $\{a, b, c\}$ has the 2-element subsets $\{a, b\}, \{a, c\}, \{b, c\}$, so that $\binom{3}{2} = 3$. But what is $\binom{69}{5}$? You see the concrete meaning of this question: How many possibilities are there to select five numbers from 69? So we need a way to calculate the binomial coefficient. The first method for this provides the following recursive description of the binomial coefficient:

▶ **Theorem 4.4** Let $n, k \in \mathbb{N}_0, n \geq k$. Then:

$$\binom{n}{k} = \begin{cases} 1 & \text{if } k = 0 \text{ or } k = n \\ \binom{n-1}{k} + \binom{n-1}{k-1} & \text{else} \end{cases} \tag{4.1}$$

This can be proven directly: For $k = 0$ and $k = n$ we have already seen this is right, so we can assume $0 < k < n$.

Let $M = \{x_1, x_2, \ldots, x_n\}$. The k-element subsets N of M fall into two types:

1st type of sets: Those with $x_n \notin N$. These correspond exactly to the k-element subsets of $M \setminus \{x_n\}$ and there are $\binom{n-1}{k}$ of them.

2nd type of sets: Those with $x_n \in N$. These correspond exactly to the $(k-1)$-element subsets of $M \setminus \{x_n\}$, so we have $\binom{n-1}{k-1}$ of them. In total, there are $\binom{n-1}{k} + \binom{n-1}{k-1}$ subsets.   □

Behind this recursive rule lies the possibility of calculating the binomial coefficient with the help of Pascal's triangle. Each coefficient results as sum of the two coefficients diagonally above:

$$\binom{0}{0} = 1$$

$$\binom{1}{0} = 1 \qquad \binom{1}{1} = 1$$

$$\binom{2}{0} = 1 \qquad \binom{2}{1} = 2 \qquad \binom{2}{2} = 1 \tag{4.2}$$

$$\binom{3}{0} = 1 \qquad \binom{3}{1} = 3 \qquad \binom{3}{2} = 3 \qquad \binom{3}{3} = 1$$

It takes a while to calculate 69 choose 5 in this way; fortunately, we can take our computer's help and write a small recursive program very quickly that implements the calculation rule (4.1). In the end, we finally get the number 11 238 513. So you really have to wait a long time for the five in Powerball.

> If you implement this program, you will notice that the calculation of 69 choose 5 takes an amazingly long time. Remember the warnings about using recursive functions from Sect. 3.3? Try to determine the runtime of the algorithm as a function of n and k. You will find more about this in the exercises for this chapter.

Those of you who already know the binomial coefficient from before may have seen another definition for it. Of course, this is equivalent to our Definition 4.3; we formulate it here as a theorem:

▶   **Theorem 4.5** Let $n, k \in \mathbb{N}_0, n \geq k$. Then:

$$\binom{n}{k} = \frac{n!}{k!(n-k)!} \tag{4.3}$$

We can use the recursion rule (4.1) to prove this amazing relationship with induction. I want to carry out the proof here because a new aspect arises: Two natural numbers appear in the proof (n and k) which both have to be dealt with. With induction, you have to specify in this case for which of the numbers this is carried out. We do an "induction on n" and then for n we have to examine all admissible k in the base case and in the induction step.

Base case: $n = 1 \, (k = 0 \text{ or } 1)$: $\binom{1}{0} = \frac{1!}{0!1!} = 1$ and $\binom{1}{1} = \frac{1!}{1!0!} = 1$ are correct.
Induction hypothesis: The assertion is correct for n and for all $0 \leq k \leq n$.
Induction step: Show: The assertion is correct for $n + 1$ and for all $0 \leq k \leq n+1$.

For $k = 0$ and $k = n + 1$ you can verify the assertion directly:

$$\binom{n+1}{0} = \frac{(n+1)!}{0!(n+1)!} = 1 \quad \text{and} \quad \binom{n+1}{n+1} = \frac{(n+1)!}{(n+1)!0!} = 1.$$

So it can be assumed that $0 < k < n + 1$. Then, according to Theorem 4.4,

$$\binom{n+1}{k} = \binom{n}{k} + \binom{n}{k-1}$$

$$\underset{\substack{\uparrow \\ \text{induction} \\ \text{hypothesis}}}{=} \frac{n!}{k!(n-k)!} + \frac{n!}{(k-1)!(n-(k-1))!}$$

$$\underset{\substack{\uparrow \\ \text{common} \\ \text{denominator}}}{=} \frac{n!(n-k+1)}{k!(n-k)!\underbrace{(n-k+1)}_{=(n-(k-1))}} + \frac{n!k}{k(k-1)!(n-(k-1))!}$$

$$= \frac{n!(n-k+1+k)}{k!(n+1-k)!} = \frac{(n+1)!}{k!((n+1)-k)!}$$

The following theorems are applications of the binomial coefficient in combinatorics:

▶ **Theorem 4.6** From a set with n different elements, k elements (order doesn't matter) can be selected in $\binom{n}{k}$ ways.

This follows directly from the Definition 4.3.

▶ **Theorem 4.7** From a set with n different elements, k elements (order does matter) can be selected in

$$n(n-1)(n-2)\cdots(n-k+1) = \frac{n!}{(n-k)!}$$

ways.

Order matters means that, for example, the selection $(1, 2, 3, 4, 5, 6)$ and the selection $(2, 1, 3, 4, 5, 6)$ are different.

Fortunately, this is not the case with the lottery.

Proof of Theorem 4.7: Each of the $\binom{n}{k}$ subsets with k elements can be arranged in $k!$ different ways. This gives a total of $\binom{n}{k}k! = \frac{n!}{k!(n-k)!}k! = \frac{n!}{(n-k)!}$ possibilities. □

Example

Determine the number of different passwords with 5 characters in length that can be formed from 26 letters and 10 digits, where no character occurs more than once: According to Theorem 4.7, this results in $36 \cdot 35 \cdot 34 \cdot 33 \cdot 32 = 45\,239\,040$. It is certainly clear to you that no password cracker program gets sweating here. ◀

You are probably already familiar with an application of the binomial coefficient in algebra: the binomial theorem. Here is probably where the name "binomial coefficient" comes from; it is related to the binomial "$x + y$" and is not intended to remind us of the mathematician Giuseppe Binomi:

▶ **Theorem 4.8: The binomial theorem** Let $x, y \in \mathbb{R}$, $n \in \mathbb{N}_0$. Then:

$$(x + y)^n = \sum_{k=0}^{n} \binom{n}{k} x^{n-k} y^k. \tag{4.4}$$

This means, for example (compare with Pascal's triangle!):

$$(x + y)^2 = x^2 + 2xy + y^2, \quad (x + y)^3 = x^3 + 3x^2 y + 3xy^2 + y^3.$$

We will often have to work with summation formulas. There are tricks and calculation rules that need to be applied again and again. Before we prove the binomial theorem, I would like to put together a few such simple rules. Here $A(k)$, $B(k)$ are any expressions depending on k and c is a number:

Calculation rules for summation formulas:

1. Combining sums:

$$\sum_{k=m}^{n} A(k) + \sum_{k=m}^{n} B(k) = \sum_{k=m}^{n} (A(k) + B(k)).$$

2. Multiplication with constant elements:

$$c \cdot \sum_{k=m}^{n} A(k) = \sum_{k=m}^{n} (c \cdot A(k)).$$

3. Index shift:

$$\sum_{k=m}^{n} A(k) = \sum_{k=m+1}^{n+1} A(k - 1).$$

4. Change of summation limits:

$$\sum_{k=m}^{n} A(k) = A(m) + \sum_{k=m+1}^{n} A(k) = \sum_{k=m}^{n-1} A(k) + A(n).$$

Rule 1 contains the commutative and associative laws of addition, rule 2 the distributive law. Rule 3 also applies analogously to other index shifts than 1, and the summation limits in rule 4 can be changed by more than 1.

These rules can also often be used when programming. A sum is usually implemented by a for loop, where the summation limits represent the initialization or the termination condition. Clever index shifting or summarizing of different loops are useful tools. For this reason, you should look at the following proof; here, these tricks are used very intensively.

We prove the binomial theorem by induction on n:

Base case: $\qquad\qquad (x+y)^0 = \binom{0}{0}x^0y^0 = 1$ always holds.

Induction hypothesis: $\quad$ Let $(x+y)^n = \sum_{k=0}^{n} \binom{n}{k}x^{n-k}y^k$.

Induction step:

$$(x+y)^{n+1} = (x+y)^n(x+y)$$

$$= \left(\sum_{k=0}^{n} \binom{n}{k} x^{n-k}y^k\right)(x+y)$$

$$\underset{\substack{\uparrow \\ \text{rules 1 and 2}}}{=} \sum_{k=0}^{n} \binom{n}{k}x^{n-k+1}y^k + \sum_{k=0}^{n} \binom{n}{k}x^{n-k}y^{k+1}$$

$$\underset{\substack{\uparrow \\ \text{index shift} \\ \text{second term}}}{=} \sum_{k=0}^{n} \binom{n}{k}x^{n-k+1}y^k + \sum_{k=1}^{n+1} \binom{n}{k-1}x^{n-k+1}y^k$$

$$\underset{\substack{\uparrow \\ \text{change summation} \\ \text{limits}}}{=} \binom{n}{0}x^{n+1}y^0 + \sum_{k=1}^{n} \binom{n}{k}x^{n-k+1}y^k + \sum_{k=1}^{n} \binom{n}{k-1}x^{n-k+1}y^k + \binom{n}{n}x^0y^{n+1}$$

$$\underset{\substack{\uparrow \\ \text{rule 1}}}{=} \binom{n}{0}x^{n+1}y^0 + \sum_{k=1}^{n}\left[\binom{n}{k-1} + \binom{n}{k}\right]x^{n-k+1}y^k + \binom{n}{n}x^0y^{n+1}$$

$$\underset{\substack{\uparrow \\ \text{recursion formula}}}{=} \binom{n+1}{0}x^{n+1}y^0 + \sum_{k=1}^{n} \binom{n+1}{k} x^{n-k+1}y^k + \binom{n+1}{n+1}x^0y^{n+1}$$

$$= \sum_{k=0}^{n+1} \binom{n+1}{k}x^{n-k+1}y^k. \qquad\qquad \square$$

If you insert 1 for x and y in the binomial theorem, you get: $\sum_{k=0}^{n} \binom{n}{k} = 2^n$. According to our definition of the binomial coefficient as the number of k-element subsets of a n-element set, this means: The number of all subsets of a n-element set is 2^n. We have already proven this in a completely different way in Theorem 4.2.

4.2   Divisibility and Euclidean Algorithm

The task of division with remainder has already arisen in the representation of integers in different numeral systems. Now we want to take a closer look at division, more precisely at the divisibility of integers.

▶   **Definition 4.9** If $a, b \in \mathbb{Z}$ and $b \neq 0$, then a is divisible by b (b divides a, we write: $b \mid a$), if there is an integer q such that $a = bq$.

▶   **Theorem 4.10: Divisibility rules**

a) If $c \mid b$ and $b \mid a$, then $c \mid a$.
   (Example: $3 \mid 6$ and $6 \mid 18 \Rightarrow 3 \mid 18$)
b) If $b_1 \mid a_1$ and $b_2 \mid a_2$, then $b_1 b_2 \mid a_1 a_2$.
   ($3 \mid 6$ and $4 \mid 8 \Rightarrow 12 \mid 48$)
c) If $b \mid a_1$ and $b \mid a_2$, then for $\alpha, \beta \in \mathbb{Z}: b \mid \alpha a_1 + \beta a_2$.
   ($3 \mid 6$ and $3 \mid 9 \Rightarrow 3 \mid (7 \cdot 6 + 4 \cdot 9)$)
d) If $b \mid a$ and $a \mid b$, then $a = b$ or $a = -b$.

The rules are all easily checked by reducing them to the definition. I only want to show the first one as an example: $c \mid b$ and $b \mid a$ imply $b = q_1 c$ and $a = q_2 b$ and therefore $a = q_2 q_1 c$, thus $c \mid a$.   □

▶   **Theorem and Definition 4.11: The Division with remainder** For two integers a, b with $b \neq 0$ there is exactly one representation $a = bq + r$ with $q, r \in \mathbb{Z}$ and $0 \leq r < |b|$. a is called *dividend*, b *divisor*, q *quotient* and r *remainder* of the division of a by b. We denote q with a/b and r with $a \bmod b$ (to read: a modulo b).

In the proof, we first assume $a, b > 0$. Now let q be the largest integer with $bq \leq a$. Then there is a $r \geq 0$ with $bq + r = a$ and it applies $r < b$, otherwise q would not have been maximum. Are a or b or both negative, the considerations are quite similar.   □

In most programming languages, there are special operators for division and remainder. In C++ and Java, for example, $q = a/b$ and $r = a\% b$.

To my regret, no mathematician was involved in the definition of the division and modulo operators in C++ and Java. For positive numbers a and b, the effect corresponds to our definition. But we sometimes have to divide with negative numbers. Also for this there is a unique solution according to the above definition for q and r. So the division of -7 and 2, for example, results in $q = -4$ and $r = 1$, since $-7 = -4 \cdot 2 + 1$. Java, on the other hand, calculates $(-7)/2 = -3$ (according to the rule: first calculate $|a|/|b|$, are a or b negative, the result is multiplied by -1) and from this according to the fact $r = a - bq$ the remainder -1. Mathematical remainders are always greater than or equal to 0! Therefore, do not use "/" and "%" on negative numbers. Fortunately, the (in the true sense of the word) negative cases can be reduced to positive ones.

▶   **Definition 4.12: The greatest common divisor (gcd)** Are $a, b, d \in \mathbb{Z}$ and $d \mid a$ and
$d \mid b$, then d is called *common divisor* of a and b. The largest positive common
divisor of a and b is called *greatest common divisor* of a and b and is denoted by
$\gcd(a, b)$.

The calculation of the greatest common divisor of two numbers is carried out using the
famous Euclidean algorithm. Quite unusually for a mathematician, Euclid was not satisfied with the existence of the greatest common divisor, but he gave a concrete calculation
procedure giving the greatest common divisor as a result. The concept of algorithm represents a central concept in computer science and the Euclidean algorithm is a prototype
for this: a calculation procedure with a beginning and an end, input values and results.
One could thus call Euclid one of the fathers of computer science.
   In order to understand the effect of the algorithm, we put a lemma first:

▶   **Lemma 4.13** Let $a, b, q \in \mathbb{Z}$. Then it holds:
   1. If $a = bq$, then $|b| = \gcd(a, b)$.
   2. If $a = bq + r$ with $0 < r < |b|$, then $\gcd(a, b) = \gcd(b, r) = \gcd(b, a \bmod b)$.

For the first part: b is of course a common divisor and b cannot have a divisor with a
larger absolute value, so $|b| = \gcd(a, b)$ applies. In this case, $a \bmod b = 0$.
   For the second part, it is enough to show that the sets of common divisors of a, b resp.
b, r coincide.
   To do this, let d be a divisor of a and b. Then, according to Theorem 4.10c), $d \mid a - bq$
follows. So d is a common divisor of b and $r = a - bq$. If d is now a common divisor of
b and r, then d is also a divisor of bq and, according to 4.10c) again, $r - bq = a$, so d is a
common divisor of a and b.   □

And now the announced theorem:

▶   **Theorem 4.14: The Euclidean algorithm** Let $a, b \in \mathbb{Z}$, $a, b \neq 0$. Then the $\gcd(a, b)$
can be determined by a continued division with remainder recursively according
to the following rule:

$$\gcd(a, b) = \begin{cases} |b| & \text{if } a \bmod b = 0 \\ \gcd(b, a \bmod b) & \text{else} \end{cases}$$

From Lemma 4.13 it follows first of all both lines are correct representations of the
$\gcd(a, b)$. When carrying out the algorithm, we repeatedly apply the second line: a is
replaced by b and b by the remainder when dividing a by b. The gcd does not change.
If finally the remainder is 0, then according to the first line the $\gcd(a, b)$ is the last

remainder different from 0. Does the termination condition actually occur at some point? Yes, because in each step the remainder really gets smaller. But since it is always positive, it will eventually reach 0.   □

Example

gctd($-42, 133$):

$$
\begin{array}{rl}
a = & b{\cdot}q \quad + r \\
-42 = & 133{\cdot}(-1){+}91 \\
133 = & 91{\cdot}1 \quad +42 \\
91 = & 42{\cdot}2 \quad + 7 \\
42 = & 7{\cdot}6 \quad + 0
\end{array}
$$

so $\gcd(-42, 133) = 7.$ ◄

In addition to the greatest common divisor, another result is obtained in the algorithm, which we will need later:

▶ **Theorem 4.15: The extended Euclidean algorithm** Let $a, b \in \mathbb{Z}$, $a, b \neq 0$ and $d = \gcd(a, b)$. Then there are integers α, β with $d = \alpha \cdot a + \beta \cdot b$.

We go through the Euclidean algorithm line by line and cleverly rearrange each line to be able to represent the remainder r_i as a combination $r_i = \alpha_i a + \beta_i b$ of a and b. If r_n is the last remainder different from 0, then we finally get $r_n = \alpha a + \beta b$:

$$
\begin{aligned}
a = bq_0 + r_0 \quad &\Rightarrow r_0 = 1a - q_0 b \quad &&= \alpha_0 a + \beta_0 b. \\
b = r_0 q_1 + r_1 \quad &\Rightarrow r_1 = 1b - r_0 q_1 \quad &&= 1b - (\alpha_0 a + \beta_0 b)q_1 \\
& &&\underset{\substack{\uparrow \\ \text{sort} \\ \text{by } a \text{ and } b}}{=} \quad -\alpha_0 q_1 a + (1 - \beta_0 q_1)b \\
& &&= \alpha_1 a + \beta_1 b \\
r_0 = r_1 q_2 + r_2 \quad &\Rightarrow r_2 = 1r_0 - r_1 q_2 \quad &&= 1(\alpha_0 a + \beta_0 b) - (\alpha_1 a + \beta_1 b)q_2 \\
& &&\underset{\substack{\uparrow \\ \text{sort} \\ \text{by } a \text{ and } b}}{=} \quad \alpha_2 a + \beta_2 b
\end{aligned}
$$

$$\vdots$$

$$
\begin{aligned}
r_{n-2} = r_{n-1} q_n + r_n \Rightarrow r_n &= 1 r_{n-2} - r_{n-1} q_n = 1(\alpha_{n-2} a + \beta_{n-2} b) - (\alpha_{n-1} a + \beta_{n-1} b)q_n \\
&= \alpha_n a + \beta_n b
\end{aligned}
$$

This proof contains also a recursive calculation possibility for α and β: From the last line you can see that α_n, β_n can be determined as the n-th elements of the sequences $\alpha_i = \alpha_{i-2} - \alpha_{i-1} q_i, \beta_i = \beta_{i-2} - \beta_{i-1} q_i$. The initial values are obtained from the first two lines: $\alpha_0 = 1, \alpha_1 = 1, \beta_0 = q_0, \beta_1 = r_0$.

Example of the application of the extended Euclidean algorithm

Calculate $\gcd(168, 133)$ and at the same time α, β with $\gcd = \alpha \cdot 168 + \beta \cdot 133$:

$$168 = 133 \cdot 1 + 35 \Rightarrow 35 \qquad\qquad\qquad = \quad 1 \cdot 168 + (-1) \cdot 133$$
$$133 = \ \ 35 \cdot 3 + 28 \Rightarrow 28 = -3 \cdot 35 + 1 \cdot 133 = (-3) \cdot 168 + \quad 4 \ \cdot 133$$
$$35 = \ \ 28 \cdot 1 + \ \ 7 \Rightarrow \ \ 7 = -1 \cdot 28 + 1 \cdot \ 35 = \quad 4 \ \cdot 168 + (-5) \cdot 133$$
$$28 = \quad \ 7 \cdot 4 + \ \ 0$$

So it is $\gcd(168, 133) = 7 = 4 \cdot 168 + (-5) \cdot 133$. ◀

The α, β would not have been found so easily by trial and error. The existence of these numbers is an amazing fact. It always works, no matter how big a, b may be.

As an immediate consequence, we can calculate that every common divisor of two numbers a and b is also a divisor of $\gcd(a, b)$:

▶   **Theorem 4.16** Let $a, b \in \mathbb{Z}$, $a, b \neq 0$ and $d = \gcd(a, b)$.
a) If $e \in \mathbb{Z}$ is a common divisor of a and b, then e is also a divisor of d.
b) If $f \in \mathbb{Z}$ can be written as a combination $f = \alpha a + \beta b$, then f is a multiple of d.

To a): It is $a = q_a e$ and $b = q_b e$. Let $d = \alpha a + \beta b$ be the representation of the greatest common divisor from Theorem 4.15. Then $d - (\alpha q_a + \beta q_b)e$ follows, so e is a divisor of d.

To b): It is $a = q_a d$ and $b = q_b d$ and from $f = \alpha a + \beta b$ it follows that $f = (\alpha q_a + \beta q_b)d$, thus f is a multiple of d. □

Statements about prime numbers are closely related to the question of divisibility. In Theorem 2.11 we have already proved that there are infinitely many prime numbers. At this point I would like to introduce some important properties of prime numbers. Let's start with the precise definition:

▶   **Definition 4.17** A natural number $p > 1$ is called a *prime number* or a *prime*, if it has no other positive divisors than 1 and itself.

It is a convention one does not count 1 to the prime numbers, 2 is obviously the only even prime number. The first prime numbers are:

$$2, 3, 5, 7, 11, 13, 17, 19, 23, \ldots$$

▷   **Theorem 4.18** Every natural number a greater than 1 can be represented as a
product of prime numbers or is itself a prime number:

$$a = p_1 \cdot p_2 \cdot p_3 \cdot \ldots \cdot p_n.$$

This representation is unique up to the order of the factors.

If one combines multiple prime factors, one obtains the representation
$a = p_1^{\alpha_1} \cdot p_2^{\alpha_2} \cdot p_3^{\alpha_3} \cdot \ldots \cdot p_m^{\alpha_m}$ with different primes p_i and exponents $\alpha_i > 0$.
    For example, $120 = 2 \cdot 2 \cdot 2 \cdot 3 \cdot 5 = 2^3 \cdot 3 \cdot 5$ and $315 = 3 \cdot 3 \cdot 5 \cdot 7 = 3^2 \cdot 5 \cdot 7$. The
proof of this is a somewhat cumbersome induction, which I would like to spare us.
    If b is a divisor of a, then there is a q with

$$a = p_1 \cdot p_2 \cdot p_3 \cdot \ldots \cdot p_n = bq.$$

If one now also represents b and q in their prime factorization, then:

$$a = p_1 \cdot p_2 \cdot p_3 \cdot \ldots \cdot p_n = bq = \underbrace{b_1 \cdot b_2 \cdot \ldots \cdot b_r}_{\text{prime factors of } b} \cdot \underbrace{q_1 \cdot q_2 \cdot \ldots \cdot q_s}_{\text{prime factors of } q}.$$

From the uniqueness of the representation it follows all factors b_i must already occur on
the left side of the equation. Therefore one gets the

▷   **Corollary 4.19** All divisors of a number a are obtained by the possible products
of their prime factors.

For example, 315 has the divisors $3, 5, 7,$ $\underbrace{9, 15, 21, 35}_{\text{all products of 2 factors}}$, $\underbrace{45, 63, 105}_{\text{all products of 3 factors}}$, 315.

    Similarly, the following statement can be derived from the prime factorization:

▷   **Corollary 4.20** If p is a prime number and a, b are natural numbers with the property $p \mid ab$. Then $p \mid a$ or $p \mid b$.

A prime number cannot be split into two other numbers as a factor, it is a divisor of one
or the other, perhaps of both numbers, if it occurs multiple times in the prime factorization. From $3 \mid 42$ and $42 = 6 \cdot 7$ it follows, for example, $3 \mid 6$ or $3 \mid 7$.

4.3   Modular Arithmetic

In Sect. 1.2 after Definition 1.9 we have called equivalent those integers for which the
difference is divisible by 5. This way, the set of integers was divided into the disjoint
equivalence classes $[0], [1], \ldots, [4]$. I will now investigate such relations in more detail
and present some applications in computer science.

▶ **Definition 4.21** Let $a, b \in \mathbb{Z}$ and $n \in \mathbb{N}$. The numbers a, b are called *congruent modulo n*, if $a - b$ is divisible by n. We write: $a \equiv b \bmod n$, or just $a \equiv b$, if it is clear which modulus m it is.

Please do not confuse the statement "$a \equiv b \bmod n$" with the number "$a \bmod n$", which denotes the remainder!

▶ **Theorem 4.22** "$\equiv$" is an equivalence relation on $\mathbb{Z}$.

This means, as we have already written down in Definition 1.8:

1. $a \equiv b \bmod n \Rightarrow b \equiv a \bmod n$ (Symmetry)
2. $a \equiv a \bmod n$ (Reflexivity)
3. $a \equiv b \bmod n, b \equiv c \bmod n \Rightarrow a \equiv c \bmod n$ (Transitivity)

We have carried out the proof for the case $n = 5$ in Sect. 1.2; in general, it goes exactly the same way.

The elements congruent to a number a form the equivalence class of a. The equivalence classes with respect to the relation "$\equiv$" are called *residue classes* modulo m. For the sake of reminder: It is $[a] = \{z \mid z \equiv a \bmod n\}$ and $b \equiv a \bmod n$ holds if and only if $[a] = [b]$. If the modulus n is not clear from the context, I write for the equivalence class $[a]_n$.

▶ **Theorem 4.23** $a, b \in \mathbb{Z}$ are congruent modulo n ($a \equiv b \bmod n$), if and only if a and b leave the same remainder when divided by n.

Proof: "$\Leftarrow$": Let $a = q_1 n + r, b = q_2 n + r \Rightarrow a - b = (q_1 - q_2)n$.

"$\Rightarrow$": From $a - b = qn$, $a = q_1 n + r_1$, $b = q_2 n + r_2$ it follows that $a - b = qn = (q_1 n - q_2 n) + (r_1 - r_2)$. Then, $(r_1 - r_2)$ must also be a multiple of n, which means $r_1 - r_2 = kn$ or $r_1 = r_2 + kn$ for a $k \in \mathbb{Z}$. But since both r_1 and r_2 are remainders out of $\{0, 1, 2, \ldots, n - 1\}$, only $k = 0$ is possible, thus $r_1 = r_2$. □

The numbers $0, 1, 2, \ldots, n - 1$ are all possible remainders modulo n. There are thus exactly the residue classes $[0], [1], \ldots, [n - 1]$. A residue class $[r]$ is represented by the remainder r. Often, we will not even distinguish between remainders and residue classes. This is a bit unusual at first, but there is nothing mysterious about it. Perhaps it is comparable to the fact that you often work with references to objects in a program, rather than with the object itself.

Calculating with Residue Classes

The set of residues resp. residue classes with respect to the number n is denoted by $\mathbb{Z}/n\mathbb{Z}$ (read: "$\mathbb{Z}$ modulo $n\mathbb{Z}$") or simply by $\mathbb{Z}_n$. The symbol $n\mathbb{Z}$ should symbolize that all the elements are equivalent that differ by a multiple of n, that is, by nz for a $z \in \mathbb{Z}$.

If a and b are remainders modulo n, we can add or multiply them. The result is usually not a remainder anymore, but we can take it modulo n and get a remainder again. In this way, we can define an addition and a multiplication on the set of residues:

▶   **Definition 4.24** For $n \in \mathbb{Z}$ and $a, b \in \mathbb{Z}/n\mathbb{Z}$, let

$$a \oplus b := (a + b) \bmod n,$$
$$a \otimes b := (a \cdot b) \bmod n. \tag{4.5}$$

The addition and multiplication on $\mathbb{Z}/n\mathbb{Z}$ and on $\mathbb{Z}$ are compatible with the modulo operation. It doesn't matter if you take two elements modulo first and then connect them or vice versa:

▶   **Theorem 4.25** For the mapping

$$\varphi : \mathbb{Z} \to \mathbb{Z}/n\mathbb{Z}$$
$$a \mapsto a \bmod n$$

it holds that

$$\varphi(a + b) = \varphi(a) \oplus \varphi(b),$$
$$\varphi(a \cdot b) = \varphi(a) \otimes \varphi(b).$$

Let's check this for the addition: If $a = nq_1 + r_1$ and $b = nq_2 + r_2$, then

$$\varphi(a + b) = (nq_1 + nq_2 + r_1 + r_2) \bmod n = (r_1 + r_2) \bmod n$$

and

$$\varphi(a) \oplus \varphi(b) = r_1 \oplus r_2 = (r_1 + r_2) \bmod n.$$

are the same. We conclude this for multiplication in a similar way.   □

The set $\mathbb{Z}/n\mathbb{Z}$ with these two operations plays an important role in discrete mathematics and also in computer science. We will work with it a lot. The compatibility of the operations with the modulo operation will also meet us again and again.

> In anticipation of Sect. 5.6: The mapping from Theorem 4.25 is a homomorphism between $\mathbb{Z}$ and $\mathbb{Z}_n$.

Now we are able to set up addition and multiplication tables for the operations, the so-called Cayley tables, after the mathematician Arthur Cayley who invented them in 1854. Two examples of this:

Examples

$n = 3$:

$\oplus$	0	1	2
0	0	1	2
1	1	2	0
2	2	0	1

$\otimes$	0	1	2
0	0	0	0
1	0	1	2
2	0	2	1

$n = 4$:

$\oplus$	0	1	2	3
0	0	1	2	3
1	1	2	3	0
2	2	3	0	1
3	3	0	1	2

$\otimes$	0	1	2	3
0	0	0	0	0
1	0	1	2	3
2	0	2	0	2
3	0	3	2	1

◀

An example of the application of modulo calculation:

Example

What is the remainder modulo 7 of $(7 \cdot 31 + 1 \cdot 28 + 4 \cdot 30)$? You can first calculate the bracket and get $365 \bmod 7 = 1$. Or you first form all remainders. Then the much simpler calculation results:

$$(7 \cdot 31 + 1 \cdot 28 + 4 \cdot 30) \bmod 7 = (0 \cdot 3 + 1 \cdot 0 + 4 \cdot 2) \bmod 7 = 8 \bmod 7 = 1. \blacktriangleleft$$

Do you see what this example is about? There are 7 months with 31 days, one with 28 (except in a leap year) and 4 with 30 days. The remainders of 0 to 6 correspond to the weekdays. For example, if today is Monday $= 0$, then you get by the above calculation: $0 + 365 \bmod 7 = 1$, so in a year it is Tuesday.

Not only for people, but also for computers, the second way of calculation is much more suitable. If, for example, you want to build an eternal calendar and calculate somewhat clumsy on which weekday in 4 billion years the sun will set forever, you will quickly get an overflow of numbers. Always calculate modulo 7, then you even get by with one byte as integer data type.

In the next section on hashing, I would like to show you an important application of the modulo calculation in computer science. You will learn about other areas of application in Sect. 5.3 on fields and Sect. 5.7 on cryptography.

4.4   Hashing

Hash Functions

In data processing, the problem often arises that data records, which are identified by a key, have to be stored or found quickly. Take a company with 5000 employees. Each employee has a 10-digit personnel number with which his data can be identified in the computer.

You could store the data in an array of 5000 elements, sorted linearly by personnel number. In the basics principles of computer science you learn the binary search for an array element then requires an average of 11.3 accesses ($\log(n + 1) - 1$).

An impossible approach is the storage in an array with 10^{10} elements, in which the index is just the personnel number. The search for a data set would require exactly one step here.

Hashing represents a compromise between these two extremes: a relatively small array with fast access. For this purpose, the long key is mapped to a short index using a so-called *hash function*.

Let K be the key space, $K \subset \mathbb{Z}$. The keys of the data to be stored come from K. Let $H = \{0, 1, \ldots, n - 1\}$ be a set of indices. These identify the storage addresses of the records in the storage area, the *hashtable*. $|H|$ is usually much smaller than $|K|$. A hash function is a mapping $h\colon K \to H, k \mapsto h(k)$. The data element with key k is then stored at index $h(k)$.

A function that has proven itself suitable for many purposes is our modulo mapping:

$$h(k) = k \bmod n.$$

The basic problem with hashing is that such a mapping h can never be injective. There are keys $k \neq k'$ with $h(k) = h(k')$. This is called a *collision*. The records for k and k' would have to be stored at the same address. In this case, special treatment is necessary, the collision resolution.

When choosing the hash function, one must make sure that such a collision occurs as rarely as possible. For example, if the last four digits of the personnel number from the example represent the year of birth and the hash function $h(k) = k \bmod 10\,000$ is chosen, then $h(k)$ is just the year of birth. Collisions occur constantly and large areas of the available array of 10,000 elements are not addressed directly. Prime numbers p often turn out to be a good choice as a modulus. Now we assume the modulus is a prime, we denote it with p. This is actually used frequently. Regularities in the key space are usually destroyed in the address space.

Collision Resolution

The closer the number of positions in the hash table and the number of records are to each other, the more likely collisions will occur. One approach to solving this is to search for another free storage location in the table according to a reproducible rule when a collision occurs. If the address $h(k)$ is already occupied, a probing sequence $(s_i(k))_{i=1,\ldots,p-1}$ is formed and the addresses $s_1(k), s_2(k), \ldots, s_{p-1}(k)$ are successively visited until a free address is found. The numbers $s_1(k), s_2(k), \ldots, s_{p-1}(k)$ have to go through all hash addresses, with the exception of $h(k)$ itself. Only then can it be guaranteed that an existing free space will also be found.

The simplest method is the *linear probing*:

$$s_i(k) = (h(k) + i) \bmod p, \quad i = 1, \ldots, p - 1.$$

Here, the next larger free address is simply chosen each time. If you reach the end of the table, you continue at the beginning. A small experiment on the computer will quickly show that this can easily lead to data clusters in the table. Data like to go where there is already data, and there are many collisions.

Such clustering is avoided with the help of the quadratic probing:

$$\left. \begin{array}{l} s_{2i-1}(k) = (h(k) + i^2) \bmod p \\ s_{2i}(k) = (h(k) - i^2) \bmod p \end{array} \right\} i = 1, \ldots, (p-1)/2.$$

The probing sequence is therefore $h(k) + 1$, $h(k) - 1$, $h(k) + 4$, $h(k) - 4$, $h(k) + 9$, $h(k) - 9$ and so on, where the value is forced into the correct range again by the modulo operation when the edge is exceeded. But what about our requirement that the probing sequence must go through all possible addresses? Let's take a look at the probing sequences for the prime numbers 5, 7, 11 and 13:

i	1		2			
i^2	1		4			
$\pm i^2 \bmod 5$	1	4	4	1		

i	1		2		3	
i^2	1		4		9	
$\pm i^2 \bmod 7$	1	6	4	3	2	5

i	1		2		3		4		5	
i^2	1		4		9		16		25	
$\pm i^2 \bmod 11$	1	10	4	7	9	2	5	6	3	8

i	1		2		3		4		5		6	
i^2	1		4		9		16		25		36	
$\pm i^2 \bmod 13$	1	12	4	9	9	4	3	10	12	1	10	3

If we start with $h(k) = 0$ for simplicity, the last line of the table always gives the probing sequence. With 7 and 11 it works well, we get a permutation of the possible addresses, with 5 and 13 it does not work. What is the common property of 7 and 11? They both leave 3 as the remainder when divided by 4. In fact, for all prime numbers p with $p \equiv 3 \bmod 4$:

$$\{i^2 \bmod p, i = 1, \ldots, (p-1)/2\} \cup \{-(i^2) \bmod p, i = 1, \ldots, (p-1)/2\}$$
$$= \{1, 2, \ldots, p-1\}$$

There are infinitely many such prime numbers. Try, for example, 9967, 33 487, 99 991. I have to postpone the proof of this theorem to the next chapter (Theorem 5.22), we have to learn more about the properties of fields first.

Since we calculate modulo p, the probing sequence always reaches all addresses, even if one does not start with $h(k) = 0$, but with any other value.

A third common method for resolving collisions is the so-called *double hashing*. In this case, for the probing sequence in the case of a collision, a second hash function $h'(k)$ is used, whereby $h'(k) \neq 0$ must always be. The sequence is then:

$$s_i(k) = (h(k) + i \cdot h'(k)) \bmod p, \quad i = 1, \ldots, p-1.$$

The sequence $i \cdot h'(k) \bmod p, i = 1, \ldots, p-1$ also reaches all adresses different from 0: Assuming there is $i \neq j$ with $i \cdot h'(k) \bmod p = j \cdot h'(k) \bmod p$, then $(i-j)h'(k) \bmod p = 0$ and thus $i - j$ or $h'(k)$ would be a divisor of p. But that's not possible if p is a prime number. So all $p - 1$ elements of the sequence are different and thus all numbers from 1 to $p - 1$ occur. For the second hash function, you can, for example, choose: $h'(k) = 1 + k \bmod (p - 1)$. This always results in a value between 1 and $p - 1$.

How many collisions occur on average with hashing methods? If we choose the array size 9967 and enter the 5000 employees from the example at the beginning, we get a filling level of about 0.5. If the company continues to grow, we get 8970 entries with a filling level of 0.9. With these levels, the following average numbers of collisions occur when the 5001st (or the 8971st) employee is entered:

Probing	linear	quadratic	double hashing
Filling level 0.5	1.57	1.24	1.01
Filling level 0.9	47.02	11.06	8.93

Of course, the array can only be filled up to exactly 100%. However, there are also methods of dynamic hashing in which the size of the hash table can be automatically adjusted to the demand during runtime.

4.5   Comprehension Questions and Exercises

Comprehension Questions

1. If you want to calculate the binomial coefficient using Theorem 4.4 in a program, you will quickly run into a problem. What is it? Do you have a solution for it?
2. Why does the Euclidean algorithm always lead to a result?

3. If the greatest common divisor of two numbers a, b is a prime number, can it be a, b have other common divisors?

4. How many even prime numbers are there?

5. How is the remainder of a division a/b calculated in mathematics, how in common programming languages (Java or C++)?

6. If the modulo mapping modulo a prime number is used as a hash function, all prime numbers can be used for linear probing, but not for quadratic probing. Why?

7. In $\mathbb{N}$ it always holds that $n! \neq 0$. Can $n! = 0$ be in $\mathbb{Z}/n\mathbb{Z}$?

Exercises

1. Show: if a, b are integers with $a \mid b$ and $b \mid a$, then $a = b$ or $a = -b$.

2. Show: if $a_1 \mid b_1$ and $a_2 \mid b_2$ are integers, then $a_1 a_2 \mid b_1 b_2$.

3. Show that the divisibility relation | (Definition 4.9) on the natural numbers is a partial order.

4. Write a recursive program to calculate $\binom{n}{k}$. Use the formula $\binom{n+1}{k} = \binom{n}{k} + \binom{n}{k-1}$.

5. A slightly more tricky induction exercise: Show for the recursive calculation of $\binom{n}{k}$ (according to exercise 4) exactly $2 \cdot \binom{n}{k} - 1$ function calls are necessary.

    You can also try this out by integrating a counter into your implementation that counts the number of function calls. In this form, the algorithm is therefore not suitable for practical purposes. If you analyze the calculation of a binomial coefficient in detail, you will notice that in the recursion many coefficients are calculated several times. With a little trick you can avoid this and improve the algorithm so that it becomes very fast after all.

6. Calculate the probability of having 6 correct numbers in the lottery if you can choose 8 numbers from 49.

7. Show by induction that for all $n \in \mathbb{N}$ it holds:
    a) $n^7 + n$ is divisible by 2,
    b) $n^3 - n$ is divisible by 6.

8. Calculate the greatest common divisor d of 456 and 269 using the Euclidean algorithm. Determine numbers α, β with $\alpha \cdot 269 + \beta \cdot 456 = d$.

9. Write in C++ or Java a modulo operator that also works mathematically correct even for negative numbers.

10. Write the Cayley multiplication tables for $\mathbb{Z}/7\mathbb{Z}$ and $\mathbb{Z}/8\mathbb{Z}$. Look closely at the rows and columns in the two tables. What do you notice?

11. Show: If $a \equiv a' \bmod m$ and $b \equiv b' \bmod m$, then $a + b \equiv a' + b' \bmod m$.

12. When working with remainders, you can exchange the operations +, respectively ·
    with the modulo operation. The same applies to exponentiation: To calculate
    $a^2 \bmod n$, it is easier to calculate $[(a \bmod n)(a \bmod n)] \bmod n$. (Why actually? Try a
    few examples.) Follow a similar procedure to calculate $a^m \bmod n$. With this knowl-
    edge you can, with the help of the algorithm given in Exercise 8 of Chap. 3 for the
    calculation of x^n, formulate a recursive algorithm for the calculation of $a^m \bmod n$.

13. Implement the Euclidean algorithm; once iteratively and once recursively.

Algebraic Structures

5

Abstract

At the end of this chapter

- you will know what an algebraic structure is,
- you will know the most important algebraic structures: groups, rings, fields,
- and many important examples thereof: permutation groups, elliptic curves, polynomial rings, the ring $\mathbb{Z}/n\mathbb{Z}$, the field $\mathbb{C}$ and the finite fields $GF(p)$ and $GF(p^n)$.
- you will know homomorphisms as the structure-preserving mappings between algebraic structures,
- and you will have learned public key cryptography as an important application of finite rings and fields.

In mathematics and computer science, we often deal with sets on which certain operations are defined. It happens again and again that such operations have similar properties on quite different sets, so one can also do similar things with them. In Chap. 2 we have already encountered something like this: For the operations $\cup$, $\cap$ and $^-$ on sets, the same laws of arithmetic apply as for $\vee$, $\wedge$ and $\neg$ on propositions. Mathematical strategy is to find and describe prototypes of such operations and properties that occur again and again, and then to form theorems about sets with these operations. These theorems are then valid in every concrete example of such a set (a model).

Supplementary Information The online version contains supplementary material available at https://doi.org/10.1007/978-3-658-40423-9_5.

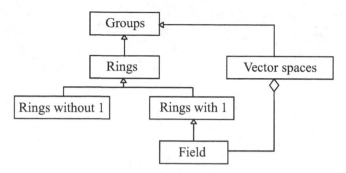

Fig. 5.1 Algebraic structures

In this and the next chapter we will get to know the most important such sets with operations, we call them algebraic structures. We formulate axioms for these structures and at the same time introduce concrete examples of the structures. The structures investigated can be classified as in the diagram in Fig. 5.1.

Maybe you know the symbols from UML (Unified Modelling Language), they fit exactly. We read the arrow ——▷ as "is a", this is the inheritance relationship: A ring is a group, a field is a ring. Only a few more properties are added to the properties of the "superclass". The arrow ——◇ denotes aggregation, we can read it as "has a" or "knows a". A vector space is a group that additionally knows and uses a field. Just as objects are concrete instances of a class, there are concrete realizations of these prototypes. Among the fields we already know from the first chapter, for example, $\mathbb{Q}$ and $\mathbb{R}$.

All structures investigated have in common that one or two operations are defined on them:

▷   **Definition 5.1** Let M be a set. A *binary operation* on the set M is a mapping

$$v : M \times M \to M$$
$$(n, m) \mapsto v(n, m)$$

Most of the time we denote the operations with $+$, $\cdot$, $\oplus$, $\otimes$ or similar symbols and then write for $v(n, m)$ for example $n + m$ or $n \cdot m$.

The operations $\cup$, $\cap$ on sets or $\vee$, $\wedge$ on propositions are such operations, as well as multiplication and addition on the real numbers or the addition and multiplication $\oplus$ and $\otimes$ between the elements of $\mathbb{Z}/m\mathbb{Z}$, which we explained in Definition 4.24:

$$\oplus : \mathbb{Z}/n\mathbb{Z} \times \mathbb{Z}/n\mathbb{Z} \to \mathbb{Z}/n\mathbb{Z} \qquad \otimes : \mathbb{Z}/n\mathbb{Z} \times \mathbb{Z}/n\mathbb{Z} \to \mathbb{Z}/n\mathbb{Z}$$
$$(a, b) \mapsto a \oplus b \qquad\qquad (a, b) \mapsto a \otimes b$$

Now we can describe the structures one after the other, we start at the top of the hierarchy with groups; these have relatively weak structural properties:

5.1 Groups

▶ **Definition 5.2: The group axioms** A *group* $(G, *)$ consists of a set G and a binary operation $*$ on G with the following properties:

(G1)   There is an element $e \in G$ with the property $a * e = e * a = a$ for all $a \in G$. e is called the *identity element* of G.

(G2)   For each $a \in G$ there is a uniquely determined element $a^{-1} \in G$ with the property $a * a^{-1} = a^{-1} * a = e$. The element a^{-1} is called the *inverse element* to a.

(G3)   For all $a, b, c \in G$ it holds that $a * (b * c) = (a * b) * c$. G is associative.

The group $(G, *)$ is called *commutative group* or *abelian group*, if in addition it holds:

(G4)   For all $a, b \in G$ it is $a * b = b * a$.

The abelian groups are named after the brilliant Norwegian mathematician Niels Hendrik Abel, who lived from 1802 to 1829. He revolutionized essential areas of modern algebra. Far away from the world centers of mathematics at that time in France and Germany, his work remained largely unnoticed or simply ignored during his lifetime. The Norwegian government established the Abel Prize in Mathematics in his honor. This was first awarded in 2003. The prize is similarly endowed as the Nobel Prizes and it is quite justified to consider it as a kind of "Nobel Prize for Mathematics".

Why is there actually no Nobel Prize for Mathematics? Mathematicians tell the story that Alfred Nobel was not well disposed towards our profession after the prominent mathematician Mittag-Leffler had stolen his girlfriend Sonja Kowalewski, also a mathematician, from him. Even if the historical content of this story is rather meager, I like it better than to assume Nobel would have underestimated the importance of mathematics for the modern sciences.

From the axioms, further calculation rules for groups can be derived. I would like to mention two simple rules here we will need more often later. The first states that certain equations can be solved, with the second one you can calculate the inverse of a product. I write down the proofs for this in detail here, later we will omit the brackets again due to associativity.

▶ **Theorem 5.3** If $(G, *)$ is a group, then for all elements $a, b \in G$ it holds:

a) There is exactly one $x \in G$ with $a * x = b$ and exactly one $y \in G$ with $y * a = b$.

b) For the inverse of $a * b$ it holds: $(a * b)^{-1} = b^{-1} * a^{-1}$.

To a): It is $x = a^{-1} * b$, because $a * (a^{-1} * b) = (a * a^{-1}) * b = e * b = b$. For uniqueness: from $a * x_1 = a * x_2 = b$, it follows $a^{-1} * (a * x_1) = a^{-1} * (a * x_2) \Rightarrow$

$(a^{-1} * a) * x_1 = (a^{-1} * a) * x_2$, thus $x_1 = x_2$. Similarly, $y = b * a^{-1}$ is uniquely determined. (x and y can be different!)

To b): It is $(a * b) * (b^{-1} * a^{-1}) = a * (b * b^{-1}) * a^{-1} = a * e * a^{-1} = a * a^{-1} = e$, thus $b^{-1} * a^{-1}$ is the inverse of $a * b$. □

Do we really have to distinguish between $a * x = b$ and $y * a = b$? Probably it is new to you there are operations which are not commutative. So far, the order did not play a role in the operations $+$ and $\cdot$. But soon we will work with such structures and therefore we have to be careful that we do not use commutativity implicitly anywhere. So, in theorem 5.3a) the statement really has to be formulated separately for multiplication from the left and from the right!

Examples of groups

1. $(G, *) = (\mathbb{R}, +)$ is a commutative group with identity element $e = 0$ and inverse $a^{-1} = -a$.
2. $(G, *) := (\mathbb{R} \setminus \{0\}, \cdot)$ is a group as well: "$\cdot$" always assigns a number different from 0 to each pair from $\mathbb{R} \setminus \{0\}$, so we have an operation. It is $e = 1$ and $a^{-1} = 1/a$.
3. In Sect. 4.3 from Definition 4.24 we formed residue classes and added and multiplied in these. $(\mathbb{Z}/n\mathbb{Z}, \oplus)$ is a commutative group:
   The identity element is 0. The inverse element to m is just $n - m$, because $m \oplus (n - m) = m + (n - m) \bmod n = 0$. The associativity and the commutativity can be traced back to the corresponding rules in $\mathbb{Z}$ for addition. For the associativity law, for example, it holds:

   $$(a \oplus b) \oplus c = (a \oplus b) + c \bmod n = (a + b) \bmod n + c \bmod n$$
   $$= a \bmod n + b \bmod n + c \bmod n,$$
   $$a \oplus (b \oplus c) = a + (b \oplus c) \bmod n = a \bmod n + (b + c) \bmod n$$
   $$= a \bmod n + b \bmod n + c \bmod n.$$

4. $(G, *) := (\mathbb{R}^+, \cdot)$ is a group as well, which you should calculate yourself! ◀

$(G, *) := (\mathbb{R}^-, \cdot)$ is not a group, however. Why not?

$\mathbb{R}^+$ is a subset of $\mathbb{R} \setminus \{0\}$. When investigating subsets of a group with the same operation, we can simplify the investigation of whether it is a group:

▶ **Theorem 5.4** If U is a subset of $(G, *)$ and G is a (commutative) group, then $(U, *)$ is a (commutative) group if and only if:

$$a, b \in U \implies a * b \in U, a^{-1} \in U. \tag{5.1}$$

U is then called a *subgroup* of G.

Proof: The condition (5.1) states $*$ is really an operation on U and that (G2) is fulfilled. Then it follows $a * a^{-1} = e \in U$ and thus (G1) is valid. The associativity or commutativity is automatically fulfilled in U, since the elements of U are also elements of G and therefore behave associatively or commutatively.   □

Examples

5. $(\mathbb{Z}, +)$ is obviously a subgroup of $(\mathbb{R}, +)$, $(\mathbb{N}, +)$ on the other hand not.

6. $(m\mathbb{Z}, +)$ is a subgroup of $(\mathbb{Z}, +)$, where $m\mathbb{Z} := \{mz \mid m \in \mathbb{Z}\}$ represents all multiples of m. Because let $a = mz_1$ and $b = mz_2$, then $a + b = m(z_1 + z_2) \in m\mathbb{Z}$ and $-a = m(-z_1) \in m\mathbb{Z}$.

7. Groups of bijective mappings: Let M be a set. The set F of bijective mappings of $M \to M$ forms a group with the composition $f \circ g$ as a binary operation: It is $e = id_M$ and f^{-1} the inverse map to f. The operation is associative, because for all $x \in M$ it applies:

$$(f \circ (g \circ h))(x) = f((g \circ h)(x)) = f(g(h(x)))$$
$$\parallel \qquad\qquad (5.2)$$
$$((f \circ g) \circ h)(x) = (f \circ g)(h(x)) = f(g(h(x)))$$

See Definition 1.15 for the composition of mappings. Such a composition is always associative.

8. Elliptic curves over the field $\mathbb{R}$ are graphs in $\mathbb{R}^2$. On the points of such a curve, one can define an addition making it to a commutative additive group. We will investigate this in more detail in Sect. 5.5.

9. Special groups of bijective mappings are the so-called *permutation groups*: Let S_n be the set of all permutations of the set $\{1, 2, \ldots, n\}$. A permutation of these numbers is nothing else but a bijective mapping. From Theorem 3.5 we know there are exactly $n!$ such mappings, according to example 7 the set of permutations with composition is a group. Let us take a closer look at these groups. ◄

Permutation groups

How can one characterize the group elements as simply as possible? First, we write the elements of the set $\{1,2,\ldots,n\}$ next to each other and below them the respective images. The group S_4, for example, contains among other the elements:

$$a = \begin{pmatrix} 1 & 2 & 3 & 4 \\ 3 & 4 & 2 & 1 \end{pmatrix}, \quad b = \begin{pmatrix} 1 & 2 & 3 & 4 \\ 3 & 1 & 2 & 4 \end{pmatrix}, \quad c = \begin{pmatrix} 1 & 2 & 3 & 4 \\ 3 & 4 & 1 & 2 \end{pmatrix}.$$

Let's look at these three elements and see what happens to 1 when we execute each of these permutations several times in succession: With a, 1 is mapped to 3, then 3 to 2, 2 to

4 and 4 finally back to 1. With b, 1 is mapped to 3, 3 to 2 and 2 back to 1, and with c, 1 is mapped to 3, 3 to 1. We call such a sequence, which ends again at the starting element, a *cycle* of the permutation. For the permutation π, we write such a cycle if $\pi^{k+1}(e) = e$, in the form: $(e, \pi(e), \pi^2(e), \ldots, \pi^k(e))$. In the example, we therefore found the cycles (1 3 2 4), (1 3 2) and (1 3). No matter which element we start with, if we look at its fate when executed several times, it ends up back at the starting element at some point; every element is therefore part of a cycle. Why? Since we only have finitely many elements available, at some point an element must occur twice. But this can only be the starting element, otherwise the permutation would not be injective. You will immediately see this if you try it out on paper.

Now we have found a simple way to write down a permutation: We start with the first element and write down the associated cycle. Then we look for an element that does not occur in the first cycle and write its cycle behind it, and so on, until no element is left. Usually, in this notation, one-cycles (for example, (4) in b) are omitted. This gives us the cycle notation. For a, b and c it is:

$$a = (1\ 2\ 3\ 4), \quad b = (1\ 3\ 2), \quad c = (1\ 3)(2\ 4).$$

Note this notation is not unique, for example (1 3)(2 4) = (4 2)(1 3).

Each individual cycle itself represents a permutation: $(e, \pi(e), \pi^2(e), \ldots, \pi^k(e))$ denotes the permutation that maps e to $\pi(e)$, $\pi(e)$ to $\pi^2(e), \ldots, \pi^k(e)$ to e and leaves the other elements unchanged. (1 3) (2 4) can therefore also be seen as the product (1 3) ∘ (2 4) of the two permutations (1 3) and (2 4) in the permutation group.

A two-cycle $(a\ b)$ is called a *transposition*. One can simply see that for a cycle it holds: $(e_1, e_2, e_3, \ldots e_k) = (e_1, e_2) \circ (e_2, e_3) \circ \cdots \circ (e_{k-1}, e_k)$. So every permutation can be written as a succession of transpositions.

Now you also know why one can sort any unordered data field using a series of element permutations.

You have to pay close attention here. According to the convention when performing mappings one after the other, we carry them out starting from the right, and it is not possible to swap the order! The S_n is the first non-commutative group we get to know. For example, (in our first notation):

$$(3\ 2) = \begin{pmatrix} 1\ 2\ 3\ 4 \\ 1\ 3\ 2\ 4 \end{pmatrix} \quad (1\ 3) = \begin{pmatrix} 1\ 2\ 3\ 4 \\ 3\ 2\ 1\ 4 \end{pmatrix}$$

and

$$(1\ 3) \circ (3\ 2) = \begin{pmatrix} 1\ 2\ 3\ 4 \\ 3\ 1\ 2\ 4 \end{pmatrix} \quad (3\ 2) \circ (1\ 3) = \begin{pmatrix} 1\ 2\ 3\ 4 \\ 2\ 3\ 1\ 4 \end{pmatrix}.$$

The operations in finite groups can be written down in Cayley tables. If we denote the six elements of the S_3 with a, b, c, d, e, f, we get the table:

$$
\begin{array}{c|cccccc}
\circ & a & b & c & d & e & f \\
\hline
a & a & b & c & d & e & f \\
b & b & c & a & e & f & d \\
c & c & a & b & f & d & e \\
d & d & f & e & a & c & b \\
e & e & d & f & b & a & c \\
f & f & e & d & c & b & a \\
\end{array}
\tag{5.3}
$$

Wait, something is missing: I'll tell you in this table a is the identical permutation, $b = (1\ 2\ 3)$ and $d = (2\ 3)$. From this you can deduce the other elements and check the correctness of the table.

▷ **Definition 5.5** The finite group $(G, *)$ with n elements is called *cyclic*, if there is an element $g \in G$ with the property

$$
G = \{g, g^2, g^3, \ldots, g^n\}.
\tag{5.4}
$$

g is then called *generator* of the group.

The powers of the element g therefore generate the whole group. Then $g^n = e$, because e must be a power of g, and if it were $g^m = e$ already for a $m < n$ then $g^{m+1} = g$, and then not all n elements of the group would be present on the right side of (5.4).

Take another look at the Cayley table (5.3): This group does not have a generator, because $b^3 = a$, $c^3 = a$, $d^2 = a$, $e^2 = a$, $f^2 = a$, where here a is the identity element. However, the sets $\{b, b^2, b^3\}, \{c, c^2, c^3\}, \{d, d^2\}$ and so on are cyclic subgroups of the S_3. You can easily check this.

5.2 Rings

▷ **Definition 5.6: The ring axioms** A *ring* $(R, \oplus, \otimes)$ consists of a set R with two binary operations $\oplus, \otimes$ on R, for which the following properties hold:

(R1) $(R, \oplus)$ is an abelian group.

(R2) For all $a, b, c \in R$, $a \otimes (b \otimes c) = (a \otimes b) \otimes c$. R is associative.

(R3) For all $a, b, c \in R$, $a \otimes (b \oplus c) = (a \otimes b) \oplus (a \otimes c)$ and $(b \oplus c) \otimes a = (b \otimes a) \oplus (c \otimes a)$. R is distributive.

The ring $(R, \oplus, \otimes)$ is called *commutative*, if in addition it holds:

(R4) For all $a, b \in R$, $a \otimes b = b \otimes a$.

Soon, we will run out of symbols for our operations. For this reason, we will proceed as you know it from object-oriented programming languages with polymorphism: We will almost exclusively use two symbols for our operations from now on, namely "+" and "·". These are polymorphic, which means depending on which objects are standing to the left and right of them, they have a different meaning. We will get used to this very quickly. Groups whose operation is denoted by "+" we call *additive groups*, groups with the operation "·" are called *multiplicative groups*. To complete the analogy to the operations of numbers we are familiar with, we will from now on call the identity element of an additive group "0" (the *0-element*), the identity element of a multiplicative group "1" (the *1-element*). The inverse to a in an additive group we will denote by $-a$ instead of a^{-1}, instead of $a + (-b)$ we will write $a - b$. For $a \cdot b$ we will write ab, and for $a \cdot b^{-1}$ in commutative groups we will often write a/b. Just as we introduce the convention "point before line" to save us many brackets.

The operations in a ring we will denote by $+$ and by $\cdot$. So a ring is a commutative additive group with an additional associative and distributive multiplication. A ring always has a 0, but not always a 1. All following examples are commutative rings. Just as with groups, there is a simple way to test if a subset of a ring is a ring itself:

▶   **Theorem 5.7** If S is a subset of the ring $(R, +, \cdot)$ then $(S, +, \cdot)$ is a ring (a subring of R) if and only if the following conditions are satisfied:

a) $(S, +)$ is a subgroup of $(R, +)$, that is $a, b \in S \Rightarrow a + b \in S, a^{-1} \in S$.
b) $a, b \in S \Rightarrow a \cdot b \in S$.

The proof is similar to that of Theorem 5.4.

Examples of rings

1. $(\mathbb{Z}, +, \cdot)$ is a ring.
2. $(m\mathbb{Z}, +, \cdot)$ is a subring of $\mathbb{Z}$, because $(m\mathbb{Z}, +)$ is a subgroup of $\mathbb{Z}$, and if $a = mz_1$, $b = mz_2$ are elements of $m\mathbb{Z}$, then $ab = m(mz_1 z_2) = mz_3 \in m\mathbb{Z}$.
3. $(\mathbb{Q}, +, \cdot), (\mathbb{R}, +, \cdot)$ are rings.
4. Let R be any ring and $(\mathcal{F}, \oplus, \otimes)$ the set of all mappings from R to R with the operations $f \oplus g$ and $f \otimes g$, which are defined for all $x \in R$ by:

$$(f \oplus g)(x) := f(x) + g(x)$$
$$(f \otimes g)(x) := f(x) \cdot g(x)$$

$(\mathcal{F}, \oplus, \otimes)$ is a ring. $\oplus$ and $\otimes$ are called pointwise addition and pointwise multiplication of mappings, respectively. Here I use the symbols $\oplus$, $\otimes$ again for a moment to avoid confusion: Two maps are added (new addition) by adding the function values at each point (old addition). The 0-element is the 0-map, that is, the map that maps each $x \in R$ to 0. The axioms can be checked by reducing them to the rules of the underlying ring. The calculation for the distributive law, for example, looks like this:

$$((f \oplus g) \otimes h)(x) = (f \oplus g)(x) \cdot h(x) = (f(x) + g(x))h(x)$$
$$= f(x)h(x) + g(x)h(x) = (f \otimes h)(x) + (g \otimes h)(x)$$
$$= ((f \otimes h) \oplus (g \otimes h))(x).$$

Be careful, the operations in $(\mathcal{F}, \oplus, \otimes)$ must not be confused with the composition of mappings, as we have studied, for example, in (5.2).

This is a very abstract example at first. More concretely, you can think of real functions, for example. Such functions can be added and multiplied. You are familiar with this from school and in the second part of the book we will do this again and again.

5. We already know $(\mathbb{Z}/n\mathbb{Z}, \oplus)$ is a group. In fact, it is even true that $(\mathbb{Z}/n\mathbb{Z}, \oplus, \otimes)$ is a commutative ring: $\otimes$ is a binary operation on $\mathbb{Z}/n\mathbb{Z}$ and just like in Example 3 after Theorem 5.3 associativity, commutativity and distributivity can be traced back to the corresponding rules in $\mathbb{Z}$. ◄

Polynomial Rings

A particularly important type of rings are the polynomial rings, the set of all polynomials with coefficients from a ring R. I formulate the following definition somewhat more general. If it still causes you problems in this form, think of R as the real numbers $\mathbb{R}$ at first, and you have the polynomials in front of you, which you have known from school for a long time.

▶ **Definition 5.8** Let R be a commutative ring and $a_0, a_1, \ldots, a_n \in R$. The map

$$f : R \to R$$
$$t \mapsto a_n t^n + a_{n-1} t^{n-1} + \cdots + a_1 t + a_0$$

is called *polynomial function* or shortly *polynomial* over R. If $a_n \neq 0$, then $\deg f := n$ is called the *degree* of f.

We must not confuse functions and function values. I denote with $f(x) = a_n x^n + a_{n-1} x^{n-1} + \cdots + a_1 x + a_0$ the function f, where x is the variable, while for a specific element $t \in R$ the value $f(t) = a_n t^n + a_{n-1} t^{n-1} + \cdots + a_1 t + a_0$ is again an element of R, the function value of $f(x)$ at the point t.

▷   **Definition 5.9** Let R be a commutative ring. The set of all polynomials with coefficients from R is denoted by $R[x]$. With the operations $p + q$ and $p \cdot q$, which are defined for $t \in R$ by

$$(p + q)(t) := p(t) + q(t)$$
$$(p \cdot q)(t) := p(t) \cdot q(t), \tag{5.5}$$

$R[x]$ is called the *polynomial ring* over R.

This ring is a subring of the ring of all mappings from R to R (Example 4 from before). The operations are defined pointwise, that is, for each element $t \in R$ individually. Now I have again denoted the polynomial operations with $+$, $\cdot$ instead of with $\oplus$, $\otimes$. In (5.5) the "$+$" and the "$\cdot$" on the left and right have different meanings! The zero polynomial of $R[x]$ is the zero map, that is, the polynomial for which all coefficients are equal to 0.

In order to see $R[x]$ is really a ring, and to make the definition meaningful at all, we have to apply the subring criterion from Theorem 5.7. Are the sum and product of two polynomial functions really again a polynomial function? We will look at this with an example. Let $p, q \in \mathbb{R}[x]$, $p = x^3 + 3x^2 + 1$, $q = x^2 + 2$. Then for all $t \in \mathbb{R}$: $(p + q)(t) = (t^3 + 3t^2 + 1) + (t^2 + 2) = t^3 + 4t^2 + 3$, and so we obtain as sum $p + q = x^3 + 4x^2 + 3$. Without going into detail (there is nothing behind it but paperwork), we have for the two polynomials $p(x) = a_n x^n + \cdots + a_1 x + a_0$ and $q(x) = b_n x^n + \cdots + b_1 x + b_0$:

$$(p + q)(x) = (a_n + b_n)x^n + (a_{n-1} + b_{n-1})x^{n-1} + \cdots + (a_0 + b_0)$$
$$= \sum_{k=0}^{n} (a_k + b_k)x^k$$
$$(p \cdot q)(x) = (a_n b_n)x^{n+n} + (a_n b_{n-1} + a_{n-1}b_n)x^{n+n-1} + \cdots + a_0 b_0 \tag{5.6}$$
$$= \sum_{k=0}^{n+n} \left(\sum_{\substack{i+j=k \\ 0 \le i,j \le n}} a_i b_j \right) x^k$$

I have written both polynomials up to the coefficient n here. If p and q have different degrees, one can simply add a few zeros for the polynomial of smaller degree and then have them in this form. The formulas look complicated, but they mean nothing more than that you can add and multiply two polynomials as if x were some number (or a ring element).

> If you ever want to implement a polynomial class, you have to use these definitions. The statement "multiply as if x were a number" will not be of any use to you then.

So now we know the sum and product of polynomial functions are again polynomial functions. To finish the subring test, it still has to be shown that the polynomials form a subgroup with addition. I leave that to you. Apply Theorem 5.4 and remember that the element a^{-1} mentioned there is now called $-a$.

5.3 Fields

▷ **Definition 5.10: The field axioms** A *field* $(K, +, \cdot)$ consists of a set K and two binary operations $+, \cdot$ on K with the following properties:

(K1) $(K, +, \cdot)$ is a commutative ring.
(K2) There is an element 1 in K with $1 \cdot a = a$ for all $a \in K \backslash \{0\}$.
(K3) For all $a \in K \backslash \{0\}$ with there is an element $a^{-1} \in K$ with $a^{-1} \cdot a = 1$.

▷ **Theorem 5.11** In a field K it holds:

a) $a \cdot 0 = 0$ for all $a \in K$.
b) Are $a, b \in K$ and $a, b \neq 0$, so it also holds $a \cdot b \neq 0$.

Proof: To a): $a \cdot 0 = a \cdot (0 + 0) = a \cdot 0 + a \cdot 0$. If one subtracts from both sides of the equation $a \cdot 0$ (more precisely: adds the inverse of $a \cdot 0$ to it), one obtains $0 = a \cdot 0$.
To b): assumed $ab = 0$. Then $b = (a^{-1}a)b = a^{-1}(ab) = a^{-1} \cdot 0 = 0$. □

From a) it follows that one may not divide by 0 in fields: $a/0$ is defined as $a \cdot 0^{-1}$, there would have to be an inverse 0^{-1} to 0. Then $0^{-1} \cdot 0 = 1$, in contradiction to a). And b) states that multiplication is an operation on $K \setminus \{0\}$. With the axioms it follows that $(K \setminus \{0\}, \cdot)$ is a group.

Examples of fields that you know are $\mathbb{Q}$ and $\mathbb{R}$. The integers $\mathbb{Z}$ do not form a field, because there are no multiplicative inverses in $\mathbb{Z}$. We will now get to know more fields.

The Field $\mathbb{C}$ of Complex Numbers

The integers $\mathbb{Z}$ arose from $\mathbb{N}$ by adding the negative numbers, because one wanted to carry out calculations like 5–7. In $\mathbb{Z}$ the problem 3/4 is not solvable. If one adds to $\mathbb{Z}$ all fractions, one obtains the rational numbers $\mathbb{Q}$. We have seen that, for example, $\sqrt{2}$ is not an element of $\mathbb{Q}$. The real numbers arise from $\mathbb{Q}$ by filling in the last gaps on the number line. In Chap. 12 we will deal with the real numbers in more detail. Unfortunately, there are still problems in $\mathbb{R}$: For example, there is no real number r mit $r^2 = -1$, because every square of a real number is positive.

Now one can extend $\mathbb{R}$ again to a field in which $\sqrt{-1}$ exists. Since there is no more space for this on the number line, one has to go into the second dimension: $\mathbb{C} := \mathbb{R}^2 = \{(a, b) \mid a, b \in \mathbb{R}\}$. The real numbers should be the x-axis in it, that is

$\mathbb{R} = \{(a, 0) \mid a \in \mathbb{R}\}$. Of course, this is not a real equality sign, we identify a with $(a, 0)$, we pretend it is the same. This also makes it clear how the operations in the set $\{(a, 0) \mid a \in \mathbb{R}\}$ look: It is $(a, 0) + (b, 0) = (a + b, 0)$ and $(a, 0) \cdot (b, 0) = (a \cdot b, 0)$.

In this set $\mathbb{C}$ we now need an addition and multiplication so that the set becomes a field. Restricted to $\mathbb{R}$ (more precisely: restricted to $\{(a, 0) \mid a \in \mathbb{R}\}$), they should of course give the operations already present there. And of course $\sqrt{-1}$ should be there, that is, there must be $(a, b) \in \mathbb{C}$ with $(a, b) \cdot (a, b) = (-1, 0)$.

I write down a definition for these operations:

▶ **Definition 5.12** $\mathbb{C} := \mathbb{R}^2 = \{(x, y) \mid x, y \in \mathbb{R}\}$ with the following binary operations for $(a, b), (c, d) \in \mathbb{R}^2$:

$$(a, b) + (c, d) := (a + c, b + d)$$

$$(a, b) \cdot (c, d) := (ac - bd, bc + ad) \tag{5.7}$$

is called *field of complex numbers*.

The definition of addition is perhaps still somewhat understandable, multiplication seems to fall from the sky. But we will see in a moment that with our just formulated requirements there is no other way. First, a few calculation examples:

$$(a, 0) + (c, 0) = (a + c, 0)$$
$$(a, 0) \cdot (c, 0) = (a \cdot c, 0)$$
$$(1, 0) \cdot (c, d) = (c, d)$$
$$(0, 1) \cdot (0, 1) = (-1, 0).$$

The first two lines show that on $\mathbb{R}$ really nothing new happens through these operations. In the third line we see that $(1, 0)$ (the 1 from $\mathbb{R}$) also represents an identity element for complex numbers, and from the last line we actually get our root of -1: it is $(0, 1)$.

▶ **Theorem 5.13** The complex numbers $(\mathbb{C}, +, \cdot)$ form a field.

We will not calculate the properties in detail; everything is elementary to carry out. The only exciting property is the existence of the multiplicative inverse. I just give the inverse and show it is correct. So let's say $(a, b) \neq (0, 0)$, then

$$(a, b) \cdot \left(\frac{a}{a^2 + b^2}, \frac{-b}{a^2 + b^2} \right) = \left(\frac{a^2}{a^2 + b^2} - \frac{b(-b)}{a^2 + b^2}, \frac{ba}{a^2 + b^2} + \frac{a(-b)}{a^2 + b^2} \right) = (1, 0)$$

and thus $(\frac{a}{a^2 + b^2}, \frac{-b}{a^2 + b^2})$ is the multiplicative inverse of (a, b). Note that $a^2 + b^2$ is always different from 0! $\qquad\square$

Now we introduce an abbreviated notation for the elements in $\mathbb{C}$: For $(a, 0)$ we simply write a again, for $(0, 1)$ we write i. This i is a new symbol, nothing more than an abbreviation. Because of $(0, b) = (b, 0)(0, 1) = b \cdot i$ and because of $(a, b) = (a, 0) + (0, b)$ we can now also write $a + bi$ for (a, b); so any complex number (a, b) can be represented in the form $a + bi$.

Let's form the product of the two numbers $a + bi$ and $c + id$, without using the Definition 5.12, but only taking into account the laws of arithmetic that must apply in a field, then we get:

$$(a + bi)(c + id) = (ac + i(bc + ad) + i^2 bd) = (ac - bd) + (bc + ad)i.$$

Compare this with (5.7); see the match? In a field in which there is a root of -1 (which we denote by i), multiplication can not look any different than written in (5.7) !

For many, complex numbers have something mysterious about them; this may be because i stands for the "imaginary unit", for something that is not real, in contrast to the real numbers, which somehow seem "tangible" to us (and "complex" also sounds so difficult!). However, complex numbers are just as real or unreal as real numbers, we are just more used to $\mathbb{R}$. Isn't it actually very suspicious that real numbers can never be represented in a calculator? We can only ever work with finite decimal numbers, but real numbers are almost all infinitely long.

It has been shown in mathematics and in the applications of mathematics that it is incredibly practical to work with $\mathbb{R}$ and the rules that apply to it. Similarly, it has been shown that many very real problems can only be reasonably solved using complex numbers. An electrical engineer would despair today without complex numbers! You will get to know some of the useful and beautiful properties and applications of complex numbers in this book.

You may be wondering if the series $\mathbb{N} \subset \mathbb{Z} \subset \mathbb{Q} \subset \mathbb{R} \subset \mathbb{C}$ can be continued? $\mathbb{C}$ is an "algebraically closed" field, there is no need for such an extension, as the following important theorem states:

▶ **Theorem 5.14: The fundamental theorem of algebra** Every polynomial $a_n x^n + a_{n-1} x^{n-1} + \cdots + a_0 \in \mathbb{C}[x]$ of degree greater than 0 has a root in $\mathbb{C}$.

Our only goal was to add $\sqrt{-1}$ to the real numbers, but a lot more has happened: Every negative number a now has a root as a root of $x^2 - a$, and even every polynomial has a root.

Although this theorem is called the "fundamental theorem of algebra", it is actually a theorem which is proved using methods of complex analysis. As simple as the theorem is to formulate, the proof is just as difficult. It takes a few semesters of mathematics studies to be able to follow it, and we just have to accept it as true here.

To conclude this introduction to complex numbers, I would like to show you two important mappings of complex numbers, which we will use again and again later.

Fig. 5.2 Absolute value and
conjugate element

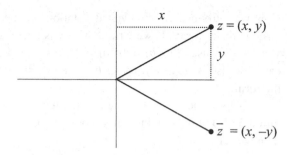

▶   **Definition 5.15** Let $z = x + iy \in \mathbb{C}$. It is

$$|z| := \sqrt{x^2 + y^2} \quad \bar{z} = x - iy$$

$|z|$ is called the *absolute value* of z and $\bar{z}$ the *conjugate element* of z (see Fig. 5.2).

Both mappings have an intuitive meaning: $|z|$ is, according to the Pythagorean theorem, just the length of the line segment from the point $(0, 0)$ to the point (x, y), $\bar{z}$ is obtained by reflecting z at the x-axis.

▶   **Theorem 5.16: Properties of absolute value and conjugate** For $z_1, z_2 \in \mathbb{C}$ it holds:

$$\begin{aligned} |z_1 + z_2| &\leq |z_1| + |z_2|, \\ |z_1 z_2| &= |z_1||z_2|, \end{aligned} \tag{5.8}$$

$$\begin{aligned} \overline{z_1 + z_2} &= \overline{z_1} + \overline{z_2}, \\ \overline{z_1 \cdot z_2} &= \overline{z_1} \cdot \overline{z_2}, \\ z_1 \cdot \overline{z_1} &= |z_1|^2. \end{aligned}$$

Only the first part of (5.8) is a little tricky, the other properties are easy to calculate. Equation (5.8) is, by the way, also valid for the absolute value in the real numbers, so we don't have to learn anything new.

The Field GF(p)

Now let's look at the rings $\mathbb{Z}/n\mathbb{Z}$. The field axioms (K1) and (K2) are fulfilled. The question is whether each element which is different from 0 has a multiplicative inverse. For $\mathbb{Z}/4\mathbb{Z}$ and $\mathbb{Z}/5\mathbb{Z}$ we write down the multiplication tables:

$$
\begin{array}{c|cccc}
n=4 & 0 & 1 & 2 & 3 \\
\hline
0 & 0 & 0 & 0 & 0 \\
1 & 0 & 1 & 2 & 3 \\
2 & 0 & 2 & 0 & 2 \\
3 & 0 & 3 & 2 & 1 \\
\end{array}
\qquad
\begin{array}{c|ccccc}
n=5 & 0 & 1 & 2 & 3 & 4 \\
\hline
0 & 0 & 0 & 0 & 0 & 0 \\
1 & 0 & 1 & 2 & 3 & 4 \\
2 & 0 & 2 & 4 & 1 & 3 \\
3 & 0 & 3 & 1 & 4 & 2 \\
4 & 0 & 4 & 3 & 2 & 1 \\
\end{array}
$$

In the case of $n = 4$, the element 2 has no inverse, $\mathbb{Z}/4\mathbb{Z}$ can therefore not be a field. In the case of $n = 5$, we see that in each row (except the first) the 1 occurs, and that means nothing else than that each number has an inverse: $1^{-1} = 1, 2^{-1} = 3, 3^{-1} = 2, 4^{-1} = 4$. $\mathbb{Z}/5\mathbb{Z}$ is therefore a field. If you look closely at the right table, you will notice: in each row and in each column each element occurs exactly once. This is not a coincidence, it has to be like this, because according to Theorem 5.3 the equations $ax = b$ for $a, b \neq 0$ are uniquely solvable.

5 is a prime number and actually $\mathbb{Z}/n\mathbb{Z}$ is a field if and only if n is a prime number. We can prove this surprising statement with our elementary number theoretical knowledge. The following theorem serves as a preparation:

▷   **Theorem 5.17** An element $a \in \mathbb{Z}/n\mathbb{Z}$ has a multiplicative inverse if and only if $\gcd(a, n) = 1$.

Let $\gcd(a, n) = 1$. By the extended Euclidean algorithm (Theorem 4.15) there are $\alpha, \beta \in \mathbb{Z}$ with $\alpha a + \beta n = 1$. Then $b := \alpha \bmod n$ is the inverse of a, because according to Theorem 4.25 we have $\underbrace{a \bmod n}_{a} \otimes \underbrace{\alpha \bmod n}_{b} = (a \cdot \alpha) \bmod n = (1 - \beta n) \bmod n$, thus $ab = 1$ in $\mathbb{Z}/n\mathbb{Z}$.

Conversely, if $ab = 1$ is in $\mathbb{Z}/n\mathbb{Z}$, then $1 = ab + \gamma n$ for an integer γ. A common divisor of a and n would then also have to be a divisor of 1, thus $\gcd(a, n) = 1$.   □

The Euclidean algorithm therefore also provides a method for determining the inverse elements, so to speak a division algorithm for the invertible elements of $\mathbb{Z}/n\,\mathbb{Z}$. We can derive another important property of $\mathbb{Z}/n\mathbb{Z}$ from the theorem just proved:

▷   **Theorem 5.18** The set $\mathbb{Z}/n\mathbb{Z}^* := \{a \in \mathbb{Z}/n\mathbb{Z} \mid \gcd(a, n) = 1\}$ forms   a commutative group with respect to multiplication modulo n.

The group axioms have to be checked:

Is multiplication a binary operation on $\mathbb{Z}/n\mathbb{Z}^*$? Yes, because if $\gcd(a, n) = 1$ and $\gcd(b, n) = 1$, then $\gcd(ab, n) = 1$, because a common prime divisor of ab and n would also divide a or b.

Of course, 1 is in $\mathbb{Z}/n\mathbb{Z}^*$ and Theorem 5.17 says that the inverse of a is also in $\mathbb{Z}/n\mathbb{Z}^*$. Associativity and commutativity hold true because $\mathbb{Z}/n\mathbb{Z}^*$ is a subset of the ring $\mathbb{Z}/n\mathbb{Z}$. □

If n is now a prime number, we get a field:

▶   **Theorem and Definition 5.19** $\mathbb{Z}/p\mathbb{Z}$ is a field if and only if p is a prime number. $\mathbb{Z}/p\mathbb{Z}$ is denoted by GF(p) .

$\mathbb{Z}/p\mathbb{Z}$ is a commutative ring with 1. If p is a prime number, then for $a \in \{1, 2, \ldots, p - 1\}$ it is always true that $\gcd(a, p) = 1$. So, according to Theorem 5.17, each element that is different from 0 has a multiplicative inverse and $\mathbb{Z}/p\mathbb{Z}$ is a field. If p is not a prime number, then p has divisors in $\{1, 2, \ldots, p - 1\}$ to which, according to Theorem 5.17, there is no inverse. □

> The finite fields GF(p) are called Galois fields. The name was chosen in honor of the French mathematician Évariste Galois (1811-1832), who made groundbreaking results in the study of these fields. He was a mathematical genius and a political hothead who, because of his youth and his ideas, which were far ahead of his time, was not recognized by the mathematical establishment. In the end, he died, just 20 years old, in a duel. His written legacy, which he only sketched in the night before his death, still consumes mathematicians attention today.

You can now also set up a multiplication table for $p = 9967$ (and for any other prime number) and you will notice that in each row and column (except the first) each remainder occurs exactly once. Here is a first application example:

Example

Books are identified by the ISBN, the International Standard Book Number. The ISBN10 consists of a code of 10 digits, with the last digit being a check digit. We ignore the hyphens between the individual digit blocks. The check digit is calculated as follows: If a_i is the i th digit of the ISBN, $i = 1, \ldots, 9$, then the check digit p is equal to

$$p = 1a_1 + 2a_2 + 3a_3 + 4a_4 + 5a_5 + 6a_6 + 7a_7 + 8a_8 + 9a_9 \bmod 11.$$

The first edition of this book had the ISBN 3-528-03181-6, it is

$$p = 1 \cdot 3 + 2 \cdot 5 + 3 \cdot 2 + 4 \cdot 8 + 5 \cdot 0 + 6 \cdot 3 + 7 \cdot 1 + 8 \cdot 8 + 9 \cdot 1 \bmod 11 = 6.$$

If the remainder modulo 11 is 10, the letter X is used as the check digit. Wouldn't it be easier to calculate modulo 10? The use of the prime number 11 as a modulus has the consequence that in the ISBN every digit permutation and every single wrong digit can be detected. Let us consider the permutation of the different digits a_i and a_j, $i \neq j$. We calculate this in the field GF(11).

   If we received the same check digit after this permutation, the following rule would have to apply in a field:

$$(\cdots + ia_i + \cdots + ja_j + \cdots) - (\cdots + ia_j + \cdots + ja_i + \cdots)$$
$$= (ia_i + ja_j) - (ia_j + ja_i) = 0,$$

and thus

$$0 = ia_i + ja_j - ia_j - ja_i = (i - j)(a_i - a_j).$$

Here, $i - j$ and $a_i - a_j$ in the field GF(11) are not equal to 0. But this product cannot be 0 according to Theorem 5.11b)!

You can conclude in a similar way that every single digit error can be detected. With modulo 10 not all errors and permutations could be detected. Even a smaller prime number is not enough. Look for number examples! ◀

Since 2006, the 13-digit ISBN13 is used more and more. This corresponds in structure to the global trade item number, which is printed as a barcode on many commercial products. The check digit a_{13} of this article number is calculated such that

$$(a_1 + a_3 + \ldots + a_{13}) + 3(a_2 + a_4 + \ldots + a_{12}) \bmod 10 = 0$$

is. There is no finite field involved here anymore. But the error detection is not as good as with ISBN10: Not all permutations can be detected.

The IBAN, the International Bank Account Number, has a two-digit check number modulo 97. The number 97 is the largest prime less than 100 and therefore particularly suitable for such a check number.

Now an important theorem of number theory, which is due to the famous mathematician Fermat. It can be easily proved using the field properties of GF(p). We use it immediately to justify the quadratic probing in hashing (see Sect. 4.4). Later we will need it in the derivation of the RSA encryption algorithm, which we examine in Sect. 5.6.

▶ **Theorem 5.20: Fermat's little theorem** If p is a prime number, then in GF(p) for each $a \neq 0$:

$$a^{p-1} = 1.$$

Most of the time, the theorem is formulated without the explicit use of the structure GF(p) as follows:

▶ **Theorem 5.20a: Fermat's little theorem** If p is a prime number, then for each $a \in \mathbb{Z}$, which is not a multiple of p:

$$a^{p-1} \equiv 1 \bmod p.$$

The two formulations of the theorem are equivalent.

$$p = 5 \quad a = 2: \quad 2^4 = 16 = 3 \cdot 5 + 1$$
$$a = 7: \quad 7^4 = 2401 = 480 \cdot 5 + 1.$$
$$p = 3 \quad a = 8: \quad 8^2 = 64 = 21 \cdot 3 + 1$$
$$a = 9: \quad 9^2 = 81 = 27 \cdot 3 + 0 \ (a \text{ is a multiple of } p!). \ \blacktriangleleft$$

To prove Fermat's little theorem, we now calculate in $GF(p)$ and show that for $a \neq 0$ always $a^{p-1} = 1$ holds: First, we note that the numbers $1a, 2a, 3a, \ldots, (p-1)a$ in $GF(p)$ are all different. If, for example, $ma = na$, then $(m-n)a = 0$ would have to be, but according to Theorem 5.11b) this is only possible if $m = n$. So the sets $\{1a, 2a, 3a, \ldots, (p-1)a\}$ and $\{1, 2, 3, \ldots, (p-1)\}$ contain the same elements, and therefore the products of all elements of these two sets are also equal. So $1a \cdot 2a \cdot 3a \cdot \ldots \cdot (p-1)a = 1 \cdot 2 \cdot 3 \cdot \ldots \cdot (p-1)$. In other words,

$$(p-1)! \, a^{p-1} = (p-1)!,$$

from which, after division by $(p-1)!$, the assertion $a^{p-1} = 1$ follows.   □

Unfortunately, I no longer remember where I saw this beautiful proof. I think it also deserved an entry in the "Proofs from the BOOK" that I mentioned after Theorem 2.11. Fermat, who knew nothing about Galois fields, had a lot more work to do with it. Fermat (1607–1665) was one of the first great number theorists, he left many important theorems. With one of his problems, he came back into the spotlight at the end of the 20th century: Every construction worker knows that the equation $x^2 + y^2 = z^2$ has integer solutions (for example, $3^2 + 4^2 = 5^2$). With a string that is marked in the ratio $3 : 4 : 5$, you can precisely determine a right angle. Fermat examined the question of whether the equations $x^n + y^n = z^n$ also have integer solutions for $n > 2$. Despite many attempts, it had not been possible to find such solutions until then. Around 1637, Fermat wrote in the margin of a book something like: "I possess a truly wonderful proof that $x^n + y^n = z^n$ for $n > 2$ is not solvable in $\mathbb{Z}$. The margin is just too small to write it down."

Fermat had the habit of never writing down proofs, but all his theorems turned out to be correct. However, during 350 years, the world's smartest mathematicians efforts were fruitless. They were unable to proof this claim as it was a very difficult nut to crack. Only in 1993 was "Fermat's last theorem" proven by Andrew Wiles. As simple as the problem can be formulated, so difficult is the mathematics that is contained in the proof. Whether Fermat really possessed a wonderful proof?

If you want to learn more about the nature of mathematics, I can recommend the book "Fermat's last theorem" by Simon Singh. It is also very exciting to read for non-mathematicians.

As the first application of the theorem, we can now prove the assertion that, with quadratic probing, every address is reached if for the prime number holds $p \equiv 3 \bmod 4$. See Sect. 4.4:

▶ **Theorem 5.21** If p is a prime number with $p \equiv 3 \bmod 4$, then in $\mathrm{GF}(p)$:

$$\{i^2 \mid i = 1, \ldots, (p-1)/2\} \cup \{-(i^2) \mid i = 1, \ldots, (p-1)/2\} = \mathrm{GF}(p) \setminus \{0\}.$$

If $a \neq 0$ is a square, that is $a = i^2$, then $\pm i$ are the only roots of a: Assume $i^2 = j^2$. Then $0 = i^2 - j^2 = (i+j)(i-j)$, that is $i = j$ or $i = -j$.

Since p is a prime number, $(p-1)/2$ is a natural number. Further, according to Fermat's little theorem, for all $a \neq 0$:

$$a^{p-1} = (a^{\frac{p-1}{2}})^2 = 1,$$

and therefore $a^{\frac{p-1}{2}} = \pm 1$. Since $p \equiv 3 \bmod 4$, it follows that $\frac{p+1}{4} \in \mathbb{N}$ and for $i = a^{\frac{p+1}{4}}$,

$$i^2 = (a^{\frac{p+1}{4}})^2 = a^{\frac{p+1}{2}} = a^{\frac{p-1}{2}+1} = a^{\frac{p-1}{2}} \cdot a = \pm a.$$

Therefore, $a = i^2$ or $a = -(i^2)$. If i is greater than $(p-1)/2$, we replace i with $-i = p - i$ and the statement remains true. So every element different from 0 is a square or the negative of a square of a number between 1 and $(p-1)/2$. □

In the course of the proof, we also found a simple algorithm for extracting roots in the fields $\mathrm{GF}(p)$ with $p \equiv 3 \bmod 4$. For all squares a,

$$\sqrt{a} = \pm a^{\frac{p+1}{4}}.$$

5.4 Polynomial Division

In the ring of integers, we were able to carry out division with remainder, which we used to determine the greatest common divisor of two numbers using the Euclidean algorithm. In polynomial rings over a field K, there is also such a division algorithm with important applications:

▶ **Theorem and Definition 5.22: Polynomial division** Let K be a field, $K[X]$ the polynomial ring over K. Then division with remainder can be carried out in $K[X]$, that is:

For $f, g \in K[X]$ with $g \neq 0$ there are $q, r \in K[X]$ with $f = g \cdot q + r$ and $\deg r < \deg g$.

The remainder $r(x)$ is then denoted by $f(x) \bmod g(x)$: $r(x) = f(x) \bmod g(x)$. If $r(x) = 0$, we say $g(x)$ *is a divisor of* $f(x)$ or $f(x)$ *is divisible by* $g(x)$.

Before we constructively find the polynomials q and r, some immediate consequences of this theorem:

▶ **Theorem 5.23** If $f \in K[X]$ has the root x_0, (that is, $f(x_0) = 0$), then f is divisible by $(x - x_0)$ without remainder. "The root can be split out."

Proof: For $g = x - x_0$, there is according to Theorem 5.23 $q(x)$, $r(x)$ with $f(x) = (x - x_0)q(x) + r(x)$ and $\deg r < lt$; $1 = \deg g$. So $r(x) = a_0 x^0 = a_0$ is constant. If one puts in x_0, then $0 = f(x_0) = 0 \cdot q(x_0) + a_0$ and thus $a_0 = 0$. □

▶ **Theorem 5.24** A polynomial $f \in K[X]$ of degree n that is different from 0 has at most n roots.

Proof: Assume f has more than n roots. By splitting out the first n roots, one obtains $f(x) = (x - x_1)(x - x_2) \cdots (x - x_n)q(x)$ and $q(x)$ must be constant, otherwise it would be $\deg f > n$. If now $f(x_{n+1}) = 0$, then one of the factors on the right must be 0 (Theorem 5.11, we are in a field!), so x_{n+1} already occurs among the $x_1, \ldots, x_n$. □

A first important consequence of Theorem 5.14, the fundamental theorem of algebra, is:

▶ **Theorem 5.25** Every polynomial $f(x) \in \mathbb{C}[X]$ can be factorized as:

$$f(x) = a_n(x - x_0)(x - x_1) \cdots (x - x_n).$$

Because the roots can be successively split out. Each quotient of degree greater than 0 has another root according to the fundamental theorem. □

Now to execute the polynomial division, we work out the algorithm. First, we remember how we learned to divide natural numbers in school:

$$365 : 7 = 52 \quad \text{Remainder 1 that is: } 365 = 7 \cdot 52 + 1$$

$$\underline{-35}$$
$$15 \qquad\qquad \text{How many times does}$$
$$\underline{-14} \qquad\qquad \text{7 fit into 36?}$$

We now do the same with a polynomial in $\mathbb{R}[X]$:

$$(2x^5 + 5x^3 + x^2 + 3x + 1) : (2x^2 + 1) = x^3 + 2x + \tfrac{1}{2} \quad \text{remainder: } (x + \tfrac{1}{2})$$

$-g \cdot x^3$: $\quad \underline{-(2x^5 + x^3)}$

f_1: $\qquad\qquad 4x^3 + x^2 + 3x + 1$

$-g \cdot 2x$: $\quad \underline{-(4x^3 + 2x)}$

f_2: $\qquad\qquad\qquad x^2 + x + 1$

$-g \cdot \tfrac{1}{2}$: $\qquad\qquad \underline{-(x^2 + \tfrac{1}{2})}$

f_3: $\qquad\qquad\qquad\qquad x + \tfrac{1}{2}$

How many times does $2x^2$ fit into $2x^5$? (examine only the first terms of the polynomials!)

This means:

$$(2x^5 + 5x^3 + x^2 + 3x + 1) = (2x^2 + 1)(x^3 + 2x + \tfrac{1}{2}) + (x + \tfrac{1}{2}).$$

With this example, you can also see that K really has to be a field; in $\mathbb{Z}[X]$ this polynomial division does not work, because $\tfrac{1}{2} \notin \mathbb{Z}$!

Why does this work? The calculation shows that

$$f(x) = f_1(x) + g(x)x^3,$$
$$f_1(x) = f_2(x) + g(x)2x,$$
$$f_2(x) = f_3(x) + g(x)\tfrac{1}{2}.$$

We insert step by step and get

$$f(x) = \left(\left(f_3(x) + g(x)\tfrac{1}{2}\right) + g(x)2x\right) + g(x)x^3 = \underbrace{f_3(x)}_{r(x)} + g(x)\underbrace{(x^3 + 2x + \tfrac{1}{2})}_{q(x)}$$

Similarly, we can derive this for polynomials $f(x) = a_n x^n + \cdots + a_0$, $g(x) = b_m x^m + \cdots + b_0$ with arbitrary coefficients from K, we are just much longer busy with the writing of the indices. I would like to spare us that here.

It is important to know that this division not only works with coefficients in $\mathbb{R}$, but with any field, for example with our just discovered fields $GF(p)$. Let's do an example in $GF(3)[X]$: The coefficients that come into question here are only 0, 1 and 2. Because we have to do a lot in the field $GF(3)$, I will write down the Cayley tables for this again:

+	0	1	2		·	0	1	2
0	0	1	2		0	0	0	0
1	1	2	0		1	0	1	2
2	2	0	1		2	0	2	1

$$
\begin{array}{ll}
(2x^4 \quad\quad + 2x^2 + \ x + 1) : (x + 2) = 2x^3 + 2x^2 + x + 2 & \\
① \ \underline{-(2x^4 + \ x^3)} & \\
② \quad\quad\quad 2x^3 + 2x^2 + \ x + 1 & \\
\quad\quad \underline{-(2x^3 + \ x^2)} & \\
\quad\quad\quad\quad\quad\quad x^2 + \ x + 1 & (5.9) \\
\quad\quad\quad\quad \underline{-(x^2 + 2x)} & \\
③ \quad\quad\quad\quad\quad\quad\quad 2x + 1 & \\
\quad\quad\quad\quad\quad\quad -(2x + 1) & \\
\end{array}
$$

Notes on the calculation:

① $\quad x \cdot 2x^3 = 2x^4,\ 2 \cdot (2x^3) = (2 \cdot 2)x^3 = 1 \cdot x^3$.
② $\quad 0 \cdot x^3 - 1 \cdot x^3 = 0 \cdot x^3 + 2x^3 = 2x^3$ (because $-1 = 2!$).
③ $\quad x - 2x = x + x = 2x$.

Let's do a test run:

$$(x+2)(2x^3 + 2x^2 + x + 2) = 2x^4 + 2x^3 + x^2 + 2x + x^3 + x^2 + 2x + 1$$
$$= 2x^4 + 2x^2 + x + 1.$$

Horner's Method

In technical applications it is often necessary to calculate the function values of polynomials quickly. We will see in the second part of the book that many functions, such as sine, cosine, exponential function or logarithm, can be approximated by polynomials. Whenever you type $\sin(x)$ on your calculator, such a polynomial is evaluated at the point x. For this we need an algorithm that requires as few operations as possible. Let's look at the following polynomial:

$$f(x) = 8x^7 + 3x^6 + 2x^5 - 5x^4 + 4x^3 - 3x^2 + 2x - 7.$$

If we want to evaluate this polynomial at a point $x = b$ and just start calculating, we need $7 + 6 + 5 + 4 + 3 + 2 + 1 = 28$ multiplications and 7 additions.

Horner's method reduces this effort considerably. We calculate one after the other:

$$c_0 := 8$$
$$c_1 := c_0 \cdot b + 3 \quad = 8 \cdot b + 3$$
$$c_2 := c_1 \cdot b + 2 \quad = 8 \cdot b^2 + 3b + 2$$
$$c_3 := c_2 \cdot b - 5 \quad = 8 \cdot b^3 + 3b^2 + 2b - 5$$
$$\vdots$$
$$c_7 := c_6 \cdot b - 7 \quad = 8 \cdot b^7 + 3b^6 + 2b^5 - 5b^4 + 4b^3 - 3b^2 + 2b - 7 = f(b) \tag{5.10}$$

c_7 is therefore the desired function value. For its calculation we need 7 multiplications and 7 additions. This method can be used not only for the calculation of function values, but also for factoring:

▶   **Theorem 5.26: Horner's method** Let K be a field, $f(x) = a_n x^n + \cdots + a_1 x + a_0 \in K[X]$ a polynomial of degree n and $b \in K$. For the recursively defined sequence of numbers

$$c_0 := a_n, \quad c_k := c_{k-1} b + a_{n-k}, \quad k = 1, \ldots, n$$

$f(b) = c_n$ holds.

If $f(b) = 0$, then the quotient $f(x)/(x - b) = q(x)$ is obtained by

$$q(x) = c_0 x^{n-1} + c_1 x^{n-2} + \cdots + c_{n-2} x + c_{n-1}.$$

The example (5.10) shows how the statement $f(b) = c_n$ comes about. Of course, one proves this in the general case by mathematical induction. To see the correctness of the second statement, we only have to test it. For this we use that $f(b) = c_n = 0$:

$$q(x)(x-b) = \underbrace{c_0}_{a_n} x^n + \underbrace{(c_1 - c_0 b)}_{a_{n-1}} x^{n-1} + \underbrace{(c_2 - c_1 b)}_{a_{n-2}} x^{n-2} + \cdots$$

$$+ \underbrace{(c_{n-1} - c_{n-2} b)}_{a_1} x - \underbrace{c_{n-1} b}_{c_n - a_0 = -a_0}$$

$$= f(x) \qquad \qquad \square$$

The following scheme can be used to calculate the coefficients c_i simply: First, we write the coefficients of the polynomial in the first line of a three-line table. Starting from the left, we fill in the other two lines column by column: the second line contains 0 in the first column, in the column i for $i > 1$ the intermediate result $c_{i-1} \cdot b$, so that in the third line the sum of the first two lines is the element c_i:

a_{n-i}:	a_n	a_{n-1}	a_{n-2}	...	a_1	a_0
$+$	0	$c_0 \cdot b$	$c_1 \cdot b$	...	$c_{n-2} \cdot b$	$c_{n-1} \cdot b$
c_i:	c_0	c_1	c_2	...	c_{n-1}	c_n

Let's try it with the example (5.9): The polynomial $2x^4 + 2x^2 + x + 1 \in GF(3)[X]$ has root 1, it is therefore divisible by $(x + (-1)) = x + 2$. From Horner's method we get the coefficients c_i:

a_{n-i}:	2	0	2	1	1
$+$		0	$2 \cdot 1 = 2$	$2 \cdot 1 = 2$	$1 \cdot 1 = 1$	$2 \cdot 1 = 2$
c_i:	2	2	1	2	0

and thus the quotient $2x^3 + 2x^2 + x + 2$, just as in (5.9).

Residue Classes in the Polynomial Ring, The Field GF(p^n)

The fact that division with remainder is possible in the polynomial ring $K[X]$ allows us to carry out similar calculations in $K[X]$ as we did in Chap. 4 with integers. So you can form residue classes modulo a polynomial and carry out the Euclidean algorithm. Calculating with remainders is analogous to $\mathbb{Z}$ and so we will discover new interesting rings and fields.

First, I would like to list a few definitions and theorems that are almost literal translations of our results with the integers:

▶   **Definition 5.27** If K is a field and $f(x) \in K[X]$ is a polynomial of degree n, then $f(x)$ is called *irreducible*, if in every representation $f(x) = g(x)h(x)$ with polynomials $g(x), h(x)$ it holds: $\deg g = 0$ or $\deg g = n$.

If $f(x)$ is divisible by $g(x)$ and k is a field element, then $f(x)$ is also divisible by $k \cdot g(x)$. But except for this multiplicative factor, an irreducible polynomial $f(x)$ only has itself and the polynomial $1 \cdot x^0 = 1$ as divisors. Do you see the analogy to the prime numbers in $\mathbb{Z}$?

▶ **Definition 5.28** Let $f(x), g(x) \in K[X]$. The polynomial $d(x) \in K[X]$ is called *greatest common divisor* of f and g, if d is a common divisor of f and g and if d has the maximum degree with this property. We write $d(x) = \gcd(f(x), g(x))$.

Compare this to the Definition 4.12 in Sect. 4.2. Please note: In contrast to the greatest common divisor in $\mathbb{Z}$, the greatest common divisor of two polynomials is not uniquely determined, but only up to a multiplicative factor from K: If $d(x)$ is a greatest common divisor and $k \in K$, then $k \cdot d(x)$ is also a greatest common divisor. Often, in the polynomial, the highest coefficient is normalized to 1 and then called *the* greatest common divisor.

▶ **Theorem 5.29: The Euclidean algorithm for** $K[X]$ Let $f(x), g(x) \in K[X]$, $f(x), g(x) \neq 0$. Then a $\gcd(f(x), g(x))$ can be determined recursively by continued division with remainder according to the following rule:

$$\gcd(f(x), g(x)) = \begin{cases} g(x) & \text{if } f(x) \bmod g(x) = 0 \\ \gcd(g(x), f(x) \bmod g(x)) & \text{else} \end{cases}$$

This is a literal translation of Theorem 4.14, and the proof proceeds analogously: at each step, the degree of the remainder becomes at least 1 smaller, so at some point the remainder must be 0. The last non-zero remainder is a greatest common divisor.

For the following calculations with residue classes, the extended Euclidean algorithm is of particular importance, It can also be transferred together with its proof from Theorem 4.15:

▶ **Theorem 5.30: The extended Euclidean algorithm for** $K[X]$ Let $f(x), g(x) \in K[X]$, $f(x), g(x) \neq 0$ and $d(x) = \gcd(f(x), g(x))$. Then there are polynomials $\alpha(x), \beta(x) \in K[X]$ with $d = \alpha f + \beta g$.

In the polynomial ring $K[X]$ we will now form residue classes modulo a polynomial and calculate with them. We proceed in the same way as in Sec. 4.3:

▶ **Definition 5.31** Let $f(x), g(x), p(x) \in K[X]$. The polynomials $f(x)$ and $g(x)$ are called *congruent modulo* p if $f - g$ is divisible by p. In symbols: $f(x) \equiv g(x) \bmod \mathrm{p}(x)$.

▶ **Theorem 5.32** "$\equiv$" is an equivalence relation on $K[X]$.

As in the integers, one finds that two polynomials are equivalent if they leave the same remainder modulo $p(x)$. The residue class of a polynomial can be represented by the corresponding residue. Which residues are possible when dividing by the polynomial $p(x)$

with degree n? These are exactly all polynomials with degree less than n. All these polynomials with degree less than n therefore form a natural system of representatives of the residue classes, just as the numbers $\{0, 1, \ldots, n-1\}$ represent the residue classes modulo n in $\mathbb{Z}$.

We denote the set of residue classes or representatives by $K[X]/p(x)$ (read: $K[X]$ modulo $p(x)$). Similar to Definition 4.24 we can now define addition and multiplication on the set of residues modulo $p(x)$:

▶ **Definition 5.33** Let $f(x), g(x) \in K[X]$ be residues modulo $p(x)$. Then

$$f(x) \oplus g(x) := (f(x) + g(x)) \bmod p(x)$$
$$f(x) \otimes g(x) := (f(x) \cdot g(x)) \bmod p(x).$$

Examples

1. In $\mathbb{Q}[X]$ the polynomial $p(x) = x^3 + 1$ is given. The set of residues consists of the polynomials with degree less than 3: $\mathbb{Q}[X]/p(x) = \{ax^2 + bx + c \mid a, b, c \in \mathbb{Q}\}$.
   The addition in $\mathbb{Q}[X]/p(x)$ is simple: The sum of two polynomials with degree less than 3 again has a degree less than 3, so it remains in the set. When multiplying, the degree can become greater than 2, then we have to take the remainder. So for example $(2x^2 + 1) \otimes (x + 2) = 2x^3 + 4x^2 + x + 2 \bmod (x^3 + 1)$. Division with remainder gives $2x^3 + 4x^2 + x + 2 = 2(x^3 + 1) + (4x^2 + x)$, so the remainder is $4x^2 + x$:

   $$(2x^2 + 1) \otimes (x + 2) = 4x^2 + x.$$

2. In GF(2) the polynomial $x^2 + 1$ is given. The possible residues are the polynomials $0, 1, x, x + 1$. For example:

   $$(x + 1)x = x^2 + x = (x^2 + 1) + x + 1 \equiv (x + 1) \bmod (x^2 + 1)$$
   $$(x + 1)(x + 1) = x^2 + 1 \equiv 0 \bmod (x^2 + 1).$$

   $$(5.11)$$

   Since we only have four residues, we can give a complete multiplication table:

$\otimes$	0	1	x	$x+1$
0	0	0	0	0
1	0	1	x	$x+1$
x	0	x	1	$x+1$
$x+1$	0	$x+1$	$x+1$	0

   The polynomials of degree less than n in GF$(p)[X]$ are often abbreviated simply as the n-tuple of their coefficients from GF(p), here $0 = (0,0)$, $1 = (0,1)$, $x = (1,0)$

and $x + 1 = (1, 1)$. In this notation, the Cayley tables in $\mathrm{GF}(2)/(x^2 + 1)$ look as follows:

$\oplus$	$(0,0)$	$(0,1)$	$(1,0)$	$(1,1)$
$(0,0)$	$(0,0)$	$(0,1)$	$(1,0)$	$(1,1)$
$(0,1)$	$(0,1)$	$(0,0)$	$(1,1)$	$(1,0)$
$(1,0)$	$(1,0)$	$(1,1)$	$(0,0)$	$(0,1)$
$(1,1)$	$(1,1)$	$(1,0)$	$(0,1)$	$(0,0)$

$\otimes$	$(0,0)$	$(0,1)$	$(1,0)$	$(1,1)$
$(0,0)$	$(0,0)$	$(0,0)$	$(0,0)$	$(0,0)$
$(0,1)$	$(0,0)$	$(0,1)$	$(1,0)$	$(1,1)$
$(1,0)$	$(0,0)$	$(1,0)$	$(0,1)$	$(1,1)$
$(1,1)$	$(0,0)$	$(1,1)$	$(1,1)$	$(0,0)$

It is not difficult to check that the structure $(K[X]/p(x), \oplus, \otimes)$ forms a ring, just like $(\mathbb{Z}/n\mathbb{Z}, \oplus, \otimes)$. The 0-element is the zero polynomial and there is also a 1-element, which is the polynomial $1 = 1 \cdot x^0$. Can a field sometimes arise from this construction? At least in the second example from above this is not the case. You can see, for example, the product of two elements different from 0 is again 0. $\mathbb{Z}/n\mathbb{Z}$ is a field if n is a prime number. Let's try it here with an irreducible polynomial $p(x)$! Another example:

3. In $\mathrm{GF}(2)[X]$ the polynomial $x^2 + x + 1$ is irreducible, because every product of two polynomials of degree 1 in $\mathrm{GF}(2)[X]$ is different from $x^2 + x + 1$. As sets, $\mathrm{GF}(2)[X]/(x^2 + 1)$ and $\mathrm{GF}(2)[X]/(x^2 + x + 1)$ are equal. But how do the Cayley tables look now? Nothing changes with the addition, but the multiplication gives different results. So now, for example (compare with (5.11))

$$(x + 1)x = x^2 + x = (x^2 + x + 1) + 1 \equiv 1 \bmod (x^2 + x + 1)$$
$$(x + 1)(x + 1) = x^2 + 1 = (x^2 + x + 1) + x \equiv x \bmod (x^2 + x + 1)$$

and in total we get for the multiplicative connection table:

$\otimes$	$(0,0)$	$(0,1)$	$(1,0)$	$(1,1)$
$(0,0)$	$(0,0)$	$(0,0)$	$(0,0)$	$(0,0)$
$(0,1)$	$(0,0)$	$(0,1)$	$(1,0)$	$(1,1)$
$(1,0)$	$(0,0)$	$(1,0)$	$(1,1)$	$(0,1)$
$(1,1)$	$(0,0)$	$(1,1)$	$(0,1)$	$(1,0)$

The 1-element is $(0, 1)$. See that here again every element different from 0 has a multiplicative inverse? This is the property that was still missing for the field. ◀

In fact, the following theorem holds (analogous to Theorem 5.19):

▷ **Theorem 5.34** Let K be a field and $p(x) \in K[X]$ a polynomial. Then $K[X]/p(x)$ is a field if and only if $p(x)$ is irreducible.

The proof is exactly the same as in the case of integers with the help of the extended Euclidean algorithm: Let $p(x)$ be irreducible and $f(x)$ a remainder modulo $p(x)$. Then

$p(x)$ and $f(x)$ have no common divisor with a degree greater than 0. This would have to be a divisor of $p(x)$ with a degree less than n. This is not possible because of the irreducibility of $p(x)$. In particular, $1 = \gcd(p, f)$, and according to Theorem 5.31 there are polynomials $\alpha(x), \beta(x) \in K[X]$ with $1 = \alpha \cdot f + \beta \cdot p$, that is $\alpha \cdot f = 1 - \beta \cdot p(x)$. So we have: $\alpha \cdot f \equiv 1 \bmod p(x)$ and $\alpha \bmod p(x)$, the representative of the residue class α, is the multiplicative inverse of $f(x)$. $\qquad\square$

The field K is actually contained in the new field $K[X]/p(x)$ as a subfield: The elements $k \in K$ correspond exactly to the residues $k \cdot x^0$, the constant polynomials.

Example

In $\mathbb{R}$ the polynomial $x^2 + 1$ is irreducible, because there is no root of -1. The polynomials in $\mathbb{R}[X]/(x^2 + 1)$ form the set $\{a + bx \mid a, b \in \mathbb{R}\}$. Let's multiply two such polynomials together:

$$
\begin{aligned}
(a + bx)(c + dx) &= ac + (ad + bc)x + bdx^2 \\
&= bd \cdot (x^2 + 1) + (ac - bd) + (ad + bc)x \\
&\equiv (ac - bd) + (ad + bc)x \bmod (x^2 + 1).
\end{aligned}
$$

Please compare this with the addition and multiplication rule of the field $\mathbb{C}$ from Definition 5.12. It is the same rule.

In fact, $\mathbb{R}[X]/(x^2 + 1) = \mathbb{C}$! The polynomial x has taken on the role of i and really is $x^2 \bmod (x^2 + 1) = (x^2 + 1) - 1 \equiv -1 \bmod (x^2 + 1)$, so $x^2 = -1$. ◄

I think, when you see it for the first time, your head has to spin. Mathematicians always like to look for roots of polynomials. So they have made from $\mathbb{Q}$ the field $\mathbb{R}$, to find a root of $x^2 - 2$, and from $\mathbb{R}$ the field $\mathbb{C}$, to be able to solve $x^2 + 1 = 0$. And now we have found a way to give every irreducible polynomial a root: We form the associated polynomial ring modulo this irreducible polynomial, the result is again a field, and in that just this irreducible polynomial has a root! Surprisingly, this not only works in the just seen example, but always: Is f in $K[X]$ irreducible, then the remainder x is root of the polynomial f in $K[X]/f$. Every field can be extended so that an irreducible polynomial has roots in the field extension. A great piece from the mathematical magic box. A large mathematical theory is based on this, the theory of fields.

In computer science, the finite fields are particularly interesting: If one starts from a field $GF(p)$ and finds an irreducible polynomial $f(x)$ of degree n in this field, then $GF(p)/f(x)$ is again a field. The elements are just all polynomials of the form $a_{n-1}x^{n-1} + a_{n-2}x^{n-2} + \ldots + a_1 x + a_0$ with $a_i \in GF(p)$. If you write the polynomials as a sequence of coefficients again, then $GF(p)/f(x) = \{(a_{n-1}, \ldots, a_1, a_0) \mid a_i \in GF(p)\}$. So this is a field with p^n elements. We call this $GF(p^n)$:

▶   **Definition 5.35: The fields** $GF(p^n)$ Is p a prime number, n a natural number
and $f(x)$ an irreducible polynomial of degree n in $GF(p)[X]$, then the field
$GF(p)[X]/f(x)$ is called $GF(p^n)$.

One would assume that for different irreducible polynomials of degree n, the corresponding residue class fields are different, just as different structures arose in the previous examples 2 and 3. To precisely identify the field $GF(p^n)$, one would therefore have to specify
the irreducible polynomial f. If one wants to perform concrete operations on the field elements, that is correct. But surprisingly, for different irreducible polynomials f and g, the
corresponding fields are practically the same. In anticipation of Sect. 5.6: There is an isomorphism between the fields. In this respect, the designation $GF(p^n)$ for *the* field with p^n
elements is justified.
    By the way, one can prove that there are no more finite fields: Every finite field is of the
form $K = GF(p^n)$ for a prime number p and a natural number n.

In computer science, a lot is calculated with sequences of 0s and 1s.
    Such a sequence, for example the set of all bytes, can now be equipped with a field
structure in a simple way. For example, the polynomial $x^8 + x^4 + x^3 + x + 1$ is irreducible in $GF(2)$. This allows one to form $GF(2^8)$ and obtain a field structure on the set of
bytes. Such structures play an important role in coding theory and cryptography. For
example, the encryption algorithm AES (Advanced Encryption Standard) uses internally
multiplications in $GF(2^8)$ with the polynomial mentioned above. Another very current
encryption mode, GCM (Galois Counter Mode), uses multiplications in $GF(2^{128})$ with
the irreducible polynomial $x^{128} + x^7 + x^2 + x + 1$.

Using Polynomial Division for Error Detection

When data is transported over a line, physical influences can always lead to errors, for
example due to "thermal noise", "crosstalk" between different lines and many other
things. Therefore, it is necessary to check the data for correctness at the receiver. The
simplest form of such an error detection is the attachment of a so-called "check bit "
to a transmitted data block. For example, you can append the sum of the bits modulo 2
(more precisely, we can say: the sum in $\mathbb{Z}/2\mathbb{Z}$) to each byte. At the receiver, the check bit
is recalculated, in the case of a transmission error the check bit does not match anymore
and the data has to be requested again:

$$10011011|1 \quad \longrightarrow \quad 11011011|1$$
$$\underset{\text{checkbit}}{\uparrow} \qquad\qquad \underset{\text{error}}{\uparrow} \quad \underset{\substack{\text{checkbit no} \\ \text{longer correct}}}{\uparrow}$$

This error detection is not very effective: You have to transmit $1/8$, i.e. 12.5%, more data,
which reduces the effective transmission rate and, if 2 bits are flipped in the byte, the
error is already overlooked.

The polynomial division in GF(2)[X] we have just learned, provides a much more efficient method. We proceed as follows:

1. The bits of a code word are interpreted as coefficients of a polynomial in GF(2)[X], for example

$$10011011 \stackrel{\triangle}{=} x^7 + x^4 + x^3 + x + 1$$

   The code word to be transmitted is the polynomial $f(x)$. Usually, much longer data words are transmitted.
2. A fixed polynomial $g(x) \in$ GF(2)[X] is used as divisor. $g(x)$ is called the *generator polynomial*. It is deg $g = n$.
3. Determine the remainder when dividing $f(x)$ by $g(x)$: $r(x) = f(x) \bmod g(x)$. It is deg $r(x) < n$.
4. Append the remainder $r(x)$ to the polynomial $f(x)$ and transmit $f(x), r(x)$. The remainder $r(x)$ is used for error detection.
5. The receiver divides the received polynomial again by $g(x)$ and receives a remainder $r'(x)$. If an error has occurred during the transmission, usually the difference $r(x) - r'(x) \neq 0$.

An error is only not recognized if the erroneous polynomial when divided by $g(x)$ gives the same remainder as $f(x)$. From the type of the difference of the remainders (how "large" or "small" they are), it is often possible to infer the error and correct it. The question of what the words "large", "small" mean in this context is a topic of coding theory.

This type of error detection is called "Cyclic Redundancy Check" (CRC), it is used in many data transmission protocols. For example, in the Ethernet protocol, frames up to 1514 bytes in length are transmitted, $f(x)$ therefore has a degree less than or equal to 12 112 here.

The generator polynomial

$$x^{32} + x^{26} + x^{23} + x^{22} + x^{16} + x^{12} + x^{11} + x^{10} + x^8 + x^7 + x^5 + x^4 + x^2 + x^1 + 1$$

is used for error checking, so 4 bytes are used for error checking, that's just 2.6 promille!

With the help of this polynomial, all 1 and 2 bit errors, any odd number of errors and all error bursts up to 32 bit length can be safely detected.

A lot of brainpower goes into the clever choice of the generator polynomial, which must have "good" properties. Algorithms for polynomial division modulo 2 can be implemented very efficiently in software as well as in hardware, so that even with large data rates this type of protection is not a problem.

5.5   Elliptic Curves

Elliptic Curves over the Field of Real Numbers

If K is a field and $a, b \in K$, then the set $\{(x, y) \in K^2 \mid y^2 = x^3 + ax + b\}$ is a subset of K^2. We would like to define a group structure on this set. For this purpose, we add one more point to this set, which we denote by $\mathcal{O}$ and call the "point at infinity", you will see why in a moment. The set $E = \{(x, y) \in K^2 \mid y^2 = x^3 + ax + b\} \cup \{\mathcal{O}\}$ is called an elliptic curve. Let's first take the real numbers $\mathbb{R}$ as the field. The rule $y^2 = x^3 + ax + b$ is not a mapping, because there is no unique calculation rule $y = f(x)$. For illustration, we can divide the curve (without the point $\mathcal{O}$) into two parts: $y_1 = \sqrt{x^3 + ax + b}$ and $y_2 = -\sqrt{x^3 + ax + b}$. Both parts are real functions, the union of the graphs of these functions is $E \setminus \{\mathcal{O}\}$. We can already read something from this: the curve is symmetrical to the x-axis, there are gaps in the domain of definition, because the value under the square root must not be less than 0. If $x^3 + ax + b = 0$, then it has at least one root. Possible are one, two or three real roots. If the polynomial has 2 roots, it can be split into three linear factors, and one of the roots must occur twice in it. Such curves are not suitable for our purposes. You can avoid double roots if you additionally require for E that $4a^3 + 27b^2 \neq 0$.

Typical graphs of elliptic curves can be seen in Figs. 5.3 and 5.4. Fig. 5.3 shows the curve for $y^2 = x^3 - 20x + 5$, Fig. 5.4 the curve for $y^2 = x^3 - 10x + 15$.

We want to define an addition on the set of points of E, which makes the curve a group. The point $\mathcal{O}$ is to be the identity element. The easiest way to explain this addition is geometrically. If P, Q are points of $E \setminus \{\mathcal{O}\}$, we draw the connecting line. If P and Q are not exactly vertically aligned, this line has exactly one more intersection point with E, we will calculate it right away. We reflect this intersection point at the x-axis and obtain $P+Q$, see Fig. 5.5 for this. What about the special case? If P and R lie vertically, then the connecting line g no longer intersects the curve, we say "g goes through the point at

Abb. 5.3 The elliptic curve
$y^2 = x^3 - 20x + 5$

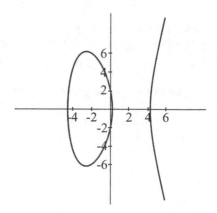

Fig. 5.4 The elliptic curve
$y^2 = x^3 - 10x + 15$

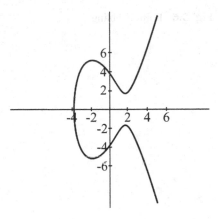

Fig. 5.5 Point addition

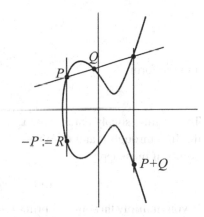

infinity $\mathcal{O}$" and we define $P + R = \mathcal{O}$. It now follows that $-P = R$ must be. The inverse of an element is therefore obtained by reflection at the x-axis, see also Fig. 5.5.

This way we have defined a sum for each of two different points of E. There is still the addition of a point to itself. What is $P + P$? How do we proceed here?

If we move the point Q closer and closer to P in Fig. 5.5, the line through P and Q will eventually become the tangent to the curve through P. This also intersects E in exactly one more point, which we reflect again across the x-axis, and obtain $P + P$, see Fig. 5.6. Again, there is one exception: if the tangent is perpendicular, then $R + R = \mathcal{O}$ and thus $-R = R$.

Now we need to convert the geometric representation into algebraic formulas. So let E be given by $y^2 = x^3 + ax + b$. Every line, except the perpendiculars, has the form $y = mx + t$, where m is the slope of the line. The connecting line g of two points $P = (x_P, y_P), Q = (x_Q, y_Q)$ has the slope

Fig. 5.6 Point doubling

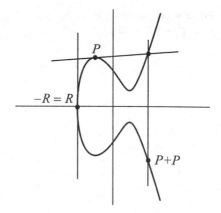

$$m = \frac{y_Q - y_P}{x_Q - x_P}$$

and the form

$$y = m(x - x_P) + y_P. \tag{5.12}$$

To test this, simply enter x_P or x_Q into this line equation, you will get y_P and y_Q. So it's the line through P and Q. Where is the third intersection of E with g? For the common points, it must apply:

$$(m(x - x_P) + y_P)^2 = x^3 + ax + b. \tag{5.13}$$

If you multiply this out and bring it to one side, you will get a third-degree polynomial in x:

$$h(x) = x^3 - m^2 x^2 + \alpha x + \beta \tag{5.14}$$

The values of the coefficients α and β are irrelevant, so I didn't even calculate them.

There are formulas for solving a third-degree polynomial, but they look pretty ugly. Fortunately, however, we already know two roots of the polynomial, namely x_P and x_Q, because P and Q are intersections of E and g! So we can split two roots off the polynomial $h(x)$, and the rest is a linear polynomial:

$$h(x) = (x - x_P)(x - x_Q)(x - x_R) \tag{5.15}$$

So there is exactly one more root x_R. How do we get its value? Expand (5.15) again, you get $(-x_P - x_Q - x_R)$ as the coefficient of x^2. The comparison with (5.14) yields $-m^2 = -x_P - x_Q - x_R$, respectively

$$x_R = m^2 - x_P - x_Q = \left(\frac{y_Q - y_P}{x_Q - x_P}\right)^2 - x_P - x_Q. \tag{5.16}$$

The y value of x_R is most easily obtained by insertion into the line (5.12). But we must not forget to reflect at the x-axis:

$$y_R = -m(x_R - x_P) - y_P$$

Now to the special case of addition $P + P$. We can assume $y_P \neq 0$ here, otherwise we would have the perpendicular tangent and $P + P = \mathcal{O}$. To calculate the tangent equation, one needs the derivative of the curve at the point x_P. For this I have to anticipate the second part of the book, see Sect. 15.1: If $y = f(x) = \sqrt{x^3 + ax + b}$, then the tangent to f at P has the slope $f'(x_P)$. To calculate the derivative, we need the chain rule from Theorem 15.5 and the rules for exponentiation from Definition 14.13 and Theorem 14.14:

$$f'(x_P) = \frac{1}{2\sqrt{x_P^3 + ax_p + b}}(3x_P^2 + a) = \frac{3x_P^2 + a}{2y_P} = m'$$

The tangent equation is as before $y = m'(x - x_P) + y_P$. Also the determination of the further intersection point of g with E takes place as in the case of $P \neq Q$: In (5.13) one has to calculate with the slope m'. In (5.15) x_P is now a double root and we get

$$x_R = m'^2 - 2x_P = \left(\frac{3x_P^2 + a}{2y_P}\right)^2 - 2x_P \qquad (5.17)$$

and for the y value, reflected again at the x-axis:

$$y_R = -m'(x_R - x_P) - y_P.$$

Now we can calculate the sum for all points of the curve. The sum $P + \mathcal{O}$ should of course be P. Does this make E a group? $\mathcal{O}$ is the identity element and each element has an inverse. The construction also automatically results in the commutativity of the operation. The associativity of the connection is still missing. Calculating this is a long and tedious task, which I would like to avoid.

Elliptic Curves Over Finite Fields

In the construction of the operation on an elliptic curve, we have made very intensive use of the properties of the real numbers: we have taken roots, formed derivatives and calculated tangents. Something like this only works to a limited extent in other fields. But if you look at the results of the calculations in (5.16) and (5.17), you will see that only the operations possible in every field are used here. So we can try to extend the definition to general fields. There is one exception: if in the field K it is true that $1 + 1 = 0$ or $1 + 1 + 1 = 0$, then the formula (5.17) breaks down, because then either $2 = 0$ or $3 = 0$. In all other cases our attempt works.

Elliptic curves over finite fields are used in cryptography. Here it is very important to have efficient algorithms for calculating the sum of two points. Now there are very fast implementations of the field operations for the Galois fields $GF(2^n)$. Unfortunately, we cannot use them, because in them $1 + 1 = 0$ is true. If you look at the equation of degree 3, which defines an elliptic curve, in a different way, the concept of groups can also be extended to elliptic curves over the fields $GF(2^n)$. Such curves are also used in applications. I don't want to go into that here.

▶   **Definition and Theorem 5.36** Let K be a field in which $1 + 1 \neq 0$ and $1 + 1 + 1 \neq 0$ is true. Let $a, b \in K$ and let $4a^3 + 27b^2 \neq 0$. The element $\mathcal{O}$ is called "point at infinity". Then the set

$$E = \{(x, y) \in K^2 \mid y^2 = x^3 + ax + b\} \cup \{\mathcal{O}\}$$

is called an *elliptic curve* over K. With the addition of the points $P = (x_P, y_P)$, $Q = (x_Q, y_Q)$, which is defined as follows
- For $P \neq Q$, $x_P \neq x_Q$:

$$P + Q := (x_R, -m(x_R - x_P) - y_P),$$

$$x_R = m^2 - x_P - x_Q, \quad m = \frac{y_Q - y_P}{x_Q - x_P},$$

- For $P = Q$, $y_P \neq 0$:

$$P + P := (x_R, -m(x_R - x_P) - y_P),$$

$$x_R = m^2 - 2x_P, \quad m = \frac{3x_P^2 + a}{2y_P},$$

- Otherwise:

$$P + Q := \mathcal{O},$$

is E a commutative group with the identity element $\mathcal{O}$.

These elliptic curves and the addition on them are not as intuitive as in the case of real numbers. In Fig. 5.7 you can see the elliptic curve to the equation $y^2 = x^3 + 6x + 9$ over the field $GF(13)$. It has 17 points including the $\mathcal{O}$: For 8 elements of $GF(13)$, $x^3 + 6x + 9$ is a square. If z is a root of it, then $-z = 13 - z$ is too. Therefore, the "curve" is symmetrical like in the real case: Above the horizontal axis $13/2 = 6.5$, the additive inverses of the lower half are located. The connecting line of two points can also be interpreted intuitively, see exercise 14 in this chapter. This line intersects the curve in one other point. There is no geometric interpretation for the tangent anymore, but we can still calculate $P + P = 2P$ like in the real case. For example, $(1, 9) + (1, 9) = 2 \cdot (1, 9) = (8, 7)$ and $3 \cdot (1, 9) = (7, 2)$. Check this using the Definition 5.36.

In Sect. 5.7 on cryptography, we will see how elliptic curves are used in encryption algorithms.

Fig. 5.7 An elliptic curve over GF(13)

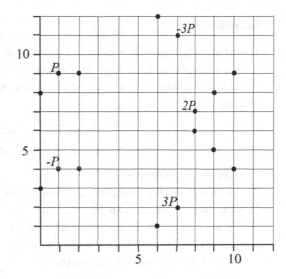

5.6 Homomorphisms

We have already seen several times that mappings are an important concept in mathematics. When comparing different things, one usually tries to establish a mapping between these things in order to find similarities and differences. In the context of algebraic structures, mappings are interesting which preserve the structure. Such operation-compatible mappings are called homomorphisms. If there is a bijective homomorphism between two structures, they are practically indistinguishable; it does not matter whether one investigates the one or the other.

▶ **Definition 5.37: Homomorphism** If G and H are groups and $\varphi\colon G \to H$ is a mapping with the property

$$\varphi(a \cdot b) = \varphi(a) \cdot \varphi(b) \quad \text{for all } a, b \in G,$$

then φ is called a *(group) homomorphism*. If R and S are rings and $\varphi\colon R \to S$ is a mapping with the property

$$\varphi(a + b) = \varphi(a) + \varphi(b), \quad \varphi(a \cdot b) = \varphi(a) \cdot \varphi(b) \quad \text{for all } a, b \in R,$$

then φ is called a *(ring) homomorphism*.
 A bijective homomorphism is called an *isomorphism*.

Attention: Even if I do not mark the symbols differently anymore: "+" and "·" on the left and on the right side of the equation mean different operations! Left in G (or in R), right in H (or in S).

Examples

1. We already know an important homomorphism. The operations on the set of residue classes modulo n from Definition 4.24 are defined in such a way that they are compatible with the operations on $\mathbb{Z}$. This means nothing else than that the mapping

$$\varphi \colon \mathbb{Z} \to \mathbb{Z}/n\mathbb{Z}$$
$$z \mapsto z \bmod n$$

   is a ring homomorphism.

2. When introducing complex numbers, I said $\mathbb{R}$ is contained in $\mathbb{C}$. For this we have identified $a \in \mathbb{R}$ with $(a, 0) \in \mathbb{C}$. Now we can express precisely what this identification means: $\varphi \colon \mathbb{R} \to \mathbb{C}$, $a \mapsto (a, 0)$ is an injective homomorphism and $\varphi \colon \mathbb{R} \to \varphi(\mathbb{R})$, $a \mapsto (a, 0)$ is even an isomorphism. This means $\mathbb{R}$ and $\varphi(\mathbb{R}) \subset \mathbb{C}$ have the same structure; we simply say $\mathbb{R} \subset \mathbb{C}$.

3. $\varphi \colon (\mathbb{Z}, +) \to (m\mathbb{Z}, +)$, $z \mapsto mz$ is:

a mapping	(if $z \in \mathbb{Z}$, then always $\varphi(z) \in m\mathbb{Z}$)
a homomorphism	($\varphi(z_1 + z_2) = m(z_1 + z_2) = mz_1 + mz_2 = \varphi(z_1) + \varphi(z_2)$)
surjective	(the image of mz is just z)
injective	(if $z_1 \neq z_2$, then also $mz_1 \neq mz_2$)

   and thus an isomorphism. $(\mathbb{Z}, +)$ and $(m\mathbb{Z}, +)$ are practically indistinguishable as groups.

4. For $\varphi \colon (\mathbb{Z}, +, \cdot) \to (m\mathbb{Z}, +, \cdot)$, $z \mapsto mz$ it holds:

$$\varphi(z_1 z_2) = m(z_1 z_2) \neq mz_1 mz_2 = \varphi(z_1)\varphi(z_2).$$

   So this is not a ring isomorphism: As rings, $\mathbb{Z}$ and $m\mathbb{Z}$ have different structures. For example, in $\mathbb{Z}$ there is a 1-element, in $m\mathbb{Z}$ not!

5. $\overline{} \colon \mathbb{C} \to \mathbb{C}$, $z \mapsto \bar{z}$ is a ring isomorphism, since $\mathbb{C}$ is a field, we also say field isomorphism. The mapping is bijective, the homomorphism property we have already seen in Theorem 5.16. This isomorphism has the special property that it leaves all real numbers unchanged. ◄

▶ **Definition 5.38** Let $(G, +)$, $(H, +)$ be groups or rings, $\varphi \colon G \to H$ a homomorphism. Then

$$\ker \varphi := \{x \in G \mid \varphi(x) = 0\} \qquad \text{the kernel of } \varphi,$$
$$\operatorname{im} \varphi := \{y \in H \mid \exists x \in G \text{ mit } \varphi(x) = y\} \quad \text{the image of } \varphi.$$

In particular, the kernel, the set of all elements that are mapped to 0, plays an important role in the use of homomorphisms and often simplifies calculations. We can already see this in the following theorems:

▶ **Theorem 5.39** For a homomorphism $\varphi: G \to H$, always holds:

a) $\varphi(0) = 0$ (0 is always an element of the kernel).
b) $\varphi(-a) = -\varphi(a)$.
c) For all $a, b \in G$, it holds that $\varphi(a - b) = \varphi(a) - \varphi(b)$.   □

Proof:
a) $\varphi(0) = \varphi(0 + 0) = \varphi(0) + \varphi(0) \Rightarrow (-\varphi(0)$ on both sides$)\ 0 = \varphi(0)$.
b) $\varphi(a) + \varphi(-a) = \varphi(a + (-a)) = 0 \Rightarrow \varphi(-a) = -\varphi(a)$.
c) $\varphi(a - b) = \varphi(a + (-b)) = \varphi(a) + \varphi(-b) = \varphi(a) + (-\varphi(b)) = \varphi(a) - \varphi(b)$.   □

▶ **Theorem 5.40** A homomorphism is injective if and only if ker $\varphi = \{0\}$.

Proof: "$\Rightarrow$" Let φ be injective. Because of $\varphi(0) = 0$, $0 \in$ ker φ, and because the mapping is injective, this is the only preimage of 0.

"$\Leftarrow$" Let ker $\varphi = \{0\}$. Assume there is $a \neq b$ with $\varphi(a) = \varphi(b)$. Then $\varphi(a) - \varphi(b) = 0 = \varphi(a - b) \Rightarrow a - b \in$ ker φ in contradiction to the assumption.   □

Here is an application of Theorem 5.40 right away. You know the rule for solving quadratic equations. A formula for it is: For real numbers p, q, the equation $x^2 + px + q$ has the roots

$$x_{1/2} = -\frac{p}{2} \pm \sqrt{\left(\frac{p}{2}\right)^2 - q}. \tag{5.18}$$

We call the expression under the square root the discriminant D. You know that there are 0, 1, or 2 real solutions for this, depending on whether the discriminant is less than, equal to, or greater than 0. With our knowledge of complex numbers, we can now say that this equation always has two complex solutions in the case $D < 0$, namely $-\frac{p}{2} \pm i\sqrt{-D}$. A similar rule applies not only to quadratic equations, but to the roots of all real polynomials:

▶ **Theorem 5.41** If $f(x) \in \mathbb{R}[X]$ and $z = a + bi \in \mathbb{C}$ is a root of the polynomial in $\mathbb{C}$, then $a - bi$ is also a root of $f(x)$.

Proof: Let $\varphi: \mathbb{C} \to \mathbb{C}, z \mapsto \bar{z}$ be the isomorphism. Then we have:

$$\varphi(f(z)) = \varphi(a_n z^n + \cdots + a_0)$$
$$\underset{\substack{\uparrow \\ \varphi \text{ is} \\ \text{homo.}}}{=} \varphi(a_n)\varphi(z)^n + \cdots + \varphi(a_0)$$
$$\underset{\substack{\uparrow \\ a_i \in \mathbb{R}}}{=} a_n\varphi(z)^n + \cdots + a_0 = f(\varphi(z)),$$

and because of $f(z) = 0$ we have $\varphi(f(z)) = 0$, thus also $f(\varphi(z)) = f(a - bi) = 0$.   □

This proof is based on a brilliant discovery that revolutionized algebra at the time: There is a close connection between roots of polynomials and field isomorphisms. Here we have used it to find a new root. For a long time it was an unsolved problem which polynomials can be solved by formulas similar to (5.18). In the formulary you will find formulas for solving equations of the third and fourth degree. It could be proven that it is impossible for polynomials of the 5th degree to give a solution formula (even though we know that the solutions exist). This proof only succeeded because it was possible to translate the question into the investigation of properties of certain groups of field homomorphisms.

5.7   Cryptography

Cryptography is the science of encrypting and decrypting data. Especially in the age of the information society and e-commerce, cryptography is gaining more and more importance. Results of number theory play a major role in the construction and analysis of cryptographic algorithms and in particular the rings $\mathbb{Z}/n\mathbb{Z}$ and the fields GF(p) and GF(p^n) occur again and again. I would like to explain some basic concepts of cryptography here and present some important algorithms as examples.

Most of you have already encrypted messages as children and passed them along the school benches. A simple method was already used by Caesar. For this you write the alphabet twice one below the other, once opposite the other shifted:

```
ABCDEFGHIJKLMNOPQRSTUVWXYZ

DEFGHIJKLMNOPQRSTUVWXYZABC
```

The encryption rule is: Replace A with D, B with E and so on. We see from this: the encryption is a function that is applied to a plaintext and that is reversible. Here the function is: "Shift the plaintext 3 characters to the right in the alphabet".

In computer science, messages are always encoded as bit strings. For example, the message MAUS in ASCII code: 01001101, 01000001, 01010101, 01010011. Combined into an `integer` results in the 4 byte long number

$$01001101\ 01000001\ 01010101\ 01010011,$$

as a decimal number this is 1 296 127 315. Encrypting the MAUS therefore means encrypting the number 1 296 127 315. This takes us over into the field of mathematics: Encrypting means applying a reversible function to a number.

$$f_K : \{\text{plaintexts}\} \rightarrow \{\text{ciphertexts}\}$$

$$N \mapsto f_K(N)$$

In real applications, the function is always dependent on a key K. It is practically impossible to keep an algorithm secret, much easier the key. The secret that ensures that no unauthorized person can decipher the message is hidden in this key. In the Caesar code, the algorithm is "shift to the right", the key is "3". Of course, the algorithm is very simple and there are so few keys that you can try them all. This must not be possible.

Therefore, common algorithms use keys that are random numbers of 128- bit to 256-bit length. The data is also not encrypted byte by byte or "integer" wise, as in the example with the MAUS, larger data blocks are processed at once. Today, 64- bit portions are often common, AES (Advanced Encryption Standard), which has replaced the old DES algorithm, encrypts blocks with 128- bit size and offers key lengths of 128 or 256 bits. These keys cannot all be tried by any computer in the world to crack a code.

There are two basic encryption principles that cryptographers work with today: *encryption with secret keys* and *encryption with public keys*.

Encryption with Secret Keys

Each of two communication partners (in cryptographic literature these are always Alice and Bob) share a secret key K. If Alice wants to send the message M to Bob, she forms:

$$C := f_K(M).$$

The function is designed such that from the knowledge of K the inverse function f_K^{-1} can be derived. Then Bob (and no one else) can decrypt the message:

$$M = f_K^{-1}(C).$$

This is called *symmetric encryption*. The main problems with these method are two things:

1. The keys must be kept secret between the partners themselves. At first it looks as if you are chasing your own tail: To exchange secret messages, you have to exchange secret messages.
2. With n communication partners, everyone has to share a key with everyone. For this you need $n(n-1)/2$ keys. How are these keys managed and distributed?

Help promises the principle of encryption with public keys:

Encryption with Public Keys

Different keys are used here to encrypt and decrypt. Encryption can be done by anyone, so this key does not have to be kept secret. Only the recipient of the message can decrypt, so the decryption key must remain secret.

Each communication partner needs a key pair in this system, a public and a matching private key. Bob, for example, has the keys K_{EB} and K_{DB}. Here E stands for "encryption key", D stands for "decryption key". The corresponding functions I denote with f_{EB} and f_{DB}. Now comes the essential difference to symmetric encryption: The functions f_{EB} and

f_{DB} are indeed inverse to each other, but it must be ensured that under no circumstances is it possible to derive f_{DB} from f_{EB}.

Now if Alice sends Bob a message M, she must first obtain his public key for encryption: she either has him send it to her beforehand or she retrieves it from a key database. She then sends to Bob:

$$C := f_{EB}(M).$$

Only Bob can reverse this encryption by computing:

$$M = f_{DB}(C).$$

Encryption using public and private keys is called *asymmetric encryption* or *public-key-encryption*. Even if it initially looks as if all the problems of symmetric methods are thereby eliminated, here too there is a fly in the ointment: one must be certain that the public keys cannot be manipulated during transport. Otherwise Charly could replace Bob's public key with his own and play it to Alice. She thinks she is encrypting for Bob, but the message can be read by Charly. This problem can be handled with the help of so-called *certificates*: the authenticity of public keys and the assignment of a key to a person are confirmed by an independent third party, the certification authority, in a certificate; this too using cryptographic methods. However, one must expend some effort here.

In addition, asymmetric methods are so computationally complex that real-time encryption of large data sets is impossible. Symmetric methods are orders of magnitude faster.

In practice, the advantages of both methods are combined. It is typical to use public keys to encrypt small data sets, for example symmetric keys, which can then be securely exchanged between two partners as *session keys*. The user data is then encrypted with these. This happens, for example, every time you request a protected document with your internet browser.

The concept of public encryption is well understandable. It is much more difficult to find mathematical algorithms that realize it. In fact, algorithms were not published until the 1970s. The main problem that must always be solved is the construction of secure and fast functions that are reversible but for which the inverse function cannot be calculated. There are essentially two classes of difficult mathematical problems on which today's algorithms are based. For both of these problem classes, I will present an algorithm to you.

The first such problem is, it is very easy to multiply two large prime numbers, but it is very difficult to decompose a large number into its prime factors in human time (the *factorization problem*).

The second problem is, to a given natural number a and to a prime number p one can very easily calculate $a^x \bmod p$. But there is no known method how to a given y find a x with $y \equiv a^x \bmod p$ that is significantly faster than calculating all powers (the problem of the *discrete logarithm*).

Interestingly, for neither of the two classes there is a proof that the underlying problem is not solvable quickly after all. Only no one has found a way yet. If tomorrow someone discovers a fast method for factoring large numbers, half of the algorithms become unusable. There is some consolation in the fact that at least one can fall back on the other half.

The RSA Algorithm

The probably best-known public-key method is the *RSA algorithm*, named after its inventors (or discoverers?) Rivest, Shamir and Adleman, who published it in 1978. The RSA is based on the factorzation problem.

To generate a key pair, you need two large prime numbers p and q. Currently, most prime numbers are chosen in the order of 1024 bits. Then $n = p \cdot q$ is formed. This n has a size of 2048 bits, which is more than a 600-digit decimal number. Now one looks for an invertible number e in $\mathbb{Z}/(p-1)(q-1)\mathbb{Z}^*$, that is, an e with $\gcd(e, (p-1)(q-1)) = 1$ (see Theorem 5.18). Finally, d is the multiplicative inverse of e in $\mathbb{Z}/(p-1)(q-1)\mathbb{Z}^*$. This way we have everything we need for encryption:

- p, q large prime numbers
- $n = p \cdot q$
- e, d with $ed = 1 + k(p-1)(q-1)$.

Now we can define the keys, with e standing for encrypt, d for decrypt:

the public key: $n, e,$
the private key: $n, d.$

The secret of the private key lies of course in the number d, not in n. The security is based on the assumption that it is not possible to determine d from the knowledge of n and e without decomposing n into its prime factors p and q. By the way, this has not been proven yet either.

Now to the algorithm itself. This is of an incredibly simplicity. The message M is a number between 1 and $n - 1$. Then

$$C = M^e \bmod n \quad (M \text{ is encrypted to } C),$$
$$M = C^d \bmod n \quad (C \text{ is decrypted to } M). \tag{5.19}$$

Try this out with not too large prime numbers! If you use the algorithm from the formula (3.4) in Sect. 3.3 for exponentiation and take the modulo after each step, you avoid number overflows. In Java there is a class BigInteger included, so that you can play with longer numbers. If you use a mathematics tool like Sage, you can also work with really large prime numbers.

Why is (5.19) true? I would like to prove this.

So let's say $C = M^e \bmod n$. Because taking the modulo and exponentiating are interchangeable, $C^d \bmod n = (M^e)^d \bmod n = M^{ed} \bmod n$. So it must be calculated that $M = M^{ed} \bmod n$ is true.

First we show $M \bmod p = M^{ed} \bmod p$:

There is an integer α so that $ed = 1 + (p-1)\alpha$ is true. Then according to Fermat's little Theorem (5.20a):

$$M^{ed} \bmod p = M^{1+(p-1)\alpha} \bmod p = M \bmod p \cdot \underbrace{(M^{p-1} \bmod p)^\alpha}_{=1} = M \bmod p.$$

This also holds for the special case $\gcd(M,p) \neq 1$ from Theorem 5.20a, because then p is a divisor of M, and therefore also a divisor of M^{ed}, and therefore $M \bmod p = M^{ed} \bmod p = 0$.

We show $M \bmod q = M^{ed} \bmod q$ in the same way.

$M - M^{ed}$ is therefore divisible by both p and q. Then $M - M^{ed}$ is also divisible by $pq = n$, which means $M - M^{ed} \equiv 0 \bmod n$ Since M is a remainder $\bmod n$, according to Theorem 4.23 $M = M^{ed} \bmod n$ must be true.   □

The Diffie-Hellman Algorithm

The first public-key method to become known was the *Diffie-Hellman algorithm*, which was published in 1976. Diffie-Hellman (DH) is an algorithm used to exchange a secret key between Alice and Bob. In this respect, its operation is somewhat different from the public-key methods I described at the beginning of this section. The underlying mathematics uses the field GF(p) where p is a very large prime number. Its security is based on the discrete logarithm problem. Without proof, I would like to mention that the multiplicative group of a finite field is always cyclic (see Definition 5.5). A generator $a \in \mathrm{GF}(p) \setminus \{0\}$ is needed for the algorithm. There are usually many generators, but it is difficult to check whether a given element has this property when p is large. Fortunately, the numbers p and a generator a only need to be determined once, they remain the same for all time and for all users. If Alice and Bob now want to exchange a key, they each choose a random number q_A and q_B less than $p - 1$. These are their private key parts, which they keep to themselves. They then exchange messages:

$$\text{Alice to Bob:} \quad r_A = a^{q_A} \bmod p,$$
$$\text{Bob to Alice:} \quad r_B = a^{q_B} \bmod p.$$

r_A and r_B are the public key parts of Alice and Bob. There is no known method to calculate q_A from r_A or q_B from r_B that is much faster than trial and error. If p is large enough, this is impossible. Now Alice and Bob can calculate a common number K from the messages received:

$$\text{Alice:} \quad r_B^{q_A} \bmod p = a^{q_B q_A} \bmod p =: K,$$
$$\text{Bob:} \quad r_A^{q_B} \bmod p = a^{q_A q_B} \bmod p =: K.$$

Do you see that we used the field properties of GF(p) here? Without the respective private parts, K is not computable, only Alice and Bob therefore share this information. K can now be used as a key or as a starting value for a key generator, so that the following communication can be encrypted.

The *ElGamal algorithm* is an encryption method that builds directly on the DH key exchange. Alice wants to encrypt the message M, which is a number less than p. After the DH key exchange, Alice sends Bob the number

$$C = KM \bmod p.$$

Without the knowledge of K, no one can calculate M. However, if Bob reads Theorem 5.19, he can determine the inverse element GF(p) in K^{-1} and receives

$$M = K^{-1}C \bmod p.$$

Diffie-Hellman and ElGamal look very attractive at first glance, but I have to pour some water into the wine right away: Alice and Bob must be sure that the exchange of public keys is not hacked by a third party who substitutes their own key parts. This is not so easy, but the Diffie-Hellman algorithm is widespread, it must be embedded in a corresponding protocol. ElGamal has the disadvantage that after each message M, the key K must be changed, so that the transmitted text is effectively twice as long as the clear text. If this did not happen, an attacker with the knowledge of a single matching pair M, C of clear text and key text could crack the encryption. But cryptographic algorithms must be prepared for this attack. However, the ElGamal algorithm also has its areas of application.

The Diffie-Hellman Algorithm with Elliptic Curves

The security of the Diffie-Hellman algorithm is based on the problem of the discrete logarithm in the multiplicative group of a finite field. This logarithm problem also exists in other groups. The currently most important are the elliptic curves. First, it is a bit confusing when we talk about exponentiation and logarithms in an additive group. We have to translate the terms into the language of addition: exponentiation is a continued multiplication, which in an additive group corresponds to a continued addition. In elliptic curves over a finite field, it is very easy to add a point P n times to itself: $nP = P + P + \ldots + P$. This n-fold addition can be implemented recursively, similar to exponentiation in GF(p) or in $\mathbb{R}$. But there is no known algorithm that, given P and $Q = nP$, finds the number n. This would correspond to taking the logarithm.

Now the Diffie-Hellman algorithm can be formulated for elliptic curves. As system parameters one needs a finite field K, an elliptic curve E over K and a so called base point P on E. The construction of the elliptic curve is very demanding. As a field one usually chooses GF(p) or GF(2^m) for a large prime number p or for a large m. The curve itself must contain many points and the base point P must have the property that the order of P, that is the smallest n for which $nP = \mathcal{O}$ has only very large factors. And last but not least, there are some elliptic curves in which the logarithm is fast anyway. Curves that are used in cryptography must be analyzed very carefully. Such curves are specified in security standards. These can then be used in encryption protocols.

> In 1999, the National Institute of Standards and Technology (NIST) in the USA standardized elliptic curves that had been developed by the National Security Agency (NSA). The design criteria were not disclosed and in the following years, especially after the revelations of Edward Snowden in 2013, the suspicion arose in the community of cryptologists that curves were generated here in which there is a back door that would allow the NSA to decrypt. In many protocols, these NIST curves are therefore now avoided.

For key exchange, Alice and Bob choose as private key a number q_A or q_B which is smaller than the order of P.

The public keys are then $q_A P$ and $q_B P$. Both Alice and Bob can then calculate the shared secret $q_B(q_A P) = q_A(q_B P)$ from their private key and the public key of the partner.

Even if it is difficult to find suitable elliptic curves for the Diffie-Hellman method, they are still very interesting for applications. The reason lies in the performance and in the key lengths of the algorithms. In the age of the Internet of Things, in which soon every light switch will want to communicate with its lamp in encrypted form, suddenly storage space sizes and computing power become interesting again. Algorithms over elliptic curves have significantly shorter key lengths and faster run times than standard algorithms with comparable security.

Key Generation

Until around the turn of the century, numbers in the order of 512 bits were chosen as the module n for the RSA algorithm. In 1999 it was possible for the first time to break such a code, that is, to decompose such a number n. It took about half a year and a total of 35.7 CPU years of computing time. For cryptologists, the algorithm was no longer viable in this form, and modules of at least 1024 bits in length were switched to. Since 2011, the BSI (German Federal Office for Information Security) has recommended a module length of 2048 bits.

When generating the key, one then needs prime numbers in the order of 1024 bits. Are there enough of them? Yes, in this number range there are more than 10^{300} useful prime numbers, more than enough for every atom in the universe (see Exercise 9 in Chap. 14).

But how do you find them? A simple prime number test consists in factoring the number. But we can't invest 35 computer years in that. There are indeed faster test algorithms, in 2002 even an algorithm was discovered in polynomial time (the AKS algorithm). But in practice this is still much too slow.

> The discovery of a polynomial prime number test caused great excitement in mathematics. And at first it almost looks as if this algorithm could nibble at the security of cryptographic algorithms that are based on the factorization problem. Fortunately, the factorization of a number and the test for the primality seem to be independent problems. And there are no

progress in factorization that could be dangerous for cryptographic applications—at least as long as there are no quantum computers.

So what do you do if you want to find a large prime number in a few seconds? There are various prime number tests that can determine that a number is at least probably prime. I will show you the *Miller-Rabin Prime Number Test*, which is widely used in applications.

Fermat's little Theorem 5.20 first gives us a negative criterion: If p is the candidate to be tested for its primality and a is a number with $a \not\equiv 0 \bmod p$, then at least $a^{p-1} \bmod p = 1$ must be. If this property is violated for any a, then p is composite. Conversely, if for many such a always $a^{p-1} \bmod p = 1$ holds, then we could hope that p is probably prime with certain probability. Any number a for which $a^{p-1} \bmod p = 1$ is fulfilled, we call a *witness* for the primality of p.

We assume that p is prime and try to confirm it by as many witnesses as possible. As potential witnesses we take numbers $a \in \{2, 3, \ldots, p-1\}$. We calculate in the field $\mathbb{Z}/p\mathbb{Z}$ and of course we want to calculate as efficiently as possible.

In order to $a^{p-1} = 1$ (i.e. $a^{p-1} \bmod p = 1$), it is sufficient to know that $a^{(p-1)/2}$ is equal to $+1$ or equal to $-1 (= p - 1)$, because $\pm 1^2 = 1$. If $(p-1)/2$ is an even number, we can compute $a^{(p-1)/4}$ and it follows from $a^{(p-1)/4} = \pm 1$ that $a^{p-1} = 1$. Continuing in this way, we can save some powers. The algorithm in detail:

1. Find the largest power of two in $p - 1$, i.e. $p - 1 = 2^k q$, q is odd.
2. Choose a random $a \in \{2, 3, \ldots, p-1\}$.
3. Calculate a^q. If $a^q = \pm 1$, then $a^{p-1} = 1$ and thus a is a witness for the property of being prime.
4. If $a^q \neq \pm 1$ and $a^{q \cdot 2}, a^{q \cdot 2^2}, a^{q \cdot 2^3}, \ldots, a^{q \cdot 2^{k-1}} = a^{(p-1)/2}$ eventually reaches the value -1, then $a^{p-1} = 1$, a is a witness for the property of being prime.
5. Otherwise, $a^{p-1} \neq 1$ and thus p is definitely not prime.

Why is it not necessary to consider the alternative $a^{q \cdot 2^m} = +1$ in step 4? In the field $\mathbb{Z}/p\mathbb{Z}$, the polynomial $x^2 - 1 = 0$ has exactly two roots, namely ± 1. So if, for example, $a^{q \cdot 2} = +1$, then $a^q = \sqrt{a^{q \cdot 2}} = \pm 1$ must have been too. Accordingly, one excludes the higher powers.

If p is prime, we can find many witnesses for the primality of p in this way. This would not help us if we could not prove that of all potential witnesses at most one fourth is willing to lie about p, that is, to claim that p is prime when it is not. So if a randomly chosen a is a witness for p, then p is not prime with a probability of less than $1/4$!

There are only a few composite numbers q for which it holds true for all a with $\gcd(a, q) = 1$ that $a^{q-1} \bmod q = 1$. These numbers are called Carmichael numbers, the first five of which are 561, 1105, 1729, 2465, 2821. From my argumentation one could conclude that these numbers would falsely be recognized as prime during testing. Deep-seated results from number theory however show that the Miller-Rabin test surprisingly also detects Carmichael numbers as composite: The Miller-Rabin test is stronger than the check whether for all $a < q$ it holds true $a^{q-1} \bmod q = 1$. If q is composite, then for example there can be numbers $a < q$ with the property $a^{(q-1)/2} \bmod q \neq \pm 1$, even though $a^{q-1} \bmod q = 1$! Look for number examples!

Let us now carry out a certain number of such prime number tests, for example 30. If p is not a prime number, then it holds true:

The probability that the 1st test provides a witness is smaller than $1/4$.

The probability that the 2nd test provides a witness is smaller than $1/4$.

The probability that the 1st and 2nd test provide witnesses is smaller than $(1/4) \cdot (1/4)$.

The probability that all 30 tests provide witnesses is smaller than $1/4^{30} = 1/2^{60}$.

In Part III of the book we will learn why one can multiply the probabilities in this way and how one can determine the probability with the help of such a test that a randomly chosen number is prime if it has passed 30 tests. This is different from $1 - 1/2^{60}$! See the second example after Theorem 19.8 in Sect. 19.3.

If one has found two numbers p and q, for example, which have passed 30 tests, then it is assumed that they are prime and thus the RSA key can be generated. Is one taking a risk with this? The probability that I will have a five twice in a row in Powerball is greater than the probability of not catching a prime number. It just doesn't happen!

I have heard that computer scientists refer to such numbers as "prime numbers of industrial quality". A real mathematician's hair stands on end! But it works, and thus such a prime number search is justified. Of course, such "prime numbers"—just like industrial diamonds—are not as valuable as the real thing.

Random Numbers

The generation of a cryptographic key always starts with a random number. In RSA you need a whole series of candidates to test them for their primality. These candidates must be chosen randomly. Random numbers also play an important role in other areas of computer science, for example for simulations. In this book I used them in a Monte Carlo method for integration, see the example after Calculation Rule 22.12 in Sec. 22.3. How can you find such numbers?

Even if it does not always look like it in everyday use of the computer: Computers work deterministically, that is, their outputs are predetermined by the inputs and the programs and reproducible. This would be fatal in key generation: A deterministic process would always generate the same keys.

Real random numbers are difficult to implement in the computer. They use, for example, physical processes such as thermal noise in a semiconductor or data entered by the user at random, such as mouse movements or time gaps between keyboard input. However, for many applications so-called *pseudorandom numbers* are used. A pseudorandom number generator outputs a reproducible sequence of numbers that look random after entering a starting value. It is not trivial to assess the quality of such a sequence of numbers. It must at least pass a whole series of statistical tests before the corresponding random number generator can be used. In Sec. 22.4 we will deal a little with such tests. Generators used in cryptography must meet another requirement in addition to these statistical tests: Even if all the numbers in the sequence are known up to a certain point, it must not be possible to predict the next number.

The modulo operation is suitable for generating random numbers. A common method is to specify numbers a, b, m and after entering the starting value z_0 to calculate the $(i + 1)$-th random number from the i-th as follows:

$$z_{i+1} = (az_i + b) \bmod m.$$

This algorithm is called *linear congruence generator*. Of course, not all numbers a, b and m are suitable. Try it out to see what happens if $b = 0$ is and a is a divisor of m. However, if the numbers are chosen wisely, such a generator will generate pseudo-random numbers that are well distributed statistically. The numbers are between 0 and $m - 1$, with a good choice the generator has the maximum period m. There are long lists of suitable values in the literature. I will give you two to try out: $a = 106$, $b = 1283$, $m = 6075$ and $a = 2416$, $b = 374\,441$, $m = 1\,771\,875$.

When implementing the second selection, you must make sure that you do not produce an overflow. How big does your data type have to be at least?

Linear congruence generators are not to be used in cryptography because they are predictable. However, the brainpower that has already been invested in the development of secure algorithms can be used a second time in the field of cryptography: Random numbers arise from continued encryption. There are a whole range of methods for this. For example, the RSA generator is based on the security of the RSA algorithm. As there, large prime numbers p and q are chosen, $n = p \cdot q$ and e, d are calculated with $e \cdot d \equiv 1 \bmod (p - 1)(q - 1)$. Then

$$x_{i+1} = x_i^e \bmod n$$

is calculated. x_0 is the starting value here. Anyone who can calculate the number x_{i+1} from x_i without knowing e can also crack RSA. The generator is supposed to generate a sequence of zeros and ones, so only the last bit of x_{i+1} is chosen as the random number.

The generated random number sequence depends on the starting value. If I want to generate a cryptographic key and choose a long value with four bytes as the starting value for the generator, I may have the best generator and choose the longest conceivable keys, but I still only get 2^{32} different keys. No problem for a hacker who knows how I generate my keys. A good encryption system therefore needs a good random number generator in addition to the algorithm and long, truly random starting values. The latter is unfortunately occasionally underestimated. In fact, attacks on key generation mechanisms are often more promising today than attacks on the encryption algorithms themselves.

5.8   Comprehension Questions and Exercises

Comprehension questions

1. A ring does not necessarily have to have a 1-element. Take another look at the examples in Sect. 5.2. Can you find a ring without 1?
2. What is $\sqrt{-25}$ in $\mathbb{C}$?
3. Is it true in $\mathbb{C}$ that $x \cdot \bar{x} = 0 \Leftrightarrow x = 0$?
4. Why is it always true in the field $\mathbb{Z}/p\mathbb{Z}$ that $(p - 1)(p - 1) = 1$?
5. If you have a ring homomorphism from R to S, is it automatically also a homomorphism of the additive groups of R and S?
6. If you want to generate keys for a cryptographic algorithm, you need a random number generator. Why are the random number generators that are standard in programming languages or operating systems not suitable for this purpose?
7. What is the great importance of elliptic curves in cryptography?
8. Which two unsolved mathematical problems are the basis for most public-key algorithms ?

Exercises

1. Set up an addition and multiplication table for $\mathbb{Z}/6\mathbb{Z}$. Show that $\mathbb{Z}/6\mathbb{Z} \setminus \{0\}$ does not form a group with multiplication.
2. Assign the elements of the table (5.3) to the elements of the S_3.
3. Show that $(\mathbb{R}^+, \cdot)$ forms a group.
4. Show that in the field $\mathbb{C}$ it holds: $z \cdot \bar{z} = |z|^2$, $z^{-1} = \frac{\bar{z}}{|z|^2}$.
5. Let $z = 4 + 3i$, $w = 6 + 5i$. Express z^{-1} and $\frac{w}{z}$ in the form $a + bi$. (For this, use exercise 4.)
6. Express $\frac{1+i}{2-i}$ in the form $a + bi$.
7. In the field $\mathbb{C}$ of complex numbers let $z = \frac{\sqrt{2}}{2} + \frac{\sqrt{2}}{2}i$. Calculate and draw in the Complex plane z, z^2, z^3, z^4, z^5.
8. Carry out division with remainder for the following polynomials and then make the test in each case:
   a) in $\mathbb{Z}/2\mathbb{Z}[x]$: $(x^8 + x^6 + x^2 + x)/(x^2 + x)$,
   b) in $\mathbb{Z}/2\mathbb{Z}[x]$: $(x^5 + x^4 + x^3 + x + 1)/(x^3 + x^2 + 1)$,
   c) in $\mathbb{Z}/5\mathbb{Z}[x]$: $(4x^3 + 2x^2 + 1)/(2x^2 + 3x)$.
9. Show in $\mathbb{Z}/n\mathbb{Z}[x]$, n not a prime, there are polynomials of degree two with more than two distinct roots.

10. Show that for $n \in \mathbb{N}$ the set $(\mathbb{R}^n, +)$ forms a group with the following addition:
    $(a_1, a_2, \ldots, a_n) + (b_1, b_2, \ldots, b_n) := (a_1 + b_1, a_2 + b_2, \ldots, a_n + b_n)$.

11. Show that the following mappings are homomorphisms. $\mathbb{R}^2$ or $\mathbb{R}^3$ are the groups from the last exercise. Calculate $\ker f$ and $\ker g$.

    a) $f: \mathbb{R}^3 \to \mathbb{R}^2, (x, y, z) \mapsto (x + y, y + z)$

    b) $g: \mathbb{R}^2 \to \mathbb{R}^3, (x, y) \mapsto (x, x + y, y)$

12. Show that in the Definition 5.2, Axiom (G2) the requirement of uniqueness can be dispensed with: In a group, the inverse a^{-1} of an element a is uniquely determined.

13. Show that in an ISBN number, each individual digit error and also the transposition of the check digit with one of the preceding digits can be detected.

14. In $\mathbb{R}^2$, the set of (x, y) with $y = mx + t$ forms a line. m is the slope and t is the y-intercept. Now investigate lines over the field GF(p), that is, in GF(p)^2:

    a) Show that each line in GF(p)^2 contains exactly p points.

    b) Use a mathematical tool to plot the points of the line for some prime numbers and for some m, t. For example, take $p = 31$, $m = 1, 2, 13, 16(= \frac{1}{2}), 29(= -2)$ and $t = 0.5$. Compare the drawings with the corresponding lines in $\mathbb{R}^2$.

Vector Spaces

6

Abstract

In this chapter you will learn

- the operations on the classical vector spaces $\mathbb{R}^2$, $\mathbb{R}^3$ and $\mathbb{R}^n$,
- the vector space as the central algebraic structure of linear algebra,
- the linear mappings as the homomorphisms of vector spaces,
- the concepts of linear independence, basis and dimension of vector spaces, and how they are related,
- the calculation with coordinates with respect to different bases of a vector space.

Vector spaces are an algebraic structure that is particularly important to us. This is partly because the space we live in can be regarded as a vector space. For example, we constantly work with the points (the vectors) of space and compute movements in it in graphical data processing and robotics. To represent spatial objects in a plane, for example on a screen, we have to carry out mappings from the three-dimensional to the two-dimensional space and also examine movements in the plane.

It will turn out that vector spaces are also a powerful tool in other areas of mathematics. So we will solve linear equations with their help, later we will also get to know applications in analysis.

For this reason, I dedicate a chapter to the structure of the vector space. In the following chapters we will deal with vector space applications.

Supplementary Information The online version contains supplementary material available at https://doi.org/10.1007/978-3-658-40423-9_6.

6.1   The Vector Spaces $\mathbb{R}^2$, $\mathbb{R}^3$ and $\mathbb{R}^n$

$\mathbb{R}^2 = \{(x, y) \mid x, y \in \mathbb{R}\}$ as well as $\mathbb{R}^3 = \{(x, y, z) \mid x, y, z \in \mathbb{R}\}$ are already known to us. The elements of $\mathbb{R}^2$ and $\mathbb{R}^3$ are called vectors. This allows us to designate the points of a plane, respectively the points of space, with coordinates. If one wants to assign the points of space to the vectors of $\mathbb{R}^3$, one defines an origin in space (the point (0,0,0)), as well as three coordinate axes through this point. Usually we choose three axes that are perpendicular to each other in pairs (Fig. 6.1).

The coordinates (x, y, z) then designate the point one obtains when one goes x length units in the direction of the first axis, y units in the direction of the second and z units in the direction of the third axis. The axes must also have a direction (forward) specified. In a drawing, this can be done by specifying the points $(1, 0, 0)$, $(0, 1, 0)$ and $(0, 0, 1)$. This is the *Cartesian coordinate system*. So we get a bijective mapping between $\mathbb{R}^3$ and the space we live in, at least if we assume that our space extends infinitely far in every direction.

We often have this idea of a coordinate system in mind when we calculate with the elements of $\mathbb{R}^3$.

Just as $\mathbb{R}^2$ and $\mathbb{R}^3$, you can form the $\mathbb{R}^n = \{(x_1, x_2, \ldots, x_n) \mid x_i \in \mathbb{R}\}$. This space is no longer as intuitively meaningful, but it is often very practical for calculations. You know arrays from programming: The instruction

```
float[]x = new float[25];
```

creates nothing else in Java than an element of $\mathbb{R}^{25}$.

It has proved to be particularly practical in physics to identify the vector (x, y, z) of $\mathbb{R}^3$ with an arrow that has the direction and length of the line segment from the origin to (x, y, z). This way, every vector has a magnitude (the length) and a direction that is determined by the endpoint. Velocity, acceleration and forces can thus be interpreted as vectors.

You can add forces. You know the force parallelogram that results when two forces act on a body in different directions. The resulting force corresponds in size and direction to the diagonal. What happens to the coordinates? You can follow this in a sketch in $\mathbb{R}^2$ (Fig. 6.2), they just add up.

Fig. 6.1 Cartesian coordinates in $\mathbb{R}^3$

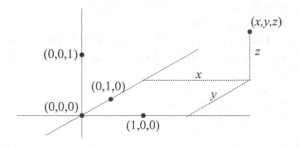

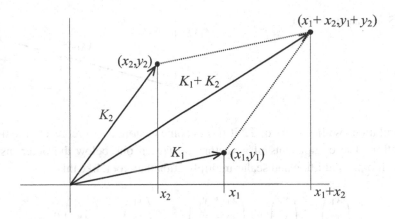

Fig. 6.2 Parallelogram of force

It is very easy to check that $(\mathbb{R}^n, +)$ with the component-wise addition $(a_1 + b_1, a_2 + b_2, \ldots, a_n + b_n) := (a_1, a_2, \ldots, a_n) + (b_1, b_2, \ldots, b_n)$ is an additive group with $(0, 0, \ldots, 0)$, the origin, as the 0-element (compare exercise 10 in Chap. 5).

The arrow that represents a vector does not necessarily have to be drawn with the beginning in the origin; any arrow of the same length and direction denotes the same vector. So you can also view the addition as the combining of the arrows. Similarly, subtraction can be represented geometrically (Fig. 6.3).

Another important operation of vectors is the change in length (Fig. 6.4). This corresponds to multiplication by a real number, a quantity without direction (a so-called scalar). This operation can also be easily transferred to coordinate notation: $\lambda(x_1, x_2, \ldots, x_n) = (\lambda x_1, \lambda x_2, \ldots, \lambda x_n)$.

For a multiplication of vectors we currently have no meaningful interpretation. I refer you to Chap. 10.

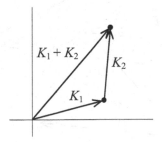
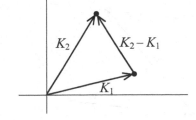

Fig. 6.3 Vector addition and subtraction

Fig. 6.4 Scalar multiplication

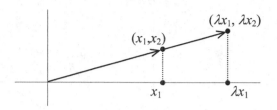

In calculations with vectors of $\mathbb{R}^n$ it has become customary to represent them in column notation. The components of a vector are written one below the other instead of next to each other. Addition and scalar multiplication then look like this:

$$
\begin{pmatrix} x_1 \\ x_2 \\ \vdots \\ x_n \end{pmatrix} + \begin{pmatrix} y_1 \\ y_2 \\ \vdots \\ y_n \end{pmatrix} = \begin{pmatrix} x_1 + y_1 \\ x_2 + y_2 \\ \vdots \\ x_n + y_n \end{pmatrix}, \quad \lambda \begin{pmatrix} x_1 \\ x_2 \\ \vdots \\ x_n \end{pmatrix} = \begin{pmatrix} \lambda x_1 \\ \lambda x_2 \\ \vdots \\ \lambda x_n \end{pmatrix}.
$$

If one does not consider individual vectors, but sets of vectors, these describe formations or shapes in $\mathbb{R}^2$, $\mathbb{R}^3$, ..., $\mathbb{R}^n$. Lines and planes will be particularly important to us: Are u, v vectors in the $\mathbb{R}^n$, then the set $g := \{u + \lambda v \mid \lambda \in \mathbb{R}\}$ is the line in the direction of the vector v through the endpoint of the vector u, and the set $E = \{u + \lambda v + \mu w \mid \lambda, \mu \in \mathbb{R}\}$ represents the plane that goes through u and is such that the arrows belonging to v and w just lie in the plane. Make these two constructions clear to you by trying out some values for λ, μ (Fig. 6.5).

> In my explanation of the plane I cheated a little: What happens if v and w have the same direction? Try to draw the set E for this special case.

The surface of more complex objects in $\mathbb{R}^3$ can be approximated by polygons, flat surfaces bounded by line segments.

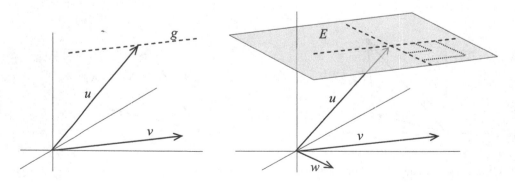

Fig. 6.5 Line and plane

6.2   Vector Spaces

Similar to groups, rings and fields in Chap. 5, I now want to define an abstract algebraic structure by putting together typical properties of $\mathbb{R}^2$ and $\mathbb{R}^3$. We thus obtain the axioms of the vector space and from now on we will only use these axioms. The resulting theorems then apply to all vector spaces, in particular of course also to the already known examples.

▶ **Definition 6.1: The vector space axioms** Let K be a field. A *vector space* V with scalars from K consists of a commutative group $(V, +)$ and a scalar multiplication $\cdot : K \times V \to V, (\lambda, v) \mapsto \lambda \cdot v$, so that for all $v, w \in V$ and for all $\lambda, \mu \in K$ it holds:

- (V1) $\lambda(\mu v) = (\lambda\mu)v$,
- (V2) $1 \cdot v = v$,
- (V3) $\lambda(v + w) = \lambda v + \lambda w$,
- (V4) $(\lambda + \mu)v = \lambda v + \mu v$.

We call a vector space with scalar field K a K-vector space or vector space over K. If $K = \mathbb{R}$ or $K = \mathbb{C}$, we speak of a real or complex vector space. This will almost always be the case with us.

> For $\mathbb{R}^n$, these properties are quite elementary to check. The surprising thing is that these few rules are enough to characterize vector spaces. We will find that the vector spaces $\mathbb{R}^n$ are important prototypes of vector spaces.

So the definition of a vector space always involves a field. In order not to confuse the elements of the structures, one usually denotes the scalars with small Greek letters, for example with λ, μ, ν, and the vectors with small Latin letters, such as u, v, w. Physicists often paint arrows over the vectors ($\vec{v}$) to be quite clear. I will only do this if there is a risk of confusion, such as between the $0 \in K$ and the $\vec{0} \in V$, the 0-element of the additive group V, the zero-vector. In $\mathbb{R}^n$ it is $\vec{0} = (0, 0, \dots, 0)$.

▶ **Theorem 6.2: Simple calculation rules for vector spaces**

- (V5) $\lambda \cdot \vec{0} = \vec{0}$ for all $\lambda \in K$,
- (V6) $0 \cdot v = \vec{0}$ for all $v \in V$,
- (V7) $(-1) \cdot v = -v$ for all $v \in V$.

To get started with vector spaces, I want to prove the first property, it is not difficult and the other properties are shown analogously:

Proof of (V5):
$$\lambda \cdot \vec{0} \underset{\substack{\uparrow \\ \vec{0}+\vec{0}=\vec{0}}}{=} \lambda \cdot (\vec{0} + \vec{0}) \underset{\substack{\uparrow \\ (V3)}}{=} \lambda \cdot \vec{0} + \lambda \cdot \vec{0}.$$

Subtracting $\lambda\vec{0}$ on both sides, $\vec{0} = \lambda\vec{0}$ remains.   □

Examples of vector spaces

1. $\mathbb{R}^2$, $\mathbb{R}^3$, $\mathbb{R}^n$ are $\mathbb{R}$-vector spaces.
2. Similarly, for $n \in \mathbb{N}$, the set $\mathbb{C}^n$ with component-wise addition and scalar multiplication (as in $\mathbb{R}^n$) is a $\mathbb{C}$-vector space.
3. Let us take $\mathbb{Z}/2\mathbb{Z}$ as the underlying field, the field that only contains the elements 0 and 1. The vector space $(\mathbb{Z}/2\mathbb{Z})^n$ consists of all n-tuples of zeros and ones. We can therefore regard an integer in its binary representation as an element of the $\mathbb{Z}_2^{32}$. This vector space plays an important role in coding theory: The code words are vectors. If errors are to be detected in the code, not all elements of the $\mathbb{Z}_2^n$ may be elements of the code: The code consists of a subset of the $\mathbb{Z}_2^n$ that is chosen so that, in the event of errors, an element outside the code is likely to occur. *Linear codes* are subsets of the $\mathbb{Z}_2^n$ that are themselves vector spaces, i.e. subspaces of the $\mathbb{Z}_2^n$. With methods of vector space arithmetic, it can be quickly determined from a given element of the $\mathbb{Z}_2^n$ whether it is a valid code word or not.
4. Analogously, for every field K the set K^n is a K-vector space.
5. In many areas of mathematics, function spaces are of great importance: The set $\mathcal{F}$ of all mappings from $\mathbb{R}$ to $\mathbb{R}$ with pointwise defined operations

$$(f + g)(x) := f(x) + g(x),$$
$$(\lambda \cdot f)(x) := \lambda \cdot f(x)$$

is a $\mathbb{R}$-vector space. We have already encountered this set in Sect. 5.2 as the 4th example of a ring. A vector here is an entire function. The properties (V1) to (V4) can be traced back to the corresponding properties in $\mathbb{R}$ just as in the ring example. The zero vector $\vec{0}$ consists of the zero function, the function that has the value 0 everywhere.
6. Accordingly, one can form many other vector spaces of functions: the vector space of all polynomials, the vector space of all continuous functions, the vector space of all convergent sequences of numbers, and many more.
   In the K-vector space of polynomials $K[X]$, for example, polynomials are added as I have written in Sect. 5.2 in (5.6). Scalar multiplication looks like this:

$$\lambda(a_n x^n + a_{n-1} x^{n-1} + \cdots + a_1 x + a_0)$$
$$= \lambda a_n x^n + \lambda a_{n-1} x^{n-1} + \cdots + \lambda a_1 x + \lambda a_0.$$

The zero vector is the zero polynomial, the polynomial whose coefficients are all 0. ◄

The theory of vector spaces is a powerful tool for such function spaces. Everything we learn about vector spaces can later be used for functions, polynomials, and sequences.

▶   **Definition 6.3** Let V be a K-vector space and $U \subset V$. If U with the operations of V is itself a K-vector space, then U is called a *subspace* of V.

▶   **Theorem 6.4** Let V be a K-dimensional vector space and $U \subset V$ a non-empty subset. Then U is a subspace of V if and only if the following conditions are satisfied.

(U1) $u + v \in U$ for all $u, v \in U$,
(U2) $\lambda u \in U$ for all $\lambda \in K, u \in U$.

Proof: First, $(U, +)$ is a subgroup of V by Theorem 5.4 from Sect. 5.1, because for all $u, v \in U$ it holds:

$$u + v \in U, \quad -u \underset{\underset{(V7)}{\uparrow}}{=} (-1)u \underset{\underset{(U2)}{\uparrow}}{\in} U.$$

From (U2) it follows that multiplication with scalars is a mapping from $K \times U \to U$, as it must be, and the properties (V1) to (V4) are satisfied for the elements of U, since they already hold in V.

Conversely, of course, in any vector space the rules (U1) and (U2) apply, so the "if and only if" in the theorem also applies.   □

A simple negative criterion consists in testing whether the zero vector is contained in U : Every subspace must contain $\vec{0}$. Therefore, a line in $\mathbb{R}^3$ that does not go through the origin cannot be a subspace.

Examples of subspaces

1. V itself is a subspace of the vector space V.
2. $\{\vec{0}\}$ is a subspace of the vector space V, and there is no smaller subspace.
3. Linear codes are subspaces of the vector space $(\mathbb{Z}/2\mathbb{Z})^n$.
4. Are lines $g = \{u + \lambda v \mid \lambda \in \mathbb{R}\}$ subspaces in $\mathbb{R}^3$? In any case, $\vec{0} \in g$ must apply to this. Such lines can be written in the form $g = \{\lambda v \mid \lambda \in \mathbb{R}\}$, with $v \in \mathbb{R}^3$. If now $u_1, u_2 \in g$, that is $u_1 = \lambda_1 v$ and $u_2 = \lambda_2 v$, then $u_1 + u_2 = \lambda_1 v + \lambda_2 v = (\lambda_1 + \lambda_2)v \in g$ and $\lambda u_1 = \lambda(\lambda_1 v) = (\lambda \lambda_1)v \in g$ and thus the lines through the origin are subspaces of $\mathbb{R}^3$.
5. Not only in $\mathbb{R}^3$, but in every K-vector space V, the sets $\{\lambda v \mid \lambda \in K\}$ are subspaces for all $v \in V$. ◀ ·

This last example can be generalized:

▶   **Definition 6.5** Let V be a K-vector space. Let $n \in \mathbb{N}$, $u_1, u_2, u_3, \ldots, u_n \in V$ and $\lambda_1, \lambda_2, \lambda_3, \ldots, \lambda_n \in K$. The sum

$$\lambda_1 u_1 + \lambda_2 u_2 + \lambda_3 u_3 + \cdots + \lambda_n u_n$$

is called *linear combination* of $u_1, u_2, u_3, \ldots, u_n$. Let

$$\text{span}(u_1, u_2, u_3, \ldots, u_n) := \{\lambda_1 u_1 + \lambda_2 u_2 + \lambda_3 u_3 + \cdots + \lambda_n u_n | \lambda_i \in K\}.$$

If M is an infinite subset of V, let

$$\text{span}\, M := \{\lambda_1 u_1 + \lambda_2 u_2 + \lambda_3 u_3 + \cdots + \lambda_n u_n \,|\, n \in \mathbb{N}, \lambda_i \in K, u_i \in M\}.$$

So $\text{span}(u_1, u_2, u_3, \ldots, u_n)$ is the set of all possible linear combinations of $u_1, u_2, u_3, \ldots, u_n$ and $\text{span}\, M$ is the set of all finite linear combinations with elements from M. You can easily check:

▶   **Theorem 6.6** Let V be a vector space and let $M \subset V$. Then span M is a subspace of V.

Lines and planes through the origin in $\mathbb{R}^3$ are special cases of this theorem. The proof proceeds analogously, as in Example 4.

6.3   Linear Mappings

In Sect. 5.6 we dealt with homomorphisms of groups and rings. The structure-preserving mappings (the homomorphisms) between vector spaces are called linear mappings. In particular, linear mappings of a vector space to itself will be particularly important to us. With such mappings, for example, we can describe movements in space. Bijective linear mappings (the isomorphisms) are a means of determining when vector spaces can be considered essentially equal. For example, there is no essential difference between $\mathbb{R}^2$ and $U := \{(x_1, x_2, 0) \,|\, x_1, x_2 \in \mathbb{R}\} \subset \mathbb{R}^3$. In this example, we will specify a bijective linear mapping between $\mathbb{R}^2$ and U and thus consider the two vector spaces as "equal". First the definition, which differs somewhat from that of the group or ring homomorphism, since we now have to take into account the scalar multiplication:

▶   **Definition 6.7** Let U and V be vector spaces over K. A mapping $f : U \to V$ is called a *linear mapping* if for all $u, v \in U$ and for all $\lambda \in K$ it holds that

$$f(u + v) = f(u) + f(v)$$
$$f(\lambda u) = \lambda f(u). \tag{6.1}$$

The vector spaces U and V are called isomorphic if there is a bijective linear mapping $f : U \to V$. We write $U \cong V$ for this.

The first of the conditions states that a linear mapping is in particular also a homomorphism of the underlying additive groups of U and V. Note in the second condition that on the left the scalar multiplication is carried out in U, on the right in V, both times with the same element $\lambda \in K$. K remains fixed under the linear mapping. There are no linear mappings between vector spaces with different scalar fields, so there is, for example, no linear mapping between $\mathbb{C}^7$ and $\mathbb{R}^5$.

Examples of linear mappings

1. $f : \mathbb{R}^2 \to \mathbb{R}^3, \begin{pmatrix} x_1 \\ x_2 \end{pmatrix} \mapsto \begin{pmatrix} x_1 \\ x_2 \\ 0 \end{pmatrix}$ is linear, because

$$f\left(\begin{pmatrix} x_1 \\ x_2 \end{pmatrix} + \begin{pmatrix} y_1 \\ y_2 \end{pmatrix} \right) = f\left(\begin{pmatrix} x_1 + y_1 \\ x_2 + y_2 \end{pmatrix} \right) = \begin{pmatrix} x_1 + y_1 \\ x_2 + y_2 \\ 0 \end{pmatrix},$$

$$f\left(\begin{pmatrix} x_1 \\ x_2 \end{pmatrix} \right) + f\left(\begin{pmatrix} y_1 \\ y_2 \end{pmatrix} \right) = \begin{pmatrix} x_1 \\ x_2 \\ 0 \end{pmatrix} + \begin{pmatrix} y_1 \\ y_2 \\ 0 \end{pmatrix} = \begin{pmatrix} x_1 + y_1 \\ x_2 + y_2 \\ 0 \end{pmatrix},$$

and in the same way we show $f\left(\lambda \begin{pmatrix} x_1 \\ x_2 \end{pmatrix} \right) = \lambda f\left(\begin{pmatrix} x_1 \\ x_2 \end{pmatrix} \right)$.

The mapping f is obviously injective, because different elements have different images, but not surjective: $(0, 0, 1)$ for example has no preimage. However, if one restricts the target set to the image:

$$f : \mathbb{R}^2 \to U, \quad \begin{pmatrix} x_1 \\ x_2 \end{pmatrix} \mapsto \begin{pmatrix} x_1 \\ x_2 \\ 0 \end{pmatrix}, \quad U := \left\{ \begin{pmatrix} x_1 \\ x_2 \\ 0 \end{pmatrix} \, \middle| \, x_1, x_2 \in \mathbb{R} \right\} \subset \mathbb{R}^3,$$

we obtain an isomorphism.

2. Let $u, v \in \mathbb{R}^3$ and $f : \mathbb{R}^2 \to \mathbb{R}^3, \begin{pmatrix} x_1 \\ x_2 \end{pmatrix} \mapsto x_1 u + x_2 v$. f is linear. Let us compute it for scalar multiplication, addition goes similarly:

$$f\left(\lambda \begin{pmatrix} x_1 \\ x_2 \end{pmatrix} \right) = f\left(\begin{pmatrix} \lambda x_1 \\ \lambda x_2 \end{pmatrix} \right) = \lambda x_1 u + \lambda x_2 v = \lambda (x_1 u + x_2 v) = \lambda f\left(\begin{pmatrix} x_1 \\ x_2 \end{pmatrix} \right).$$

3. Check for yourself that the mappings

$$f : \mathbb{R}^3 \to \mathbb{R}^2, \quad \begin{pmatrix} x_1 \\ x_2 \\ x_3 \end{pmatrix} \mapsto \begin{pmatrix} x_1 \\ x_2 \end{pmatrix}, \qquad g : \mathbb{R}^3 \to \mathbb{R}^3, \quad \begin{pmatrix} x_1 \\ x_2 \\ x_3 \end{pmatrix} \mapsto \begin{pmatrix} 2x_1 + x_3 \\ 0 \\ -x_2 \end{pmatrix}$$

are linear. Show that f is surjective, but not injective and g is neither surjective nor injective.

4. A counter-example: $f: \mathbb{R} \to \mathbb{R}, x \mapsto x + 1$ is not a linear mapping of the vector space $\mathbb{R} = \mathbb{R}^1$, because $1 = f(0 + 0) \neq f(0) + f(0) = 1 + 1$.

Remember Theorem 5.39: In a homomorphism, $f(0) = 0$ must always be true. This also applies to linear mappings!

5. Something non-linear: $f: \mathbb{R}^2 \to \mathbb{R}^2, \begin{pmatrix} x_1 \\ x_2 \end{pmatrix} \mapsto \begin{pmatrix} x_1 \\ (x_1 + x_2)^2 \end{pmatrix}$. A numerical example is enough to refute (6.1): It is

$$f\left(2 \begin{pmatrix} 1 \\ 1 \end{pmatrix}\right) = f\left(\begin{pmatrix} 2 \\ 2 \end{pmatrix}\right) = \begin{pmatrix} 2 \\ 16 \end{pmatrix} \neq 2f\left(\begin{pmatrix} 1 \\ 1 \end{pmatrix}\right) = 2 \begin{pmatrix} 1 \\ 4 \end{pmatrix} = \begin{pmatrix} 2 \\ 8 \end{pmatrix}.$$

At this example you can see where the name "linear mapping" comes from. At the end of this chapter (in Theorem 6.23) we will see that the images of linear mappings can always only be linear combinations of the original coordinates. Squares, roots and similar things have no place in this theory. ◄

▶   **Theorem 6.8** If $f: U \to V$ is an isomorphism, then the inverse mapping $g := f^{-1}$ is linear and thus an isomorphism.

Proof: We have to check (6.1) for g: Let $v_1 = f(u_1)$, $v_2 = f(u_2)$ be. Since g is inverse to f, $u_1 = g(v_1)$, $u_2 = g(v_2)$ and thus

$$g(v_1 + v_2) = g(f(u_1) + f(u_2)) = g(f(u_1 + u_2)) = u_1 + u_2 = g(v_1) + g(v_2)$$
$$g(\lambda v_1) = g(\lambda f(u_1)) = g(f(\lambda u_1)) = \lambda u_1 = \lambda g(v_1). \qquad \qquad \square$$

▶   **Theorem 6.9** If $f: U \to V$ and $g: V \to W$ are linear mappings, then the composition $g \circ f: U \to W$ is linear.

This can be easily shown in a similar way as in Theorem 6.8.

The terms kernel and image, which we have already briefly discussed in the context of homomorphisms in Definition 5.39, are transferred literally to linear mappings:

▶   **Definition 6.10** Let U, V vector spaces, $f: U \to V$ a linear mapping. Then

$$\ker f := \{u \in U \,|\, f(u) = 0\} \qquad \qquad \text{the } kernel \text{ of } f,$$
$$\operatorname{im} f := \{v \in V \,|\, \exists u \in U \text{ with } f(u) = v\} \quad \text{the } image \text{ of } f.$$

▶   **Theorem 6.11** The linear mapping $f: U \to V$ is injective if and only if $\ker f = \{0\}$.

We have already given the proof of this in Theorem 5.40.

Fig. 6.6 Kernel and image

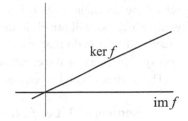

Examples

1. $f\colon \mathbb{R}^2 \to \mathbb{R}^2,\ \begin{pmatrix} x_1 \\ x_2 \end{pmatrix} \mapsto \begin{pmatrix} x_1 - 2x_2 \\ 0 \end{pmatrix}$. The kernel consists of all (x_1, x_2) with the

   property $x_1 - 2x_2 = 0$, that is, with $x_1 = 2x_2$. The image consists of all (x_1, x_2) with

   $x_2 = 0$. Look at the drawing in (Fig. 6.6).

2. $f\colon \mathbb{R}^3 \to \mathbb{R}^2,\ \begin{pmatrix} x_1 \\ x_2 \\ x_3 \end{pmatrix} \mapsto \begin{pmatrix} x_1 - x_2 + x_3 \\ 2x_1 - 2x_2 + 2x_3 \end{pmatrix}$. The kernel consists of the set of

   (x_1, x_2, x_3), for which it holds:

$$x_1 - x_2 + x_3 = 0$$
$$2x_1 - 2x_2 + 2x_3 = 0$$

For example, $(1, 1, 0)$ and $(0, 1, 1)$ are solutions of these equations, but by far not all: $(0, 0, 0)$ or $(1, 2, 1)$ are further solutions. How can we completely specify these? ◄

A simple, but very important theorem helps us to determine the structure of such solutions in the future:

▶ **Theorem 6.12** If $f\colon U \to V$ is a linear mapping, then $\ker f$ and $\operatorname{im} f$ are subspaces of U and V, respectively.

I show that the kernel is a subspace and use the subspace criterion from Theorem 6.4 for this:

(U1): $f(u) = \vec{0}, f(v) = \vec{0} \Rightarrow f(u + v) = f(u) + f(v) = \vec{0}$, that means $u + v \in \ker f$.

(U2): $f(\lambda u) = \lambda f(u) = \lambda \vec{0} = \vec{0}$, that means $\lambda u \in \ker f$.

Similarly, we conclude for the image of f. □

If you look at the two equations from Example 2 again, you will now see an interesting relation for the first time, which we will examine in more detail below: The solution set of the equation system is a vector space, namely a subspace of $\mathbb{R}^3$. Our next big goal, which we will attack in the rest of this chapter, will be to describe such subspaces exactly. If we can do that, we are able to completely specify the solutions of linear equation systems, such as the equation system from Example 2.

Theorem 6.12 can be generalized:

▷ **Theorem 6.13** Let $f: U \to V$ be a linear mapping, U_1 a subspace of U and V_1 a subspace of V. Then:

$$f(U_1) = \{f(u) \mid u \in U_1\} \text{ is a subspace of } V,$$
$$f^{-1}(V_1) = \{u \in U \mid f(u) \in V_1\} \text{ is a subspace of } U.$$

For $U_1 = U$ or $V_1 = \{\vec{0}\}$ we just get Theorem 6.12. The proof is very similar.   □

6.4   Linear Independence

We want to describe subspaces of vector spaces in more detail. First, we will take a closer look at the generating sets. Let's look at Example 2 after Theorem 6.11 again: We had seen that the solution set of the two equations

$$x_1 - x_2 + x_3 = 0$$
$$2x_1 - 2x_2 + 2x_3 = 0$$

is a subspace U of $\mathbb{R}^3$. $u_1 = (1,1,0)$, $u_2 = (0,1,1)$, $u_3 = (0,0,0)$, $u_4 = (1,2,1)$ were elements of this solution space. With these elements, the sums and multiples of the elements are also in U and in general every linear combination of these 4 vectors, since U is a vector space. Do we get all elements of U by such linear combinations? Is maybe $U = \text{span}(u_1, u_2, u_3, u_4)$? Then the space U would be completely determined by the specification of the 4 solutions. We will see that this is correct; for each subspace of $\mathbb{R}^3$ we can specify vectors that generate (span) this space by linear combinations. In our example, u_1 to u_4 are such generating vectors.

Further, $u_4 = u_1 + u_2$, so u_4 is already a linear combination of u_1 and u_2 and thus $\text{span}(u_1, u_2, u_3, u_4) = \text{span}(u_1, u_2, u_3)$. Also u_3 can be left out, it does not contribute to the span.

u_2 cannot be written as a linear combination of u_1: $u_2 \neq \lambda u_1$ for all $\lambda \in K$, and so $\text{span}(u_1, u_2) \supsetneq \text{span}(u_1)$. Similarly, for all $\lambda \in K$ we also have $u_1 \neq \lambda u_2$. Neither u_1 nor u_2 can therefore be removed from the generating set.

A vector u is called linearly dependent from a set of other vectors if it can be represented as a linear combination of these vectors:

▶ **Definition 6.14** Let V be a K-vector space and $u, v_1, v_2, v_3, \ldots, v_n \in V$. The vector u is called *linearly dependent* from $v_1, v_2, v_3, \ldots, v_n$ if there are scalars $\lambda_1, \lambda_2, \ldots \lambda_n \in K$ with

$$u = \lambda_1 v_1 + \lambda_2 v_2 + \ldots + \lambda_n v_n$$

Otherwise, u is called *linearly independent* from $v_1, v_2, v_3, \ldots, v_n$.

A set of vectors is called linearly dependent if one of them is linearly dependent on the others. This statement can be formulated in the following way, which initially looks a bit strange:

▶ **Definition 6.15** The vectors $v_1, v_2, v_3, \ldots, v_n$ of the K-vector space V are called *linearly independent*, if for every linear combination of the vectors it holds:

$$\lambda_1 v_1 + \lambda_2 v_2 + \cdots + \lambda_n v_n = \vec{0} \quad \Rightarrow \quad \lambda_1, \lambda_2, \ldots, \lambda_n = 0. \tag{6.2}$$

The vectors are called *linearly dependent*, if they are not linearly independent.

A (possibly infinite) system B of vectors is called *linearly independent*, if every finite selection of vectors from B is linearly independent.

Does (6.2) really mean that none of the vectors is linearly dependent from the others? For example, is v_1 dependent from the other vectors, then there are scalars with $v_1 = \lambda_2 v_2 + \ldots + \lambda_n v_n$ and thus $\vec{0} - (-1)v_1 + \lambda_2 v_2 + \ldots + \lambda_n v_n$ is a linear combination in which not all coefficients are 0. And if there is a linear combination of the form (6.2) in which not all $\lambda_i = 0$, for example, $\lambda_1 \neq 0$, then

$$v_1 = \frac{\lambda_2}{-\lambda_1} v_2 + \frac{\lambda_3}{-\lambda_1} v_3 + \cdots + \frac{\lambda_n}{-\lambda_1} v_n,$$

follows, so v_1 is dependent on the other vectors.

You often read the abbreviation w.l.o.g. or WLOG in connection with proofs in mathematics books. This means "without loss of generality". The above consideration makes it clear what is meant by this: It does not have to be v_1 linearly dependent on the other vectors, it can also be v_2, v_3 or any other v_i. But this does not change the proof, it goes analogously, since v_1 is not distinguished from the other vectors by anything. The problem is completely symmetrical. We can simply assume "w.l.o.g." of the proof that v_1 is our special element. This trick often makes proofs much easier, as it can avoid many case distinctions in this way. But this means should be used carefully and be absolutely sure that there is really no restriction.

Caution: If a set of vectors is linearly dependent, this does not mean that each of the vectors can be represented as a linear combination of the others: $(1, 1, 0), (2, 2, 0)$ and $(0, 1, 1)$ are linearly dependent, since $(2, 2, 0) = 2 \cdot (1, 1, 0) + 0 \cdot (0, 1, 1)$, but $(0,1,1)$ can not be linearly combined from the two other vectors.

The formula (6.2) looks abstract and inconvenient compared to the intuitive explanation "one of the vectors is linearly dependent from the others" at first sight. But we will see that the opposite is the case. Equation (6.2) is very well suited to test a set of vectors for linear independence. We will often use this rule.

Are there vector spaces with infinitely many linearly independent vectors? For this an

---

Example

Let $V = \mathbb{R}[X]$ be the $\mathbb{R}$-vector space of polynomials with coefficients from $\mathbb{R}$ (compare Example 6 in Sect. 6.2 after Theorem 6.2). Then the system $B = \{1, x, x^2, x^3, \dots\}$ is linearly independent.

Assume this is not the case. Then there is a finite subset of these vectors that can be linearly combined so that the zero polynomial results:

$$\lambda_1 x^{m_1} + \lambda_2 x^{m_2} + \lambda_3 x^{m_3} + \cdots + \lambda_n x^{m_n} = 0.$$

On the left side there is a real polynomial, on the right side the zero polynomial. Now let's look at the polynomials again as functions for a moment. Then we know from Theorem 5.25 that the polynomial on the left side has only finitely many roots. But the polynomial on the right has the value 0 for all $x \in \mathbb{R}$, it is the zero polynomial after all. Obviously this is not possible, so the polynomials in B are linearly independent. ◀

6.5   Basis and Dimension of Vector Spaces

The span of a set B of vectors forms a vector space V. If the set B is linearly independent, it is called a basis of V. We will see in this section that every vector space has a basis and derive properties of such bases.

▷   **Definition 6.16** Let V be a vector space. A subset B of V is called a basis of V, if the following conditions are satisfied:

(B1)   span $B = V$ (B generates V).
(B2) B is a linearly independent set of vectors.

Examples

1. $\left\{ \begin{pmatrix} 1 \\ 0 \end{pmatrix}, \begin{pmatrix} 0 \\ 1 \end{pmatrix} \right\}$ forms a basis of $\mathbb{R}^2$:

   To (B1): $\begin{pmatrix} x \\ y \end{pmatrix} = x \begin{pmatrix} 1 \\ 0 \end{pmatrix} + y \begin{pmatrix} 0 \\ 1 \end{pmatrix}$, so the two vectors generate the $\mathbb{R}^2$.

   To (B2):   From $\lambda \begin{pmatrix} 1 \\ 0 \end{pmatrix} + \mu \begin{pmatrix} 0 \\ 1 \end{pmatrix} = \begin{pmatrix} \lambda \\ \mu \end{pmatrix} = \vec{0} = \begin{pmatrix} 0 \\ 0 \end{pmatrix}$ follows $\lambda, \mu = 0$, so they are according to (6.2) linearly independent.

2. Similarly, it follows:

$$\left\{ \begin{pmatrix} 1 \\ 0 \\ 0 \\ \vdots \\ 0 \end{pmatrix}, \begin{pmatrix} 0 \\ 1 \\ 0 \\ \vdots \\ 0 \end{pmatrix}, \begin{pmatrix} 0 \\ 0 \\ 1 \\ \vdots \\ 0 \end{pmatrix}, \dots, \begin{pmatrix} 0 \\ 0 \\ 0 \\ \vdots \\ 1 \end{pmatrix} \right\} \subset \mathbb{R}^n \text{ is a basis of } \mathbb{R}^n.$$

This basis is called the *standard basis* of $\mathbb{R}^n$.

3. $\left\{ \begin{pmatrix} 2 \\ 3 \end{pmatrix}, \begin{pmatrix} 3 \\ 4 \end{pmatrix} \right\}$ is a basis of $\mathbb{R}^2$:

To (B1):   Let $\begin{pmatrix} x \\ y \end{pmatrix} \in \mathbb{R}^2$. Find λ, μ with the property $\lambda \begin{pmatrix} 2 \\ 3 \end{pmatrix} + \mu \begin{pmatrix} 3 \\ 4 \end{pmatrix} = \begin{pmatrix} x \\ y \end{pmatrix}$.

For this, the two equations have to be solved:

$$\text{(I)} \quad \lambda 2 + \mu 3 = x,$$
$$\text{(II)} \quad \lambda 3 + \mu 4 = y.$$

Solve for λ and μ, and you will get

$$\lambda = -4x + 3y,$$
$$\mu = 3x - 2y.$$

To (B2):   From $\lambda \begin{pmatrix} 2 \\ 3 \end{pmatrix} + \mu \begin{pmatrix} 3 \\ 4 \end{pmatrix} = \begin{pmatrix} 0 \\ 0 \end{pmatrix}$ the equations

$$\lambda 2 + \mu 3 = 0$$
$$\lambda 3 + \mu 4 = 0$$

arise and you can calculate that they only have the solution $\lambda = 0, \mu = 0$.

4. $\{1, x, x^2, x^3, \dots\}$ forms a basis of the vector space $\mathbb{R}[X]$: We already know that the vectors are linearly independent, but are they also generating? Each $p(x) \in \mathbb{R}[X]$ has the form

$$p(x) = a_n x^n + a_{n-1} x^{n-1} + \dots + a_1 x + a_0.$$

But that's already everything: The coefficients of the linear combination are just the coefficients of the polynomial! You can see from this example that there are vector spaces with infinite bases. ◄

We have already found bases for many of the vector spaces we know. But does every vector space have a basis? $\mathbb{R}^n$ has a basis with n elements. Can we generate $\mathbb{R}^n$ maybe with less than n elements, if we only choose the vectors clever enough? Or could it be we find more than n linearly independent vectors in $\mathbb{R}^n$? We will see: This is not the case. Every vector space has a basis, and different bases of a vector space always have the same number of elements.

The ideas behind the proof of these assertions are easy to understand and constructive. The precise implementation of the proof is quite tedious, though. I would like to forego this and only sketch the way.

▶   **Theorem 6.17** Every vector space $V \neq \{\vec{0}\}$ has a basis.

The proof for this consists of an induction, which we could carry out in the case of vector spaces with finite bases: We construct a basis B by starting with a vector different from $\vec{0}$. As long as the set B does not yet generate the vector space V, we can always add a vector v from V, so that B remains linearly independent. In the case of vector spaces that have infinite bases, the proof exceeds the knowledge you have acquired so far.

Next, we investigate the question of whether any two bases of a vector space have the same number of elements. Our goal is the assertion: If a vector space has a finite basis, then every other basis has the same number of elements.

It follows from this, of course: If a vector space has an infinite basis, then every other basis of the space has infinitely many elements.

So let's now restrict ourselves to vector spaces that have a finite basis. The core of the proof is in the technical theorem:

▶   **Theorem 6.18: The exchange theorem** Let $B = \{b_1, b_2, \ldots, b_n\}$ be a basis of V, $x \in V$ and $x \neq \vec{0}$. Then there is a $b_i \in B$, so that also $\{b_1, \ldots, b_{i-1}, x, b_{i+1}, \ldots, b_n\}$ is a basis of V.

Since the proof of this statement is simple and very typical for the calculations with linearly independent vectors, I would like to present it here:

Since B is a basis,

$$x = \lambda_1 b_1 + \lambda_2 b_2 + \cdots + \lambda_n b_n, \tag{6.3}$$

applies, where not all coefficients $\lambda_i = 0$. For example (w.l.o.g.!), $\lambda_1 \neq 0$. I claim that then $\{x, b_2, b_3, \ldots, b_n\}$ forms a basis of V. To prove this, (B1) and (B2) from Definition 6.16 have to be checked:

(B1):   Let $v \in V$. There is $\mu_1, \mu_2, \ldots \mu_n$ with the property

$$v = \mu_1 b_1 + \mu_2 b_2 + \cdots + \mu_n b_n. \tag{6.4}$$

From the representation (6.3) of x we get

$$b_1 = \frac{1}{\lambda_1} x - \frac{\lambda_2}{\lambda_1} b_2 - \frac{\lambda_3}{\lambda_1} b_3 - \cdots - \frac{\lambda_n}{\lambda_1} b_n.$$

If we set this in the representation (6.4) of v, we get v as a linear combination of $x, b_2, \ldots, b_n$. This makes the set generating.

(B2): Let $\mu_1 x + \mu_2 b_2 + \cdots \mu_n b_n = \vec{0}$. We show all coefficients are 0, that means (6.2) is fulfilled. We replace x according to (6.3) and get

$$\mu_1(\lambda_1 b_1 + \lambda_2 b_2 + \cdots + \lambda_n b_n) + \mu_2 b_2 + \cdots + \mu_n b_n$$
$$= \mu_1 \lambda_1 b_1 + (\mu_1 \lambda_2 + \mu_2)b_2 + \cdots + (\mu_1 \lambda_n + \mu_n)b_n = \vec{0}.$$

Since the b_i form a basis, all coefficients are 0 here. Because of $\lambda_1 \neq 0$, first $\mu_1 = 0$ and then $\mu_2, \mu_3, \ldots, \mu_n = 0$ follow one after the other. So the set is linearly independent. □

As a result, we get the announced assertion. Its proof also consists of an induction, which is quite tricky, however. I will only present the idea.

▶ **Theorem 6.19** If the vector space V has a finite basis B with n elements, then every other basis of V also has n elements.

With the help of the exchange theorem, we can first derive the following assertion: If $B = \{b_1, b_2, \ldots, b_n\}$ and further $B' = \{b'_1, b'_2, \ldots, b'_m\}$ is another basis of V with m elements, then $m \leq n$ applies. For if m were greater than n, the elements of the basis B could be successively exchanged for n elements of B', while the basis property is retained. But then the further $m - n$ elements of B' can no longer be linearly independent from the n vectors that have already been exchanged.

If we now exchange the roles of B and B', the same argumentation results in $n \leq m$.

The number of elements in a basis of a vector space is therefore an important characteristic of the space. We call this number the dimension. This is the number of vectors that are needed to generate the vector space.

▶ **Definition 6.20** Let $B = \{b_1, b_2, \ldots, b_n\}$ be a basis of the vector space V. Then the number n is called *dimension* of V. We write $n = \dim V$. The zero space $\{\vec{0}\}$ is assigned the dimension 0.

It is only after Theorem 6.19 that we are allowed to write this definition, now the concept of dimension is "well-defined".

▶ **Theorem 6.21** In a vector space of dimension n, every linearly independent set with n elements is already a basis.

Because otherwise one could extend it like in the proof of Theorem 6.17 to a basis. But a basis cannot have more than n elements. □

Since we already know bases of $\mathbb{R}^2$, $\mathbb{R}^3$, $\mathbb{R}^n$, we also know the dimensions of these spaces: $\mathbb{R}^n$ has the dimension n and we will never be able to find $n + 1$ linearly independent vectors in it, just as we will never be able to generate it from $n - 1$ elements.

We can also make precise assertions about possible subspaces with our new knowledge. Let's take the $\mathbb{R}^3$ as an example: We already know some subspaces: Lines through the origin $g = \{\lambda u \mid \lambda \in \mathbb{R}\}$, $u \in \mathbb{R}^3 \setminus \{\vec{0}\}$ have dimension 1. The vector u is a basis of the subspace g. Planes through the origin are given by $E = \{\lambda u + \mu v \mid \lambda, \mu \in \mathbb{R}\}$, $u, v \in \mathbb{R}^3 \setminus \{\vec{0}\}$. E only represents a plane if u and v are linearly independent, and then u and v also form a basis of the subspace E. Are there any other subspaces? The zero space $\{\vec{0}\}$ is the extreme case. Let U be any subspace of $\mathbb{R}^3$. Does U contain at least one vector different from $\vec{0}$, then U already contains a line through the origin. If U is not a line through the origin, then U contains at least two linearly independent vectors and thus includes an entire plane through the origin. If U is not a plane through the origin itself, then a further vector must be contained outside the plane. But this is then linearly independent, U itself contains three linearly independent vectors and $U = \mathbb{R}^3$ applies. The subspaces of the $\mathbb{R}^3$ are therefore exactly $\{\vec{0}\}$, all lines and all planes through the origin and $\mathbb{R}^3$ itself.

All subspaces of the $\mathbb{R}^n$ can be classified in the same way.

Take a look at example 2 after Theorem 6.11 again: We there found that the solution set of the two linear equations

$$
\begin{aligned}
x_1 - x_2 + x_3 &= 0 \\
2x_1 - 2x_2 + 2x_3 &= 0
\end{aligned}
$$

is a subspace of the $\mathbb{R}^3$. We guessed some solutions: $(1, 1, 0)$, $(0, 1, 1)$, $(1, 2, 1)$. But we didn't know the structure of the space and the complete solution set yet. Now we can find out: $(1, 1, 0)$ and $(0, 1, 1)$ are linearly independent, so the solution space has at least dimension 2. $(1, 1, 1)$ is, for example, not a solution, so the dimension cannot be 3 (otherwise it would be $\mathbb{R}^3$). This means that the set of solutions consists of the plane spanned by $(1, 1, 0)$ and $(0, 1, 1)$:

$$
L = \left\{ \lambda \begin{pmatrix} 1 \\ 1 \\ 0 \end{pmatrix} + \mu \begin{pmatrix} 0 \\ 1 \\ 1 \end{pmatrix} \,\middle|\, \lambda, \mu \in \mathbb{R} \right\}.
$$

Of course you can also enter other basis vectors here, the plane always remains the same.

In Chap. 8 we will deal extensively with the connection between solutions of linear equations and the subspaces of $\mathbb{R}^n$.

6.6   Coordinates and Linear Mappings

With the development of the concepts of basis and dimension, we have made a great leap forward in our further work with vector spaces. This is mainly due to the fact that we can now describe vectors by their coefficients (the coordinates) in a basis representation.

At least in finite-dimensional vector spaces, these are only finitely many data and the calculation with them is simple.

In the $\mathbb{R}^n$ we have already learned coordinates: For

$$v = \begin{pmatrix} x_1 \\ x_2 \\ x_3 \end{pmatrix} \in \mathbb{R}^3$$

the x_1, x_2, x_3 are the coordinates of v. How do coordinates look in arbitrary vector spaces with respect to a basis B? The $\mathbb{R}^3$ with its standard basis helps us with the construction. For it is

$$v = \begin{pmatrix} x_1 \\ x_2 \\ x_3 \end{pmatrix} = x_1 \begin{pmatrix} 1 \\ 0 \\ 0 \end{pmatrix} + x_2 \begin{pmatrix} 0 \\ 1 \\ 0 \end{pmatrix} + x_3 \begin{pmatrix} 0 \\ 0 \\ 1 \end{pmatrix}.$$

That is: If v is represented as a linear combination of the basis, then the coefficients of this linear combination are just the coordinates of v. We now carry out this construction in general:

▶ **Theorem and Definition 6.22** If $B = \{b_1, b_2, \ldots, b_n\}$ is a basis of the vector space V, then there are uniquely determined elements $x_1, x_2, \ldots, x_n$ with the property

$$v = x_1 b_1 + x_2 b_2 + \cdots + x_n b_n.$$

The x_i are called coordinates of v with respect to B. We write

$$v = \begin{pmatrix} x_1 \\ x_2 \\ \vdots \\ x_n \end{pmatrix}_B.$$

Proof: Of course there is such a representation, since B is a basis. So it's just the uniqueness that has to be shown. Assume there are different representations of v:

$$v = x_1 b_1 + x_2 b_2 + \cdots + x_n b_n,$$
$$v = y_1 b_1 + y_2 b_2 + \cdots + y_n b_n.$$

Then by forming the difference we get

$$\vec{0} = (x_1 - y_1) b_1 + (x_2 - y_2) b_2 + \cdots + (x_n - y_n) b_n$$

and from that because of the linear independence of the basis vectors $x_i - y_i = 0$, that is $x_i = y_i$ for all i. □

These coordinates are of course basis-dependent, indeed they are even dependent on the order of the basis vectors. If you change the basis or the order of the basis vectors, you will get other coordinates.

Let's calculate the coordinates of the basis vectors themselves: It is

$$b_1 = 1b_1 + 0b_2 + 0b_3 + \cdots + 0b_n,$$
$$b_2 = 0b_1 + 1b_2 + 0b_3 + \cdots + 0b_n,$$

$$\vdots$$

$$b_n = 0b_1 + 0b_2 + 0b_3 + \cdots + 1b_n.$$

and from that we get

$$b_1 = \begin{pmatrix} 1 \\ 0 \\ 0 \\ \vdots \\ 0 \end{pmatrix}_B, \quad b_2 = \begin{pmatrix} 0 \\ 1 \\ 0 \\ \vdots \\ 0 \end{pmatrix}_B, \quad \ldots, \quad b_n = \begin{pmatrix} 0 \\ 0 \\ 0 \\ \vdots \\ 1 \end{pmatrix}_B.$$

Oops, that's certainly familiar to you. That's exactly how the coordinates of the basis of the $\mathbb{R}^n$ looked like, which we once called the standard basis. Now we have made a very amazing discovery. In the calculation with coordinates, one basis is as good as any other. The basis vectors always have the coordinates of the standard basis.

There is also no basis that is in any way distinguished in the $\mathbb{R}^n$: If, for example, we work with the vectors of the $\mathbb{R}^3$ to describe the space we live in, we first look for a basis, that is, an origin and three vectors of certain length that do not lie in one plane. Then we calculate with coordinates relative to this basis. This randomly chosen basis is our "standard basis".

It is also important that this basis choice can be made so freely. For example, if one describes the movements of a robot arm, one usually places a coordinate system in each joint, with which one describes the movements of this joint exactly. In Sect. 10.3 we will examine this application case in more detail.

However, it is quite possible and often necessary to establish the connection between different bases computationally. If you calculate with coordinates of a basis C and the coordinates of another basis B are given with respect to this basis, the coordinates of a vector v can be given with respect to both bases. For this purpose, an

---

Example

In the $\mathbb{R}^2$ the basis $B = \{b_1, b_2\}$ is given. The vectors $c_1 = \begin{pmatrix} 2 \\ 2 \end{pmatrix}_B$, $c_2 = \begin{pmatrix} 1 \\ 2 \end{pmatrix}_B$ are linearly independent. If $v = \begin{pmatrix} x \\ y \end{pmatrix}_B$ is given, we now want to calculate the coordinates

$v = \begin{pmatrix} \lambda \\ \mu \end{pmatrix}_C$ with respect to the basis $C = \{c_1, c_2\}$: If $v = \lambda c_1 + \mu c_2$, then λ and μ are these coordinates. But

$$\begin{pmatrix} x \\ y \end{pmatrix}_B = v = \lambda c_1 + \mu c_2 = \lambda \begin{pmatrix} 2 \\ 2 \end{pmatrix}_B + \mu \begin{pmatrix} 1 \\ 2 \end{pmatrix}_B.$$

From this we get the determining equations for λ and μ:

$$x = 2\lambda + \mu,$$
$$y = 2\lambda + 2\mu.$$

Solved for λ, μ we get:

$$\lambda = x - \frac{1}{2}y,$$
$$\mu = -x + y.$$

Now we can convert the coordinates. For example, we get

$$\begin{pmatrix} 2 \\ 1 \end{pmatrix}_B = \begin{pmatrix} \frac{3}{2} \\ -1 \end{pmatrix}_C, \quad \begin{pmatrix} 2 \\ 2 \end{pmatrix}_B = \begin{pmatrix} 1 \\ 0 \end{pmatrix}_C, \quad \begin{pmatrix} 1 \\ 0 \end{pmatrix}_B = \begin{pmatrix} 1 \\ -1 \end{pmatrix}_C, \quad \begin{pmatrix} 0 \\ 1 \end{pmatrix}_B = \begin{pmatrix} -\frac{1}{2} \\ 1 \end{pmatrix}_C.$$

The last two calculated points are just the coordinates of the basis B with respect to the basis C. ◄

In Fig. 6.7 I have entered the two bases and the coordinates of the calculated points with respect to both bases. Determine the coordinates for other points, graphically and mathematically.

Our new ability to calculate in vector spaces with coordinates can now be applied to linear mappings. The hard work we did in connection with the concepts of basis and dimension pays off: Some important and beautiful results are now becoming apparent.

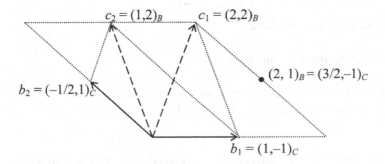

Fig. 6.7 Coordinates with change of basis

First, we note that every linear mapping of a finite-dimensional vector space can already be completely described by finitely many data, namely the images of a basis. This will be very useful in the next chapter.

▶   **Theorem 6.23** Let U and V be vector spaces over K, $b_1, b_2, \ldots, b_n$ be a basis of U and $v_1, v_2, \ldots, v_n$ be elements of V. Then there is exactly one linear mapping $f : U \to V$ with the property $f(b_i) = v_i$ for all $i = 1, \ldots, n$.

Let us first investigate the existence of the mapping: For $u = x_1 b_1 + x_2 b_2 + \cdots + x_n b_n$ we define

$$f(u) = x_1 v_1 + x_2 v_2 + \cdots + x_n v_n. \tag{6.5}$$

The mapping is linear, because for

$$u = x_1 b_1 + x_2 b_2 + \cdots + x_n b_n$$
$$v = y_1 b_1 + y_2 b_2 + \cdots + y_n b_n$$
$$u + v = (x_1 + y_1) b_1 + (x_2 + y_2) b_2 + \cdots + (x_n + y_n) b_n$$

we have:

$$f(u) = x_1 v_1 + x_2 v_2 + \cdots + x_n v_n$$
$$f(v) = y_1 v_1 + y_2 v_2 + \cdots + y_n v_n$$
$$f(u) + f(v) = (x_1 + y_1) v_1 + (x_2 + y_2) v_2 + \cdots + (x_n + y_n) v_n = f(u + v).$$

Similarly,

$$\lambda f(u) = \lambda(x_1 v_1 + x_2 v_2 + \cdots + x_n v_n) = (\lambda x_1) v_1 + (\lambda x_2) v_2 + \cdots + (\lambda x_n) v_n = f(\lambda u).$$

Now for uniqueness: Let g be another mapping with this property. Then, because of the linearity of g

$$g(u) = g(x_1 b_1 + x_2 b_2 + \cdots + x_n b_n) = x_1 g(b_1) + x_2 g(b_2) + \cdots + x_n g(b_n)$$
$$= x_1 v_1 + x_2 v_2 + \cdots + x_n v_n = f(u)$$

and thus f and g agree everywhere.   ☐

We come to a highlight of vector space theory:

▶   **Theorem 6.24** Let U and V be finite-dimensional vector spaces over K. Then U and V are isomorphic if and only if they have the same dimension:

$$U \cong V \quad \Leftrightarrow \quad \dim U = \dim V.$$

Why is this theorem so important? It immediately follows from this: Every n-dimensional vector space is isomorphic to K^n, in particular, every finite-dimensional real vector space is isomorphic to a $\mathbb{R}^n$.

Thus we now know the structure of all finite-dimensional vector spaces in the universe; there are (up to isomorphism) only the vector spaces K^n.

This is a moment when the true mathematician first leans back in his chair and enjoys life. It is rare that it is possible to classify all possible instances of a given algebraic structure. This is a highlight of the theory, then one knows the structure exactly. Here we have at least succeeded in classifying all finite-dimensional vector spaces.

Flip back to the beginning of this chapter: We started with $\mathbb{R}^2$, $\mathbb{R}^3$ and $\mathbb{R}^n$. From this we crystallized out the vector space properties and generally defined the structure of a vector space. Now we find (at least for real vector spaces of finite dimension) there is nothing but the $\mathbb{R}^n$. So why the whole thing, are we not just going around in circles?

We have gained several things: On the one hand, if we come across a structure in some application that we see has the properties of a real vector space, then we now know: This structure "is" the $\mathbb{R}^n$, we can use everything we know about $\mathbb{R}^n$. Of course, there is still the great theory of infinite-dimensional vector spaces, which builds on what we have developed in this chapter, but which I have only briefly mentioned here. And finally, the process that led to Theorem 6.24 gave us a whole toolbox we can use in the future when working with vectors. Our knowledge of bases, dimensions, linear dependence, coordinates and other things will be immensely important in the next chapters. So even here the journey was the goal.

For the proof of the Theorem 6.24: As an equivalence proof it consists of two parts. We start with the direction from left to right:

For this purpose, let $f: U \to V$ be an isomorphism, $b_1, b_2, \ldots, b_n$ a basis of U. We show that then the images $v_1, v_2, \ldots, v_n$ of $b_1, b_2, \ldots, b_n$ form a basis of V. Thus the dimensions of the spaces are equal.

The v_i are linearly independent, as from

$$\vec{0} = \lambda_1 v_1 + \lambda_2 v_2 + \cdots + \lambda_n v_n = f(\lambda_1 b_1 + \lambda_2 b_2 + \cdots + \lambda_n b_n)$$

follows because of the injectivity (only $f(\vec{0}) = \vec{0}$!) also $\lambda_1 b_1 + \lambda_2 b_2 + \cdots + \lambda_n b_n = \vec{0}$. But since B is a basis, then all $\lambda_i = 0$, and thus the v_i are linearly independent.

The v_i generate V: Since f is surjective, there is for every $v \in V$ a $u \in U$ with $f(u) = v$, thus is

$$v = f(u) = f(x_1 b_1 + x_2 b_2 + \cdots + x_n b_n) = x_1 f(b_1) + x_2 f(b_2) + \cdots + x_n f(b_n)$$
$$= x_1 v_1 + x_2 v_2 + \cdots x_n v_n.$$

Now to the other direction: Let $\dim U = \dim V$ and let $u_1, u_2, \ldots, u_n$ be a basis of U and $v_1, v_2, \ldots, v_n$ be a basis of V. According to Theorem 6.23 there are then two linear mappings which map the basis vectors to each other:

$$f: U \to V, \quad f(u_i) = v_i,$$
$$g: V \to U, \quad g(v_i) = u_i.$$

These are inverse to each other and thus they are bijective linear mappings, that is isomorphisms, between U and V. $\qquad \square$

This proof contains two sub-statements, which are worth writing down:

▶ **Theorem 6.25** An isomorphism between finite dimensional vector spaces maps a basis to a basis.

And from the second part one can read:

▶ **Theorem 6.26** A linear mapping between two vector spaces of equal dimension, which maps a basis to a basis, is an isomorphism.

In Chap. 8 I will go into the determination of solutions of linear equation systems. For this, the following theorem is important, with which I would like to conclude this chapter:

▶ **Theorem 6.27** Let U, V vector spaces, dim $U = n$ and $f: U \to V$ a linear mapping. Then

$$\dim \ker f + \dim \operatorname{im} f = \dim U.$$

I do not want to carry out the proof. For this one chooses a basis of the kernel, this can be extended to a basis of U. Then one shows that the images of this extension are precisely a basis of the image space. □

In short: The more is mapped onto $\vec{0}$ in a linear map, the smaller the image becomes. The limiting cases are still easy to follow: If the kernel is equal to $\{\vec{0}\}$, then f is injective, the image is isomorphic to U and has dimension n. Conversely, if the image $= \{\vec{0}\}$, then the kernel $= U$, so it has dimension n. But the Theorem goes even further: For example, for linear mappings $f: \mathbb{R}^3 \to \mathbb{R}^3$: If the kernel has dimension 1, then the image has dimension 2 and vice versa.

6.7   Comprehension Questions and Exercises

Comprehension questions

1. Is a line in $\mathbb{R}^3$, which does not go through the origin, a vector space?
2. Are there any vector spaces that are a proper subspace of $\mathbb{R}^2$ and a proper superset of the x-axis?
3. Is $\mathbb{Q}^3$ a subspace of $\mathbb{R}^3$?
4. Are $\mathbb{R}^4$ and $\mathbb{C}^2$ isomorphic as vector spaces? Can there be a linear mapping between $\mathbb{R}^4$ and $\mathbb{C}^2$?
5. Are there vector spaces with bases that have different numbers of elements?

6. True or false: If you choose from an infinite-dimensional vector space V a finite number of elements, then the span of these elements forms a finite-dimensional subspace of V.
7. Can a surjective linear map from $\mathbb{R}^n$ to $\mathbb{R}^n$ always be inverted?

Exercises

1. Show that the mappings
   a) $f: \mathbb{R}^3 \to \mathbb{R}^2, (x, y, z) \mapsto (x + y, y + z)$
   b) $g: \mathbb{R}^2 \to \mathbb{R}^3, (x, y) \mapsto (x, x + y, y)$
   are linear. Calculate $\ker f$ and $\ker g$.
2. The equation $y = 3x + 4$ can be interpreted as a line in $\mathbb{R}^2$:
$$g = \{(x, y) \mid y = 3x + 4\}, \quad a, b \in \mathbb{R}^2.$$
   Find vectors a, b in $\mathbb{R}^2$ with $g = \{a + \lambda b \mid \lambda \in \mathbb{R}\}$.
3. Every graph of a line $y = mx + c$ has a representation in the form $g = \{a + \lambda b \mid \lambda \in \mathbb{R}\}$. Determine vectors a, b for this representation. Is there a representation of every line g in $\mathbb{R}^2$ as a graph of a line $y = mx + c$?
4. Check the vector space conditions (V1) to (V4) for $\mathbb{R}^3$.
5. Check whether $\mathbb{R}^2$ with the usual addition and with the scalar multiplication $\lambda \begin{pmatrix} x \\ y \end{pmatrix} = \begin{pmatrix} \lambda x \\ y \end{pmatrix}$ is a vector space.
6. Find a linear mapping $f: \mathbb{R}^2 \to \mathbb{R}^2$ for which $\ker f = \operatorname{im} f$ applies.
7. What do you say about exercise 6 if I replace $\mathbb{R}^2$ each time by $\mathbb{R}^5$?
8. The vectors $\begin{pmatrix} 3 \\ 5 \end{pmatrix}$ and $\begin{pmatrix} 2 \\ 4 \end{pmatrix}$ form a basis B of $\mathbb{R}^2$. Let $\begin{pmatrix} x \\ y \end{pmatrix} \in \mathbb{R}^2$.
   Calculate the coordinates of $\begin{pmatrix} x \\ y \end{pmatrix}$ in the basis B.
9. Show: If u, v are linearly independent vectors in V, then so are $u + v$ and $u - v$ (make a sketch in $\mathbb{R}^2$!).

Matrices

7

Abstract

The use of coordinates and matrices in linear algebra lays the foundation for algorithms in many areas of computer science. By the end of this chapter, you will know

- the relation between matrices and linear mappings,
- important linear mappings in the $\mathbb{R}^2$ and their representing matrices: stretchings, rotations, reflections,
- the matrix multiplication and its interpretation as a composition of linear mappings,
- matrices and matrix operations in the K^n, where K can be any field, the rank of a matrix.

7.1   Matrices and Linear Mappings in $\mathbb{R}^2$

In the last chapter we saw that every linear mapping is already completely determined by the images of the basis vectors of a vector space. Matrices are suitable for describing such a mapping. I do not want to cause illegibility by the abundance of coordinates right from the start, so I will develop the basic concepts for matrices first in the vector space $\mathbb{R}^2$. We calculate with coordinates with respect to a basis and investigate linear mappings from $\mathbb{R}^2$ to $\mathbb{R}^2$.

Supplementary Information The online version contains supplementary material available at https://doi.org/10.1007/978-3-658-40423-9_7.

From Theorem 6.23 it follows that every linear mapping $f: \mathbb{R}^2 \to \mathbb{R}^2$ has the form:

$$f\left(\begin{pmatrix} x_1 \\ x_2 \end{pmatrix}\right) = \begin{pmatrix} a_{11}x_1 + a_{12}x_2 \\ a_{21}x_1 + a_{22}x_2 \end{pmatrix}.$$

To see this, we first determine the images of the basis for the linear mapping f. If these are

$$f\left(\begin{pmatrix} 1 \\ 0 \end{pmatrix}\right) = \begin{pmatrix} a_{11} \\ a_{21} \end{pmatrix} \quad \text{and} \quad f\left(\begin{pmatrix} 0 \\ 1 \end{pmatrix}\right) = \begin{pmatrix} a_{12} \\ a_{22} \end{pmatrix}, \tag{7.1}$$

then for every vector $x \in \mathbb{R}^2$:

$$f(x) = f\left(\begin{pmatrix} x_1 \\ x_2 \end{pmatrix}\right) = f\left(x_1 \begin{pmatrix} 1 \\ 0 \end{pmatrix} + x_2 \begin{pmatrix} 0 \\ 1 \end{pmatrix}\right)$$

$$= x_1 \begin{pmatrix} a_{11} \\ a_{21} \end{pmatrix} + x_2 \begin{pmatrix} a_{12} \\ a_{22} \end{pmatrix} = \begin{pmatrix} a_{11}x_1 + a_{12}x_2 \\ a_{21}x_1 + a_{22}x_2 \end{pmatrix}.$$

We introduce a shorthand notation: Instead of

$$\begin{pmatrix} a_{11}x_1 + a_{12}x_2 \\ a_{21}x_1 + a_{22}x_2 \end{pmatrix} \quad \text{we write:} \quad \begin{pmatrix} a_{11} & a_{12} \\ a_{21} & a_{22} \end{pmatrix} \begin{pmatrix} x_1 \\ x_2 \end{pmatrix}.$$

Here $\begin{pmatrix} x_1 \\ x_2 \end{pmatrix} = x$ is a vector, $\begin{pmatrix} a_{11} & a_{12} \\ a_{21} & a_{22} \end{pmatrix} = A$ we call matrix. For $\begin{pmatrix} a_{11} & a_{12} \\ a_{21} & a_{22} \end{pmatrix} \begin{pmatrix} x_1 \\ x_2 \end{pmatrix}$ we write briefly Ax.

▶ **Definition 7.1** A quadruple of real numbers

$$\begin{pmatrix} a_{11} & a_{12} \\ a_{21} & a_{22} \end{pmatrix}$$

is called 2×2-*matrix* (read: "two by two matrix"). The set of all 2×2-matrices with real coefficients we call $\mathbb{R}^{2 \times 2}$.

In the matrix $A = \begin{pmatrix} a_{11} & a_{12} \\ a_{21} & a_{22} \end{pmatrix}$ the $\begin{pmatrix} a_{11} \\ a_{21} \end{pmatrix}$, $\begin{pmatrix} a_{12} \\ a_{22} \end{pmatrix}$ are called *columns* or *column vectors* and (a_{11}, a_{12}), (a_{21}, a_{22}) are called *rows* or *row vectors*. The first index is called *row index*, it remains constant in a row, the second index is called *column index*, it remains constant in a column. a_{ij} is thus the element in the i-th row and in the j-th column. Occasionally the matrix A is also denoted by (a_{ij}).

Our previous knowledge about the connection between matrices and linear mappings can now be formulated in the

▶ **Theorem 7.2** For every linear mapping $f: \mathbb{R}^2 \to \mathbb{R}^2$ there is exactly one matrix $A \in \mathbb{R}^{2 \times 2}$ with the property $f(x) = Ax$. Conversely, every matrix $A \in \mathbb{R}^{2 \times 2}$ defines a linear mapping by the rule $f: \mathbb{R}^2 \to \mathbb{R}^2$, $x \mapsto Ax$.

We have already seen that every linear mapping determines such a matrix. Conversely, for a matrix A the mapping $f: \mathbb{R}^2 \to \mathbb{R}^2$, $x \mapsto Ax$ is also linear: It is exactly the linear mapping according to Theorem 6.23, which maps $\begin{pmatrix} 1 \\ 0 \end{pmatrix}$ to $\begin{pmatrix} a_{11} \\ a_{21} \end{pmatrix}$ and $\begin{pmatrix} 0 \\ 1 \end{pmatrix}$ to $\begin{pmatrix} a_{12} \\ a_{22} \end{pmatrix}$. □

There is thus a bijective relationship between the set of linear mappings of $\mathbb{R}^2$ and the set of matrices $\mathbb{R}^{2 \times 2}$. We will make intensive use of this relationship and not distinguish between matrices and mappings anymore. A matrix "is" a linear mapping and a linear mapping "is" a matrix. So I will often speak of the linear mapping $A: \mathbb{R}^2 \to \mathbb{R}^2$, where A is a matrix from $\mathbb{R}^{2 \times 2}$. $A(x)$ is then the image of x under A, which is the same as Ax.

However, when making this identification, one should note that the matrix depends on the basis. With respect to another basis, the corresponding matrix usually looks quite different. But first we fix a basis, so no problems arise from this.

If you look again at the (7.1), you will notice the columns of the matrix are just the images of the basis vectors. You should remember this.

Now we finally want to look at some concrete linear mappings of $\mathbb{R}^2$ to itself. Imagine a section of the $\mathbb{R}^2$ as a computer screen. If I work with a drawing program and want to draw a rectangle, I can first drag a prototype of the rectangle onto the screen. Usually it is marked in some places by small rectangles that show that I can grab and edit the rectangle with the mouse. By various operations, I can generate from the rectangle, the shapes 1-6 found in Fig. 7.1. The prototype is transformed into figure 1, figure 1 into figure 2 and so on.

A change of the figure means a mapping of the object points in $\mathbb{R}^2$. The computer has to calculate the new coordinates of the object from the old ones and draw them on the screen. With one exception, all these mappings are linear. We will now look for the linear mappings that can generate these figures. The origin is always marked, the x_1-axis goes to the right, the x_2-axis goes up.

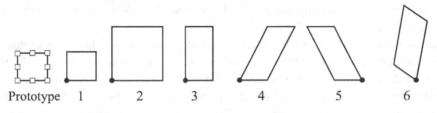

Prototype 1   2   3   4   5   6

Fig. 7.1 Linear mappings of a square

Examples of linear mappings

1. Let $A = \begin{pmatrix} 1 & 0 \\ 0 & 1 \end{pmatrix}$. Remember: The images of the basis are the columns of the matrix. So here $A\left(\begin{pmatrix} 1 \\ 0 \end{pmatrix}\right) = \begin{pmatrix} 1 \\ 0 \end{pmatrix}$, $A\left(\begin{pmatrix} 0 \\ 1 \end{pmatrix}\right) = \begin{pmatrix} 0 \\ 1 \end{pmatrix}$. This mapping fixes the basis and thus all points. It is the identity mapping. The matrix A is called the identity matrix I. This corresponds to the transition from the prototype to (1): Nothing has happened.

2. Let $A = \begin{pmatrix} \lambda & 0 \\ 0 & \lambda \end{pmatrix}$. For $(x_1, x_2) \in \mathbb{R}^2$ we have $Ax = \begin{pmatrix} \lambda & 0 \\ 0 & \lambda \end{pmatrix}\begin{pmatrix} x_1 \\ x_2 \end{pmatrix} = \begin{pmatrix} \lambda x_1 \\ \lambda x_2 \end{pmatrix} = \lambda \begin{pmatrix} x_1 \\ x_2 \end{pmatrix} = \lambda x$.

   Every vector is stretched by the factor λ. This is the transition from (1) to (2). The mapping is called *stretching*.

3. For the mapping $A = \begin{pmatrix} \lambda & 0 \\ 0 & 1 \end{pmatrix}$ the basis vector $\begin{pmatrix} 0 \\ 1 \end{pmatrix}$ remains unchanged, $\begin{pmatrix} 1 \\ 0 \end{pmatrix}$ is stretched by the factor λ. This turns the square (2) into a rectangle (3). In Fig. 7.1 we have $\lambda < 1$, the x_1-axis is shortened.

4. The transition from (3) to (4) becomes a bit more difficult: This is a *shear*. The first basis vector remains unchanged, but the further one goes from the x_1-axis upwards, the more the figure is pulled to the right. Fortunately, we only have to determine the image of the second basis vector, and we can read it from figure (4): $\begin{pmatrix} 0 \\ 1 \end{pmatrix}$ becomes $\begin{pmatrix} \lambda \\ 1 \end{pmatrix}$, the length of the x_2-coordinate remains unchanged. This gives us the matrix $A = \begin{pmatrix} 1 & \lambda \\ 0 & 1 \end{pmatrix}$.

5. The transition from (4) to (5) represents a *reflection* over the x_2-axis. Try to understand for yourself that the matrix $A = \begin{pmatrix} -1 & 0 \\ 0 & 1 \end{pmatrix}$ belongs to this. ◀

For the transition from (5) to (6) I have to go a bit further. The matrix will look a bit more complicated. This is a *rotation*, one of the most important movements of all. In Sect. 10.2 we will find out that movements of bodies in space only consist of rotations and translations, there are no other movements that do not bend the bodies. To represent rotation matrices we need the trigonometric functions cosine and sine.

In their geometric definition, cosine and sine describe side ratios in the right-angled triangle, see Fig. 7.2.

Here $\sin \alpha = b/c$ (opposite /hypotenuse) and $\cos \alpha = a/c$ (adjacent /hypotenuse). These ratios are only dependent on the angle α and not on the size of the triangle. If we draw a circle with radius 1 around the origin (the unit circle) in the two-dimensional Cartesian coordinate system and look at the coordinates of a point on this circle in the first quadrant, we get a right-angled triangle with hypotenuse 1, adjacent x_1 and opposite x_2. Thus the point on the circle has the coordinates $(x_1, x_2) = (\cos \alpha, \sin \alpha)$. If we also

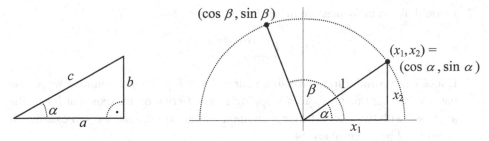

Fig. 7.2 Cosine and sine on the unit circle

Fig. 7.3 Rotation

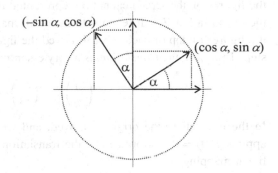

interpret the points of the unit circle in the other quadrants in this way, we can now also define sine and cosine for angles greater than 90°: If β is the angle between the x_1-axis and the vector (x_1, x_2) on the unit circle, we set $\cos \beta := x_1$ and $\sin \beta := x_2$. Cosine and sine can also become negative.

Let us now look at what happens when we rotate at the origin by the angle α:

More examples

6. From Fig. 7.3 one can read that the vector $\begin{pmatrix} 1 \\ 0 \end{pmatrix}$ is rotated into the vector $\begin{pmatrix} \cos \alpha \\ \sin \alpha \end{pmatrix}$, and $\begin{pmatrix} 0 \\ 1 \end{pmatrix}$ becomes $\begin{pmatrix} -\sin \alpha \\ \cos \alpha \end{pmatrix}$.

The matrix of the rotation by the angle α is thus:

$$R_\alpha = \begin{pmatrix} \cos \alpha & -\sin \alpha \\ \sin \alpha & \cos \alpha \end{pmatrix}.$$

For this mapping applies:

$$\begin{pmatrix} x_1 \\ x_2 \end{pmatrix} \mapsto \begin{pmatrix} \cos \alpha \, x_1 - \sin \alpha \, x_2 \\ \sin \alpha \, x_1 + \cos \alpha \, x_2 \end{pmatrix}.$$

Note that we always perform rotations counterclockwise.

7. I would like to introduce you to one last mapping matrix:

$$M_\alpha = \begin{pmatrix} \cos\alpha & \sin\alpha \\ \sin\alpha & -\cos\alpha \end{pmatrix}.$$

It looks very similar to the rotation matrix. From Fig. 7.4 you can read where the basis vectors are mapped. This mapping is a *reflection* on the axis that forms the angle $\alpha/2$ with the x_1-axis. I cannot compute this at the moment, we will calculate it in Sect. 9.2 on eigenvalues. ◄

But what is the exception I mentioned at the beginning of the examples? What change of the figures on the screen cannot be represented by a linear mapping? I cheated a bit with the origin in Fig. 7.1: It remains fixed for all mappings. But in order not to have to draw the figures on top of each other, I moved the figure a bit further to the right from step to step. This shift, the *translation*, is a very essential operation:

$$\begin{pmatrix} x_1 \\ x_2 \end{pmatrix} \mapsto \begin{pmatrix} x_1 \\ x_2 \end{pmatrix} + \begin{pmatrix} a_1 \\ a_2 \end{pmatrix}.$$

In the translation, the origin is moved, and we know that for linear mappings always applies $f(\vec{0}) = \vec{0}$. Unfortunately, the translation requires a special treatment, it is not a linear mapping.

Composition of linear mappings

If A, B are two linear mappings, $A, B \in \mathbb{R}^{2\times2}$, then the composition is possible and $B \circ A$ is again a linear mapping. Which matrix C belongs to it? If we calculate the images of the basis, we obtain the columns of the matrix C. For this, let

$$A = \begin{pmatrix} a_{11} & a_{12} \\ a_{21} & a_{22} \end{pmatrix}, \quad B = \begin{pmatrix} b_{11} & b_{12} \\ b_{21} & b_{22} \end{pmatrix}, \quad C = \begin{pmatrix} c_{11} & c_{12} \\ c_{21} & c_{22} \end{pmatrix}.$$

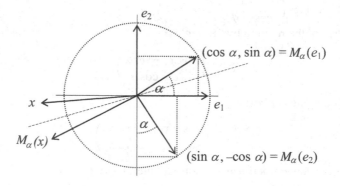

Fig. 7.4 Reflection

It follows that

$$\begin{pmatrix} c_{11} \\ c_{21} \end{pmatrix} = (B \circ A) \begin{pmatrix} 1 \\ 0 \end{pmatrix} = B \left(A \left(\begin{pmatrix} 1 \\ 0 \end{pmatrix} \right) \right) = B \left(\begin{pmatrix} a_{11} \\ a_{21} \end{pmatrix} \right) = \begin{pmatrix} b_{11}a_{11} + b_{12}a_{21} \\ b_{21}a_{11} + b_{22}a_{21} \end{pmatrix},$$

$$\begin{pmatrix} c_{12} \\ c_{22} \end{pmatrix} = (B \circ A) \begin{pmatrix} 0 \\ 1 \end{pmatrix} = B \left(A \left(\begin{pmatrix} 0 \\ 1 \end{pmatrix} \right) \right) = B \left(\begin{pmatrix} a_{12} \\ a_{22} \end{pmatrix} \right) = \begin{pmatrix} b_{11}a_{12} + b_{12}a_{22} \\ b_{21}a_{12} + b_{22}a_{22} \end{pmatrix}.$$

No one can remember this. But there is a trick, with which one can easily determine the matrix C from A and B. For this, you write the matrix A not next to C, but as shown in (7.2), right above it. In the field to the right of B, the matrix C is then created. For the calculation of an entry in the matrix C, exactly the elements that are in the same row of B and in the same column of A are needed. To calculate the element c_{ij}, the elements of the i-th row of B have to be multiplied with those of the j-th column of A and then added up:

$$\begin{pmatrix} & a_{11} & & a_{12} \\ & a_{21} & & a_{22} \\ \begin{pmatrix} b_{11} & b_{12} \\ b_{21} & b_{22} \end{pmatrix} & \begin{pmatrix} b_{11}a_{11} + b_{12}a_{21} & b_{11}a_{12} + b_{12}a_{22} \\ b_{21}a_{11} + b_{22}a_{21} & b_{21}a_{12} + b_{22}a_{22} \end{pmatrix}. \end{pmatrix}$$

(7.2)

Written as a formula, for all i, j:

$$c_{ij} = \sum_{k=1}^{2} b_{ik}a_{kj}.$$

From now on, we write for the composition $B \circ A$, shortly BA, and call this operation *matrix multiplication*.

Examples

1.
$$\begin{pmatrix} 4 & 3 \\ 2 & 1 \end{pmatrix} \qquad \begin{pmatrix} 1 & 2 \\ 2 & 4 \end{pmatrix}$$
$$\begin{pmatrix} 1 & 2 \\ 2 & 4 \end{pmatrix} \begin{pmatrix} 8 & 5 \\ 16 & 10 \end{pmatrix} \qquad \begin{pmatrix} 4 & 3 \\ 2 & 1 \end{pmatrix} \begin{pmatrix} 10 & 20 \\ 4 & 8 \end{pmatrix}.$$

   You can see from this that in general $AB \neq BA$. The order matters for the composition of linear mappings!

2. For a rotation by $45°$, $\cos \alpha = \sin \alpha$ (Fig. 7.5). From the Pythagorean theorem, we know $(\cos \alpha)^2 + (\sin \alpha)^2 = 1$ and from this we obtain here $\cos \alpha = \sin \alpha = 1/\sqrt{2}$. Thus we get the following rotation matrix:

$$R_{45°} = \begin{pmatrix} 1/\sqrt{2} & -1/\sqrt{2} \\ 1/\sqrt{2} & 1/\sqrt{2} \end{pmatrix}.$$

Fig. 7.5 Rotation by 45°

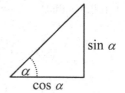

$\cos \alpha$

If we now rotate twice in a row by 45°, we obtain:

$$\begin{pmatrix} 1/\sqrt{2} & -1/\sqrt{2} \\ 1/\sqrt{2} & 1/\sqrt{2} \end{pmatrix} \begin{pmatrix} 1/\sqrt{2} & -1/\sqrt{2} \\ 1/\sqrt{2} & 1/\sqrt{2} \end{pmatrix} = \begin{pmatrix} 0 & -1 \\ 1 & 0 \end{pmatrix} = (R_{45°})^2.$$

This corresponds exactly to a rotation by 90°: $\begin{pmatrix} 1 \\ 0 \end{pmatrix} \mapsto \begin{pmatrix} 0 \\ 1 \end{pmatrix}, \begin{pmatrix} 0 \\ 1 \end{pmatrix} \mapsto \begin{pmatrix} -1 \\ 0 \end{pmatrix}$.

3.
$$\begin{pmatrix} 1 & 0 \\ 0 & 1 \end{pmatrix} \begin{pmatrix} a & b \\ c & d \end{pmatrix} = \begin{pmatrix} a & b \\ c & d \end{pmatrix}.$$

This result is not surprising: the identity mapping does nothing. For the identity matrix I and any other matrix A, it always holds: $IA = AI = A$.

4.
$$\begin{pmatrix} 1/\lambda & 0 \\ 0 & 1/\lambda \end{pmatrix} \begin{pmatrix} \lambda & 0 \\ 0 & \lambda \end{pmatrix} = \begin{pmatrix} 1 & 0 \\ 0 & 1 \end{pmatrix}.$$

A stretching by λ and then by $1/\lambda$ results in the identity.

5. If you take the matrix $R = R_{45°}$ from example 2, you can calculate R^7 and then it holds $R \cdot R^7 = R^8 = I$, this is the rotation by 360°.

   The examples 4 and 5 show us that there are matrices A that have an inverse matrix B, with $BA = E$. But this is not always the case:

6.
$$\begin{pmatrix} 1 & 3 \\ 0 & 0 \end{pmatrix} \begin{pmatrix} a & b \\ c & d \end{pmatrix} = \begin{pmatrix} a+3c & b+3d \\ 0 & 0 \end{pmatrix}.$$

The matrix $\begin{pmatrix} 1 & 3 \\ 0 & 0 \end{pmatrix}$ has no inverse, because no matter what you multiply it with, the second row always reads $(0,0)$, and so I can never come out. This is because the corresponding linear mapping is not bijective and therefore cannot have an inverse mapping. ◀

7.2   Matrices and Linear Mappings from $K^n \to K^m$

After our preparations in the $\mathbb{R}^2$, we now examine general matrices and general linear mappings. K stands for any field, usually it will be the field of real numbers for us. We start with the definition of a $m \times n$-matrix.

▶   **Definition 7.3** The rectangular scheme of elements of the field K

$$\begin{pmatrix} a_{11} & a_{12} & \cdots & a_{1n} \\ a_{21} & a_{22} & & a_{2n} \\ \vdots & & \ddots & \\ a_{m1} & a_{m2} & & a_{mn} \end{pmatrix}$$

is called a $m \times n$-*matrix*. We denote the set of all $m \times n$-matrices with elements from K by $K^{m \times n}$. Here, m is the number of rows and n is the number of columns of the matrix. The terms *column index, row index, column vector, row vector* are transferred literally from Definition 7.1.

If $x = \begin{pmatrix} x_1 \\ \vdots \\ x_n \end{pmatrix} \in K^n$ and $b = \begin{pmatrix} b_1 \\ \vdots \\ b_m \end{pmatrix} \in K^m$, we introduce a shorthand notation: For

$$\begin{pmatrix} a_{11}x_1 + a_{12}x_2 + \cdots + a_{1n}x_n \\ a_{21}x_1 + a_{22}x_2 + \cdots + a_{2n}x_n \\ \vdots \\ a_{m1}x_1 + a_{m2}x_2 + \cdots + a_{mn}x_n \end{pmatrix} = \begin{pmatrix} b_1 \\ b_2 \\ \vdots \\ b_m \end{pmatrix} \tag{7.3}$$

we write from now on:

$$\underbrace{\begin{pmatrix} a_{11} & a_{12} & \cdots & a_{1n} \\ a_{21} & a_{22} & & a_{2n} \\ \vdots & & \ddots & \\ a_{m1} & a_{m2} & & a_{mn} \end{pmatrix}}_{\text{matrix } A} \underbrace{\begin{pmatrix} x_1 \\ x_2 \\ \vdots \\ x_n \end{pmatrix}}_{\text{vector } x} = \underbrace{\begin{pmatrix} b_1 \\ b_2 \\ \vdots \\ b_m \end{pmatrix}}_{\text{vector } b}, \quad \text{therefore } Ax = b. \tag{7.4}$$

We will first study the connection between matrices and linear mappings as in the two-dimensional case. But before that, I would like to point out an interesting fact: In (7.3) there are m linear equations of the form

$$a_{i2}x_1 + a_{i2}x_2 + \cdots + a_{in}x_n = b_i$$

with the n unknowns $x_1, x_2, \ldots, x_n$. The notation $Ax = b$ also stands as a shorthand for a system of linear equations. In the next chapter, we will systematically solve such systems of equations using this matrix notation.

But now to the linear mappings: The proof of the following theorem goes exactly as in Theorem 7.2:

▶ **Theorem 7.4** For every linear mapping $f: K^n \to K^m$ there is exactly one matrix $A \in K^{m \times n}$ with the property $f(x) = Ax$ for all $x \in K^n$. Conversely, every matrix $A \in K^{m \times n}$ defines a linear mapping $f: K^n \to K^m$ by the rule

$$f: K^n \to K^m, \quad x \mapsto Ax = \begin{pmatrix} a_{11}x_1 + a_{12}x_2 + \cdots + a_{1n}x_n \\ a_{21}x_1 + a_{22}x_2 + \cdots + a_{2n}x_n \\ \vdots \\ a_{m1}x_1 + a_{m2}x_2 + \cdots + a_{mn}x_n \end{pmatrix}.$$

Again, the columns of the matrix contain the images of the basis vectors, which, as we know, completely determine the mapping. For example,

$$\begin{pmatrix} a_{11} & a_{12} & \cdots & a_{1n} \\ a_{21} & a_{22} & & a_{2n} \\ \vdots & & \ddots & \\ a_{m1} & a_{m2} & & a_{mn} \end{pmatrix} \begin{pmatrix} 1 \\ 0 \\ \vdots \\ 0 \end{pmatrix} = \begin{pmatrix} a_{11}1 + a_{12}0 + \cdots + a_{1n}0 \\ a_{21}1 + a_{22}0 + \cdots + a_{2n}0 \\ \vdots \\ a_{m1}1 + a_{m2}0 + \cdots + a_{mn}0 \end{pmatrix} = \begin{pmatrix} a_{11} \\ a_{21} \\ \vdots \\ a_{m1} \end{pmatrix}.$$

One has to be careful not to confuse m and n: For a mapping from $K^n \to K^m$ one needs a matrix from $K^{m \times n}$. The number of rows of the matrix, m, determines the number of elements in the image vector, i.e. the dimension of the image space. But I can console you, we will mostly deal with the case $m = n$.

As in the two-dimensional case, we will identify the matrices and their corresponding linear mappings in the future.

The identity mapping $id: K^n \to K^n$ belongs to the identity matrix

$$c = \begin{pmatrix} 1 & 0 & \cdots & 0 \\ 0 & 1 & & \\ \vdots & & \ddots & 0 \\ 0 & & 0 & 1 \end{pmatrix}.$$

Note that the identity matrix is always square! There is no identity mapping from K^m to K^n, if n and m are different.

Examples of linear mappings

1. $A = \begin{pmatrix} 1 & 0 & 0 \\ 0 & \cos \alpha & -\sin \alpha \\ 0 & \sin \alpha & \cos \alpha \end{pmatrix}$. It maps:

$$\begin{pmatrix} x_1 \\ 0 \\ 0 \end{pmatrix} \mapsto \begin{pmatrix} x_1 \\ 0 \\ 0 \end{pmatrix} \quad \text{and} \quad \begin{pmatrix} 0 \\ x_2 \\ x_3 \end{pmatrix} \mapsto \begin{pmatrix} 0 \\ \cos \alpha\, x_2 - \sin \alpha\, x_3 \\ \sin \alpha\, x_2 + \cos \alpha\, x_3 \end{pmatrix}.$$

The x_1-axis remains fixed and the x_2-x_3-plane is rotated by the angle α. Compare this with the rotation in the $\mathbb{R}^2$ (Example 6 in Sect. 7.1). Overall, the space is rotated by A around the x_1-axis. Find the matrices yourself that describe rotations around the x_2- or around the x_3-axis.

2. We want to determine the matrix for the mapping

$$f \colon \mathbb{R}^3 \to \mathbb{R}^2, \quad \begin{pmatrix} x_1 \\ x_2 \\ x_3 \end{pmatrix} \mapsto \begin{pmatrix} x_1 + \frac{1}{2}x_3 \\ x_2 + \frac{1}{2}x_3 \end{pmatrix}$$

. The matrix must therefore be from $\mathbb{R}^{2\times 3}$. We look for the images of the basis:

$$f\left(\begin{pmatrix} 1 \\ 0 \\ 0 \end{pmatrix}\right) = \begin{pmatrix} 1 \\ 0 \end{pmatrix}, \quad f\left(\begin{pmatrix} 0 \\ 1 \\ 0 \end{pmatrix}\right) = \begin{pmatrix} 0 \\ 1 \end{pmatrix}, \quad f\left(\begin{pmatrix} 0 \\ 0 \\ 1 \end{pmatrix}\right) = \begin{pmatrix} \frac{1}{2} \\ \frac{1}{2} \end{pmatrix},$$

from this follows

$$A = \begin{pmatrix} 1 & 0 & \frac{1}{2} \\ 0 & 1 & \frac{1}{2} \end{pmatrix}.$$

You see: If you write the images of the basis next to each other, a matrix of the right size automatically comes out. You don't have to pay attention to the number of rows and columns anymore.

What does this mapping do? Let us imagine the x_1-x_2-plane as our drawing plane. All points of this plane remain unchanged. Everything that lies in front of or behind it is projected onto this plane and shifted a bit. From a three-dimensional cube of edge length 1, which has a corner at the origin, the shape in Fig. 7.6 (which I can draw in $\mathbb{R}^2$) is obtained. ◄

Since the screens of our computers will probably only produce two-dimensional representations for some more years, we have to map all three-dimensional objects that we want to view on the screen in some way into the $\mathbb{R}^2$. Here you have learned a simple mapping. A common projection is the one shown in Fig. 7.7, which gives a somewhat more realistic view. This is also generated by a linear mapping. Try to find the matrix for this yourself.

In graphical data processing, many other projections are common, for example central projections. Not all of them can be represented by linear mappings. You will find an exercise on this at the end of the chapter.

Fig. 7.6 Projection 1

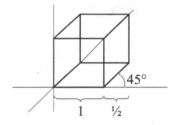

Fig. 7.7 Projection 2

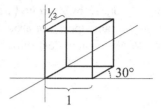

Matrix multiplication and composition of linear mappings

If $A: K^n \rightarrow K^m$ and $B: K^m \rightarrow K^r$ are linear mappings, then the composition is possible and also $B \circ A: K^n \rightarrow K^r$ is linear. The corresponding matrix C is called the product of the matrices B and $A: C = BA$. The matrix C can be calculated in the same way as in the two-dimensional case described in (Eq. 7.2):

Let $B \in K^{r \times m}$ and $A \in K^{m \times n}$. Then $C = BA \in K^{r \times n}$:

$$
\begin{pmatrix}
b_{11} & \cdots & b_{1m} \\
\vdots & & \vdots \\
b_{i1} & \cdots & b_{im} \\
\vdots & & \vdots \\
b_{r1} & \cdots & b_{rm}
\end{pmatrix}
\begin{pmatrix}
a_{11} & \cdots & a_{1j} & \cdots & a_{1n} \\
\vdots & & \vdots & & \vdots \\
a_{m1} & \cdots & a_{mj} & \cdots & a_{mn}
\end{pmatrix}
$$

$\longrightarrow b_{i1}a_{1j} + \cdots + b_{im}a_{mj}$ $\left.\right\}$ r rows

$\underbrace{\qquad\qquad\qquad}_{n \text{ columns}}$

As a formula, for all $i = 1, \ldots, r$ and for all $j = 1, \ldots, n$:

$$c_{ij} = \sum_{k=1}^{m} b_{ik}a_{kj}. \qquad (7.5)$$

When you multiply two matrices, you don't have to think long about row and column numbers, you will see: If the matrices don't fit together, the procedure fails and, if it works, the right size comes out automatically.

Example

$$
\begin{pmatrix}
0 & 3 & 4 & -1 \\
2 & 0 & 1 & 1 \\
0 & 1 & 0 & -2
\end{pmatrix}
$$

$$
\begin{pmatrix}
0 & 3 & 4 \\
2 & 0 & 1 \\
0 & 1 & 0
\end{pmatrix}
\begin{pmatrix}
6 & 4 & 3 & -5 \\
0 & 7 & 8 & -4 \\
2 & 0 & 1 & 1
\end{pmatrix} \blacktriangleleft
$$

The multiplication of a matrix with a vector, $Ax = b$, which we introduced in (7.4), is nothing but a special case of this matrix multiplication. We can also regard the column vector $x \in K^n$ as a matrix $x \in K^{n \times 1}$, and then we get:

$$\underbrace{\begin{pmatrix} a_{11} & a_{12} & \cdots & a_{1n} \\ a_{21} & a_{22} & & a_{2n} \\ \vdots & & \ddots & \\ a_{m1} & a_{m2} & & a_{mn} \end{pmatrix}}_{A \in K^{m \times n}} \underbrace{\begin{pmatrix} x_1 \\ x_2 \\ \vdots \\ x_n \end{pmatrix}}_{x \in K^{n \times 1}} = \underbrace{\begin{pmatrix} a_{11}x_1 + a_{12}x_2 + \cdots + a_{1n}x_n \\ a_{21}x_1 + a_{22}x_2 + \cdots + a_{2n}x_n \\ \vdots \\ a_{m1}x_1 + a_{m2}x_2 + \cdots + a_{mn}x_n \end{pmatrix}}_{b \in K^{m \times 1}} = \begin{pmatrix} b_1 \\ b_2 \\ \vdots \\ b_m \end{pmatrix}. \quad (7.6)$$

But note that this only works if we consistently use column vectors and multiply the vector from the right.

I would like to define a few more operations for matrices that we will need later:

▶ **Definition 7.5** For two matrices $A = (a_{ij})$, $B = (b_{ij}) \in K^{m \times n}$ and $\lambda \in K$, let

$$A + B = \begin{pmatrix} a_{11} + b_{11} & \cdots & a_{1n} + b_{1n} \\ \vdots & \ddots & \vdots \\ a_{m1} + b_{m1} & \cdots & a_{mn} + b_{mn} \end{pmatrix}, \quad \lambda A = \begin{pmatrix} \lambda a_{11} & \cdots & \lambda a_{1n} \\ \vdots & \ddots & \vdots \\ \lambda a_{m1} & \cdots & \lambda a_{mn} \end{pmatrix}.$$

So the components are simply added or multiplied by the factor λ. With these definitions, $K^{m \times n}$ becomes obviously a $m \cdot n$ dimensional K-vector space. If we think of the components not in the rectangle but written one after the other, we have just the K^{mn}. For the operations just defined, all the rules that we derived for vector spaces apply. There are a few more that are related to the multiplication:

▶ **Theorem 7.6: Rules for matrices** For matrices that fit together by their size, the following rules hold:

$$A + B = B + A,$$
$$(A + B)C = AC + BC,$$
$$A(B + C) = AB + AC,$$
$$(AB)C = A(BC),$$
$$AB \neq BA \quad \text{(generally)}.$$

The first rule follows already from the vector space property that we just noticed. The distributive laws can be easily verified with the help of the multiplication rule (7.5). If you want to check the associativity law in this way, you can get terribly tangled up with the indices. But there is another way! Let us remember the double nature of matrices:

They are also linear mappings, and mappings are always associative, as we saw in (5.2) in Sect. 5.1 after Theorem 5.4. So there is nothing to do here. The last rule or rather non-rule you already know. Here it is again as a reminder.   □

We now focus for a moment on square matrices, i.e. on matrices that describe linear mappings of a space into itself. We already know that some of these matrices are invertible. In the following theorems, I would like to summarize what we can say so far about invertible matrices. In doing so, we will harvest some fruits from Chap. 6:

▶   **Theorem 7.7: On the existence of an inverse matrix** Let $A \in K^{n \times n}$ be the matrix of a linear mapping. Then the following conditions are equivalent:

(I1)   There is a matrix $A^{-1} \in K^{n \times n}$ with the property $A^{-1}A = AA^{-1} = I$.
(I2)   The mapping $A \colon K^n \to K^n$ is bijective.
(I3)   The columns of the matrix A form a basis of the K^n.
(I4)   The columns of the matrix A are linearly independent.

(I1) says nothing else than that there is an inverse mapping for A. See for this Theorem 1.19 in Sect. 1.3. Thus, A is bijective. And since every bijective mapping f of a set onto itself has an inverse mapping g for which $f \circ g = g \circ f = id$ holds, (I1) and (I2) are equivalent.
     Theorem 6.25 states that a bijective linear mapping maps a basis onto a basis. The columns of the matrix are precisely the images of the basis. Conversely, if the columns form a basis, this means that A maps a basis onto a basis and from Theorem 6.26 it follows that A is an isomorphism, i.e. bijective. Thus we have the equivalence of (I2) and (I3).
     From (I4) of course (I3) follows and vice versa, n linearly independent vectors in an n-dimensional vector space always form a basis (Theorem 6.21), so (I3) and (I4) are also equivalent.   □

▶   **Theorem 7.8** The set of invertible matrices in $K^{n \times n}$ forms a group with respect to multiplication.

If you look up the group axioms in Definition 5.2 in sect. 5.1, you will find that we have already computed (G1) to (G3). The only thing still missing is that the set is closed under the operation. Is the product of two invertible matrices again invertible? Yes, and we can also specify the inverse matrix for AB. It follows from Theorem 5.3b that $B^{-1}A^{-1}$ is the inverse matrix for AB.   □

The order is very crucial here! $A^{-1}B^{-1}$ is in general different from $B^{-1}A^{-1}$ and then is not inverse to AB.

Computing the inverse of a matrix is usually very difficult. But we will soon learn an algorithm for it. For a 2×2-matrix we can still do the calculation completely by hand: Let $A = \begin{pmatrix} a & b \\ c & d \end{pmatrix}$ be given. If an inverse matrix $\begin{pmatrix} e & f \\ g & h \end{pmatrix}$ exists, then it must hold:

$$\begin{pmatrix} a & b \\ c & d \end{pmatrix} \begin{pmatrix} e & f \\ g & h \end{pmatrix} = \begin{pmatrix} ae+bg & af+bh \\ ce+dg & cf+dh \end{pmatrix} = \begin{pmatrix} 1 & 0 \\ 0 & 1 \end{pmatrix}.$$

We thus obtain four equations for the 4 unknowns e, f, g, h. These can be solved and we obtain:

$$\begin{pmatrix} e & f \\ g & h \end{pmatrix} = \frac{1}{ad-bc} \begin{pmatrix} d & -b \\ -c & a \end{pmatrix}. \tag{7.7}$$

This works of course only if $ad - bc \neq 0$. And indeed: Exactly when $ad - bc \neq 0$, the inverse matrix exists and has the form given in (7.7).

7.3   The Rank of a Matrix

Let us return to the not necessarily square matrices of the $K^{m \times n}$. We have already seen several times that the column vectors of the matrices play a special role. We now want to deal with the space that these vectors span. First, an important term:

▶   **Definition 7.9** The *rank* of a matrix A is the maximum number of linearly independent column vectors in the matrix.

The rank is thus the dimension of the space spanned by the column vectors. We get a first insight into the meaning of the term in the following theorem. Recall that $Ax = b$ not only means the image of x under the linear mapping A, but also the shorthand for a linear system of equations (compare (7.3) and (7.4) after Definition 7.3). From the rank of the matrix we can then read how large the solution space of the system of equations $Ax = 0$ is. Later we will also conclude from this the solutions of the system $Ax = b$.

▶   **Theorem 7.10** If $f \colon K^n \to K^m$ is a linear mapping with associated matrix $A \in K^{m \times n}$ with the columns $(s_1, s_2, \ldots, s_n)$, then:

a) $\operatorname{im} f = \operatorname{span}\{s_1, s_2, \ldots, s_n\}$.
b) $\ker f = \{x \mid Ax = 0\}$ is the set of solutions of the system of equations $Ax = 0$.
c) $\dim \operatorname{im} f = \operatorname{rank} A$.
d) $\dim \ker f = n - \operatorname{rank} A$.

For a): Since the s_i are all in the image and the image is a vector space, we have of course $\operatorname{im} f \supset \operatorname{span}\{s_1, s_2, \ldots, s_n\}$. On the other hand, we have seen in Theorem 7.4 that

$$f(x) = Ax = \begin{pmatrix} a_{11}x_1 + a_{12}x_2 + \cdots + a_{1n}x_n \\ a_{21}x_1 + a_{22}x_2 + \cdots + a_{2n}x_n \\ \vdots \\ a_{m1}x_1 + a_{m2}x_2 + \cdots + a_{mn}x_n \end{pmatrix} = s_1x_1 + s_2x_2 + \cdots s_nx_n$$

holds, so every image element $f(x)$ can be written as a linear combination of the column vectors and thus we also have $\operatorname{im} f \subset \operatorname{span}\{s_1, s_2, \ldots, s_n\}$.

Point b) is exactly the definition of the kernel. Interesting here is the interpretation: The kernel of the linear mapping is the solution set of the corresponding linear system of equations. Point c) is an immediate consequence of a), since the rank is precisely the dimension of $\operatorname{span}\{s_1, s_2, \ldots, s_n\}$.

For d) we use Theorem 6.27: The sum of the dimensions of kernel and image gives the dimension of the domain of definition.   □

Let us calculate the rank for a few matrices:

Examples

1. If $A \in K^{n \times n}$ is invertible, then Theorem 7.7 immediately implies $\operatorname{rank} A = n$.

2. $A = \begin{pmatrix} 1 & 1 & 1 \\ 0 & 1 & 1 \\ 0 & 0 & 1 \end{pmatrix}$ has rank 3: In the span of the first column vector, the second and third component are always equal to 0, so the second column is not in the span. Likewise, in the span of the first two columns, the third component is always equal to 0 and therefore the third column is linearly independent of the first two columns.

3. This can be generalized: A square $n \times n$-matrix is called an *upper triangular matrix*, if all elements below the diagonal are equal to 0. The rank of any upper triangular matrix, in which all diagonal elements have the value 1, is n:

$$\operatorname{rank} \underbrace{\begin{pmatrix} 1 & * & \cdots & * \\ 0 & 1 & \ddots & \vdots \\ \vdots & \ddots & \ddots & * \\ 0 & \cdots & 0 & 1 \end{pmatrix}}_{n \text{ columns}} = n.$$

4. $\begin{pmatrix} 1 & 0 & 1 \\ 0 & 1 & 1 \\ 2 & 0 & 2 \end{pmatrix}$ has rank 2: $s_1 + s_2 = s_3$.

5. $\begin{pmatrix} 1 & 2 & 4 \\ 2 & 4 & 8 \\ 3 & 6 & 12 \end{pmatrix}$ has rank 1, because the second and third column are multiples of the first column.

6. $\begin{pmatrix} 1 & 3 & 0 & 4 \\ 2 & 5 & 1 & 0 \\ 3 & 8 & 1 & 4 \end{pmatrix}$. By looking at it, we don't see anything at first. ◀

Let us swap for a moment the role of columns and rows and calculate the maximum number of linearly independent rows of the matrices, the "row rank": for example 1 we can't say anything yet. In example 2 and 3 we get with the same argument again 3 and n respectively. In example 4 the third row is twice the first, so the row rank is 2. In example 5 the rows are also multiples of each other, so row rank 1. And in the last example we now see that the third row is the sum of the first two: so row rank 2.

Of course you notice that in cases 2 to 5 always row rank = column rank. And of course this is not a coincidence, but it is always the case. That's why in example 1 the rows also form a basis, and that's why in example 6 the "true" rank is also 2.

I find this fact so astonishing that I think one can only believe it when one has checked it a few times. Why don't you try to combine the 3rd and 4th column vector of example 6 linearly from the first two. Somehow it has to work!

Of course we need a theorem here, and the proof is not very difficult, but a rather tricky index fiddling:

▶ **Theorem 7.11** For any $m \times n$-matrix A the rank is equal to the maximum number of linearly independent rows of the matrix: "column rank = row rank".

This statement is often very useful, as we have already seen in example 6: One can always determine the rank that is easier to calculate.

In the following proof, I will write field elements in lowercase and all vectors in uppercase. Let

$$A = \begin{pmatrix} a_{11} & \cdots & a_{1n} \\ \vdots & \ddots & \vdots \\ a_{m1} & \cdots & a_{mn} \end{pmatrix},$$

$R_1, R_2, \ldots, R_m$ be the rows and $C_1, C_2, \ldots C_n$ the columns of A. Let the row rank be r and $B_1 = (b_{11}, \ldots, b_{1n})$, $B_2 = (b_{21}, \ldots, b_{2n})$, $\cdots$, $B_r = (b_{r1}, \ldots, b_{rn})$ be a basis of the space spanned by the rows. The row vectors can therefore be linearly combined from the vectors $B_1, \ldots, B_r$:

$$R_1 = k_{11}B_1 + k_{12}B_2 + \cdots + k_{1r}B_r$$
$$R_2 = k_{21}B_1 + k_{22}B_2 + \cdots + k_{2r}B_r$$
$$\vdots \tag{7.8}$$
$$R_m = k_{m1}B_1 + k_{m2}B_2 + \cdots + k_{mr}B_r$$

Each of the m equations from (7.8) is a vector equation, which can also be written in coordinate notation. I write for example the first one of them:

$$\underbrace{\begin{pmatrix} a_{11} \\ a_{12} \\ \vdots \\ a_{1i} \\ \vdots \\ a_{1n} \end{pmatrix}}_{R_1} = k_{11} \underbrace{\begin{pmatrix} b_{11} \\ b_{12} \\ \vdots \\ b_{1i} \\ \vdots \\ b_{1n} \end{pmatrix}}_{B_1} + k_{12} \underbrace{\begin{pmatrix} b_{21} \\ b_{22} \\ \vdots \\ b_{2i} \\ \vdots \\ b_{2n} \end{pmatrix}}_{B_2} + \cdots + k_{1r} \underbrace{\begin{pmatrix} b_{r1} \\ b_{r2} \\ \vdots \\ b_{ri} \\ \vdots \\ b_{rn} \end{pmatrix}}_{B_r} \tag{7.9}$$

From each of the rows of (7.8) we now pick out the i-th component (for the row R_1 I have marked this in (7.9)), and so we get a new set of m equations:

$$a_{1i} = k_{11}b_{1i} + k_{12}b_{2i} + \cdots + k_{1r}b_{ri}$$
$$a_{2i} = k_{21}b_{1i} + k_{22}b_{2i} + \cdots + k_{2r}b_{ri}$$
$$\vdots \tag{7.10}$$
$$a_{mi} = k_{m1}b_{1i} + k_{m2}b_{2i} + \cdots + k_{mr}b_{ri}.$$

On the left side of the $=$-signs in (7.10) stands just the i-th column vector C_i and we can write (7.10) as a new vector equation:

$$C_i = \begin{pmatrix} a_{1i} \\ a_{2i} \\ \vdots \\ a_{mi} \end{pmatrix} = \underbrace{\begin{pmatrix} k_{11} \\ k_{21} \\ \vdots \\ k_{m1} \end{pmatrix}}_{K_1} b_{1i} + \underbrace{\begin{pmatrix} k_{12} \\ k_{22} \\ \vdots \\ k_{m2} \end{pmatrix}}_{K_2} b_{2i} + \cdots + \underbrace{\begin{pmatrix} k_{1r} \\ k_{2r} \\ \vdots \\ k_{mr} \end{pmatrix}}_{K_r} b_{ri}. \tag{7.11}$$

Now it is done: We have in (7.11) the column C_i linearly combined from the newly defined vectors $K_1, K_2, \ldots, K_r$. What we have done in (7.9) with the index i, we can also do with all other indices from 1 to n, and so we get that every column vector C_i can be linearly combined from the vectors $K_1, K_2, \ldots, K_r$. This means that the dimension of the column space is in any case $\leq r$ and thus column rank $\leq$ row rank.

The problem is symmetric, however: If we swap columns and rows in the proof, we get just as well: row rank $\leq$ column rank. This finally means that column rank = row rank, and from now on there is only *one* rank of a matrix.   □

I conclude with a theorem for square matrices:

▶   **Theorem 7.12** A matrix $A \in K^{n \times n}$ is invertible if and only if rank $A = n$.

This is an immediate consequence of Theorem 7.7: If rank $A = n$, then the columns form a basis, A is invertible. And if A has an inverse, then again the columns form a basis, so rank $A = n$.   □

7.4   Comprehension Questions and Exercises

Comprehension questions

1. True or false: If A and B are arbitrary matrices, then the multiplication is not commutative, but the products $A \cdot B$ and $B \cdot A$ can always be calculated.
2. If the n basis vectors in $\mathbb{R}^n$ are stretched by different factors, is the resulting mapping a linear mapping?
3. Why can a simple translation by a fixed vector in the $\mathbb{R}^3$ not be described by a linear mapping?
4. Can two non-square matrices be multiplied so that a (square) identity matrix results? If so: What does this mean for the linear mappings associated with the matrices?
5. Is there a difference between the multiplication $\langle$matrix$\rangle \cdot \langle$vector$\rangle$ and $\langle$matrix$\rangle \cdot \langle$single column matrix$\rangle$?
6. Why is matrix multiplication associative?
7. Is the set of equal-sized square matrices with the multiplication a group? Is the set of equal-sized matrices with the addition a group?

Exercises

1. In Fig. 7.8 you see a central projection sketched, with the help of which objects of the space can be mapped into the $\mathbb{R}^2$. The points of the $\mathbb{R}^3$, which are to be projected, are connected with the projection center, which is located at the point $(0, 0, 1)$.
   The projection plane is the x-y-plane. The intersection point of the connecting line with the projection plane gives the point to be represented. Calculate where the point (x, y, z) is mapped to by this projection. Is this mapping defined for all points of the $\mathbb{R}^3$? Is it a linear mapping?

Fig. 7.8 Central projection

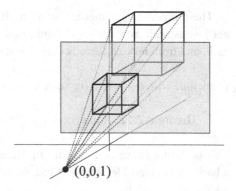

$(0,0,1)$

2. The vectors $\begin{pmatrix} 3 \\ 5 \end{pmatrix}$ and $\begin{pmatrix} 2 \\ 4 \end{pmatrix}$ form a basis B of $\mathbb{R}^2$. Let $\begin{pmatrix} x \\ y \end{pmatrix} \in \mathbb{R}^2$. Calculate the coordinates of $\begin{pmatrix} x \\ y \end{pmatrix}$ with respect to the basis B.

3. Determine the matrix of the mapping that maps $\begin{pmatrix} 1 \\ 0 \end{pmatrix}$ to $\begin{pmatrix} 3 \\ 5 \end{pmatrix}$ and $\begin{pmatrix} 0 \\ 1 \end{pmatrix}$ to $\begin{pmatrix} 2 \\ 4 \end{pmatrix}$ and calculate the inverse of this matrix.

4. Determine the matrices of the mappings
   a) $f : \mathbb{R}^3 \to \mathbb{R}^2, (x, y, z) \mapsto (x + y, y + z)$
   b) $g : \mathbb{R}^2 \to \mathbb{R}^3, (x, y) \mapsto (x, x + y, y)$.

5. Perform the following matrix multiplications:

   a) $\begin{pmatrix} -2 & 3 & 1 \\ 6 & -9 & -3 \\ 4 & -6 & -2 \end{pmatrix} \cdot \begin{pmatrix} 3 & 1 & 1 \\ 2 & 0 & 1 \\ 0 & 2 & -1 \end{pmatrix}$

   b) $\begin{pmatrix} 1 & 0 & 0 \\ 0 & 1 & 0 \\ 4 & 0 & 0 \end{pmatrix} \cdot \begin{pmatrix} 1 & 0 & 0 \\ 0 & 1 & 0 \\ 4 & 0 & 0 \end{pmatrix}$

   c) $\begin{pmatrix} 1 & 1 \\ 1 & 1 \end{pmatrix} \cdot \begin{pmatrix} 1 & 2 \\ 0 & 4 \end{pmatrix}$

   d) $\begin{pmatrix} 1 & 1 \\ 1 & 1 \end{pmatrix} \cdot \begin{pmatrix} 0 & 3 \\ 1 & 3 \end{pmatrix}$

6. Determine the matrices and their rank of the following linear mappings.

   a) $f : \mathbb{R}^2 \to \mathbb{R}^2, \begin{pmatrix} x_1 \\ x_2 \end{pmatrix} \mapsto \begin{pmatrix} x_1 - 2x_2 \\ 0 \end{pmatrix}$.

   b) $f : \mathbb{R}^3 \to \mathbb{R}^2, \begin{pmatrix} x_1 \\ x_2 \\ x_3 \end{pmatrix} \mapsto \begin{pmatrix} x_1 - x_2 + x_3 \\ 2x_1 - 2x_2 + 2x_3 \end{pmatrix}$.

7. Show that for matrices $A, B \in K^{n \times m}$ and $C \in K^{m \times r}$ holds:

$$(A + B)C = AC + BC.$$

Gaussian Algorithm and Linear Equations

8

Abstract

If you have worked through this chapter,

- you know what a system of linear equations is and can interpret the solutions geometrically,
- you can write a system of linear equations in matrix notation,
- you master the Gaussian algorithm and can apply it to
- solve systems of linear equations and determine the inverse of matrices.

Systems of linear equations occur in many engineering and economic problems. In the last chapters we came across systems of linear equations in connection with matrices and linear mappings. It now turns out that the methods we have learned there are suitable for developing systematic solution methods for such systems of equations. The focus is on the Gaussian algorithm. This is a method with which a matrix is transformed in such a way that the solutions of the associated system of linear equations can be read off directly.

8.1 The Gaussian Algorithm

Our most important application of this algorithm will be solving equations. But we will also compute other things with it. If we apply the algorithm to a matrix, we can read at the end:

Supplementary Information The online version contains supplementary material available at https://doi.org/10.1007/978-3-658-40423-9_8.

- the rank of the matrix A,
- the solutions of the system of equations $Ax = 0$, later also the solutions of $Ax = b$,
- with a small extension: the inverse of a matrix,
- and—in anticipation of the next chapter—the determinant of the matrix.

The transformation of the matrices is carried out with the help of elementary row transformations:

▶   **Definition 8.1: Elementary row transformations** The following operations on matrices are called *elementary row transformations*:

1. Interchange two rows.
2. Add λ times row r_j to row r_i $(i \neq j)$.
3. Multiply a row by the scalar $\lambda \neq 0$.

Let's start with determining the rank of a matrix. For this it holds:

▶   **Theorem 8.2** Elementary row transformations do not change the rank of a matrix.

In the proof it is again shown how useful it is that the rank of a matrix can also be determined by the rows:

a) is clear, because the row space does not change.

To b): Show that $\operatorname{span}(r_1, r_2, \ldots, r_i, \ldots, r_m) = \operatorname{span}(r_1, r_2, \ldots, r_i + \lambda r_j, \ldots, r_m)$:
"⊂": for $u \in \operatorname{span}(r_1, r_2, \ldots, r_i, \ldots, r_m)$ it holds:

$$u = \lambda_1 r_1 + \cdots + \lambda_i r_i + \cdots + \lambda_j r_j + \cdots + \lambda_m r_m$$
$$= \lambda_1 r_1 + \cdots + \lambda_i (r_i + \lambda r_j) + \cdots + (\lambda_j - \lambda_i \lambda) r_j + \cdots + \lambda_m r_m$$
$$\in \operatorname{span}(r_1, \ldots, r_i + \lambda r_j, \ldots, r_j, \ldots, r_m).$$

And vice versa:

"⊃": $u = \lambda_1 r_1 + \cdots + \lambda_i (r_i + \lambda r_j) + \cdots + \lambda_m r_m \in \operatorname{span}(r_1, r_2, \ldots, r_i, \ldots, r_m)$.
Part c) is similar, I do not present it.   □

For the time being, we only need the transformations a) and b) to determine the rank.

Now we come to the Gaussian algorithm. Do you remember that you could immediately read the rank from an upper triangular matrix with ones on the diagonal? We will try to transform the given matrix A into something similar.

While you read the description of this algorithm, you should solve it on paper at the same time. Use the following matrix as an exercise:

$$\begin{pmatrix} 1 & 2 & 0 & 1 \\ 1 & 2 & 2 & 3 \\ 4 & 8 & 2 & 6 \\ 3 & 6 & 4 & 8 \end{pmatrix} \tag{8.1}$$

We begin the algorithm with the element a_{11} of the $m \times n$-matrix A:

$$\begin{pmatrix} a_{11} & \cdots & a_{1n} \\ \vdots & \ddots & \vdots \\ a_{m1} & \cdots & a_{mn} \end{pmatrix}.$$

If $a_{11} \neq 0$, we subtract (a_{i1}/a_{11}) times the first row from the i-th row for all $i > 1$.
This makes the first element of row i a 0:

$$a_{i1} \mapsto a_{i1} - (a_{i1}/a_{11})a_{11} = 0, \tag{8.2}$$

and then the matrix A looks as follows:

$$\begin{pmatrix} a_{11} & a_{12} & \cdots & a_{1n} \\ 0 & a'_{22} & \cdots & a'_{2n} \\ \vdots & \vdots & \ddots & \vdots \\ 0 & a'_{m2} & \cdots & a'_{mn} \end{pmatrix} \tag{8.3}$$

If $a_{11} = 0$, we first look for a row whose first element is not 0. If there is such a row, we swap it with the first row and then carry out operation (8.2). After that, A also has the form (8.3), only with a different first row.

If all elements of the first column are 0, we examine the second column and begin the algorithm in this column with the element a_{12}, that is, we try again to make the elements below a_{12} to 0, possibly after a row swap. If this does not work again, we take the third column, and so on. If the matrix is not the zero matrix, the attempt will be successful at some point.

In any case, the matrix then also has the form (8.3), possibly with some zero columns on the left.

Now the first step is over. We go one column to the right and one row down. We call the element at this point b and start the process again, now with the element b.

It may be the case again that $b = 0$ and we have to swap the current row with another row or go one column further to the right. Here you have to be careful: we are only allowed to swap rows that are below b, otherwise something would get mixed up again in the columns to the left of b, which we have just nicely transformed.

The process ends when we arrive at the last line or when we would leave the matrix when progressing to the next element (to the right or to the right and down), that is, when there are no more columns.

Finally, the matrix has a shape that looks similar to that in (8.4):

$$\begin{pmatrix} a & * & & & & \cdots & & & * \\ 0 & b & * & & & \cdots & & & * \\ 0 & 0 & 0 & c & * & & \cdots & & * \\ 0 & 0 & 0 & 0 & d & * & \cdots & & * \\ \vdots & & \vdots & & 0 & 0 & \cdots & & 0 \\ & & & & & & & & \vdots \\ 0 & & & \cdots & & & & & 0 \end{pmatrix}. \tag{8.4}$$

The numbers $a, b, c, d, \ldots$ are field elements different from 0. The entries designated by $*$ have any arbitrary values.

In each row, the first element different from 0 is called the *leading coefficient* of the row. Lines without leading coefficients can only be at the bottom of the matrix.

If you have carried out the process with the exercise matrix (8.1), you have obtained the matrix

$$\begin{pmatrix} 1 & 2 & 0 & 1 \\ 0 & 0 & 2 & 2 \\ 0 & 0 & 0 & 1 \\ 0 & 0 & 0 & 0 \end{pmatrix} \tag{8.5}$$

as a result and have run through each alternative of the algorithm at least once. The algorithm ends in this case because there is no further element to the right of 1.

A matrix in the form (8.4) is called a matrix in *row-echelon form*. You see the staircase that I have drawn. The number of steps in this staircase is the rank of the matrix or, expressed more distinguished:

▶   **Theorem 8.3** The rank of a matrix in row-echelon form is exactly the number of leading coefficients.

We already know the argument from the upper triangular matrices in Example 3 in Sect. 7.3. We examine the columns in which leading coefficients are located. The i-th such column is always linearly independent of the previous $i-1$ columns with leading coefficients, since the i-th coordinate was 0 in all previous column vectors. So the columns with leading coefficients are linearly independent.

However, the dimension of the column space cannot be greater than the number of leading coefficients: If there are k leading coefficients, there are not more than k linearly independent rows, and as we know the dimension of the row space is equal to that of the column space.   □

In the literature, the leading coefficients are often normalized to 1 during the execution of the Gaussian algorithm. This may facilitate the calculation by hand, but it means a whole series of more multiplications. But since today nobody solves linear equation systems manually anymore, it is important to pay attention to numerical efficiency. Therefore, the algorithm will be implemented without this.

You may have noticed while reading the algorithm that it contains a recursion. It was quite tedious to describe the process with all possible special cases. In the following recursive formulation you will see again what a powerful tool we have in our hands. It goes very quickly:

I call the recursive function gauss(i, j) and write it down in pseudocode. The function has as parameters the row and column index. It starts with $i = j = 1$, in each step either j, or i and j is increased. The procedure ends when i is equal to the number of rows or j is greater than the number of columns:

> **gauss(i, j):**
>
> if i = number of rows or j > number of columns:
>
>   end.
>
> if $a_{ij} = 0$:
>
>   look for $a_{kj} \neq 0$, $k > i$, if there is none: **gauss$(i, j + 1)$**, end.
>
>   swap row k and row i
>
>   subtract for all $k > i$ from the k-th row (a_{kj}/a_{ij}) times the i-th row.
>
> **gauss$(i + 1, j + 1)$**, end.

Is that not incredible? This algorithm can be implemented in a few lines!

In addition to some permutations, the algorithm consists of a single operation:

$$\text{subtract for all } k > i \text{ from the } k\text{-th row } (a_{kj}/a_{ij}) \text{ times the } i\text{-th row.} \tag{8.6}$$

The element a_{ij}, by which is divided, is called the *pivot element*. We will deal with the importance of this element at the end of Sect. 18.1.

We want to calculate an example:

Example

$$\begin{pmatrix} 2 & 2 & 0 & 2 \\ 4 & 6 & 4 & 7 \\ 5 & 6 & 2 & 7 \\ 2 & 3 & 2 & 4 \end{pmatrix} \begin{matrix} \\ \text{II} - 2 \cdot \text{I} \\ \text{III} - 5/2 \cdot \text{I} \\ \text{IV} - \text{I} \end{matrix} \mapsto \begin{pmatrix} 2 & 2 & 0 & 2 \\ 0 & 2 & 4 & 3 \\ 0 & 1 & 2 & 2 \\ 0 & 1 & 2 & 2 \end{pmatrix} \begin{matrix} \\ \\ \text{III} - 1/2 \cdot \text{II} \\ \text{IV} - 1/2 \cdot \text{II} \end{matrix}$$

$$\mapsto \begin{pmatrix} 2 & 2 & 0 & 2 \\ 0 & 2 & 4 & 3 \\ 0 & 0 & 0 & 1/2 \\ 0 & 0 & 0 & 1/2 \end{pmatrix} \begin{matrix} \\ \\ \\ \text{IV} - \text{III} \end{matrix} \mapsto \begin{pmatrix} 2 & 2 & 0 & 2 \\ 0 & 2 & 4 & 3 \\ 0 & 0 & 0 & 1/2 \\ 0 & 0 & 0 & 0 \end{pmatrix}.$$

This is the row-echelon form; the matrix has three leading coefficients and therefore rank 3. ◀

If you carry out the algorithm on paper, it is very useful to write down the row transformations carried out in each case.

▶   **Theorem 8.4** An elementary row transformation of the matrix $A \in K^{m \times n}$ corresponds to the multiplication with a certain invertible matrix from the left.

I simply give the matrices that correspond to the row transformations, it is then easy to calculate that they really do what they are supposed to do:

Type a): If you multiply A from the left with the matrix I', which you obtain when you swap in the $m \times m$-identity matrix I the i-th and j-th row, the rows i and j are exchanged in the matrix A.

Type b): I'' arises from the identity matrix by adding to the i-th row λ times the j-th row. Only the element i_{ij} (a 0) is replaced by λ. If you multiply A from the left with I'', then λ times the j-th row of A is added to the i-th row.

Type c): Finally, if you multiply the i-th row of the unit matrix by λ, and multiply A from the left with the resulting matrix I''', the i-th line of A is multiplied by λ.

The $m \times m$-matrices I', I'' and I''' are themselves derived from I by elementary row transformations. So their rank does not change, it is m, and thus the matrices are invertible according to Theorem 7.12.   □

This statement is not of importance for concrete calculations; elementary row transformations can be carried out faster than with matrix multiplications. However, the theorem is an important proof tool. With its help we will see in a moment how we can determine the inverse of a matrix with elementary transformations.

8.2   Calculating the Inverse of a Matrix

A square $n \times n$-matrix can only be invertible if it has rank n. The row-echelon form of such a matrix therefore has the following shape:

$$\begin{pmatrix} a_{11} & * & & * \\ 0 & a_{22} & * & \\ & 0 & \ddots & * \\ 0 & & 0 & a_{nn} \end{pmatrix}$$

where all diagonal elements are different from 0. If we multiply each row i with $1/a_{ii}$, then the diagonal will have 1s everywhere.

Starting from the right, we can now, similarly to the Gaussian algorithm, make the elements in the upper right half to 0:

We subtract for all $i = 1, \ldots, n-1$ from the i-th row a_{in} times the last row. Now the last column has a row of zeros above the 1. Now we go to the second to last column and subtract from the i-th row for $i = 1, \ldots, n-2$ now $a_{i,n-1}$ times the second to last row. Everything above the element $a_{n-1,n-1}$ will become 0. The last column remains unchanged, since $a_{n-1,n}$ is already 0.

We continue in this form until we reach the left edge of the matrix, and thus we have finally converted the original matrix into the identity matrix by elementary row operations.

This description of the algorithm can also be formulated very briefly recursively.

It is important in this part of the transformation that you start from the right. Try once what happens if you want to make the elements above the diagonal to 0, starting from the left.

What did we gain from this transformation?

▶   **Theorem 8.5** If we transform an invertible $n \times n$-matrix A into the identity matrix by elementary row operations, then the inverse matrix A^{-1} of A is obtained if we apply these operations to the identity matrix I in the same order.

Theorem 8.4 helps us with the proof: Let $D_1, \ldots, D_k$ be the matrices that correspond to the row operations performed on A, then we have:

$$I = D_k D_{k-1} \cdots D_1 A = (D_k D_{k-1} \cdots D_1) A.$$

So

$$A^{-1} = D_k D_{k-1} \cdots D_1 = D_k D_{k-1} \cdots D_1 I. \qquad \square$$

When calculating the inverse, it is best to write A and I next to each other and transform both matrices at the same time, writing the newly obtained matrices under A or EI:

$$
\begin{array}{ccc|ccc l}
\multicolumn{3}{c}{A} & \multicolumn{3}{c}{I} & \\
1 & 0 & 2 & 1 & 0 & 0 & \\
2 & -1 & 3 & 0 & 1 & 0 & \text{II} - 2 \cdot \text{I} \\
4 & 1 & 8 & 0 & 0 & 1 & \text{III} - 4 \cdot \text{I} \\[4pt]
1 & 0 & 2 & 1 & 0 & 0 & \\
0 & -1 & -1 & -2 & 1 & 0 & \\
0 & 1 & 0 & -4 & 0 & 1 & \text{III} + \text{II} \\[4pt]
1 & 0 & 2 & 1 & 0 & 0 & \\
0 & -1 & -1 & -2 & 1 & 0 & \text{II} \cdot (-1) \\
0 & 0 & -1 & -6 & 1 & 1 & \text{III} \cdot (-1) \\[4pt]
1 & 0 & 2 & 1 & 0 & 0 & \text{I} - 2 \cdot \text{III} \\
0 & 1 & 1 & 2 & -1 & 0 & \text{II} - \text{III} \\
0 & 0 & 1 & 6 & -1 & -1 & \\[4pt]
1 & 0 & 0 & -11 & 2 & 2 & \\
0 & 1 & 0 & -4 & 0 & 1 & \\
0 & 0 & 1 & 6 & -1 & -1 &
\end{array}
$$

It is recommended to always make the test at the end of such a calculation:

$$\begin{pmatrix} -11 & 2 & 2 \\ -4 & 0 & 1 \\ 6 & -1 & -1 \end{pmatrix}$$

$$\begin{pmatrix} 1 & 0 & 2 \\ 2 & -1 & 3 \\ 4 & 1 & 8 \end{pmatrix} \begin{pmatrix} 1 & 0 & 0 \\ 0 & 1 & 0 \\ 0 & 0 & 1 \end{pmatrix}.$$

8.3   Systems of Linear Equations

We investigate a system of linear equations with m-equations in n unknowns. This system is described by a matrix $A \in K^{m \times n}$, the result vector $b \in K^m$, and the vector of unknowns $x \in K^n$:

$$Ax = b.$$

If $b \neq 0$, such a system of linear equations is called *inhomogeneous system*, $Ax = 0$ is called the corresponding *homogeneous system*.

We call

$$\text{sol}(A, b) = \{x \in K^n \mid Ax = b\} \tag{8.7}$$

the solution set of the linear equation system $Ax = b$.

The matrix (A, b), which arises when we add the column A to b, is called the *augmented matrix* of the system.

In the course of the solution process, it will prove to be useful to interpret the matrix A as a linear mapping. In the language of linear mappings, $\text{sol}(A, b)$ is precisely the set of $x \in K^n$, which are mapped by the linear mapping A to b.

▶   **Theorem 8.6**

a) $\text{sol}(A, 0) = \ker A$, in particular this solution set is a subspace of the K^n.
b) The following three statements are equivalent:
   - $Ax = b$ has at least one solution,
   - $b \in \text{im} A$,
   - $\text{rank} A = \text{rank}(A, b)$.
c) If w is a solution of $Ax = b$, then the general solution set is obtained by
   $$\text{sol}(A, b) = w + \ker A := \{w + x \mid x \in \ker A\}.$$

The first part is known: $\text{sol}(A, 0)$ is precisely the set of x, which are mapped to 0, that is, the kernel of A.

For the second part: If $Ax = b$ has a solution w, then $Aw = b$, thus $b \in \text{im}\, A$, and conversely, if $b \in \text{im}\, A$, then every preimage of b is a solution of the equation system.

In Theorem 7.10 we have seen that the image is generated by the columns of the matrix. Therefore, if $b \in \text{im}\, A$, then the rank of (A, b) cannot be greater than that of A, the vector b can be linearly combined from the columns of A. Now if $\text{rank}\, A = \text{rank}(A, b)$, then b can be combined from the columns of A again, and since the columns generate the image, b is also in the image.

For part c) of the theorem: We first show that $w + \ker A \subset \text{sol}(A, b)$. For this purpose, let $y \in \ker A$. Since A is a linear mapping, $A(w + y) = Aw + Ay = Aw + 0 = b$, thus $w + y \in \text{sol}(A, b)$.

For the other direction, let $v \in \text{sol}(A, b)$. Then $A(v - w) = Av - Aw = b - b = 0$. So $v - w \in \ker A$ and $v = w + (v - w) \in w + \ker A$.   □

By reversing part 2 of the theorem, we can now see that a system of linear equations is not solvable if $\text{rank}(A, b)$ is greater than $\text{rank}\, A$. However, a homogeneous system always has at least one solution: the zero vector.

Above all, the last part of theorem 8.6 helps us to compute all solutions of the system of linear equations $Ax = b$: It is enough to solve the homogeneous system $Ax = 0$ completely and then calculate one single particular solution of the inhomogeneous system.

Geometrical Interpretation of Systems of Linear Equations

The solutions of homogeneous systems of linear equations are vector spaces: the zero vector, lines and planes through the origin, and so on. The solutions of inhomogeneous systems of linear equations are, in general, not subspaces. Nevertheless, they can be interpreted geometrically: They are lines, planes or spaces of higher dimension shifted from the origin.

The dimension of the solution space can be read from the rank of the matrix:

▶   **Theorem 8.7**

$$\dim \text{sol}(A, 0) = \text{number of unknowns} - \text{rank}\, A. \tag{8.8}$$

Because we already know:

$$\text{rank of matrix } A = \text{dimension of im}\, A$$
$$\text{number of unknowns} = \text{dimension of domain of definition}$$
$$\text{dimension of solution space} = \text{dimension of ker}\, A$$

and according to Theorem 6.27 we have $\dim \text{im}\, A + \dim \ker A = \dim (\text{domain of definition})$. Now plug in the other interpretations of these numbers and you will get equation (8.8).   □

Examples

1. Given the system of equations $x_1 - x_2 = 0$, $x_3 = 1$.

    It is $A = \begin{pmatrix} 1 & -1 & 0 \\ 0 & 0 & 1 \end{pmatrix} \in \mathbb{R}^{2 \times 3}$, $b = \begin{pmatrix} 0 \\ 1 \end{pmatrix}$.

    The rank of the matrix is 2. Since we have 3 unknowns, the kernel has dimension 1. A solution of the homogeneous system of equations $x_1 - x_2 = 0$, $x_3 = 0$ different from 0 is $(1, 1, 0)$. The kernel thus consists of the set $\{\lambda(1, 1, 0) \mid \lambda \in \mathbb{R}\}$ and forms a line through the origin. An inhomogeneous solution is, for example, $(0, 0, 1)$, and thus the solution set is the line $g = \{(0, 0, 1) + \lambda(1, 1, 0) \mid \lambda \in \mathbb{R}\}$, see Fig. 8.1.

2. Let $A = (1\ 0\ 1) \in \mathbb{R}^{1 \times 3}$, $b = 1 \in \mathbb{R}^1$. This also determines a linear equation system, although a somewhat unusual one. It consists only of the one equation $x_1 + x_3 = 1$, but still has three unknowns. If you take it very precisely, you should write $1x_1 + 0x_2 + 1x_3 = 1$.

    The rank of the matrix is 1, the dimension of the kernel therefore 2. We are looking for two linearly independent solutions of the homogeneous equation $x_1 + x_3 = 0$, for example $(0, 1, 0)$ and $(1, 0, -1)$ as well as any solution of the inhomogeneous system, for example $(1, 0, 0)$. This gives us all solutions as

$$\text{sol}(A, b) = \{(1, 0, 0) + \lambda(0, 1, 0) + \mu(1, 0, -1) \mid \lambda, \mu \in \mathbb{R}\},$$

a plane in $\mathbb{R}^3$ (see Fig. 8.2). From the result you can see that for x_2 all real numbers are solutions. Of course this must be the case if it does not appear in the equation at all. ◄

For higher dimensions, this geometric interpretation is unfortunately difficult to imagine and can no longer be drawn.

Ray Tracing, Part 1

An example of application from graphical data processing: An important method in the representation of spatial objects on the screen is *ray tracing*. Each object point is

Fig. 8.1 Line as solution set

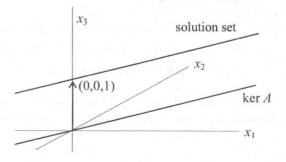

Fig. 8.2 Plane as solution set

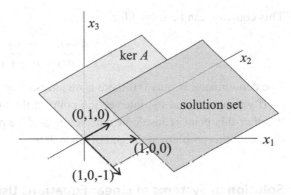

projected onto the projection plane by a central projection, in the center of which the observer's eye is located. In ray tracing, rays are now shot from the center in the direction of the object and the first intersection point with the object is determined. The color and brightness of the object determine the representation of the object on the projection plane (Fig. 8.3). The problem that occurs again and again is the calculation of the intersection point of a line with an object. Complex objects are often approximated by flat surface pieces, in this case it is necessary to determine as quickly as possible the inter section point of a line with a plane, which contains such a surface piece.

In the last example we saw the solution set of a linear equation of the form

$$ax_1 + bx_2 + cx_3 = d \tag{8.9}$$

represents a plane. In Sect. 10.1 after Theorem 10.5 we will learn how to find such an equation for a given plane. Also lines could be described as solution sets of systems of linear equations, but for the above problem there is a better possibility: We choose the parameter representation of a line, as we have learned in Sect. 6.1:

$$g = \left\{ \begin{pmatrix} u_1 \\ u_2 \\ u_3 \end{pmatrix} + \lambda \begin{pmatrix} v_1 \\ v_2 \\ v_3 \end{pmatrix} \,\middle|\, \lambda \in \mathbb{R} \right\} = \left\{ \begin{pmatrix} u_1 + \lambda v_1 \\ u_2 + \lambda v_2 \\ u_3 + \lambda v_3 \end{pmatrix} \,\middle|\, \lambda \in \mathbb{R} \right\}.$$

Here (u_1, u_2, u_3) is the projection center and (v_1, v_2, v_3) is the direction of the projection beam. We set the components of the line in the linear equation (8.9) and obtain

$$a(u_1 + \lambda v_1) + b(u_2 + \lambda v_2) + c(u_3 + \lambda v_3) = d.$$

Fig. 8.3 Ray Tracing 1

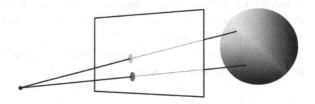

This equation can be solved for λ:

$$\lambda = \frac{d - au_1 - bu_2 - cu_3}{av_1 + bv_2 + cv_3}.$$

The denominator is only 0 if there is no intersection with the plane.

If you have found the intersection point of the line with the plane, you have to check whether this point is inside or outside the surface piece that bounds the object. We will deal with this problem at the end of Chap. 9.9.

Solution of Systems of Linear Equations Using the Gaussian Algorithm

Now, however, finally to the specific method for solving linear equations. Theorem 8.9 shows us that the Gaussian algorithm can also be used here. As preparation:

▶   **Lemma 8.8** If $B \in K^{m \times m}$ is an invertible matrix and $A \in K^{m \times n}$, $b \in K^m$, then

$$\mathrm{sol}(A, b) = \mathrm{sol}(BA, Bb).$$

For if $Av = b$, then $(BA)v = B(Av) = Bb$, thus v is also a solution of the equation system $BAx = Bb$.

Conversely, if $BAv = Bb$, then from the invertibility of B:

$$Av = B^{-1}(BAv) = B^{-1}(Bb) = b$$

and thus also $Av = b$.   □

▶   **Theorem 8.9** If the matrix A' is obtained from A by elementary row operations and the vector b' is obtained from b by the same row operations, then

$$\mathrm{sol}(A', b') = \mathrm{sol}(A, b).$$

So it has to be shown that $Ax = b$ if and only if $A'x = b'$. The elementary row operations are equivalent to a series of multiplications with invertible matrices $D = D_k D_{k-1} \cdots D_1$, and it is $A' = DA$, $b' = Db$. The product of invertible matrices is invertible again (since the invertible matrices form a group!) and thus according to Lemma 8.8 the assertion is true.   □

If we want to solve a linear equation system $Ax = b$, we first use the Gaussian algorithm to bring the augmented matrix (A, b) into row-echelon form, as we have learned when computing the rank of a matrix.

There are different, quite similar methods on how to proceed. Below I will introduce the recipe which seems simplest to me.

First, as described in the caculation of the inverse of a matrix in Sect. 8.2, we can normalize all leading coefficients to 1 by elementary row operations and then, starting from the right, make the elements above the leading coefficients to 0. The columns without leading coefficients are not of interest to us at the moment. This transforms our augmented matrix (A, b) into a matrix that looks approximately like this:

$$(A', b') = \begin{pmatrix} 1 & * & 0 & 0 & * & \cdots & 0 & * \\ 0 & 0 & 1 & 0 & * & \cdots & 0 & * \\ 0 & \cdots & 0 & 1 & * & \cdots & 0 & * \\ 0 & \cdots & & & 0 & \cdots & 1 & * \\ 0 & \cdots & 0 & 0 & 0 & \cdots & 0 & 0 \end{pmatrix}$$

This form of the matrix is called reduced echelon form: In this form, all leading coefficients are equal to 1 and 0 is above the leading coefficient everywhere. The reduced echelon form is ideally suited to immediately give the solutions of the equation system $Ax = b$. The complete procedure is carried out in 4 steps:

Step 1: Bring the augmented matrix into reduced echelon form. Now you can already check whether the rank of the augmented matrix is equal to the rank of the matrix, that is, whether there is a solution at all.

Step 2: Set the unknowns that belong to the columns without leading coefficients as free variables (for example λ_1, λ_2, λ_3, etc.).

Step 3: Each row of the matrix now represents an equation for one of the other unknowns. These unknowns can be expressed by the λ_i and by a b_j, one of the elements from the last column of the matrix.

Step 4: Now the solution set can be written in the form:

$$\text{sol}(A, b) = \left\{ \begin{pmatrix} c_1 \\ c_2 \\ \vdots \\ c_m \end{pmatrix} + \lambda_1 \begin{pmatrix} u_{11} \\ u_{21} \\ \vdots \\ u_{m1} \end{pmatrix} + \cdots + \lambda_k \begin{pmatrix} u_{1k} \\ u_{2k} \\ \vdots \\ u_{mk} \end{pmatrix} \;\middle|\; \lambda_i \in \mathbb{R} \right\}.$$

Here $c = (c_1, c_2, \ldots, c_m)$ is a special solution of the inhomogeneous equation system and the vectors $(u_{1i}, u_{2i}, \ldots, u_{mi})$ form a basis of the solution space of the homogeneous system. c contains the elements of b' and possibly some zeroes.

Think about why the vectors resulting from this process $(u_{1i}, u_{2i}, \ldots, u_{mi})$ are always linearly independent!

The best way to understand this is with an example. I will give the matrix A in reduced echelon form, and we will carry out steps 2 through 4:

$$(A, b) = \begin{pmatrix} 1 & 0 & 3 & 0 & 0 & 8 & | & 2 \\ 0 & 1 & 2 & 0 & 0 & 1 & | & 4 \\ 0 & 0 & 0 & 1 & 0 & 5 & | & 6 \\ 0 & 0 & 0 & 0 & 1 & 4 & | & 0 \end{pmatrix}.$$
$$\phantom{(A, b) = \begin{pmatrix}}x_1\ x_2\ x_3\ x_4\ x_5\ x_6$$

Under the matrix, I've written the unknowns that belong to each column. When you get to this point, you should first think about the existence and number of solutions. The rank of the matrix A is 4, because A has 4 leading coefficients. The rank of (A, b) is also 4, the extension doesn't add any new leading coefficients. So by Theorem 8.6, the system of equations has at least one solution, and we need to continue. From Theorem 8.7, we can read off the dimension of the solution space: it's equal to the number of unknowns minus the rank, so 2.

Now for step 2: The columns without leading coefficients are columns 3 and 6, so we set $x_3 = \lambda$, $x_6 = \mu$.

In step 3, we get from the four equations by solving for the remaining unknowns the results:

$$x_1 = -3\lambda - 8\mu + 2, \quad x_2 = -2\lambda - \mu + 4, \quad x_4 = -5\mu + 6, \quad x_5 = -4\mu.$$

We can summarize this result in one vector equation in step 4. Here you can see, how two 0's enter the result vector at position 3 and 6:

$$\begin{pmatrix} x_1 \\ x_2 \\ x_3 \\ x_4 \\ x_5 \\ x_6 \end{pmatrix} = \begin{pmatrix} -3\lambda - 8\mu + 2 \\ -2\lambda - \mu + 4 \\ \lambda \\ -5\mu + 6 \\ -4\mu \\ \mu \end{pmatrix} = \begin{pmatrix} 2 \\ 4 \\ 0 \\ 6 \\ 0 \\ 0 \end{pmatrix} + \lambda \begin{pmatrix} -3 \\ -2 \\ 1 \\ 0 \\ 0 \\ 0 \end{pmatrix} + \mu \begin{pmatrix} -8 \\ -1 \\ 0 \\ -5 \\ -4 \\ 1 \end{pmatrix},$$

and then we already have our solution set in the desired form:

$$\text{sol}(A, b) = \left\{ \begin{pmatrix} 2 \\ 4 \\ 0 \\ 6 \\ 0 \\ 0 \end{pmatrix} + \lambda \begin{pmatrix} -3 \\ -2 \\ 1 \\ 0 \\ 0 \\ 0 \end{pmatrix} + \mu \begin{pmatrix} -8 \\ -1 \\ 0 \\ -5 \\ -4 \\ 1 \end{pmatrix} \ \middle| \ \lambda, \mu \in \mathbb{R} \right\}.$$

As an alternative solution method, you can directly calculate all unknowns from the row echelon form by starting with the last row and working backwards, plugging in the unknowns. Here too, you must set as variables the unknowns that belong to columns without leading coefficients. This saves you the step of transforming the matrix into reduced echelon form. Even if this method looks simpler, you usually don't need fewer operations than in the method presented above.

8.4   Comprehension Questions and Exercises

Comprehension Questions

1. What are the elementary row operations in a matrix?
2. Swapping two rows does not change the rank of a matrix. Can swapping two columns change the rank?
3. Can an invertible matrix become a non-invertible matrix when performing elementary row operations?
4. What is the pivot element in the Gaussean algorithm?
5. A system of linear equations is determined by the matrix A and the result vector b. How can you read the number of unknowns and the number of equations from the matrix A?
6. Let $Ax = 0$ be a homogeneous system of linear equations. How can you determine the dimension of the solution space from the rank of A and the number of unknowns?
7. The solution set of a homogeneous system of linear equations is always a vector space. What is the solution of an inhomogeneous system of linear equations?
8. The solution set of a linear equation in $\mathbb{R}^3$ describes a plane in $\mathbb{R}^3$. Why? What is the geometric shape of the solution set of two linear equations in $\mathbb{R}^3$? Explain the relation to the planes, which are determined by the two individual equations.

Exercises

1. Determine the rank of each of the following matrices:

$$
\begin{pmatrix} 1 & 3 & -2 & 5 & 4 \\ 1 & 4 & 1 & 3 & 5 \\ 1 & 4 & 2 & 4 & 3 \\ 2 & 7 & -3 & 6 & 13 \end{pmatrix}, \quad
\begin{pmatrix} 1 & 2 & -3 & -2 & -3 \\ 1 & 3 & -2 & 0 & -4 \\ 3 & 8 & -7 & -2 & -11 \\ 2 & 1 & -9 & -10 & -3 \end{pmatrix}, \quad
\begin{pmatrix} 1 & 1 & 2 \\ 4 & 5 & 5 \\ 5 & 8 & 1 \\ -1 & -2 & 2 \end{pmatrix}, \quad
\begin{pmatrix} 2 & 1 \\ 3 & -7 \\ -6 & 1 \\ 5 & -8 \end{pmatrix}.
$$

2. Which subspaces of $\mathbb{R}^3$ are determined by the solutions of the following homogeneous systems?

   a) $2x_1 - 2x_2 + x_3 = 0$

   b) $2x_1 - 2x_2 + x_3 = 0$

   $\qquad\qquad x_2 + 3x_3 = 0$

   c) $x_1 + x_2 + x_3 = 0$

   $\qquad x_1 + \qquad x_3 = 0$

   $\qquad 2x_1 + x_2 + 2x_3 = 0$

3. Determine the inverses of the following matrices, if they exist:

$$
\begin{pmatrix} 3 & 0 & 1 \\ 1 & 0 & 1 \\ 0 & 1 & 0 \end{pmatrix}, \quad
\begin{pmatrix} 3 & 0 & 1 \\ 6 & 0 & 2 \\ 0 & 1 & 0 \end{pmatrix}, \quad
\begin{pmatrix} 5 & 0 & 0 \\ -1 & 2 & 0 \\ 4 & 1 & 3 \end{pmatrix}.
$$

4. Determine whether the following system of linear equations is solvable, and if so, determine the solution set:

$$x_1 + 2x_2 + 3x_3 = 1$$
$$4x_1 + 5x_2 + 6x_3 = 2$$
$$7x_1 + 8x_2 + 9x_3 = 3$$
$$5x_1 + 7x_2 + 9x_3 = 4$$

5. Solve the following system of linear equations:

$$2x_1 + x_2 + x_3 = a_1$$
$$5x_1 + 4x_2 - 5x_3 = a_2$$
$$3x_1 + 2x_2 - x_3 = a_3$$

where once $a_1 = 5$, $a_2 = -1$, $a_3 = 3$, and once $a_1 = 1$, $a_2 = -1$, $a_3 = 1$.

6. What is the order of the Gaussian algorithm when solving the following system of equations:

$$
\begin{pmatrix}
a_{11} & a_{12} & 0 & 0 & 0 & \cdots & 0 \\
a_{21} & a_{22} & a_{23} & 0 & 0 & \cdots & 0 \\
0 & a_{32} & a_{33} & a_{34} & 0 & \cdots & 0 \\
0 & 0 & a_{43} & a_{44} & a_{45} & \ddots & \vdots \\
0 & 0 & 0 & \ddots & \ddots & \ddots & 0 \\
\vdots & \vdots & \vdots & \ddots & a_{n-1,n-2} & a_{n-1,n-1} & a_{n-1,n} \\
0 & 0 & 0 & 0 & 0 & a_{n,n-1} & a_{nn}
\end{pmatrix}
\begin{pmatrix}
x_1 \\ x_2 \\ x_3 \\ \vdots \\ \\ \\ x_n
\end{pmatrix}
=
\begin{pmatrix}
b_1 \\ b_2 \\ b_3 \\ \vdots \\ \\ \\ b_n
\end{pmatrix}
$$

Show that this system is always uniquely solvable if for all i holds $|a_{ii}| > |a_{i-1,i}| + |a_{i+1,i}|$. Such a matrix is called "strictly diagonally dominant."

Eigenvalues, Eigenvectors and Change of Basis

9

Abstract

At the end of this chapter

- you will know what the determinant of a matrix is and can compute it,
- you have learned the terms eigenvalue and eigenvector of matrices,
- and can calculate eigenvalues and eigenvectors with the help of the characteristic polynomial of a matrix,
- you have calculated eigenvalues and eigenvectors of some important linear mappings in the $\mathbb{R}^2$ and interpreted the results geometrically,
- you can carry out a change of basis,
- and know what the orientation of vector spaces means.

9.1 Determinants

Determinants are a characteristic of matrices that is often useful for investigating certain properties of matrices. We need determinants primarily in connection with eigenvalues and eigenvectors, which we will discuss in the second part of the chapter. But they also form an important tool for solving systems of linear equations. Determinants are only defined for square matrices, so all matrices in this chapter are square. K is supposed to be some field, but you can usually just think of the real numbers.

Supplementary Information The online version contains supplementary material available at https://doi.org/10.1007/978-3-658-40423-9_9.

Let's look at a system of linear equations with two equations and two unknowns again:

$$ax + by = e,$$
$$cx + dy = f,$$

and solve for x and y. Two equations are just about manageable. We get:

$$x = \frac{ed - bf}{ad - bc}, \quad y = \frac{af - ce}{ad - bc}.$$

These fractions only make sense if the denominator $ad - bc$ is different from zero. In this case, the system has the specified unique solution, and if $ad - bc = 0$, then there is no unique solution. Check for yourself that in the case $ad - bc = 0$, the rank of the coefficient matrix is less than 2.

$ad - bc$ is the determinant of the matrix $\begin{pmatrix} a & b \\ c & d \end{pmatrix}$, and we have already learned two assertions that can be read from it: If it is not equal to 0, then the matrix has rank 2 and the system of equations $Ax = y$ is uniquely solvable for all $y \in K^2$.

Every square $n \times n$-matrix has a determinant. This is a mapping from $K^{n \times n} \to K$. The definition is quite technical, but we can't get around it.

For matrices from $K^{2 \times 2}$ and from $K^{3 \times 3}$, up to the sign, the determinants have an intuitive geometric interpretation: In $K^{2 \times 2}$, they give the area of the parallelogram spanned by the row vectors, in $K^{3 \times 3}$, the volume of the body spanned by the row vectors. This body is called parallelepiped.

So far, I have always investigated column vectors in connection with matrices. Since I want to calculate determinants using the elementary row transformations from Definition 8.1, the use of row vectors is now appropriate. We will see later that the same results also apply to column vectors: The determinant also gives the volume of the body spanned by the column vectors.

You can also define a n-dimensional volume in the K^n, and in principle the determinant does nothing else: It represents a generalized volume function for the "hyper"-parallelogram spanned by n vectors in the K^n. I will also speak of the area as a volume, namely a two-dimensional volume.

In three steps we want to come to such a determinant function and above all to an algorithm for the calculation. In the matrix A we denote the row vectors with $r_1, r_2, \ldots, r_n$ and write $A = (r_1, r_2, \ldots, r_n)$. In the first step I will collect some properties that one expects from such a volume function.

▶ **Definition 9.1** A *determinant* is a function $\det \colon K^{n \times n} \to K$ with the following properties:

(D1) $\det I = 1$.

(D2) If the rows are linearly dependent, the determinant is 0.

(D3) For all $\lambda \in K$, all $v \in K^n$ and for $i = 1, \ldots, n$ it holds:

$$\det(r_1, \ldots, \lambda r_i, \ldots, r_n) = \lambda \det(r_1, \ldots, r_i, \ldots, r_n),$$
$$\det(r_1, \ldots, r_i + v, \ldots, r_n) = \det(r_1, \ldots, r_i, \ldots, r_n) + \det(r_1, \ldots, \underset{\underset{i}{\uparrow}}{v}, \ldots, r_n).$$

The last property can also be interpreted in this way: If we keep all rows except the i-th fixed and only let this i-th row variable, we get a mapping that is linear in the i-th component, just like a linear mapping.

Are these reasonable volume properties? (D1) represents the normalization so to speak: the unit square or the unit cube has the volume 1. Of course it is also reasonable to demand this from the "n-dimensional unit cube".

To (D2): In the K^2, if two vectors are collinear, the parallelogram degenerates to a line and has no area, just as in the K^3 three linearly dependent vectors do not span a real space, it is flattened and has no volume. We also wish this property for a volume of the K^n.

(D3) is not quite as easy to understand, but it is certainly correct in K^2 and in K^3. In the two-dimensional case it still can be drawn (Fig. 9.1).

You can see in the left image the grey area is larger by the factor λ than the dotted one, in the right image the dotted and grey area are equal; what has been added to the grey area above has just been taken away below. The same relations apply in K^3. If you want, you can calculate them in coordinates. For the volume function in K^n we simply demand the two properties from (D3) should apply.

Of course there are other characteristic properties of a volume. But it turns out the three mentioned conditions are already sufficient to carry out the second step, which consists in the

▶ **Theorem 9.2** There is exactly one determinant function det: $K^{n \times n} \to K$ with the properties from Definition 9.1.

But I would like to postpone the proof of this theorem for a while and come directly to the third step. This represents a specific calculation procedure under the assumption that there is a determinant function. Once again, the Gaussian algorithm will help us here. First we investigate what effect elementary row operations have on the determinant:

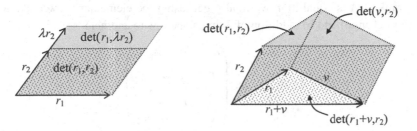

Fig. 9.1 Determinant operations

▶ **Theorem 9.3: The effect of elementary row operations on determinants**

a) Interchanging two rows reverses the sign of the determinant.
b) Adding λ times one row to another does not change the determinant.
c) If a row is multiplied by a factor $\lambda \in K$, then:

$$\det(r_1, \ldots, \lambda r_i, \ldots, r_n) = \lambda \det(r_1, \ldots, r_i, \ldots, r_n).$$

We break down the properties from behind: c) is identical to the first part of (D3). Now to b): It holds

$$\det(r_1, \ldots, r_i + \lambda r_j, \ldots, r_n) \underset{(\text{D3})}{=} \det(r_1, \ldots, r_i, \ldots, r_n) + \det(r_1, \ldots, \underset{i}{\lambda r_j}, \ldots, r_j, \ldots, r_n)$$

$$\underset{(\text{D2})}{=} \det(r_1, \ldots, r_i, \ldots, r_n) + 0.$$

a) we can reduce to b) and c) again, because interchanging rows can be composed from the following series of operations of type b) and c):

$$\det(\cdots, r_i, \cdots, r_j, \cdots) = \quad (i\text{ - th} - j\text{ - th row}) \text{ determinant doesn't change}$$
$$\det(\cdots, r_i - r_j, \cdots, r_j, \cdots) = \quad (j\text{ - th} + i\text{ - th row}) \text{ determinant doesn't change}$$
$$\det(\cdots, r_i - r_j, \cdots, r_i, \cdots) = \quad (i\text{ - th} - j\text{ - th row}) \text{ determinant doesn't change}$$
$$\det(\cdots, -r_j, \cdots, r_i, \cdots) = \quad (i\text{ - th row} \cdot (-1)) \text{ determinant changes sign}$$
$$-\det(\cdots, r_j, \cdots, r_i, \cdots)$$

You might think that something is wrong with the sign in this Theorem. Isn't a volume always positive? Actually, the determinant is something like an oriented, signed volume. We will investigate what this orientation is all about at the end of this chapter.

Now we can apply the Gaussian algorithm to a matrix A again. For most operations, the determinant does not change, but if it does, we have to keep track of it.

The procedure ends immediately if we encounter a column without a leading coefficient during our conversions. Because this can only happen in a $n \times n$-matrix if the rank is $< n$. That just means that the rows are linearly dependent and thus the determinant is zero.

If the rank of A is equal to n, we finally get, using the elementary row operations of type a) (interchange) and b) (addition of λ times one row to another):

$$\det A = \alpha \det \begin{pmatrix} a'_{11} & * & \cdots & * \\ 0 & a'_{22} & * & \vdots \\ \vdots & 0 & \ddots & * \\ 0 & \cdots & 0 & a'_{nn} \end{pmatrix}. \tag{9.1}$$

Here $\alpha = \pm 1$, depending on how many interchanges we have made, and the leading coefficients are all unequal 0. By a series of further operations of type c) (multiplication of a row by the factor λ) we get

$$\det A = \alpha a'_{11} a'_{22} \cdots a'_{nn} \cdot \det \begin{pmatrix} 1 & * & \cdots & * \\ 0 & 1 & * & \vdots \\ \vdots & 0 & \ddots & * \\ 0 & \cdots & 0 & 1 \end{pmatrix}.$$

This matrix can be transformed into the identity matrix by operations of type b). The determinant does not change anymore, and since the determinant of the identity matrix is equal to 1, we can read off the determinant from (9.1): It holds

$$\det A = \alpha a'_{11} a'_{22} \cdots a'_{nn}. \tag{9.2}$$

In this case, by the way, the determinant is not equal to zero, because all factors are unequal 0. We already know that in the case $\operatorname{rank} A < n$ the determinant $= 0$ is, and so we get:

▶ **Theorem 9.4** Let $A \in K^{n \times n}$. Then the following statements are equivalent:

$$\det A \neq 0$$
$$\Leftrightarrow \quad \operatorname{rank} A = n$$
$$\Leftrightarrow \quad \ker A = \{\vec{0}\}, \text{ that is } \dim \ker A = 0$$
$$\Leftrightarrow \quad \text{columns (rows) are linearly independent,}$$

or:

$$\det A = 0$$
$$\Leftrightarrow \quad \operatorname{rank} A < n$$
$$\Leftrightarrow \quad \dim \ker A > 0$$
$$\Leftrightarrow \quad \text{columns (rows) are linearly dependent.}$$

The relation between rank and dimension of the kernel, which is stated here again, has already been shown in Theorem 7.10. We also already know the relation between the rank and the linear dependence of columns or rows.

Finally, some examples. I will introduce a shortened notation:

$$\begin{vmatrix} a_{11} & \cdots & a_{1n} \\ \vdots & \ddots & \vdots \\ a_{m1} & \cdots & a_{mn} \end{vmatrix} := \det \begin{pmatrix} a_{11} & \cdots & a_{1n} \\ \vdots & \ddots & \vdots \\ a_{m1} & \cdots & a_{mn} \end{pmatrix}.$$

Let's calculate the determinant of a 4×4-matrix:

$$\begin{vmatrix} 1 & 2 & 0 & 0 \\ 1 & 2 & 2 & 4 \\ 2 & 3 & 1 & 2 \\ 0 & 2 & 1 & 4 \end{vmatrix} \begin{matrix} \\ \text{II} - \text{I} \\ \text{III} - 2 \cdot \text{I} \\ \\ \end{matrix} = \begin{vmatrix} 1 & 2 & 0 & 0 \\ 0 & 0 & 2 & 4 \\ 0 & -1 & 1 & 2 \\ 0 & 2 & 1 & 4 \end{vmatrix} \begin{matrix} \\ \text{II} \leftrightarrow \text{III} \\ \\ \\ \end{matrix} = (-1) \begin{vmatrix} 1 & 2 & 0 & 0 \\ 0 & -1 & 1 & 2 \\ 0 & 0 & 2 & 4 \\ 0 & 2 & 1 & 4 \end{vmatrix} \begin{matrix} \\ \\ \\ \text{IV} + 2 \cdot \text{II} \end{matrix}$$

$$= (-1) \begin{vmatrix} 1 & 2 & 0 & 0 \\ 0 & -1 & 1 & 2 \\ 0 & 0 & 2 & 4 \\ 0 & 0 & 3 & 8 \end{vmatrix} \begin{matrix} \\ \\ \\ \text{IV} - 3/2 \cdot \text{III} \end{matrix}$$

$$= (-1) \begin{vmatrix} 1 & 2 & 0 & 0 \\ 0 & -1 & 1 & 2 \\ 0 & 0 & 2 & 4 \\ 0 & 0 & 0 & 2 \end{vmatrix} = (-1) \cdot 1 \cdot (-1) \cdot 2 \cdot 2 = 4.$$

And a 3×3-determinant:

$$\begin{vmatrix} 1 & 0 & 2 \\ 2 & -2 & 3 \\ 4 & 1 & 8 \end{vmatrix} \begin{matrix} \\ \text{II} - 2 \cdot \text{I} \\ \text{III} - 4 \cdot \text{I} \end{matrix} = \begin{vmatrix} 1 & 0 & 2 \\ 0 & -2 & -1 \\ 0 & 1 & 0 \end{vmatrix} \begin{matrix} \\ \\ \text{III} + 1/2 \cdot \text{II} \end{matrix}$$

$$= \begin{vmatrix} 1 & 0 & 2 \\ 0 & -2 & -1 \\ 0 & 0 & -1/2 \end{vmatrix} = 1 \cdot (-2) \cdot (-1/2) = 1.$$

◀

For 3×3-matrices there is an easier way to calculate, the *Rule of Sarrus*: It is best to remember this by writing the two first columns of the matrix A next to it to the right and then proceeding according to the following rule:

Sum the products of the main diagonals (from top left to bottom right) and subtract the products of the secondary diagonals (from top right to bottom left). That is:

$$\begin{pmatrix} a_{11} & a_{12} & a_{13} \\ a_{21} & a_{22} & a_{23} \\ a_{31} & a_{32} & a_{33} \end{pmatrix} \begin{matrix} a_{11} & a_{12} \\ a_{21} & a_{22} \\ a_{31} & a_{32} \end{matrix}$$

$$\det A = a_{11}a_{22}a_{33} + a_{12}a_{23}a_{31} + a_{13}a_{21}a_{32} - a_{13}a_{22}a_{31} - a_{11}a_{23}a_{32} - a_{12}a_{21}a_{33}. \tag{9.3}$$

Using the example from above we get:

$$\begin{vmatrix} 1 & 0 & 2 \\ 2 & -2 & 3 \\ 4 & 1 & 8 \end{vmatrix} \begin{matrix} 1 & 0 \\ 2 & -2 \\ 4 & 1 \end{matrix}$$

$$\rightarrow \det = 1 \cdot (-2) \cdot 8 + 0 \cdot 3 \cdot 4 + 2 \cdot 2 \cdot 1 - 2 \cdot (-2) \cdot 4 - 1 \cdot 3 \cdot 1 - 0 \cdot 2 \cdot 8 = 1.$$

You already know from the beginning of the chapter for 2×2-matrices the rule

$$\begin{vmatrix} a & b \\ c & d \end{vmatrix} = ad - bc, \tag{9.4}$$

so it also applies here: "Product of main diagonal minus product of secondary diagonal". But please: never use it for 4×4 or other larger matrices! The base case cannot be continued with an induction step; it is simply wrong.

Now I want to go back to step 2 of our program to construct a determinant function. We still don't even know if there is a determinant at all. We calculate something, but it is not at all clear whether with other transformations maybe something else comes out, whether we have a function here at all and, if so, whether it fulfills our three requirements (D1), (D2), (D3). I would spare you the existence part of Theorem 9.2 if it were not constructive at the same time: It contains another method for calculating determinants, which often leads to the goal faster than the Gaussian algorithm and which one should therefore know. It is the calculation method of the expansion along the first row:

▶ **Definition 9.5** Let $A \in K^{n \times n}$. Let A_{ij} be the $(n-1) \times (n-1)$-matrix, which results from A when the i-th row and the j-th column are deleted.

With this fixed we can now recursively define the determinant function:

▶ **Definition 9.6: Expansion along the first row** For a 1×1-matrix $A = (a)$ we set $\det A := a$.

For a $n \times n$-matrix $A = (a_{ij})$ we define

$$\det A := a_{11} \det A_{11} - a_{12} \det A_{12} + a_{13} \det A_{13} - \cdots \pm a_{1n} \det A_{1n}.$$

Try it out right away for 2×2- and 3×3-matrices. The same result comes out, as we have calculated in (9.3) and (9.4).

In order for us to be able to write this definition at all, we have to check that this is really a determinant function, that is, the properties (D1), (D2), (D3) are fulfilled. This is a mathematical induction, which is not very difficult, but also not very exciting. I will therefore omit it. To finish the proof of Theorem 9.2, only the uniqueness of the function is missing. But this can be seen easily with the help of the Gaussian algorithm: Are det and det′ determinant functions, then we calculate det A and det′ A by both times trans-forming the matrix with the same transformations into an upper triangular matrix. As seen in (9.1) and (9.2), the value of the determinant is thus uniquely determined, and it is det $A = $ det′ A.

We apply the procedure from Definition 9.6 to the 3×3 determinant that we have already calculated twice:

$$\begin{vmatrix} 1 & 0 & 2 \\ 2 & -2 & 3 \\ 4 & 1 & 8 \end{vmatrix} = 1 \cdot \begin{vmatrix} -2 & 3 \\ 1 & 8 \end{vmatrix} + 2 \cdot \begin{vmatrix} 2 & -2 \\ 4 & 1 \end{vmatrix} = 1 \cdot (-19) + 2 \cdot 10 = 1.$$

I would like to introduce two more important theorems about determinants, the proofs of which are rather lengthy, without us giving us any essential new insights. I will also omit these proofs.

▶   **Theorem 9.7: The multiplication theorem for determinants** For matrices $A, B \in K^{n \times n}$ it holds:

$$\det AB = \det A \cdot \det B.$$

▶   **Definition 9.8** If $A \in K^{n \times n}$, then the *transposed matrix* A^T to A is created by reflecting A on the main diagonal: the i-th column of A is the i-th row of A^T.

▶   **Theorem 9.9** For $A \in K^{n \times n}$ it holds det $A = $ det A^T.

From Theorem 9.9 we can immediately draw some consequences:

    Just as we carried out the expansion of a matrix along the first row in Definition 9.6, we can expand the determinant along each row and column. For this, the coefficients of this row or column are multiplied by the corresponding sub-determinants and alternately added and subtracted. The sign of each summand results from the following chessboard pattern:

$$\begin{pmatrix} + & - & + & \cdots & \pm \\ - & + & - & & \mp \\ + & - & + & & \vdots \\ \vdots & & & \ddots & \\ \pm & \mp & \cdots & & \pm \end{pmatrix}.$$

Written out as a formula, the expansion along the i-th row is:

$$\det A = \sum_{j=1}^{n} (-1)^{i+j} a_{ij} \det A_{ij},$$

and correspondingly the expansion along the j-th column:

$$\det A = \sum_{i=1}^{n} (-1)^{i+j} a_{ij} \det A_{ij}.$$

The expansion along the i-th row can be traced back to the expansion along the first row by previously exchanging the rows so that the i-th row comes to the first position and the rows in front of the row i are each pushed down one position. Check that this sign pattern results.

If we now carry out a transposition of the matrix before the expansion along the i-th row, we obtain the expansion along the i-th column.

Let's take the 3×3-matrix, which has already been used three times, as an example. This time we expand along the 2nd column, and again the same comes out:

$$\begin{vmatrix} 1 & 0 & 2 \\ 2 & -2 & 3 \\ 4 & 1 & 8 \end{vmatrix} = (-2) \cdot \begin{vmatrix} 1 & 2 \\ 4 & 8 \end{vmatrix} - 1 \cdot \begin{vmatrix} 1 & 2 \\ 2 & 3 \end{vmatrix} = (-2) \cdot 0 - 1 \cdot (-1) = 1.$$

Expansion of a determinant along a row or column is only interesting if there are many zeros in this row or column, so that only a few sub-determinants have to be calculated. If this is not the case, it is better to use Gauss' method.

The determinant is a volume function: it gives the volume of the body spanned by the rows. If we transpose the matrix, we see that the determinant also gives the volume of the body generated by the columns.

Now we interpret the matrix as a linear mapping again. We know that the columns of the matrix are exactly the images of the standard basis under this mapping. The mapping transforms the n-dimensional unit cube into the body determined by the columns of the matrix. Since the volume of the unit cube is 1, the determinant thus gives something like a distortion factor of space by the mapping. A large determinant means: the space is inflated, a small determinant means: the space shrinks. In many applications, linear mappings that preserve angles and lengths are particularly important. These also leave volumes unchanged, so they have determinant ± 1. We will investigate such mappings in Sect. 10.2.

9.2   Eigenvalues and Eigenvectors

In Sect. 7.1 we have learned about some linear mappings that are important in graphical data processing, for example stretching, shearing, rotation and reflection. Many of these mappings can be described very easily: the reflection, for example, has an axis that remains fixed, the vector perpendicular to it just changes its direction. With the shear, there is also a fixed vector. The rotation is characterized by an angle, usually no vector remains fixed.

If the coordinate system of the vector space is not chosen very cleverly, it can be difficult to see which mapping is behind a matrix.

We now want to deal with the problem of finding the vectors that remain fixed or only change their length for a given matrix.

The inverse problem is also important for data processing: If, for example, we want to move an object in space or determine the coordinates of a robot gripper, we need to know which types of movements occur at all: With how many different types of matrices do we have to deal with if we want to exhaust all possibilities?

These are some applications of the theory of eigenvalues and eigenvectors. In mathematics, especially in numerics, there are many more. In Chap. 11 we will use eigenvalues to calculate the centrality of nodes in graphs, they will appear again in Part III of the book when we talk about Markov chains and principal component analysis. A useful tool for calculating these matrix characteristics are the determinants, which we got to know in the first part of this chapter.

▶   **Definition 9.10** Let $A: K^n \to K^n$ be a linear mapping. $\lambda \in K$ is called *eigenvalue* of A if there is a vector $v \neq 0$ with the property $Av = \lambda v$. The vector $v \in K^n$ is called *eigenvector* corresponding to the eigenvalue $\lambda \in K$ if $Av = \lambda v$ holds.

Eigenvectors of a linear mapping are therefore vectors that retain their direction under the mapping while only changing their length, where λ is the stretching factor.

Well, one direction change is possible: If $\lambda < 0$, the eigenvector changes its sign and points exactly in the opposite direction. Please allow me to consider this reversal of direction as a stretch and continue to say that eigenvectors retain their direction.

The requirement $v \neq 0$ in the definition is important for the existence of an eigenvalue: If one were to allow $v = 0$, then every λ would be an eigenvalue, because $A0 = \lambda 0$ always holds, of course. But once an eigenvalue λ has been found, $v = 0$ is an eigenvector of λ by our definition.

Examples in $\mathbb{R}^2$

1. First of all, the matrix that describes a rotation around the origin. We met it in Example 6 from Sect. 7.1:

$$D_\alpha = \begin{pmatrix} \cos\alpha & -\sin\alpha \\ \sin\alpha & \cos\alpha \end{pmatrix}.$$

Except for the origin, every point is rotated, so no vector $v \neq 0$ remains fixed and therefore there are no eigenvalues and no eigenvectors.

Wait, there are a few exceptions: With a rotation by 0°, 360°, and multiples thereof, every vector is mapped to itself: Every vector is an eigenvector with eigenvalue 1. With a rotation by 180°, every vector is exactly reversed and is therefore an eigenvector with eigenvalue -1.

2. I introduced the reflection matrix to you in Example 7 from Sect. 7.1. There I claimed that this is a reflection without verifying it. Now we want to do this by determining the eigenvalues and eigenvectors. So we are looking for $(x, y) \in \mathbb{R}^2$ and $\lambda \in \mathbb{R}$ with the property

$$\begin{pmatrix} \cos\alpha & \sin\alpha \\ \sin\alpha & -\cos\alpha \end{pmatrix} \begin{pmatrix} x \\ y \end{pmatrix} = \lambda \begin{pmatrix} x \\ y \end{pmatrix}.$$

We reformulate this matrix equation a little:

$$\begin{array}{ll} \cos\alpha\, x + \sin\alpha\, y = \lambda x \\ \sin\alpha\, x - \cos\alpha\, y = \lambda y \end{array} \quad \Rightarrow \quad \begin{array}{ll} (\cos\alpha - \lambda)x + & \sin\alpha\, y = 0 \\ \sin\alpha\, x + (-\cos\alpha - \lambda)y = 0 \end{array} \qquad (9.5)$$

Now we have a system of linear equations for x and y. Unfortunately, a third unknown has crept in, namely λ. And for these three unknowns the system is no longer linear. So we have to come up with something other than the usual Gaussian algorithm.

When can the system of equations (9.5) have a solution (x,y) different from $(0,0)$? We can read this from the coefficient matrix:

$$A = \begin{pmatrix} \cos\alpha - \lambda & \sin\alpha \\ \sin\alpha & -\cos\alpha - \lambda \end{pmatrix}.$$

Because according to Theorem 9.4 the existence of a non-trivial solution is equivalent to the determinant of A being equal to zero. So there can only be solutions if

$$(\cos\alpha - \lambda)(-\cos\alpha - \lambda) - (\sin\alpha)^2 = \lambda^2 - (\cos\alpha)^2 - (\sin\alpha)^2 = \lambda^2 - 1 = 0.$$

Look at this: x and y have disappeared from our equation by magic, and we can calculate λ. The solutions $\lambda_{1/2} = \pm 1$ are the possible eigenvalues of the reflection matrix. We can insert these values into (9.5) and obtain for each eigenvalue a very ordinary system of linear equations that we can solve in our sleep:

$$\lambda = +1: \quad \begin{array}{ll} (\cos\alpha - 1)x + & \sin\alpha\, y = 0 \\ \sin\alpha\, x + (-\cos\alpha - 1)y = 0 \end{array}$$

$$\lambda = -1: \quad \begin{array}{ll} (\cos\alpha + 1)x + & \sin\alpha\, y = 0 \\ \sin\alpha\, x + (-\cos\alpha + 1)y = 0 \end{array}$$

In the first system of equations we get, for example, as a solution $x = \cos\alpha + 1$, $y = \sin\alpha$. If we plot this (Fig. 9.2), we see that this vector lies exactly on the line that forms the angle $\alpha/2$ with the x-axis. Eigenvalue $+1$ says that this axis remains the same point by point.

Calculate the eigenvector for the eigenvalue -1 yourself. You will notice that it is perpendicular to the first eigenvector. This vector is reflected. Through the images of the two eigenvectors, the linear mapping is completely determined. It really is a reflection. ◄

Fig. 9.2 Eigenvectors of the
reflection

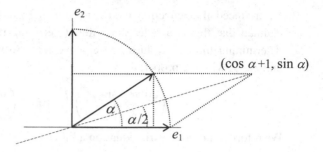

We will derive a general method for computing eigenvalues, similar to the one used for the reflection matrix. But first we have to make a few preparations:

▶ **Definition 9.11** If λ is an eigenvalue of the linear mapping A, then the set of eigenvectors to λ

$$T_\lambda := \{v \in K^n \,|\, Av = \lambda v\}$$

is called *eigenspace* of the eigenvalue λ.

A simple arithmetic trick will help us determine the eigenvalues: Because of $Iv = v$, it is actually true that $\lambda v = \lambda I v$, and so we can write, using the known matrix operations:

$$Av = \lambda v \Leftrightarrow Av = \lambda I v \Leftrightarrow Av - \lambda I v = \vec{0} \Leftrightarrow (A - \lambda I)v = \vec{0}. \tag{9.6}$$

$(A - \lambda I)$ is a linear mapping. The transformation in (9.6) says that v is an eigenvector to λ if and only if v lies in the kernel of the linear mapping $A - \lambda I$:

▶ **Theorem 9.12** The eigenspace T_λ for the eigenvalue λ is the kernel of the mapping $(A - \lambda I)$.

In the example of the reflection matrix from above, we just examined this matrix $A - \lambda I$.

Since kernels of linear mappings are subspaces, we conclude:

▶ **Theorem 9.13** The eigenspace T_λ for an eigenvalue is a subspace.

A number λ is an eigenvalue if and only if there is an eigenvector $\neq 0$, that is, if the kernel of $(A - \lambda I)$ contains more than 0, it must have dimension greater than or equal to 1. Theorem 9.4 then says that $\det(A - \lambda I) = 0$:

▶ **Theorem 9.14** λ is an eigenvalue of the linear mapping A if and only if $\det(A - \lambda I) = 0$.

Theorems 9.12 and 9.14 give us the recipe for calculating eigenvalues and eigenvectors: We determine the eigenvalues λ according to Theorem 9.14 and with the help of Theorem 9.12 we find the corresponding eigenspaces.

First of all, let us see what $\det(A - \lambda I)$ looks like for real 2×2- and 3×3-matrices:

$$A = \begin{pmatrix} a & b \\ c & d \end{pmatrix}, \quad A - \lambda I = \begin{pmatrix} a - \lambda & b \\ c & d - \lambda \end{pmatrix}.$$

$$\det(A - \lambda I) = (a - \lambda)(d - \lambda) - bc = \lambda^2 - (a + d)\lambda + (ad - bc).$$

This is a quadratic equation for λ. To calculate roots use the formula:

$$\lambda_{1/2} = \frac{a + d}{2} \pm \sqrt{\left(\frac{a + d}{2}\right)^2 - ad + bc}.$$

As you know, there are sometimes 2, sometimes 1, and sometimes no solutions for this in $\mathbb{R}$: Linear mappings in $\mathbb{R}^2$ have 0, 1, or 2 real eigenvalues.

We've seen examples of all these cases already. Can you match them up?

Now for the 3×3-matrices:

$$A = \begin{pmatrix} a & b & c \\ d & e & f \\ g & h & i \end{pmatrix}, \quad A - \lambda E = \begin{pmatrix} a - \lambda & b & c \\ d & e - \lambda & f \\ g & h & i - \lambda \end{pmatrix},$$

$$\det(A - \lambda E) = (a - \lambda)(e - \lambda)(i - \lambda) + bfg + cdh$$
$$- c(e - \lambda)g - db(i - \lambda) - hf(a - \lambda).$$

If we multiply this expression out and sort by powers of λ, we get something in the form:

$$\det(A - \lambda E) = -\lambda^3 + \alpha\lambda^2 + \beta\lambda + \gamma, \quad \alpha, \beta, \gamma \in \mathbb{R},$$

so a polynomial of degree 3 in λ. In general, we have:

▶ **Theorem and Definition 9.15** For $A \in K^{n \times n}$, $\det(A - \lambda E)$ is a polynomial in λ of degree n. This polynomial is called the *characteristic polynomial* of A.

The roots of this polynomial are the eigenvalues of the matrix A.

The fact that a $n \times n$-matrix always results in a polynomial of degree n can, for example, be derived from the determinant representation in Definition 9.6 by mathematical induction.

I don't want to hide that it is a bit laborious to write this proof down cleanly. Much easier it is, if one uses an alternative (and of course equivalent) definition of the determinant, which you can find in almost every mathematics book: The determinant of an $n \times n$-matrix is always a sum of n-fold products of permutations of the matrix elements. In the case of 2×2- and 3×3-matrices you have seen this already. But the proof of Theorem 9.15 this is the only place where this definition would be useful in this book, I have dispensed with the quite technical introduction. But if you read about this form of the determinant elsewhere: it is the same.

This theorem has an immediate, very surprising consequence for real 3×3-matrices, even for all real $n \times n$-matrices, provided that n is odd:

▶ **Theorem 9.16** For odd n every matrix from $\mathbb{R}^{n \times n}$ has at least one real eigenvalue.

For the proof you need to briefly flip to the second part of the book: From Bolzano's theorem it follows that every real polynomial of odd degree has at least one root (see Theorem 14.23 in Sect. 14.3).   □

Think about what this means: You cannot find a linear mapping in the $\mathbb{R}^3$ which does not have at least one vector that keeps its direction. If a soccer ball flies across the field during the game, it goes through linear mappings and translations. If the referee puts the ball back on the kickoff spot after the break, the translations cancel out, only a linear mapping remains. That means there is an axis in the soccer ball that has exactly the same position as at the beginning of the game. The ball has only rotated a bit around this axis.

  This circumstance makes it much easier to follow the movement of bodies on a computer.

  For complex numbers the fundamental theorem of algebra (Theorem 5.14) results in:

▶ **Theorem 9.17** Every matrix from $\mathbb{C}^{n \times n}$ has at least one eigenvalue.

We know that every complex polynomial of degree n decomposes into linear factors, so it has n roots. But beware, that does not mean that every complex $n \times n$-matrix n has eigenvalues: roots can also occur multiple times, but only the different roots provide different eigenvalues.

And now another finding follows with our knowledge of polynomials: According to Theorem 5.25 in Sect. 5.4 a polynomial of degree n has at most n roots, so it holds:

▶ **Theorem 9.18** A $n \times n$-matrix has at most n different eigenvalues.

Examples

1. $A = \begin{pmatrix} 3 & -1 \\ 1 & 1 \end{pmatrix}$, $\det(A - \lambda E) = \lambda^2 - 4\lambda + 4 = (\lambda - 2)^2$,

   has one eigenvalue, namely $\lambda = 2$. To determine the corresponding eigenvectors, we now have to solve the system of linear equations $(A - 2E)x = 0$:

   $$\begin{pmatrix} 3-2 & -1 \\ 1 & 1-2 \end{pmatrix} \begin{pmatrix} x_1 \\ x_2 \end{pmatrix} = \begin{pmatrix} 1 & -1 \\ 1 & -1 \end{pmatrix} \begin{pmatrix} x_1 \\ x_2 \end{pmatrix} = \begin{pmatrix} 0 \\ 0 \end{pmatrix}.$$

   A good control possibility to check whether you have determined the eigenvalues correctly always consists in determining the rank of the corresponding matrix. In any case, this must be smaller than the number of rows, in this case therefore at most 1. This is true.

In general, this system of equations will now be solved again using the Gaussian algorithm. But here we immediately see that $(1, 1)$ is a basis of the solution space. The matrix A therefore has the eigenvector $(1, 1)$ (and of course all multiples thereof) to the eigenvalue $+2$.

2. $A = \begin{pmatrix} 1 & -1 \\ 2 & -1 \end{pmatrix}$, $\det(A - \lambda E) = \lambda^2 + 1$.

This matrix has no real eigenvalues, but complex ones.

3. $A = \begin{pmatrix} 2 & 2 \\ 0 & 1 \end{pmatrix}$, $\det(A - \lambda E) = (2 - \lambda)(1 - \lambda)$,

has the eigenvalues 2 and 1. The following two systems of equations have to be solved:

$$\begin{pmatrix} 0 & 2 \\ 0 & -1 \end{pmatrix} \begin{pmatrix} x_1 \\ x_2 \end{pmatrix} = \begin{pmatrix} 0 \\ 0 \end{pmatrix} \quad \text{and} \quad \begin{pmatrix} 1 & 2 \\ 0 & 0 \end{pmatrix} \begin{pmatrix} x_1 \\ x_2 \end{pmatrix} = \begin{pmatrix} 0 \\ 0 \end{pmatrix}.$$

Basis of the eigenspace to 2 is the vector $(1, 0)$, basis of the eigenspace to 1 is for example $(2, -1)$.

Draw, similar to the examples in Sect. 7.1, what these mappings do with a square.

4. Finally, an example of a 3×3-matrix. I don't go through all the steps in detail, please work on paper:

$$A = \begin{pmatrix} 5 & -1 & 2 \\ -1 & 5 & 2 \\ 2 & 2 & 2 \end{pmatrix}, \quad \det(A - \lambda E) = (-\lambda)(\lambda^2 - 12\lambda + 36).$$

The first zero is $\lambda_1 = 0$, further we get $\lambda_{2/3} = 6 \pm \sqrt{36 - 36} = 6$. We therefore receive two eigenvalues and have to solve the two systems of equations $(A - 6E)x = 0$ and $(A - 0E)x = 0$:

$$A - 6E = \begin{pmatrix} -1 & -1 & 2 \\ -1 & -1 & 2 \\ 2 & 2 & -4 \end{pmatrix}.$$

This matrix has rank 1, the dimension of the solution space is therefore 2: for example, $(-1, 1, 0)$ and $(2, 0, 1)$ form a basis of the eigenspace to the eigenvalue 6.

$$A - 0E = A = \begin{pmatrix} 5 & -1 & 2 \\ -1 & 5 & 2 \\ 2 & 2 & 2 \end{pmatrix}.$$

The matrix has rank 2, the dimension of the solution space is 1 and a basis of the eigenspace is for example $(1, 1, -2)$.

Make the test! ◄

In the last example, we found one eigenspace of dimension 2 and one of dimension 1. Could the second eigenspace have also had dimension 2? No, because more than three linearly independent eigenvectors do not fit into the $\mathbb{R}^3$, and the following result applies:

▶ **Theorem 9.19** Eigenvectors different from $\vec{0}$ belonging to different eigenvalues are linearly independent.

The eigenvectors of different eigenspaces have therefore nothing to do with each other.

I think this theorem is obvious: If an eigenvector were in two different eigenspaces, he would not even know where to map. I spare us the precise induction proof. We can draw a conclusion from the theorem that will be interesting in the next section:

▶ **Corollary 9.20** If the matrix $A \in K^{n \times n}$ n has eigenvalues, it has a basis of eigenvectors.

Because to each eigenvalue belongs at least one eigenvector different from 0.   □

As you have seen in the last example, a matrix with fewer eigenvalues can also have a basis of eigenvectors: $(-1,1,0)$, $(2,0,1)$ and $(1,1,-2)$ form a basis of the $\mathbb{R}^3$.

9.3   Change of Basis

The matrix of a reflection at the x_1-axis in the $\mathbb{R}^2$ has a very simple shape: The basis vector e_1 remains fixed, e_2 is turned over. As a matrix, this results in:

$$\begin{pmatrix} 1 & 0 \\ 0 & -1 \end{pmatrix}.$$

The matrix with respect to the reflection at another axis looks much more complicated. Of course, this is because in this special case the basis vectors have not been bent, they are eigenvectors of the reflection.

A key goal of the calculation of eigenvectors is to find bases with respect to which the matrix of a linear mapping looks as simple as possible. If you can choose eigenvectors as basis vectors, you will achieve this goal. If, for example, we choose in the last example before Theorem 9.19 as a basis the three found eigenvectors $(-1, 1, 0)$, $(2, 0, 1)$ and $(1,1,-2)$, then the matrix of the corresponding mapping with respect to this basis has the shape

$$\begin{pmatrix} 6 & 0 & 0 \\ 0 & 6 & 0 \\ 0 & 0 & 0 \end{pmatrix},$$

because the columns of the matrix are the images of the basis and the basis vectors are only stretched.

In Sect. 6.6 we have already seen that vectors have different coordinates with respect to different bases. In a two-dimensional example, we have calculated in Theorem 6.22 how the coordinates of a vector can be calculated in different bases. We now have to do this a little more systematically: If a vector is given with respect to the basis B_1, we want to determine its coordinates with respect to the basis B_2. Furthermore, let A be the matrix of a linear mapping with respect to B_1, we want to know the matrix of this mapping with respect to the basis B_2.

Recall that if $B = \{b_1, b_2, \ldots, b_n\}$ is a basis of the K^n, then every $v \in K^n$ can be written as a linear combination of the basis vectors: $v = \lambda_1 b_1 + \lambda_2 b_2 + \cdots + \lambda_n b_n$. We then write

$$v = \begin{pmatrix} \lambda_1 \\ \vdots \\ \lambda_n \end{pmatrix}_B .$$

The basis vectors have the coordinates

$$b_1 = \begin{pmatrix} 1 \\ 0 \\ \vdots \\ 0 \end{pmatrix}_B , \quad \ldots, \quad b_n = \begin{pmatrix} 0 \\ 0 \\ \vdots \\ 1 \end{pmatrix}_B .$$

What is the relationship between the coordinates of a vector v with respect to the bases B_1 and B_2? Let $B_2 = \{b_1, b_2, \ldots, b_n\}$, and let the coordinates of the b_i with respect to B_1 and v with respect to B_1 and B_2 be given:

$$b_1 = \begin{pmatrix} b_{11} \\ \vdots \\ b_{n1} \end{pmatrix}_{B_1} , \ldots, b_n = \begin{pmatrix} b_{1n} \\ \vdots \\ b_{nn} \end{pmatrix}_{B_1} , \quad v = \begin{pmatrix} v_1 \\ \vdots \\ v_n \end{pmatrix}_{B_1} , \quad v = \begin{pmatrix} \lambda_1 \\ \vdots \\ \lambda_n \end{pmatrix}_{B_2} .$$

Then

$$v = \begin{pmatrix} v_1 \\ \vdots \\ v_n \end{pmatrix}_{B_1} = \lambda_1 b_1 + \lambda_2 b_2 + \cdots \lambda_n b_n$$

$$= \lambda_1 \begin{pmatrix} b_{11} \\ \vdots \\ b_{n1} \end{pmatrix}_{B_1} + \cdots + \lambda_n \begin{pmatrix} b_{1n} \\ \vdots \\ b_{nn} \end{pmatrix}_{B_1} = \begin{pmatrix} b_{11}\lambda_1 + \cdots + b_{1n}\lambda_n \\ \vdots \\ b_{n1}\lambda_1 + \cdots + b_{nn}\lambda_n \end{pmatrix}_{B_1} . \tag{9.7}$$

The matrix

$$
T = \begin{pmatrix} b_{11} & \cdots & b_{1n} \\ \vdots & & \vdots \\ b_{n1} & \cdots & b_{nn} \end{pmatrix}
$$

is an invertible matrix, because the columns are linearly independent and from (9.7) it follows:

$$
\begin{pmatrix} v_1 \\ \vdots \\ v_n \end{pmatrix} = T \begin{pmatrix} \lambda_1 \\ \vdots \\ \lambda_n \end{pmatrix}, \text{ and thus also } \begin{pmatrix} \lambda_1 \\ \vdots \\ \lambda_n \end{pmatrix} = T^{-1} \begin{pmatrix} v_1 \\ \vdots \\ v_n \end{pmatrix}. \tag{9.8}
$$

In (9.8) I have omitted the indexes. It is a calculation in base B_1, but here it is only about the mathematical connection between the numbers $(v_1, v_2, \cdots, v_n)$ and $(\lambda_1, \lambda_2, \cdots, \lambda_n)$. We can therefore formulate the following Theorem:

▶ **Theorem 9.21** Let f be the linear mapping that maps the basis B_1 to the basis B_2. T be the matrix of this mapping with respect to the basis B_1, the columns of T therefore contain the basis $B_2 = \{b_1, b_2, \ldots, b_n\}$ in the coordinates with respect to B_1. Then the coordinates of a vector v with respect to B_2 are obtained by multiplying the coordinates of v with respect to B_1 from the left with T^{-1}. The coordinates of v with respect to B_1 are obtained from those with respect to B_2 by multiplication from the left with T.

Now turn back to the example after Theorem 6.22 where we did exactly that without knowing at the time what a matrix is.

The matrix T is called the *transition matrix* from B_1 to B_2.

▶ **Theorem 9.22 Change of Basis** If T is the transition matrix from B_1 to B_2 and S is the transition matrix from B_2 to B_3, then TS is the transition matrix from B_1 to B_3.

This follows from (9.8), if you also write this line for the matrix S. I will leave the proof to you as an exercise.   □

Occasionally, the order of matrix multiplication causes confusion, it is exactly the reverse of the order of performing linear mappings: If you first perform the linear mapping A and then the mapping B, you get the linear mapping BA. But first performing the transition T and then the transition S, you get the transition TS.

Let's look at Example 4 again before Theorem 9.19 and transform into the basis of eigenvectors: The matrix T is:

$$T = \begin{pmatrix} -1 & 2 & 1 \\ 1 & 0 & 1 \\ 0 & 1 & -2 \end{pmatrix},$$

the inverse of which is

$$\frac{1}{6} \begin{pmatrix} -1 & 5 & 2 \\ 2 & 2 & 2 \\ 1 & 1 & -2 \end{pmatrix}.$$

Then, for example,

$$\frac{1}{6} \begin{pmatrix} -1 & 5 & 2 \\ 2 & 2 & 2 \\ 1 & 1 & -2 \end{pmatrix} \begin{pmatrix} -1 \\ 1 \\ 0 \end{pmatrix} = \begin{pmatrix} 1 \\ 0 \\ 0 \end{pmatrix},$$

because $\begin{pmatrix} -1 \\ 1 \\ 0 \end{pmatrix}_{B_1}$ is the first basis vector in the basis B_2:

$$\begin{pmatrix} -1 \\ 1 \\ 0 \end{pmatrix}_{B_1} = \begin{pmatrix} 1 \\ 0 \\ 0 \end{pmatrix}_{B_2}.$$

How do we determine the matrix of a linear mapping with respect to the basis B_2? Let $f: K^n \to K^n$ be a linear mapping, $A \in K^{n \times n}$ the corresponding matrix with respect to B_1 and $S \in K^{n \times n}$ the matrix of f with respect to B_2. Let $v, w \in K^n$ with $f(v) = w$. Further, the coordinates of v and w with respect to the bases B_1 and B_2 are given:

$$v = \begin{pmatrix} v_1 \\ \vdots \\ v_n \end{pmatrix}_{B_1}, w = \begin{pmatrix} w_1 \\ \vdots \\ w_n \end{pmatrix}_{B_1} \quad \text{and} \quad v = \begin{pmatrix} \lambda_1 \\ \vdots \\ \lambda_n \end{pmatrix}_{B_2}, w = \begin{pmatrix} \mu_1 \\ \vdots \\ \mu_n \end{pmatrix}_{B_2}.$$

Thus, for the matrices A and S:

$$A \begin{pmatrix} v_1 \\ \vdots \\ v_n \end{pmatrix} = \begin{pmatrix} w_1 \\ \vdots \\ w_n \end{pmatrix}, \quad S \begin{pmatrix} \lambda_1 \\ \vdots \\ \lambda_n \end{pmatrix} = \begin{pmatrix} \mu_1 \\ \vdots \\ \mu_n \end{pmatrix}. \tag{9.9}$$

If T is the transition matrix, then we get by Theorem 9.21:

$$T\begin{pmatrix} \lambda_1 \\ \vdots \\ \lambda_n \end{pmatrix} = \begin{pmatrix} v_1 \\ \vdots \\ v_n \end{pmatrix}, \quad T\begin{pmatrix} \mu_1 \\ \vdots \\ \mu_n \end{pmatrix} = \begin{pmatrix} w_1 \\ \vdots \\ w_n \end{pmatrix}. \tag{9.10}$$

We set (9.10) in the left half of (9.9) and thus $AT\begin{pmatrix} \lambda_1 \\ \vdots \\ \lambda_n \end{pmatrix} = T\begin{pmatrix} \mu_1 \\ \vdots \\ \mu_n \end{pmatrix}$ and from this by multiplication with T^{-1} from the left:

$$T^{-1}AT\begin{pmatrix} \lambda_1 \\ \vdots \\ \lambda_n \end{pmatrix} = \begin{pmatrix} \mu_1 \\ \vdots \\ \mu_n \end{pmatrix}. \tag{9.11}$$

The comparison of (9.11) with the right half of (9.9) finally yields $T^{-1}AT = S$, because the matrix of a linear mapping is uniquely determined. So we finally have the desired relationship:

▶ **Theorem 9.23** If T is the transition matrix from the basis B_1 to the basis B_2 and f is a linear mapping to which the matrix A belongs with respect to B_1, then the matrix $T^{-1}AT$ belongs to f with respect to B_2.

Please note that matrix multiplication is not commutative. In general, therefore, $T^{-1}AT \neq A$!

In the example above, you can check that it holds:

$$\underbrace{\frac{1}{6}\begin{pmatrix} -1 & 5 & 2 \\ 2 & 2 & 2 \\ 1 & 1 & -2 \end{pmatrix}}_{T^{-1}} \underbrace{\begin{pmatrix} 5 & -1 & 2 \\ -1 & 5 & 2 \\ 2 & 2 & 2 \end{pmatrix}}_{A} \underbrace{\begin{pmatrix} -1 & 2 & 1 \\ 1 & 0 & 1 \\ 0 & 1 & -2 \end{pmatrix}}_{T} = \underbrace{\begin{pmatrix} 6 & 0 & 0 \\ 0 & 6 & 0 \\ 0 & 0 & 0 \end{pmatrix}}_{S}.$$

▶ **Definition 9.24** Two matrices $A, B \in K^{n \times n}$ are called *similar* if there is an invertible matrix $T \in K^{n \times n}$ with $A = T^{-1}BT$.

This means nothing other than that A and B describe the same linear mapping with respect to two different bases. In particular, therefore, A and B have the same eigenvalues and the same determinant.

The similarity of matrices is, by the way, an equivalence relation.

▶ **Theorem 9.25** Let $f: K^n \to K^n$ be a linear mapping. If K^n has with respect to f a basis of eigenvectors, then to f with respect to this basis belongs the diagonal matrix

$$
\begin{pmatrix}
\lambda_1 & 0 & \cdots & 0 \\
0 & \lambda_2 & & \vdots \\
\vdots & & \ddots & 0 \\
0 & \cdots & 0 & \lambda_n
\end{pmatrix},
\tag{9.12}
$$

where in the diagonal are the eigenvalues of f.

We know this already: The columns are the images of the eigenvectors and because of $f(b_i) = \lambda_i b_i$ the assertion is true. □

▶ **Definition 9.26** The matrix $A \in K^{n \times n}$ is called *diagonalizable*, if there is a similar matrix S which has the form (9.12).

The matrix A is diagonalizable, if and only if the eigenvectors of A form a basis.

Let's examine the reflection in $\mathbb{R}^2$ again. We know by now that the corresponding matrix

$$
M_\alpha = \begin{pmatrix} \cos\alpha & \sin\alpha \\ \sin\alpha & -\cos\alpha \end{pmatrix}
$$

is diagonalizable. A basis of the diagonalized matrix consists of the axis b_1 and the vector perpendicular to it b_2. The length of b_1 and b_2 should be 1 each. The matrix M_B with respect to the new basis B has the form

$$
M_B = \begin{pmatrix} 1 & 0 \\ 0 & -1 \end{pmatrix}.
$$

What does the transition matrix look like? The linear mapping T, which maps e_1 to b_1 and e_2 to b_2, is just the rotation by the angle $\alpha/2$, and T^{-1} is the rotation back by $-\alpha/2$ (Fig. 9.3):

Fig. 9.3 Change of bsis

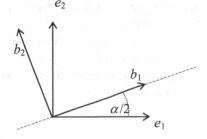

$$T = \begin{pmatrix} \cos(\alpha/2) & -\sin(\alpha/2) \\ \sin(\alpha/2) & \cos(\alpha/2) \end{pmatrix}, \quad T^{-1} = \begin{pmatrix} \cos(-\alpha/2) & -\sin(-\alpha/2) \\ \sin(-\alpha/2) & \cos(-\alpha/2) \end{pmatrix}.$$

With our knowledge of the transition matrix from Theorem 9.23 it must therefore hold:

$$\begin{pmatrix} \cos(\alpha/2) & -\sin(\alpha/2) \\ \sin(\alpha/2) & \cos(\alpha/2) \end{pmatrix} \begin{pmatrix} 1 & 0 \\ 0 & -1 \end{pmatrix} \begin{pmatrix} \cos(-\alpha/2) & -\sin(-\alpha/2) \\ \sin(-\alpha/2) & \cos(-\alpha/2) \end{pmatrix} = \begin{pmatrix} \cos\alpha & \sin\alpha \\ \sin\alpha & -\cos\alpha \end{pmatrix}.$$

Calculate it using the addition rules for cosine and sine from the formulary or from Theorem 14.21 in Sect. 14.2. It actually works out!

Orientation of Vector Spaces

Since you first saw a coordinate system, you are used to drawing the x_1-axis to the right and the x_2-axis up and not down. At first it caused me quite a bit of trouble that some drawing programs place the origin in the upper left corner of the screen; probably because the image is built up line by line from top to bottom. What is it about this top, bottom, right and left?

Man is not rotationally symmetrical and can therefore distinguish between left and right. We can sit on the tail of a vector in a plane, look in the direction of the vector and say what is to the left and what is to the right of it. It is a common convention to set up our two-dimensional coordinate system so that the basis vector e_2 is to the left of the basis vector e_1. Such a coordinate system is called positively oriented, another negatively oriented. If you exchange e_1 and e_2, the orientation just turns around.

In three dimensions it is a bit more difficult: If we sit in the origin and look in the direction of a vector, we cannot speak of left or right: We need a reference plane. This is the plane that is formed by the two basis vectors e_1 and e_2.

Can we say from this plane in the $\mathbb{R}^3$ whether its coordinate system is positive or negative? No, it depends on which side of the plane we sit on!

The convention here is: If you rotate the basis vector e_1 in the direction of e_2, e_3 should be on the side of the plane into which a screw would be screwed in during this rotation. Perhaps you also know the "right-hand rule": If you stretch your thumb, index finger and middle finger of your right hand out linearly independently, you get (in this order) a positively oriented system.

As in the $\mathbb{R}^2$ there are therefore two types of coordinate systems here: positively and negatively oriented.

How can we describe these orientations mathematically? The determinant gives us the means to do so.

▶   **Definition and Theorem 9.27** Two bases of the vector space K^n are called *equally oriented* if the determinant of the transition matrix is greater than 0. The orientation is an equivalence relation on the set of bases of K^n and divides them into two classes.

Let us call the relation "$\sim$". Then it holds:

$B_1 \sim B_1$, because the transition matrix is I and $\det I = 1$.

$B_1 \sim B_2 \Rightarrow B_2 \sim B_1$, because if T is the transition matrix from B_1 to B_2, then T^{-1} is the transition matrix from B_2 to B_1. Since $\det TT^{-1} = \det T \cdot \det T^{-1} = \det I = 1$, with $\det T$ also $\det T^{-1}$ has a positive sign.

$B_1 \sim B_2$, $B_2 \sim B_3 \Rightarrow B_1 \sim B_3$, because if is T the transition matrix from B_1 to B_2, and S the transition matrix from B_2 to B_3, so according to Theorem 9.22 TS is the transition matrix from B_1 to B_3 and $\det TS = \det T \cdot \det S > 0$.

The assertion that this equivalence relation has exactly two equivalence classes, I would like to leave as an exercise to get you used to change of basis and determinants a little more. □

The tool of the determinant does not relieve us of the task of saying which coordinate systems are positively oriented and which are not. We have to specify that. If not almost every human being had his right hand with him all the time, the prototype of a positively oriented coordinate system would have to be kept in Paris beside the prototype metre and the prototype kilogram.

But if we have fixed our prototype in the $\mathbb{R}^2$ or $\mathbb{R}^3$, does the set of positively oriented bases agree with the convention we made at the beginning of this section?

Let's think about that for the plane. We start from a positively oriented coordinate system (e_2 is to the left of e_1) and carry out a change of basis with a positive determinant. Then e_2' should be to the left of e_1' again. Let's split the transition into two parts: first e_2 is kept and e_1 is transformed into e_1', then e_1' is kept and e_2 is transformed. The base transition matrices T and S then look as follows:

$$T = \begin{pmatrix} a & 0 \\ b & 1 \end{pmatrix}, \quad S = \begin{pmatrix} 1 & c \\ 0 & d \end{pmatrix}.$$

It is $\det T = a$ and $\det S = d$. Since $\det TS = \det T \cdot \det S > 0$, both a and d are greater than 0 or both are less than 0.

Let's assume that a and d are both greater than 0. (a, b) are the coordinates of e_1' with respect to the original basis and $a > 0$ means that e_1 and e_1' are on the same side of e_2, so e_2 is to the left of e_1'. Similarly, (c, d) are the coordinates of e_2' in the coordinate system (e_1', e_2) and $d > 0$ means e_2 and e_2' are on the same side of e_1', so e_2' is also to the left of e_1' (Fig. 9.4).

Similarly, one argues if a and d are both less than 0.

Fig. 9.4 Orientation of
coordinate systems

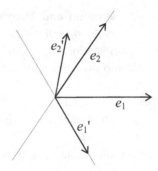

Our convention is also compatible with the determinant rule in $\mathbb{R}^3$. Is it possible to formulate something like the right-hand rule in higher-dimensional spaces? I don't think so; we can start with a triad, but in which direction should the four-dimensional screw then turn? In general, we simply work with a basis that is our standard basis and compare the orientations of other bases with this.

Now we also see what it is with the signs of a volume that the determinant provides us with: the volume of a n-dimensional parallelogram is positive if the vectors describing the edges have the same orientation as the basis of the vector space. Otherwise the volume is negative.

If you want to move or deform an object in computer graphics, you have to make sure that you apply transformations that preserve orientation, unless you want to look at the object in the mirror. Check the transformations I introduced in Sect. 7.1 from this point of view.

Ray Tracing, Part 2

In the first part of the ray tracing, at the end of Chap. 8, we dealt with the problem of determining the intersection of a line with an object. We assume that the object is bounded by a series of irregular polygons. In the first step, we determined the intersection of the line with the plane in which the polygon lies. But how do we find out whether the point is inside or outside the polygon?

First of all, we can move the problem into the $\mathbb{R}^2$: A point is inside the polygon if this is true for the projection of the polygon onto one of the coordinate planes. So we project onto a coordinate plane in which the polygon doesn't degenerate into a line. This is simply done by setting one of the three coordinates to zero.

Let's test now whether the point P is inside the polygon. First of all, we assume the polygon is convex, that is, the line segment between two points is also inside the polygon. If a reference point R is given inside the polygon, we have to check whether the line PR intersects the border of the polygon or not. The border of the polygon consists of line segments itself, and so the problem reduces to the repeated investigation of the question

Fig. 9.5 Ray Tracing 2

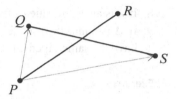

whether two line segments PR and QS intersect. For this we need a fast algorithm. The sign of the determinant and its relationship with the orientation of linearly independent vectors are a suitable tool:

PR and QS intersect exactly when Q and S are on different sides of the vector $\vec{PR}$, and P and R are on different sides of $\vec{QS}$ (Fig. 9.5). According to what we have just learned about the orientation of coordinate systems, this means:

$$\det \left(\vec{PR} \ \ \vec{PQ} \right) \cdot \det \left(\vec{PR} \ \ \vec{PS} \right) \leq 0 \quad \text{and} \quad \det \left(\vec{QS} \ \ \vec{QR} \right) \cdot \det \left(\vec{QS} \ \ \vec{QP} \right) \leq 0.$$

The vectors are the column vectors of the matrices, respectively. The special case $= 0$ occurs when one of the endpoints lies on the other line segment. To perform this check, we therefore need 8 multiplications.

The procedure can easily be extended to non-convex polygons: P lies in the polygon if the number of intersections of PR with the edge of the polygon is even, otherwise it lies outside.

9.4   Comprehension Questions and Exercises

Comprehension Questions

1. What is the geometrical meaning of the determinant of a matrix?
2. Explain the relationship between the determinant of a matrix, the rank of the matrix, and the kernel of the corresponding linear mapping.
3. In a rotation-reflection in $\mathbb{R}^3$, a plane is first reflected and then rotated around the axis perpendicular to this plane. How many eigenvectors does a rotation reflection have? What are the corresponding eigenvalues? Are there any special cases?
4. Is $\vec{0}$ an eigenvector for every eigenvalue?
5. If u is an eigenvector for the eigenvalue λ and v is an eigenvector for the eigenvalue μ $(\lambda \neq \mu)$, is $u + v$ also an eigenvector?
6. If a matrix from the $\mathbb{R}^{4\times4}$ has three eigenvalues and the dimension of the eigenspace for the first eigenvalue is 2, does there exist a basis of eigenvectors for the matrix?
7. True or false: If an eigenvalue of the matrix from the $\mathbb{R}^{3\times3}$ has an eigenspace with dimension 2, then the matrix has exactly one additional eigenvalue.

8. If λ is an eigenvalue of the matrix A from $\mathbb{R}^{2\times 2}$, then can $\operatorname{rank}(A - \lambda I) = 2$?
9. If A is a matrix from $\mathbb{R}^{n\times n}$ and $T \in \mathbb{R}^{n\times n}$ is an invertible matrix, does $T^{-1}AT$ then have the same eigenvalues as A?

Exercises

1. Show: The determinant of a 2×2-matrix $\begin{pmatrix} a & b \\ c & d \end{pmatrix}$ is exactly the area of the parallelogram determined by the two row vectors.

2. Calculate the determinants of the following matrices:

$$\begin{pmatrix} 2 & -6 & 4 & 0 \\ 4 & -12 & -1 & 2 \\ 1 & 7 & 2 & 1 \\ 0 & 10 & 3 & 9 \end{pmatrix}, \quad \begin{pmatrix} 3 & -4 & 0 & 2 \\ 0 & 7 & 6 & 3 \\ 2 & -6 & 0 & 1 \\ 5 & 3 & 1 & -2 \end{pmatrix}, \quad \begin{pmatrix} 1 & 0 & -1 & 2 \\ 2 & 1 & 0 & 1 \\ -3 & 1 & 0 & 1 \\ 2 & 2 & 0 & -1 \end{pmatrix}.$$

3. a) Calculate the eigenvalues of the matrices $\begin{pmatrix} 5 & -1 & 2 \\ -1 & 5 & 2 \\ 2 & 2 & 2 \end{pmatrix}$ and $\begin{pmatrix} 1 & -3 & 3 \\ 3 & -5 & 3 \\ 6 & -6 & 4 \end{pmatrix}$.

   b) For each eigenvalue, give a basis of the corresponding eigenspace. Are the matrices diagonalizable? If so, give the corresponding transition matrix.

4. Show that when two changes of basis are carried out in succession, the transition matrices are multiplied: If T is the transition matrix from B_1 to B_2 and S is from B_2 to B_3, then TS is the transition matrix from B_1 to B_3.

5. Show that the equivalence relation "oriented equally" partitions the set of bases of $\mathbb{R}^n$ into exactly two equivalence classes.

6. Why does a mirror swap left and right, but not top and bottom?

Dot Product and Orthogonal Mappings

10

Abstract

In this chapter you will learn:

- the dot product and norms for real vector spaces,
- the special dot product in $\mathbb{R}^2$ and $\mathbb{R}^3$,
- orthogonal mappings and their matrices,
- all types of orthogonal mappings in $\mathbb{R}^2$ and $\mathbb{R}^3$,
- homogeneous coordinates and their applications in robotics.

10.1 Dot Product

In real vector spaces, often the task arises of measuring lengths or distances, or of describing something like angles between vectors. For this, the dot product can be used. A dot product is a mapping that assigns a scalar to two vectors, in our case a real number.

Dot products are usually studied in mathematics for real or complex vector spaces. Here I would like to limit myself to real vector spaces, as these are the spaces from which our most important applications arise.

You already know how I proceed. I first collect the essential requirements for a dot product and then define a specific such product for the $\mathbb{R}^n$.

Supplementary Information The online version contains supplementary material available at https://doi.org/10.1007/978-3-658-40423-9_10.

▶   **Definition 10.1** Let V be a $\mathbb{R}$-vector space. A mapping $\langle \cdot, \cdot \rangle : V \times V \to \mathbb{R}$ is called a *dot product* (or *scalar product*) if for all $u, v, w \in V$ and for all $\lambda \in \mathbb{R}$ the following properties are satisfied:

(S1) $\langle u + v, w \rangle = \langle u, w \rangle + \langle v, w \rangle$,

(S2) $\langle \lambda u, v \rangle = \lambda \langle u, v \rangle$,

(S3) $\langle u, v \rangle = \langle v, u \rangle$,

(S4) $\langle u, u \rangle \geq 0, \langle u, u \rangle = 0 \Leftrightarrow u = 0$.

In (S4) lies the reason for our restriction to real spaces: We need to know what "≥ 0" means. For example, in $\mathbb{C}$ or $\mathrm{GF}(p)$ this is not possible. However, there is an extension of the notion to complex vector spaces.

From (S3) it follows that $\langle u, v + w \rangle = \langle u, v \rangle + \langle u, w \rangle$ and $\langle u, \lambda v \rangle = \lambda \langle u, v \rangle$ also hold. But you now know that mathematicians only include the essential requirements in axioms, nothing that can be derived from them.

The concept of norm is closely related to the concept of dot product. With the norm, we will measure the lengths of vectors:

▶   **Definition 10.2** Let V be a $\mathbb{R}$-vector space. A mapping $\| \cdot \| : V \to \mathbb{R}$ is called a *norm*, if for all $u, v \in V$ and for all $\lambda \in \mathbb{R}$ the following properties are satisfied:

(N1) $\|\lambda v\| = |\lambda| \cdot \|v\|$,

(N2) $\|u + v\| \leq \|u\| + \|v\|$ (the *triangle inequality*),

(N3) $\|v\| = 0 \Leftrightarrow v = \vec{0}$.

In Theorem 5.16 in Sect. 5.3 we have already encountered such a norm: The absolute value of a complex number is a norm if we interpret $\mathbb{C}$ as the vector space $\mathbb{R}^2$.

▶   **Theorem 10.3** If V is a vector space with a dot product, then a norm on V is defined by setting

$$\|v\| := \sqrt{\langle v, v \rangle}$$

The crux of the proof is the triangle inequality. This easily follows from the *Cauchy-Schwarz inequality*, which states that for all u, v it holds:

$$|\langle u, v \rangle| \leq \|u\| \cdot \|v\|. \tag{10.1}$$

The proof of this inequality is short but tricky, although only (S1)–(S4) and some basic properties of real numbers are used. Like the proof of Theorem 10.3 I will not carry it out here.

But finally to the specific dot product on $\mathbb{R}^n$:

▶ **Definition and Theorem 10.4** For $u = (u_1, u_2, \ldots, u_n), v = (v_1, v_2, \ldots, v_n) \in \mathbb{R}^n$ the mapping

$$\langle u, v \rangle := u_1 v_1 + u_2 v_2 + \cdots + u_n v_n$$

is a dot product and

$$\|u\| := \sqrt{\langle u, u \rangle} = \sqrt{u_1^2 + u_2^2 + \cdots + u_n^2}$$

is a norm.

All properties (S1) to (S4) can be easily calculated. (S4) follows from the fact that sums of squares in $\mathbb{R}$ are always greater than or equal to 0.   □

If in the future I speak of norm or dot product on $\mathbb{R}^n$, I always mean the mappings from Definition 10.4.

First of all, a few remarks about this theorem:

1. If $u \in \mathbb{R}^n$, $u \neq 0$, then $\frac{1}{\|u\|} u$ is a vector in the direction of u with length 1 (a normalized vector), because $\|\frac{1}{\|u\|} u\| = \frac{1}{\|u\|} \|u\| = 1$.
2. The norm of a vector in the $\mathbb{R}^2$ and $\mathbb{R}^3$ is nothing other than the usual length, which we can calculate from the coordinates using the Pythagorean theorem. You can read this from Fig. 10.1.

In this and the following theorem, it is essential for the first time that we use a Cartesian coordinate system: The Pythagorean theorem only applies to right angles, and sine and cosine are only defined in right-angled triangles. Arguments that use cosine, sine or Pythagoras are wrong, if you choose any arbitrary basis.

3. A little trick: If you consider u and v as one-column matrices, then the dot product $\langle u, v \rangle$ is nothing other than the matrix product $u^T v$, where u^T is the transposed matrix, that is, u is written as a row vector:

$$\langle u, v \rangle = u^T v. \tag{10.2}$$

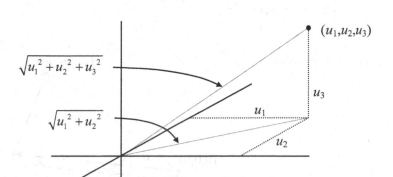

Fig. 10.1 The length of a vector

▶   **Theorem 10.5** In the $\mathbb{R}^2$ and $\mathbb{R}^3$, for $u, v \neq 0$:

$$\langle u, v \rangle = \|u\| \cdot \|v\| \cos \alpha,$$

where $0 \leq \alpha \leq 180°$ is the angle between the two vectors.

You can derive the Cauchy-Schwarz inequality (10.1), the proof of which I have withheld from you, for $\mathbb{R}^2$ and $\mathbb{R}^3$ from this formula.

For the proof: First, it is enough to examine vectors u, v with an angle $\alpha < 90°$. If the angle is more than $90°$, we calculate with the vectors u and $-v$, which then form the angle $180° - \alpha$ with each other. Since $\cos(180° - \alpha) = -\cos \alpha$, the assertion then follows from (S2).

We can further restrict ourselves in the calculation to vectors u, v of length 1 and show for such vectors: $\langle u, v \rangle = \cos \alpha$. For then, for any vectors u, v:

$$\langle u, v \rangle = \|u\| \|v\| \left\langle \frac{u}{\|u\|}, \frac{v}{\|v\|} \right\rangle = \|u\| \|v\| \cos \alpha.$$

For this situation, we can now make a drawing of the plane in which u and v lie (Fig. 10.2).

From this we can read:

$$h^2 + (\cos \alpha)^2 = 1, \tag{10.3}$$

$$h^2 + (1 - \cos \alpha)^2 = \|u - v\|^2,$$

multiplied out results in

$$h^2 + 1 - 2 \cos \alpha + (\cos \alpha)^2 = (u_1 - v_1)^2 + (u_2 - v_2)^2 + (u_3 - v_3)^2.$$

If we insert here (10.3), we get

$$1 + 1 - 2 \cos \alpha = u_1^2 - 2u_1 v_1 + v_1^2 + u_2^2 - 2u_2 v_2 + v_2^2 + u_3^2 - 2u_3 v_3 + v_3^2$$
$$2 - 2 \cos \alpha = \|u\|^2 + \|v\|^2 - 2\langle u, v \rangle$$
$$\Rightarrow \qquad \cos \alpha = \langle u, v \rangle.$$

Fig. 10.2 $\langle u, v \rangle = \|u\| \cdot \|v\| \cos \alpha$

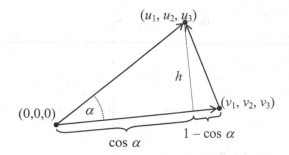

The special case $\alpha = 90°$ is still missing. But then $\|u\|^2 + \|v\|^2 = \|u - v\|^2$ (make a sketch!) And, if you insert the coordinates here, you get

$$\sum_{i=1}^{3} u_i^2 + \sum_{i=1}^{3} v_i^2 = \sum_{i=1}^{3} u_i^2 - \sum_{i=1}^{3} 2u_i v_i + \sum_{i=1}^{3} v_i^2$$

and from that $\langle u, v \rangle = 0$. □

The dot product of two vectors that are not 0 in $\mathbb{R}^2$ and $\mathbb{R}^3$ is therefore only 0 if they are perpendicular to each other. Such vectors are called *orthogonal*.

This property can be used to describe planes in $\mathbb{R}^3$ using the dot product. Let a vector $u = (u_1, u_2, u_3) \neq 0$ be given. Then the set E of all vectors that are perpendicular to u forms an plane through the origin. For

$$E = \{x \in \mathbb{R}^3 \mid \langle u, x \rangle = 0\} = \{(x_1, x_2, x_3) \mid u_1 x_1 + u_2 x_2 + u_3 x_3 = 0\},$$

and that is exactly the solution set of a homogeneous system of linear equations. Since the rank of the coefficient matrix is 1, this solution set has dimension 2, that is, a plane. The vector u is called the *normal vector* of the plane E.

If we now want to describe a plane E' perpendicular to u, which passes through the point v, we form the set

$$E' = \{x \in \mathbb{R}^3 \mid \langle u, x \rangle - \langle u, v \rangle = 0\} = \{(x_1, x_2, x_3) \mid u_1 x_1 + u_2 x_2 + u_3 x_3 = \langle u, v \rangle\}.$$

v is a special solution of this inhomogeneous system of linear equations. So, according to the theory of systems of linear equations, $E' = v + E$, the plane E shifted by v from the origin.

The following definitions, which we can formulate again independently of these special vector spaces, are based on the orthogonality relation for vectors in the $\mathbb{R}^2$ and $\mathbb{R}^3$:

▶ **Definition 10.6** In a vector space with dot product, two vectors u and v are called *orthogonal*, if $\langle u, v \rangle = 0$ holds. We then write $u \perp v$.

▶ **Definition 10.7** A basis $B = \{b_1, b_2, \ldots, b_n\}$ of a vector space with dot product is called an *orthonormal basis*, if $\|b_i\| = 1$ and $b_i \perp b_j$ hold for all i, j with $i \neq j$.

Examples

1. The standard basis of the $\mathbb{R}^n$ is an orthonormal basis.

2. $b_1 = \begin{pmatrix} \cos \alpha \\ \sin \alpha \end{pmatrix}$, $b_2 = \begin{pmatrix} -\sin \alpha \\ \cos \alpha \end{pmatrix}$ is an orthonormal basis in the $\mathbb{R}^2$. ◀

Without proof, I quote the following theorem, for which there is a constructive algorithm:

▷   **Theorem 10.8** Every subspace of the $\mathbb{R}^n$ has an orthonormal basis.

Where Does the Mouse Click?

I would like to show you a small, but clever application of the dot product in a graphics program: If you have drawn a line on the screen and want to edit it later, you have
to mark it. This is usually done by clicking on the line. Well, not quite exactly on the
line, you can't aim that precisely, but somewhere near the line. There is a region defined
around the line that represents the active area (Fig. 10.3). If you click in the (in reality
invisible) gray area, the line is marked, otherwise not.

How can you find out with the least possible effort whether the point is in the gray
area? We proceed in two steps. The first one is shown in Fig. 10.4. The endpoints of the
line form the vectors u, v, the mouse click takes place at the point m.

First we check whether m is in the marked strip. This is the case if the two angles α
and β are both acute angles, that is, if $\cos \alpha > 0$ and $\cos \beta > 0$.

Since we know that

$$\langle v - u, m - u \rangle = \|v - u\| \|m - u\| \cos \alpha, \quad \langle u - v, m - v \rangle = \|u - v\| \|m - v\| \cos \beta,$$

it reduces to checking the two relations

$$\langle v - u, m - u \rangle > 0, \quad \langle u - v, m - v \rangle > 0. \tag{10.4}$$

Fig. 10.3 Where does the
mouse click 1

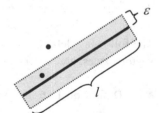

Fig. 10.4 Where does the
mouse click 2

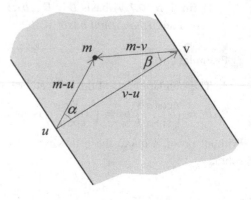

Fig. 10.5 Where does the
mouse click 3

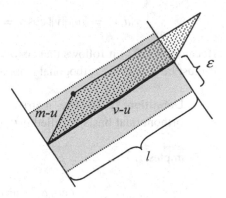

In the second step we test whether m is additionally in the rectangle shaded gray in Fig.
10.5. This is the case if the area of the dotted parallelogram is less than half the gray
area, that is, less than $l \cdot \varepsilon$.

But how do we get the dotted area? Here the determinant helps us again, whose value
indicates the size of the area:

$$\left| \det \begin{pmatrix} v - u \\ m - u \end{pmatrix} \right| < l \cdot \varepsilon. \tag{10.5}$$

You can count that (including the calculation of $l \cdot \varepsilon$) in (10.4) and (10.5) just 7 multiplications are needed to carry out the test.

10.2   Orthogonal Mappings

Particularly important linear mappings in computer graphics or robotics are mappings
that preserve lengths and angles. We will now deal with these mappings and examine
them in particular in $\mathbb{R}^2$ and $\mathbb{R}^3$. We will find there are only a few basic types of such
mappings. In this section, all bases used are orthonormal bases.

▶   **Definition 10.9** A linear mapping $f \colon \mathbb{R}^n \to \mathbb{R}^n$ is called *orthogonal*, if for all
$u, v \in \mathbb{R}^n$ it holds:

$$\langle f(u), f(v) \rangle = \langle u, v \rangle.$$

What is this definition about? For $u \in \mathbb{R}^n$ it holds

$$\|f(u)\| = \sqrt{\langle f(u), f(u) \rangle} = \sqrt{\langle u, u \rangle} = \|u\|, \tag{10.6}$$

that is, the mapping preserves the lengths of the vectors, it is *length-preserving*. Similarly, from $u \bot v$ also follows $f(u) \bot f(v)$, so right angles are also preserved. We do not
know other angles in $\mathbb{R}^n$, but in $\mathbb{R}^2$ and $\mathbb{R}^3$ we can infer even more: If α, β are the angles
between u, v or between $f(u), f(v)$, then

$$\langle u, v \rangle = \|u\| \|v\| \cos \alpha = \|f(u)\| \|f(v)\| \cos \beta = \langle f(u), f(v) \rangle.$$

Because of (10.6) it follows that $\cos \alpha = \cos \beta$. Since the angles are between 0 and 180°, it must be $\alpha = \beta$, the orthogonal mapping is also *angle-preserving*.

▷   **Definition 10.10** The matrix of an orthogonal mapping with respect to an orthonormal basis is called *orthogonal matrix*.

Examples

$$D_\alpha = \begin{pmatrix} \cos \alpha & -\sin \alpha \\ \sin \alpha & \cos \alpha \end{pmatrix}, \quad M_\alpha = \begin{pmatrix} \cos \alpha & \sin \alpha \\ \sin \alpha & -\cos \alpha \end{pmatrix}$$

are orthogonal mappings (rotation and reflection). Please calculate for yourself what $\left\langle D_\alpha \begin{pmatrix} x_1 \\ x_2 \end{pmatrix}, D_\alpha \begin{pmatrix} y_1 \\ y_2 \end{pmatrix} \right\rangle$ results in. ◀

This calculation is tedious, surprisingly, however, orthogonal matrices can also be characterized very easily:

▷   **Theorem 10.11** The following three assertions for a matrix $A \in \mathbb{R}^{n \times n}$ are equivalent:

a) A is orthogonal.
b) The columns of the matrix form an orthonormal basis.
c) It holds that $A^{-1} = A^T$.

I show by circular reasoning a) $\Rightarrow$ b) $\Rightarrow$ c) $\Rightarrow$ a). This makes the three assertions equivalent.

a) $\Rightarrow$ b): For an orthogonal mapping A it holds that $\ker A = \{0\}$, because for $u \neq 0$ it holds $\|u\| \neq 0$ and thus also $\|f(u)\| \neq 0$. Therefore, A is bijective and the columns $s_1, s_2, \ldots, s_n$ of A, which represent the images of the orthonormal basis $b_1, b_2, \ldots, b_n$, are again a basis. Furthermore,

$$\langle s_i, s_j \rangle = \langle f(b_i), f(b_j) \rangle = \langle b_i, b_j \rangle = \begin{cases} 1 & i = j \\ 0 & i \neq j, \end{cases} \tag{10.7}$$

and that means precisely that the column vectors have length 1 and are pairwise orthogonal.

b) $\Rightarrow$ c): We calculate $C = A^T A$. The rows of A^T contain, like the columns of A, the orthonormal basis. If we calculate the element c_{ij} of the product matrix, then this is just the dot product of the i-th row of A^T with the j-th column of A. So it is $c_{ij} = \langle s_i, s_j \rangle$ and according to (10.7) thus $C = I$, the identity matrix.

The inverse of an invertible matrix is however uniquely determined, therefore $A^{-1} = A^T$.

You should calculate the product $A^T A$ on paper, then you will see that there is nothing mysterious about it.

c) $\Rightarrow$ a) For this we need the trick from (10.2) and a little lemma about transposed matrices: It is $(AB)^T = B^T A^T$. I will just use it without proof. From $A^T A = I$ it then follows for all vectors u, v:

$$\langle Au, Av \rangle = (Au)^T (Av) = u^T A^T A v = u^T I v = u^T v = \langle u, v \rangle,$$

and thus A is orthogonal. $\square$

I find statement c) in particular very exciting. Can you still remember how hard it was to determine the inverse of a matrix? Of all things, for the important orthogonal matrices, we only need to transpose and we have it. Sometimes life is good to us.

From c) you can see that with A also the matrix A^T is orthogonal, because the inverse of an orthogonal mapping is again orthogonal. This has the consequence that in an orthogonal matrix not only the columns, but also the rows form an orthonormal basis.

Determinants and eigenvalues of orthogonal matrices can only take certain values:

▶   **Theorem 10.12** For an orthogonal matrix A it holds $\det A = \pm 1$, for eigenvalues λ of A it holds $\lambda = \pm 1$.

For a length- and angle-preserving mapping one would certainly not have expected anything else, but we can also write it down very quickly:

It is $1 = \det I = \det A^T A = \det A^T \cdot \det A = \det A \cdot \det A$, so $\det A = \pm 1$; and for an eigenvector to the eigenvalue λ, on the one hand $\|Av\| = \|\lambda v\|$, on the other hand, because of the orthogonality, also $\|Av\| = \|v\|$ and thus $\|\lambda v\| = \|v\|$, from which $|\lambda| \cdot \|v\| = \|v\|$ and thus, because of $v \neq \vec{0}$, also $\lambda = \pm 1$ follows. $\square$

Finally, we want to consider what happens to an orthogonal matrix under a change of basis.

▶   **Theorem 10.13** Let f be an orthogonal linear mapping with matrix A with respect to the orthonormal basis B_1 and matrix S with respect to the orthonormal basis B_2. Let T be the transition matrix from B_1 in B_2. Then A, S and T are orthogonal.

A and S are orthogonal as matrices of an orthogonal linear mapping. The transition matrix T of B_1 in B_2 is an orthogonal matrix, because the columns of the matrix are just the basis vectors of B_2. $\square$

In Chap. 9 we treated eigenvalues and eigenvectors. I now add one more important theorem from eigenvalue theory, which we can only formulate now.

▶   **Theorem 10.14** Let $A \in \mathbb{R}^{n \times n}$ be a symmetric matrix, that is $A = A^T$. Then the matrix has n eigenvectors that form an orthonormal basis.

The matrix A is therefore diagonalisable and the basis transition matrix is orthogonal.

It is easy to calculate that eigenvectors to different eigenvalues are orthogonal, see Exercise 10 in this chapter. It is just as easy to check that the characteristic polynomial of the matrix A decomposes into linear factors, that is, it has n real roots. But these roots can be multiple roots, and it is quite tricky to prove that a root of multiplicity k in this case also has an eigenspace of dimension k. I omit the proof of the theorem.

We will need the statement twice later when applying mathematical methods: first when calculating the centrality of nodes in a graph in Sect. 11.1, and second when doing principal component analysis in Sect. 22.2.

The orthogonal linear mappings in $\mathbb{R}^2$ and $\mathbb{R}^3$

Now we know a lot about the shape of orthogonal matrices, and we will use this knowledge immediately to determine all orthogonal mappings in $\mathbb{R}^2$ and $\mathbb{R}^3$.

Let's start with the $\mathbb{R}^2$. For the matrix $A = \begin{pmatrix} a & b \\ c & d \end{pmatrix}$ of such a mapping, it must hold: $a^2 + c^2 = 1$, because the column vectors have length 1. Let's take a and c as the legs of a right-angled triangle with hypotenuse 1 and call α the angle between the hypotenuse and a, then we see that

$$a = \cos \alpha, \quad c = \sin \alpha.$$

Furthermore, $a^2 + b^2 = 1$ is also true, because the row vectors also have length 1, so $(\cos \alpha)^2 + b^2 = 1$ applies. This gives $b^2 = (\sin \alpha)^2$ (because $(\cos \alpha)^2 + (\sin \alpha)^2 = 1$), and we get

$$b = \pm \sin \alpha.$$

From $c^2 + d^2 = 1$ we conclude in the same way:

$$d = \pm \cos \alpha.$$

Finally, we use the orthogonality of the columns: $ab + cd = 0$, so there is still a restriction on the signs, it turns out:

$$\begin{pmatrix} b \\ d \end{pmatrix} = \begin{pmatrix} \sin \alpha \\ -\cos \alpha \end{pmatrix} \quad \text{or} \quad \begin{pmatrix} b \\ d \end{pmatrix} = \begin{pmatrix} -\sin \alpha \\ \cos \alpha \end{pmatrix}.$$

▶ **Theorem 10.15** Every orthogonal mapping of $\mathbb{R}^2$ has the shape

$$D_\alpha = \begin{pmatrix} \cos\alpha & -\sin\alpha \\ \sin\alpha & \cos\alpha \end{pmatrix}, \quad M_\alpha = \begin{pmatrix} \cos\alpha & \sin\alpha \\ \sin\alpha & -\cos\alpha \end{pmatrix},$$

is thus a rotation or a reflection.

The situation in $\mathbb{R}^3$ is a bit more complex, but here too we can classify all orthogonal mappings. I would at least like to sketch the procedure. We know that each matrix A in $\mathbb{R}^3$ has an eigenvalue, in this case $+1$ or -1. First we carry out an orthogonal basis transition so that the corresponding eigenvector is the first basis vector. This gives us our matrix in the form:

$$A' = \begin{pmatrix} \pm1 & a_{12} & a_{13} \\ 0 & a_{22} & a_{23} \\ 0 & a_{32} & a_{33} \end{pmatrix}.$$

This is also an orthogonal matrix, in particular, for the columns s_1, s_2 holds $\langle s_1, s_2 \rangle = 0 = \pm1 a_{12} + 0 a_{22} + 0 a_{32}$. This gives $a_{12} = 0$. Similarly it follows $a_{13} = 0$. So A' has the form

$$A' = \begin{pmatrix} \pm1 & 0 & 0 \\ 0 & a_{22} & a_{23} \\ 0 & a_{32} & a_{33} \end{pmatrix}.$$

Now let's take a look at what the mapping does to the elements of the two-dimensional subspace W, which is spanned by the columns s_2 and s_3:

$$\begin{pmatrix} \pm1 & 0 & 0 \\ 0 & a_{22} & a_{23} \\ 0 & a_{32} & a_{33} \end{pmatrix} \begin{pmatrix} 0 \\ x_1 \\ x_2 \end{pmatrix} = \begin{pmatrix} 0 \\ a_{22}x_1 + a_{23}x_2 \\ a_{32}x_1 + a_{33}x_{21} \end{pmatrix}.$$

We see that the image lands again in the subspace W, that is, the restriction of A' to W is a linear mapping on W and this is of course orthogonal again. So the 2×2 submatrix of A has the form D_α or M_α from Theorem 10.15.

In the first case, A' therefore has the form

$$A' = \begin{pmatrix} \pm1 & 0 & 0 \\ 0 & \cos\alpha & -\sin\alpha \\ 0 & \sin\alpha & \cos\alpha \end{pmatrix},$$

in the second case we can simplify the matrix even further by another orthogonal basis transition: we rotate the plane W around the first coordinate axis until the second basis vector becomes the axis of reflection. The third basis vector, which is perpendicular to it, becomes the second eigenvector of the reflection, and the matrix takes the form:

$$A'' = \begin{pmatrix} \pm1 & 0 & 0 \\ 0 & 1 & 0 \\ 0 & 0 & -1 \end{pmatrix}.$$

Now we have gathered all the possibilities and we can formulate the following result as a final highlight of the section:

▶   **Theorem 10.16** For each orthogonal mapping of the $\mathbb{R}^3$ there is an orthonormal basis such that the corresponding matrix has one of the following forms:

$$\text{type 1:}\quad A = \begin{pmatrix} 1 & 0 & 0 \\ 0 & \cos\alpha & -\sin\alpha \\ 0 & \sin\alpha & \cos\alpha \end{pmatrix}, \quad B = \begin{pmatrix} -1 & 0 & 0 \\ 0 & 1 & 0 \\ 0 & 0 & -1 \end{pmatrix},$$

$$\text{type 2:}\quad C = \begin{pmatrix} -1 & 0 & 0 \\ 0 & \cos\alpha & -\sin\alpha \\ 0 & \sin\alpha & \cos\alpha \end{pmatrix}, \quad D = \begin{pmatrix} 1 & 0 & 0 \\ 0 & 1 & 0 \\ 0 & 0 & -1 \end{pmatrix}.$$

Type 1 is a rotation about an axis, type 2 is a rotation about an axis followed by a reflection in the plane perpendicular to the origin, the *rotational reflection*.

The two matrices on the right are a bit out of the ordinary, but they also fall under these types, we shouldn't have to write them down at all: With B it is a rotation around the x_2-axis by 180°, with D it is a reflection at the x_1-x_2-plane, with no rotation taking place beforehand.

There are no other orthogonal mappings in $\mathbb{R}^3$! The orientation-preserving orthogonal mappings are of type 1, as you can see from the determinant.

So if you move an object somehow in space, this results in exactly one translation and one rotation around an axis. Nothing more. Would you have believed that?

10.3   Homogeneous Coordinates

We have already come across this several times in our investigation of linear mappings: an important type of mapping does not fall under this: translation. Of course, we are constantly picking up objects with the mouse and dragging them across the screen, robots are moving objects from one place to another, but our theory cannot be used for this. Translations do not fix the origin, but linear mappings always map subspaces into subspaces, so the origin is also mapped into a subspace, in the case of bijective maps back into itself.

That's a shame, because linear mappings can be described well using matrices, and the matrix product can also be used to calculate consecutive linear mappings well. It would be nice if we could treat translations in the same way.

We use a trick for this. If the origin bothers us, we simply take it out. Let's first look at this in the plane, in space (and also in higher dimensions) it works exactly the same. Of course, we cannot simply tear a hole in the plane, but we go one dimension higher and move our x_1-x_2-plane a bit in x_3-direction, usually exactly by the value 1 (Fig. 10.6). The origin is gone already. Now the points have other coordinates: (x_1, x_2) becomes $(x_1, x_2, 1)$. We now identify this point with (x_1, x_2), and so that it is clear that I am refer-

Fig. 10.6 Homogeneous
coordinates

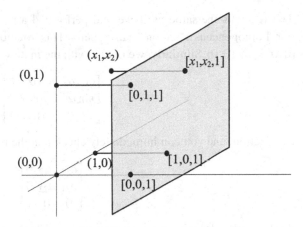

ring to the point of the plane I write $[x_1, x_2, 1]$ for it from now on and call $[x_1, x_2, 1]$ the
homogeneous coordinates of the point (x_1, x_2).

What have we gained? Now we can also specify linear mappings that shift the "origin" $[0, 0, 1]$. We have to be careful though, because we can only use such linear mappings which do not change the x_3 component, which always has to remain 1. All mappings of the following form are possible:

$$\begin{pmatrix} a_{11} & a_{12} & a_{13} \\ a_{21} & a_{22} & a_{23} \\ 0 & 0 & 1 \end{pmatrix}, \tag{10.8}$$

because

$$\begin{bmatrix} a_{11} & a_{12} & a_{13} \\ a_{21} & a_{22} & a_{23} \\ 0 & 0 & 1 \end{bmatrix} \begin{bmatrix} x_1 \\ x_2 \\ 1 \end{bmatrix} = \begin{bmatrix} y_1 \\ y_2 \\ 1 \end{bmatrix}.$$

You see: Now I have also written the matrices in square brackets. For linear mappings on homogeneous coordinates we introduce this notation. In these matrices, the third line is always equal to $[0, 0, 1]$.

But now to specific mappings in our plane: Let's look at the following matrix:

$$\begin{bmatrix} \cos\alpha & -\sin\alpha & 0 \\ \sin\alpha & \cos\alpha & 0 \\ 0 & 0 & 1 \end{bmatrix}. \tag{10.9}$$

As a mapping in the $\mathbb{R}^3$, this matrix, as we know, represents a rotation around the x_3-axis. What happens to the points in our plane? Let's try it out:

$$\begin{bmatrix} \cos\alpha & -\sin\alpha & 0 \\ \sin\alpha & \cos\alpha & 0 \\ 0 & 0 & 1 \end{bmatrix} \begin{bmatrix} x_1 \\ x_2 \\ 1 \end{bmatrix} = \begin{bmatrix} \cos\alpha\, x_1 - \sin\alpha\, x_2 \\ \sin\alpha\, x_1 + \cos\alpha\, x_2 \\ 1 \end{bmatrix}.$$

The result is the same as if we had performed a rotation in the $\mathbb{R}^2$, so we have found our homogeneous rotation matrix here. This rotation is carried out around the point $[0, 0, 1] = (0, 0)$. Similarly, we obtain with the matrix

$$\begin{bmatrix} \cos\alpha & \sin\alpha & 0 \\ \sin\alpha & -\cos\alpha & 0 \\ 0 & 0 & 1 \end{bmatrix}$$

a reflection, and you can immediately check that the matrix

$$\begin{bmatrix} a_{11} & a_{12} & 0 \\ a_{21} & a_{22} & 0 \\ 0 & 0 & 1 \end{bmatrix}$$

does nothing else with the homogeneous coordinates but the linear mapping

$$\begin{pmatrix} a_{11} & a_{12} \\ a_{21} & a_{22} \end{pmatrix}$$

in the usual x_1-x_2-plane. So we can still use all the linear mappings we have learned so far. But now to something new.

Look at the following mapping:

$$\begin{bmatrix} 1 & 0 & a \\ 0 & 1 & b \\ 0 & 0 & 1 \end{bmatrix} \begin{bmatrix} x_1 \\ x_2 \\ 1 \end{bmatrix} = \begin{bmatrix} x_1 + a \\ x_2 + b \\ 1 \end{bmatrix}.$$

Here we finally have the translation: The point $[x_1, x_2, 1] = (x_1, x_2)$ is shifted by $[a, b, 1] = (a, b)$. Even the "origin" $[0, 0, 1] = (0, 0)$ of our plane is not spared.

Of course we can still perform mappings one after the other as before, we just have to multiply the matrices. It is

$$\begin{bmatrix} 1 & 0 & a \\ 0 & 1 & b \\ 0 & 0 & 1 \end{bmatrix} \begin{bmatrix} a_{11} & a_{12} & 0 \\ a_{21} & a_{22} & 0 \\ 0 & 0 & 1 \end{bmatrix} = \begin{bmatrix} a_{11} & a_{12} & a \\ a_{21} & a_{22} & b \\ 0 & 0 & 1 \end{bmatrix}.$$

Any mapping that is generated by a matrix of the form (10.8) is thus a combination of a linear mapping and a translation. These maps are called *affine mappings*.

For example, let's carry out a rotation and then a translation:

$$\begin{bmatrix} 1 & 0 & a \\ 0 & 1 & b \\ 0 & 0 & 1 \end{bmatrix} \begin{bmatrix} \cos\alpha & -\sin\alpha & 0 \\ \sin\alpha & \cos\alpha & 0 \\ 0 & 0 & 1 \end{bmatrix} = \begin{bmatrix} \cos\alpha & -\sin\alpha & a \\ \sin\alpha & \cos\alpha & b \\ 0 & 0 & 1 \end{bmatrix}.$$

Applied to a point in our plane, we get:

$$\begin{bmatrix} \cos\alpha & -\sin\alpha & a \\ \sin\alpha & \cos\alpha & b \\ 0 & 0 & 1 \end{bmatrix} \begin{bmatrix} x_1 \\ x_2 \\ 1 \end{bmatrix} = \begin{bmatrix} \cos\alpha\, x_1 - \sin\alpha\, x_2 + a \\ \sin\alpha\, x_1 + \cos\alpha\, x_2 + b \\ 1 \end{bmatrix}. \qquad (10.10)$$

Fig. 10.7 Translation and rotation

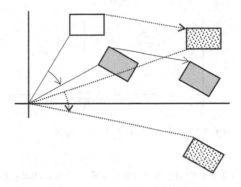

What happens if we exchange translation and rotation?

$$\begin{bmatrix} \cos\alpha & -\sin\alpha & 0 \\ \sin\alpha & \cos\alpha & 0 \\ 0 & 0 & 1 \end{bmatrix} \begin{bmatrix} 1 & 0 & a \\ 0 & 1 & b \\ 0 & 0 & 1 \end{bmatrix} = \begin{bmatrix} \cos\alpha & -\sin\alpha & \cos\alpha\,a - \sin\alpha\,b \\ \sin\alpha & \cos\alpha & \sin\alpha\,a + \cos\alpha\,b \\ 0 & 0 & 1 \end{bmatrix}.$$

Of course, this is also an affine mapping, but a different one than (10.10): Translation and rotation are not commutative, as you can see in Fig. 10.7: The gray rectangle is rotated and then moved, the dotted one is moved and then rotated.

One last example:

Example

An object is to be rotated on the screen, but not at the origin, which is probably in a corner of the screen, but at the center (a, b) of the object. However, the rotation matrix from (10.9) rotates at the origin. What can we do? Move the object to the origin, rotate it by the angle α and then move it back (Fig. 10.8). The matrix of the mapping is:

$$\begin{bmatrix} 1 & 0 & a \\ 0 & 1 & b \\ 0 & 0 & 1 \end{bmatrix} \begin{bmatrix} \cos\alpha & -\sin\alpha & 0 \\ \sin\alpha & \cos\alpha & 0 \\ 0 & 0 & 1 \end{bmatrix} \begin{bmatrix} 1 & 0 & -a \\ 0 & 1 & -b \\ 0 & 0 & 1 \end{bmatrix}$$

$$= \begin{bmatrix} \cos\alpha & -\sin\alpha & -\cos\alpha\,a + \sin\alpha\,b + a \\ \sin\alpha & \cos\alpha & -\sin\alpha\,a - \cos\alpha\,b + b \\ 0 & 0 & 1 \end{bmatrix}$$

Test that the point $[a, b, 1]$ is the fixed point of this mapping. ◀

Let's go to three-dimensional space. We proceed analogously. This time we do not move the x_1-x_2-plane in the direction of x_3, but we move the whole $\mathbb{R}^3$ in the direction of x_4. The homogeneous coordinates of a point in space are therefore $[x_1, x_2, x_3, 1]$, the matrices

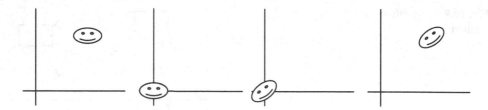

Fig. 10.8 Rotation not at the origin

of the mappings are the 4×4 matrices, in which the last row is equal to $[0, 0, 0, 1]$. For example, the following matrix defines a rotation around the x_1-axis:

$$\begin{bmatrix} 1 & 0 & 0 & 0 \\ 0 & \cos\alpha & -\sin\alpha & 0 \\ 0 & \sin\alpha & \cos\alpha & 0 \\ 0 & 0 & 0 & 1 \end{bmatrix},$$

and as translation matrix we get:

$$\begin{bmatrix} 1 & 0 & 0 & a \\ 0 & 1 & 0 & b \\ 0 & 0 & 1 & c \\ 0 & 0 & 0 & 1 \end{bmatrix}.$$

Basis Transitions in Robotics

A robot consists of several arms that are connected to each other by joints. The joints can either perform rotations or translations, and thus influence the position of the gripper (the effector) that sits at the end of this so-called kinematic chain and performs the tasks for which the robot is intended. Not only the absolute position, but also the orientation of the gripper in space is important. This depends on the position of the individual joints.

Each arm of the robot is rigidly connected to an orthogonal coordinate system that moves along with the arm during a movement. The coordinate system K_0 with the axes x_0, y_0, z_0 is fixed: It represents the position of the robot base; where its origin is placed. The last coordinate system K_n with the axes x_n, y_n, z_n is rigidly connected to the gripper. The origin and position of the base vectors of the gripper system in coordinates of the base system K_0, in world coordinates, are sought. Also for this, the description of the systems in homogeneous coordinates has proved to be suitable.

There is still some freedom in the choice of coordinate systems, which can be used to make the basis transitions as simple as possible. Here, the arm i is the connection between the i-th and $(i + 1)$-th joint, the coordinate system K_i is assigned to the i-th arm. The *Denavit-Hartenberg convention* now provides the following rules for the position of the coordinate systems:

1. The z_i-axis is placed in the direction of the movement axis of the $(i+1)$-th joint.
2. For $i > 1$, the x_i-axis is placed perpendicular to z_{i-1} (and of course to z_i).
3. The origin of K_i lies in the plane determined by z_{i-1} and x_i.
4. And finally, y_i is added so that K_i becomes a positively oriented Cartesian coordinate system.

Some special cases, such as parallel z-axes, I have not considered in this somewhat abbreviated description.

Now we want to express the coordinate system K_n by the coordinates of the system K_0. For this we carry out basis transitions.

Unlike before, our coordinate system is now defined by a basis and an origin, so we have one more data element to consider. We will use homogeneous coordinates, and thus the basis transition works quite analogously, as we have carried out in the standard coordinates in Sec. 9.3. Theorem 9.21 and 9.22 can be transferred almost literally:

▶   **Theorem 10.17** Let f be the affine mapping which maps the coordinate system K_1 to the coordinate system K_2. Let T be the matrix of this mapping with respect to the coordinate system K_1. Then the coordinates of a vector v with respect to K_2 are obtained by multiplying the coordinates of v with respect to K_1 from the left with T^{-1}. The coordinates of v with respect to K_1 are obtained from those with respect to K_2 by multiplication from the left with T.

If T is the transition matrix from K_1 to K_2 and S is the transition matrix from K_2 to K_3, then TS is the transition matrix from K_1 to K_3.

I do not want to prove this theorem, but at least give the transition matrix T. Let $K_2 = \{b_1, b_2, b_3, u\}$, where u is to be the origin of B_2. The homogeneous coordinates of K_2 with respect to K_1 are to be:

$$
b_1 = \begin{bmatrix} b_{11} \\ b_{21} \\ b_{31} \\ 1 \end{bmatrix}_{K_1}, \quad
b_2 = \begin{bmatrix} b_{12} \\ b_{22} \\ b_{32} \\ 1 \end{bmatrix}_{K_1}, \quad
b_3 = \begin{bmatrix} b_{13} \\ b_{23} \\ b_{33} \\ 1 \end{bmatrix}_{K_1}, \quad
u = \begin{bmatrix} u_1 \\ u_2 \\ u_3 \\ 1 \end{bmatrix}_{K_1}.
$$

Then the transition matrix has the form

$$
T = \begin{bmatrix}
b_{11} - u_1 & b_{12} - u_1 & b_{13} - u_1 & u_1 \\
b_{21} - u_2 & b_{22} - u_2 & b_{23} - u_2 & u_2 \\
b_{31} - u_3 & b_{32} - u_3 & b_{33} - u_3 & u_3 \\
0 & 0 & 0 & 1
\end{bmatrix}.
\tag{10.11}
$$

You can verify this by applying T to $[1, 0, 0, 1]$, $[0, 1, 0, 1]$, $[0, 0, 1, 1]$ and $[0, 0, 0, 1]$.

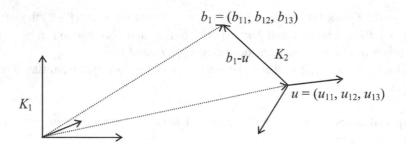

Fig. 10.9 Basis transition in robotics 1

Fig. 10.10 Basis transitions
in robotics 2

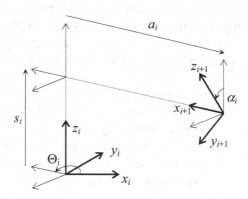

Aren't the images of the basis vectors in the columns of this matrix any more? Strictly speaking they are: $[b_{11}, b_{21}, b_{31}, 1]$ are the coordinates of the "tip" of the basis vector b_1 with respect to K_1, length and direction of the first basis vector are then given by the first column of the matrix, see Fig. 10.9.

We will now specify a number of basis transitions for the robot that transform K_0 into K_n (Fig. 10.10). First, we can generate K_{i+1} from K_i by four consecutive transitions as follows.

First, K_i is rotated around the z_i-axis by the angle Θ_i until x_i and x_{i+1} have the same direction. This is due to 2). This movement has the following transition matrix:

$$D_{\theta_i} = \begin{bmatrix} \cos\theta_i & -\sin\theta_i & 0 & 0 \\ \sin\theta_i & \cos\theta_i & 0 & 0 \\ 0 & 0 & 1 & 0 \\ 0 & 0 & 0 & 1 \end{bmatrix}.$$

The system now obtained is moved as far as possible in the direction of z_i until its origin intersects the line determined by x_{i+1}. This is possible because of 3). As a third step, the two origins can be brought into coincidence by another translation along this line. These movements have the translation matrices

$$T_{s_i} = \begin{bmatrix} 1 & 0 & 0 & 0 \\ 0 & 1 & 0 & 0 \\ 0 & 0 & 1 & s_i \\ 0 & 0 & 0 & 1 \end{bmatrix}, \quad T_{a_i} = \begin{bmatrix} 1 & 0 & 0 & a_i \\ 0 & 1 & 0 & 0 \\ 0 & 0 & 1 & 0 \\ 0 & 0 & 0 & 1 \end{bmatrix}.$$

Finally, only a further rotation around the x_i-axis is necessary to bring the vector y_i into coincidence with y_{i+1}:

$$D_{\alpha_i} = \begin{bmatrix} 1 & 0 & 0 & 0 \\ 0 & \cos\alpha_i & -\sin\alpha_i & 0 \\ 0 & \sin\alpha_i & \cos\alpha_i & 0 \\ 0 & 0 & 0 & 1 \end{bmatrix}.$$

This gives the transition matrix from K_i to K_{i+1} the following form:

$$A_{i,i+1} = D_{\theta_i} T_{s_i} T_{a_i} D_{\alpha_i} = \begin{bmatrix} \cos\theta_i & -\cos\alpha_i\sin\theta_i & \sin\alpha_i\sin\theta_i & a_i\cos\theta_i \\ \sin\theta_i & \cos\alpha_i\cos\theta_i & -\sin\alpha_i\cos\theta_i & a_i\sin\theta_i \\ 0 & \sin\alpha_i & \cos\alpha_i & s_i \\ 0 & 0 & 0 & 1 \end{bmatrix}. \quad (10.12)$$

Pay attention to the order of matrix multiplications: They belong to transitions!

Since the z_i-axis always runs along the movement axis of the $(i+1)$-th joint, a movement of this joint either results in a rotation around the $x_i - x_{i+1}$-plane or a translation in z_i-direction. This means that, depending on the type of joint, Θ_i or s_i are variable in the matrix (10.12), the other elements are robot constants.

With this transition, we therefore obtain the origin and basis vectors of K_{i+1} in the coordinates of K_i. Let's test it with the origin and two basis vectors of K_{i+1}:

$$A_{i,i+1}\begin{bmatrix} 0 \\ 0 \\ 0 \\ 1 \end{bmatrix}_{i+1} = \begin{bmatrix} a_i\cos\theta_i \\ a_i\sin\theta_i \\ s_i \\ 1 \end{bmatrix}_i, \quad A_{i,i+1}\begin{bmatrix} 1 \\ 0 \\ 0 \\ 1 \end{bmatrix}_{i+1} = \begin{bmatrix} \cos\theta_i + a_i\cos\theta_i \\ \sin\theta_i + a_i\sin\theta_i \\ s_i \\ 1 \end{bmatrix}_i,$$

$$A_{i,i+1}\begin{bmatrix} 0 \\ 0 \\ 1 \\ 1 \end{bmatrix}_{i+1} = \begin{bmatrix} a_i\cos\theta_i \\ \sin\alpha_i + a_i\sin\theta_i \\ \cos\alpha_i + s_i \\ 1 \end{bmatrix}_i.$$

Can you follow the results by looking at the drawing in Fig. 10.10? The origin and basis vectors are shifted in direction x_{i+1} and z_i. In addition to this shift, x_i is rotated in the x_1-y_1-plane by Θ_i, and z_i is rotated in the y_1-z_1-plane by α_i.

Now we can set up a whole chain of these transitions: $A_{i+1,i+2}$ expresses K_{i+2} in coordinates of K_{i+1}, so $A_{i,i+1} \circ A_{i+1,i+2}$ gives us the coordinates of basis K_{i+2} in the K_i-system, and so on. Finally, the transition

$$A = A_{0,1} \circ A_{1,2} \circ \cdots \circ A_{n-1,n}$$

provides the origin and orientation of the effector in world coordinates. From the shape of the transition matrix (10.11) you can see that the first three columns give the direction of the basis vectors and the last column gives the origin of the effector system.

Of course, the individual transitions depend on the current position of the robot joint axes, and so the transition B changes with every robot movement.

This thus poses the two problems for the programmer of a robot controller:

- To determine the position and orientation of the effector from the position of the individual joints, this is the *direct kinematic problem*, which I have briefly treated here,
- and the *inverse kinematic problem*, to find appropriate joint values for a given position of the effector.

The second is the more important, but unfortunately also by far the more difficult problem.

10.4   Comprehension Questions and Exercises

Comprehension Questions

1. Does a vector space with addition and with a dot product as multiplication form a ring?
2. Why can a dot product only be defined meaningfully for real vector spaces?
3. Explain the relationship between dot product and norm.
4. If a linear mapping in $\mathbb{R}^3$ preserves the angles between all vectors, then it also preserves the length of the vectors and is therefore an orthogonal mapping.
5. What types of orthogonal mappings are there in $\mathbb{R}^3$?
6. The rotation in $\mathbb{R}^2$ has the matrix $\begin{pmatrix} \cos\alpha & -\sin\alpha \\ \sin\alpha & \cos\alpha \end{pmatrix}$, the reflection the matrix $\begin{pmatrix} \cos\alpha & \sin\alpha \\ \sin\alpha & -\cos\alpha \end{pmatrix}$. And which mappings are described by $\begin{pmatrix} -\cos\alpha & \sin\alpha \\ \sin\alpha & \cos\alpha \end{pmatrix}$ or by $\begin{pmatrix} \cos\alpha & \sin\alpha \\ -\sin\alpha & \cos\alpha \end{pmatrix}$?
7. Although translation is not a linear mapping. But is it an "orthogonal" mapping in the sense that it preserves the length of vectors and the angle between vectors?
8. Why is it sometimes useful to calculate with homogeneous coordinates?

Exercises

1. Determine a linear equation whose solution set describes the plane determined by the points $(1, 0, 1)$, $(2, 1, 2)$, and $(1, 1, 3)$ in $\mathbb{R}^3$.
2. Determine the angle between the vectors $(2, 1, -1)$ and $(1, 2, 1)$.

3. Determine the angle between the diagonal of a cube and an edge adjacent to the diagonal.

4. Complete the vector $(1/2, 1/2, 1/\sqrt{2})$ to an orthonormal basis of $\mathbb{R}^3$.

5. Prove: If the vectors $v_1, v_2, \ldots, v_n$ are pairwise orthogonal and all different from $\vec{0}$, then they are also linearly independent.

6. What type of mapping do you get in the $\mathbb{R}^3$, if you successively carry out a rotation, 2 reflections, 2 rotations and 4 more reflections?

7. Which of the following matrices are orthogonal?

a) $\begin{pmatrix} \frac{1}{\sqrt{2}} & \frac{1}{\sqrt{2}} & 0 \\ \frac{1}{2} & \frac{1}{2} & -\frac{1}{\sqrt{2}} \\ \frac{1}{2} & \frac{1}{2} & \frac{1}{\sqrt{2}} \end{pmatrix}$

b) $\begin{pmatrix} \frac{1}{\sqrt{2}} & -\frac{1}{\sqrt{2}} \\ \frac{1}{\sqrt{2}} & \frac{1}{\sqrt{2}} \end{pmatrix}$

c) $\begin{pmatrix} \frac{1}{2} & \frac{1}{2} & \frac{1}{2} & \frac{1}{2} \\ \frac{1}{2} & -\frac{5}{6} & \frac{1}{6} & \frac{1}{6} \\ \frac{1}{2} & \frac{1}{6} & \frac{1}{6} & -\frac{5}{6} \\ \frac{1}{2} & \frac{1}{6} & \frac{5}{6} & \frac{1}{6} \end{pmatrix}$.

8. Determine the homogeneous matrix of a mapping which maps the square 1 from Fig. 10.11 with edge length 1 and origin in the lower left to the square 2.

9. Derive the triangle inequality $\|u + v\| \leq \|u\| + \|v\|$ from the Cauchy-Schwarz inequality $|\langle u, v \rangle| \leq \|u\| \cdot \|v\|$ for the norm $\|u\| := \sqrt{\langle u, u \rangle}$.

10. If A is a symmetric matrix, that is $A = A^T$, then eigenvectors to different real eigenvalues are orthogonal.

Note: If λ_1, λ_2 are eigenvalues to v_1, v_2, then show that $\lambda_1 \langle v_1, v_2 \rangle = \lambda_2 \langle v_1, v_2 \rangle$. Use the tricks for calculation: $\langle v_1, v_2 \rangle = v_1^T v_2$ and $(AB)^T = B^T A^T$.

Fig. 10.11 Exercise 8

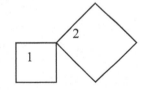

Graph Theory

11

Abstract

This chapter contains many algorithms and is particularly close to computer science. If you have worked through it

- you know the basic concepts of graph theory: nodes, edges, degree, paths, cycles, isomorphisms, networks and directed graphs,
- you know what trees and rooted trees are,
- you have constructed search trees as an application and can use trees to build the Huffman code,
- you know what breadth-first search and depth-first search mean and can find shortest paths in networks
- and you can carry out a topological sorting in directed acyclic graphs.

If you want to drive from Flensburg to Freiburg by car and do not know the way, you use a navigation system that calculates the shortest or fastest route between the two towns and leads you along this route. The device knows a list of places and connecting roads between these places. The length of the routes between the places or the approximate travel time is known.

This task is a typical problem of graph theory: Graphs are structures consisting of a number of nodes (here the places) and edges (the roads), where the edges sometimes

Supplementary Information The online version contains supplementary material available at https://doi.org/10.1007/978-3-658-40423-9_11.

carry directions (one-way streets) or weights (length, travel time). The route planner searches for the shortest path between two nodes.

Graphs occur in many areas of technology, computer science and everyday life. In addition to road networks, all kinds of networks are graphs: power distribution networks, electrical circuits, communication networks, molecule structures and much more. At the beginning of Chap. 5 you will find a graph in which the nodes are algebraic structures and the edges are relations between these. For example, class diagrams of an object-oriented design are also graphs.

In this chapter I would like to introduce you to the basic concepts of graph theory. The focus will be on trees, which play a particularly important role in computer science.

11.1 Basic Concepts of Graph Theory

At the beginning of a theory there is always a whole lot of new terms. First I would like to put them together:

▶   **Definition 11.1** A (*undirected*) *graph* G consists of a set $V = V(G)$, the *nodes* or
   vertices of G and a set $E = E(G)$ of unordered pairs $k = [x, y]$, with $x, y \in V$, the
   edges of G. We write $G = (V, E)$.

We will only deal with finite graphs, that is, with graphs in which the set V—and thus of course the set E—is finite. Furthermore, I do not want to allow two nodes to be connected with more than one edge.

If $k = [x, y]$ is an edge, then x, y *are the endpoints* of the edge, the nodes x and y are called *incident* to k. x and y are then *adjacent nodes*.

The edge $k = [x, y]$ is called a *loop*, if $x = y$. The graph G is called *complete*, if every two nodes are adjacent. $G' = (V', E')$ is called a *subgraph* of G, if $V' \subset V$ and $E' \subset E$.

Figure 11.1 shows an example. Here $V = \{x_1, x_2, x_3, x_4\}$, $E = \{k_1, k_2, k_3, k_4\}$ with $k_1 = [x_1, x_2]$, $k_2 = [x_2, x_2]$, $k_3 = [x_2, x_3]$, $k_4 = [x_3, x_4]$.

Fig. 11.1 Nodes and edges

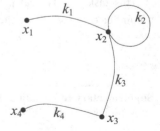

x_1 and x_2 are adjacent, x_3, x_4 are endpoints of k_4, k_2 is a loop. The graph is not complete. In Fig. 11.2 you can see a complete graph, Fig. 11.3 does not represent a graph. If you are confronted with problems in which such objects occur, you can easily turn them into a graph by, for example, adding more nodes to the multiple edges.

How can one represent graphs? First of all, drawings are an option. In Fig. 11.2, however, it is disturbing that the two diagonals intersect, even though the intersection point is not a node. Figure 11.4 shows we can also draw the edges without intersections in this graph. Is it always possible to represent graphs in such a way that edges only intersect in nodes? In the $\mathbb{R}^3$ this is possible: If V is a finite subset of the $\mathbb{R}^3$ (the nodes) and the elements of V are connected by a finite set of line segments (the edges), then we can always choose them so that they only intersect in the nodes.

In the $\mathbb{R}^2$ this is not always possible, however. Figure 11.5 shows a graph you cannot map into the $\mathbb{R}^2$ without edges intersecting. Graphs that allow a non-intersecting representation in the plane are called *planar graphs*.

A big problem in the design of circuit boards or integrated circuits is to plan the electrical connections in such a way that they can be placed with as little overlap as possible. As we have seen, this is not always possible. That is why circuit boards often consist of several layers.

Fig. 11.2 A complete graph

Fig. 11.3 No graph

Fig. 11.4 A planar graph

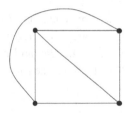

Fig. 11.5 A non-planar graph

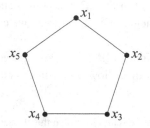
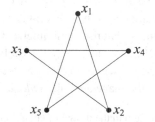

Fig. 11.6 Two representations of a graph

Drawings are an important tool. But you have to be careful with them and not read more into them than is actually there. For example, Fig. 11.6 shows the same graph twice.

Two graphs are to be considered equal if they only differ in the labeling of their nodes:

▶   **Definition 11.2** If $G = (V, E)$ and $G' = (V', E')$ are graphs, then G and G' are called *isomorphic*, if there is a bijective mapping $\varphi : V \to V'$ such that

$$[x, y] \in E \Leftrightarrow [\varphi(x), \varphi(y)] \in E'.$$

It is easy to construct an isomorphic graph to a given graph G, on the other hand it is very difficult to determine from two given graphs whether they are isomorphic. If the graphs are very large, this is practically impossible.

In Sect. 5.7 I have presented two hard mathematical problems that are used in cryptography: the factorization problem and the discrete logarithm problem. Here we have found another problem of this kind. There are cryptographic protocols that are based on it, but I do not know any widespread product for it. Maybe graph isomorphisms will become even more important if someone cracks the factorization problem?

The representation of graphs in the computer can be done using *adjacency matrices*. The nodes are numbered from 1 to n, the element a_{ij} of the adjacency matrix is 1 if the nodes i, j are adjacent, and has the value 0 otherwise. The adjacency matrix of the graph from Fig. 11.6 has the form:

$$\begin{pmatrix} 0 & 1 & 0 & 0 & 1 \\ 1 & 0 & 1 & 0 & 0 \\ 0 & 1 & 0 & 1 & 0 \\ 0 & 0 & 1 & 0 & 1 \\ 1 & 0 & 0 & 1 & 0 \end{pmatrix}.$$

The matrix of a graph is always symmetric, in the diagonal is a 1 if the respective node has a loop.

Adjacency matrices are only suitable for the implementation of graphs if they contain many different entries from 0, that is, if many edges are connected to each other. In many graphs, nodes are only connected to a few other nodes, for example, think of the connections in the route planner. For such graphs, one chooses other representations, such as the *adjacency lists*, in which a list of adjacent nodes is attached to each node.

▶ **Definition 11.3** If x is a node of the graph G, then the number of edges incident to x is called the *degree* of x. Loops are counted twice. The degree of x is denoted by $d(x)$.

In Fig. 11.1 the nodes x_1 and x_4 have degree 1, x_3 has degree 2 and x_2 has degree 4.

Let A be the adjacency matrix of a graph with nodes $x_1, x_2, \ldots, x_n$. The i-th row is $(a_{i1}, a_{i2}, \ldots, a_{in})$. a_{ij} is 1 if there is an edge from node i to node j. Therefore, the degree of node i is exactly the sum of the 1s in the i-th row of the matrix. Multiply the adjacency matrix with the n-dimensional vector $(1, 1, 1, 1, \ldots, 1)$, you get in the i-th row the value $(a_{i1} + a_{i2} + \ldots + a_{in})$, that is,

$$A \begin{pmatrix} 1 \\ 1 \\ \vdots \\ 1 \end{pmatrix} = \begin{pmatrix} d(x_1) \\ d(x_2) \\ \vdots \\ d(x_n) \end{pmatrix}.$$

▶ **Theorem 11.4** In every graph $G = (V, E)$ it holds that $\sum_{x \in V} d(x) = 2 \cdot |E|$.

Proof: Each endpoint of an edge provides exactly the contribution 1 to the degree of this point. Since each edge has exactly 2 endpoints, it contributes exactly twice the value 1 to the left side. □

▶ **Corollary 11.5** In every graph the number of nodes of odd degree is even.

Otherwise the total sum of the degrees would be odd. □

Paths in Graphs

▷   **Definition 11.6** If G is a graph and $x_1, x_2, \ldots, x_n$ are nodes of G, then a set of
edges that connects x_1 with x_n is called a *walk* from x_1 to x_n. We denote the walk
$\{[x_1, x_2], [x_2, x_3], \ldots, [x_{n-1}, x_n]\}$ with $x_1 x_2 x_3 \ldots x_n$. The walk is called *open* if
$x_1 \neq x_n$, otherwise *closed*.

A *path* from x to y is an open walk from x to y, in which all nodes are different.

A *cycle* is a closed walk, in which all nodes are different except for the starting
and ending nodes.

The graph G is called *connected*, if every two of its nodes are connected by
paths.

In Fig. 11.7 you can see a non-connected graph. In this one

$x_1 x_2 x_3 x_4 x_5 x_6 x_2 x_1$ is a closed walk,
$x_1 x_2 x_6 x_5 x_7$ is a path,
$x_2 x_3 x_4 x_5 x_6 x_2$ is a cycle.

If we also specify each node should be connected to itself, you can easily check that
the relationship between the nodes of a graph "x and y are connected by a path" is an
equivalence relation. According to Theorem 1.11 in Sect. 1.2 this means every graph
decomposes into equivalence classes, its *connected components*. The graph G in Fig.
11.7 consists of the two connected components $\{x_1, \ldots, x_7\}$ and $\{x_8, x_9\}$. These are in turn
subgraphs of G.

▷   **Theorem 11.7** Every walk of x_1 to x_n contains a path from x_1 to x_n.

You can therefore omit detours, no node has to be visited multiple times.

Fig. 11.7 Connection
components

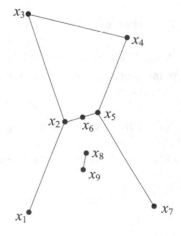

Without explicitly carrying out the proof, I give you the very intuitive method: Let $x_1 x_2 x_3 \dots x_n$ be the walk. If you start at x_1, and you encounter a node for the second time, so you have to remove all nodes in between, as well as the double node once. In the graph from Fig. 11.7 you can from the walk $x_1 x_2 x_3 x_4 x_5 x_6 x_2 x_3 x_4 x_5 x_7$ in this way obtain the path $x_1 x_2 x_3 x_4 x_5 x_7$. This is done as often as necessary until all multiple nodes are removed.   □

▶   **Definition 11.8** The number of edges of a walk or a path is called the *length* of the walk or the path.

In many cases, certain numbers are associated with the edges of a graph, for example distances, line capacities, costs or duration. If you assign such numbers to the edges, you will get networks:

▶   **Definition 11.9** A graph is called *network* if each edge $[x, y]$ is assigned a *weight* $w(x, y) \in \mathbb{R}$. In networks, the length of a path from one node to another is the sum of the weights of all the edges of the path.

The search for cost-effective paths in graphs is an important task of computer science. We will deal with this in Sect. 11.3.

Centrality of Nodes—Who is the Most Important?

In a social network, people form the nodes and relationships between people form the edges of a graph. Which people in such a network are probably the most interesting? Who are the important people in the network that you should contact yourself? I would like to assign a number to each node to denote its importance, which I call the centrality of the node.

A first approach is certainly to check who has the most contacts. The centrality is then the degree of the corresponding node. But if you want to look a little deeper, not only the number of contacts is interesting, but also the quality of the contacts: A person is not important just because he knows many other people, but because he knows many important people, that is, those who in turn know many people. Let's look at a small example (Fig. 11.8).

The adjacency matrix of the graph is

$$
\begin{pmatrix}
0 & 1 & 1 & 1 & 0 \\
1 & 0 & 1 & 0 & 0 \\
1 & 1 & 0 & 1 & 0 \\
1 & 0 & 1 & 0 & 1 \\
0 & 0 & 0 & 1 & 0
\end{pmatrix}.
$$

Fig. 11.8 Centrality of nodes

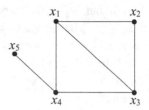

The nodes x_1, x_3 and x_4 each have degree 3, x_2 has degree 2 and the node x_5 is somewhat remote, it only has degree 1. If only the degrees are considered, the three nodes x_1, x_3 and x_4 would be equally central. But if the importance of the respective neighbors is taken into account, x_4 should have less weight. How can this be reproduced? Let's try the approach that the still unknown centrality w_i of the node x_i should be proportional to the sum of the centrality of all adjacent nodes x_i. We call the proportionality factor with μ:

$$w_i = \mu \cdot \sum_{\substack{\text{all } j \text{ for which} \\ x_j \text{ is adjacent to } x_i}} w_j \quad \underset{\substack{\uparrow \\ a_{ij}=1 \,\Leftrightarrow\, x_j \text{ adjacent to } x_i}}{=} \quad \mu \cdot (a_{i1}w_1 + a_{i2}w_2 + \ldots + a_{in}w_n).$$

The result is exactly the i-th component of the product of the matrix A with the vector $w = (w_1, w_2, \ldots, w_n)$, see (7.6). So it is

$$\begin{pmatrix} w_1 \\ w_1 \\ \vdots \\ w_n \end{pmatrix} = \mu A \begin{pmatrix} w_1 \\ w_1 \\ \vdots \\ w_n \end{pmatrix},$$

and if we set $\lambda = 1/\mu$ we get

$$Aw = \lambda w.$$

This matrix equation we have seen before: If there is such a vector w and such a factor λ, then w is just an eigenvector to the eigenvalue λ of the adjacency matrix A of the graph, see Definition 9.10 at the beginning of Sect. 9.2.

The mathematician now asks: Are there any meaningful solutions for this equation at all? Not every matrix has eigenvalues, and here all w_i should also be greater than 0 if a meaningful measure for centrality is to result. A theorem from the theory of eigenvectors helps us further: The adjacency matrix of a connected graph has at least one positive real eigenvalue. The largest of these eigenvalues has an eigenvector with only positive components. If you would like to read more about this: This result is part of the Perron-Frobenius Theorem. The matrix is symmetric, and therefore all eigenvectors are orthogonal according to Theorem 10.14. If a vector is orthogonal to the eigenvector with positive components, it must have negative components, otherwise the scalar product could not be 0. So all other eigenvectors have mixed signs in the components. Therefore, the positive eigenvector is the only reasonable candidate for the weights of the nodes of the graph. And this eigenvector always exists! Its components are called the *eigenvector centrality* of the nodes of the graph. Internet search engines use variants of the eigenvector centrality, among other things, to determine the ranking of documents in the search results: Links to documents referenced by many links are placed at the top; but these links are also referenced by many other pages.

Let's work through the example from above. With a mathematics tool, the eigenvalues are quickly calculated. The largest of them has the value 2.64, and an associated eigenvector is

$$\begin{pmatrix} 1.0 \\ 0.76 \\ 1.0 \\ 0.88 \\ 0.33 \end{pmatrix}.$$

You can see that the nodes x_1 and x_3 are now more important than x_4. This was downgraded because it is in contact with the insignificant x_5.

11.2 Trees

▶ **Definition 11.10** A graph in which each two nodes are connected by exactly one path is called a *tree*.

Fig. 11.9 shows four examples of trees. Trees can be characterized by many different properties. The following theorem gives some equivalent conditions to the definition.

▶ **Theorem 11.11** Let G be a graph with n nodes and m edges. Then the following statements are equivalent:

a) G is a tree.
b) G is a connected graph without cycles.
c) G is connected, but if you remove any edge, G splits into two connected components.
d) G is connected and $n = m + 1$ (G has one more node than edges).

We show the equivalence of a), b) and c) in a circular reasoning:

a) $\Rightarrow$ b): G is of course connected. If G had a cycle, there would be two paths between the nodes of this cycle.

b) $\Rightarrow$ c): If we remove the edge $E = [x_1, x_2]$, we get the graph G', in which x_1 and x_2 are no longer connected, G' is therefore no longer connected. Let y be any node. If y is not connected to G' in x_1, then the original path from y to x_1 contained the edge E and thus also x_2. The path to x_2 therefore still exists, y is connected to x_2. G' therefore splits into two components: The nodes of one component are all connected to x_1, the nodes of the other component are all connected to x_2.

c) $\Rightarrow$ a): Since G is connected, any two nodes are connected by a path. If there were two paths from x_1 to x_2, you could remove one of these paths without destroying the connection.

To show the equivalence of d) with a), b) and c), we first reason that after removing an edge from a tree, the resulting connected components again form trees: If, for example, y and z are nodes in such a component, then there cannot be multiple paths from y to z, since these would also be paths in the original graph G.

Now we show a) $\Rightarrow$ d) by induction on the number of nodes n. The base case ($n = 1$ and $n = 2$) can be read off from the first two trees in Fig. 11.9. Now let the assertion be satisfied for all trees with less than $n + 1$ nodes. Let G be a tree with $n + 1$ nodes. If we remove an edge we get 2 subtrees with node numbers n_1 and n_2, where $n_1 + n_2 = n + 1$ applies. For the subtrees, the number of edges is $n_1 - 1$ and $n_2 - 1$ according to the assumption. This results in the total number of edges of G: $n_1 - 1 + n_2 - 1 + 1 = n$.

We also prove d) $\Rightarrow$ a) by induction, where the base case can again be read off from Fig. 11.9. Let G be a graph with $n + 1$ nodes and n edges given. The degree of each node is at least 1, since G is connected. Then there must be nodes of degree 1, i.e. nodes that are incident with exactly one edge, because if it were for all nodes $d(x) \geq 2$, then according to Theorem 11.4 it would follow:

$$2 \cdot \underbrace{(n + 1)}_{\text{number of nodes}} \leq \sum d(x) = 2 \cdot \underbrace{n}_{\text{number of edges}} .$$

We now remove such a node x of degree 1 with its edge from G, so a smaller graph results, which again has exactly one node more than edges. This is therefore a tree according to the induction hypothesis. If we add x with its edge to this tree again, then x is also connected to every other node of the graph by exactly one path and thus G is a tree.   □

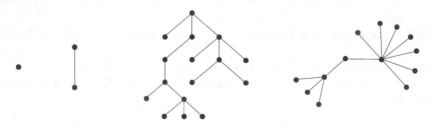

Fig. 11.9 Trees

The most surprising here is probably the property d), a very simple test criterion to determine whether a given graph is a tree or not.

Rooted Trees

So far, in a tree, all nodes and edges are equal. However, often a node in the tree is particularly distinguished, this is called the *root* and the corresponding tree *rooted tree*. In order not to overdo the analogy to botany, the root of a tree is usually drawn at the top, the edges all point downwards. The third tree in Fig. 11.9 is already shown in this way. If you imagine the nodes of a tree as electrically charged balls that are movably connected to each other by rods, you can touch any node of the tree and lift it up and get a rooted tree. So every node in a tree can be a root.

If x and y are connected by an edge and x is closer to the root than y, then x is the *parent* of y and y is a *child* of x. If there is a path from x to y, starting from x and going to the child, grandchild, great-grandchild, and so on, then x is an *ancestor* of y and y is a *descendant* of x. Nodes with descendants are called *inner nodes*, nodes without descendants are called *leaves* of the tree. The leaves of a tree are exactly the nodes different from the root with degree 1. With the relation "descendant" between the nodes of a tree, the nodes are ordered, see Definition 1.12 in Sect. 1.2.

In Fig. 11.10 x is the parent of y, and z is a descendant of x. However, z is not a descendant of y.

If the order from top to bottom is not the only one of importance, but the order of the children is also important, one speaks of the first, second, third child, and so on, and one obtains an *ordered rooted tree*. In the drawing, the children are ordered from left to right: y is the first child of x.

▶   **Theorem 11.12** In a rooted tree, each node together with all its descendants and the corresponding edges forms a rooted tree again.

Fig. 11.10 A rooted tree

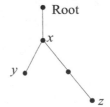

We check the definition of the tree. If x is the new root of the tree and y and z are descendants of x, then there is a path from y to x and a path from x to z. By combining, you get a walk and thus also a path from y to z. This path is unique, because otherwise there would already be several paths from y to z in the original tree.   □

This property means that trees are excellently suited to be treated recursively: The sub-trees have the same properties as the original tree, but they always become smaller. Many tree algorithms are recursive.

Example

The first task of a compiler during the translation of a program is the syntax analysis. In doing so, a *syntax tree* is generated from an expression. In such a tree, the inner nodes represent the operators, the operands of an operator are formed from the sub-trees that hang on the children of this operator. For the expression $(a+b) \cdot c$, the syntax tree from Fig. 11.11 belongs, the tree from Fig. 11.12 belongs to

$$(x\%4 == 0)\&\&\,!\,((x\%100 == 0)\&\&(x\%400\,!= 0)). \blacktriangleleft$$

If the syntax tree is successfully built, the evaluation of the expression can be done recursively: To evaluate the tree, all subtrees that belong to the children of the root have to be evaluated. Then, the operation of the root can be performed. The process ends when the root itself is an operand.

Fig. 11.11 Syntax tree 1

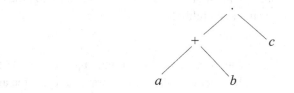

Fig. 11.12 Syntax tree 2

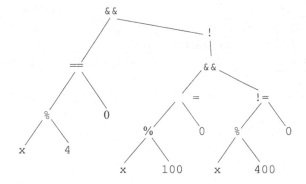

If in a rooted tree each node has at most n children, the tree is called a *n-ary rooted tree*, if each node has exactly 0 or n children, it is called a *regular n*-ary rooted tree. In the case $n = 2$ we speak of a *binary rooted tree*, the children are then also called *left* and *right children*. Now we want to take a closer look at these trees.

▶ **Theorem 11.13** In a binary regular rooted tree B holds:

a) B has exactly one node of degree 2, all other nodes have degree 1 or 3.
b) If x is a node of B, then x with all its descendants is a binary regular rooted tree again.
c) The number of nodes is odd.
d) If B has n nodes, it has $(n+1)/2$ leaves and $(n-1)/2$ inner nodes. There is therefore exactly one leaf more than inner nodes.

The properties a) and b) are clear: The root has exactly two children, so degree 2, every other node is a child and has either two children (degree 3) or no descendants (degree 1). The characterizing properties of the binary regular tree remain in all subtrees, of course.

For the third property, we can use Theorem 11.12 for induction for the first time: A minimal binary regular rooted tree is a single node, for which the statement is true. The number of nodes of the tree B is $1 +$ number of nodes of the left subtree $+$ number of nodes of the right subtree, thus odd, since the two smaller subtrees have an odd number of nodes.

d): Let p be the number of leaves of B, If B has n nodes, then B has n-1 edges and it applies according to a) and according to Theorem 11.4:

$$\sum_{x \in V} d(x) = p \cdot 1 + (n - p - 1) \cdot 3 + 1 \cdot 2 = 2|E| = 2(n - 1).$$

If you solve this equation for p, you get $p = \frac{n+1}{2}$ and for the number of inner nodes $n - p = n - \frac{n+1}{2} = \frac{n-1}{2}$. Since $\frac{n+1}{2} = \frac{n-1}{2} + 1$, there is exactly one leaf more than inner nodes. □

You can draw the games of a tennis tournament in the form of a binary regular rooted tree. The players are the leaves, the games are the inner nodes, the final represents the root. Question: If p players participate in the tournament, how many games will take place until the tournament winner is determined? Since there is exactly one leaf more than inner nodes, the number of games must be $p - 1$.

But you can answer this question quite simply without deep graph-theoretical theorems: Each game eliminates exactly one player. The winner remains: so with p players you need exactly $p-1$ games.

▶ **Definition 11.14** The *level* of a node in a rooted tree is the number of nodes of the path from the root to this node. The *height* of a rooted tree is the maximum level of the nodes.

So the root has level 1, the tree from Fig. 11.10 has height 4.

▶   **Theorem 11.15** In a binary rooted tree of height H with n nodes it applies

$$H \geq \log_2(n+1).$$

At level k of the tree, a maximum of 2^{k-1} nodes can be located (you can check this by induction if you want). So at the bottom level H, a maximum of 2^{H-1} nodes are located, which are then all leaves. If you add the inner nodes, you get $n \leq 2^{H-1} + (2^{H-1} - 1)$, that is $n \leq 2 \cdot 2^{H-1} - 1 = 2^H - 1$, and thus $\log_2(n+1) \leq H$.   □

Search Trees

Read the assertion of Theorem 11.15 the other way around: in a tree of height H, up to $2^H - 1$ nodes can be placed. For $H = 18$, for example, 2^H is about 262,000. For each of these nodes, there is a path from the root that visits at most 18 nodes. This property can be used to store data in the nodes of trees that can be accessed very quickly. These are the *search trees*. In the example with $H = 18$, you can store a larger dictionary in it.

Data records are identified by keys. We assume each key occurs only once. Each node of the search tree is assigned a record in the following way: all keys of the left subtree of p are smaller than the one of node p, all keys of the right subtree of p are greater than the one of p.

First we want to create a search tree. When entering data, we always start at the root, for each entry a new node is created as a leaf and at the same time as the root of a new subtree:

Enter the key s in the tree:
If the tree does not exist yet, create the root and insert s at the root. Otherwise:
    If s is smaller than the key of the root:
        Insert the key s in the left subtree of the root.
    If s is greater than the key of the root:
        Insert the key s in the right subtree of the root.

As you can see, this is a very simple recursive rule. I want to carry it out with a concrete example. I will assign words to the nodes of the tree, which at the same time serve as keys with their alphabetical order. Let us enter a sentence's words into my tree (Fig. 11.13).

The search for a record in such a tree is also carried out recursively: we find the data that belongs to a given key s with the following algorithm:

Fig. 11.13 A height-balanced tree

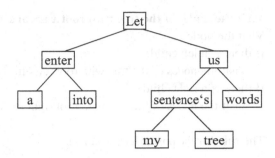

Search s starting at x:
If s is equal to the key of x, the search is finished.
If s is less than the key of x, then find s starting from the left child of x:
If s is greater than the key of x, then find s starting from the right child of x:

This will find every entered record. To cover the case that a non-existing key is searched for, the algorithm must be extended slightly. Please do this yourself.

The creation of a tree using the presented method does not always result in a tree that is as evenly balanced as shown in Fig. 11.13. If you want to convert the phone book into a tree structure in this way, the tree degenerates into an infinitely long list, and the entire list must be searched sequentially when searching. This does not provide any benefit. Trees that are particularly well suited for searching are those for which the left and right subtrees in each node differ in height by at most 1. These trees are called *height-balanced trees*. Fig. 11.13 shows such a tree. There are algorithms that create exactly such trees.

You can output all the data contained in the tree alphabetically using the rule:

Output the tree with root x alphabetically:
If there is a left child:
    Output the tree with root left child of x alphabetically.
Output the record of the node x.
If there is a right child:
    Output the tree with root right child of x alphabetically.

Try this out on the above example!

This is an algorithm that systematically visits all the nodes of a graph. Such *traversal methods* are a fundamental part of many graph algorithms. The algorithm presented describes the *inorder* traversal of a binary tree. Closely related to this are the *preorder* and *postorder* traversals. In these, the root is not visited between the subtrees, but before or after the subtrees. The preorder traversal can be described as follows:

Visit the nodes of the tree with root x according to preorder:
Visit the node x.
If there is a left child:
    Visit the nodes of the tree with root left child of x according to preorder.
If there is a right child:
    Visit the nodes of the tree with root right child of x according to preorder.

The rule for the postorder traversal is:

Visit the nodes of the tree with root x according to postorder:
If there is a left child:
    Visit the nodes of the tree with root left child of x according to postorder.
If there is a right child:
    Visit the nodes of the tree with root right child of x according to postorder.
Visit the node x.

If you output the tree from Fig. 11.13 according to preorder, you get:

> Let enter a into us sentence's my tree words

and with the postorder method you get:

> a into enter my tree sentence's words us Let

Try to implement such a tree. If you have taken the hurdle of modeling the tree data structure correctly, you will see that the insert, search, and output functions really only require a few lines of code.

The Huffman Code

As an application example of binary trees, I would like to introduce you to the Huffman code, an important algorithm for data compression. If you encode a text of the English language with the ASCII code, you need exactly one byte for each character, for the frequent "E" as well as for the rare "X". A long time ago it was discovered that one could save capacity during data transmission by replacing frequently occurring characters with short code words and rarely occurring characters with longer code words. An example of such a code is the Morse alphabet. In this, for example:

$$E = \cdot \quad T = - \quad A = \cdot- \quad H = \cdots\cdot \quad X = -\cdots-$$

With such variable length codes, however, there is a decoding problem: How, for example, is the code "·-", interpreted at the receiver? As "A" or as "ET"? The Morse code avoids this problem by using not only two characters, but also the pause in addition to "." and "-", a separator. So "· -" is equal to "ET" and "··" is equal to "A".

Is it possible to construct a code in such a way that the separator between two code words can be dispensed with? This is possible with the help of the so-called *prefix codes*: A prefix code is a code in which there is no whole code word that is a prefix (initial segment) of any other code word. You all know such a code, probably without having consciously thought about this property: the telephone number system. There is no telephone number that is a prefix of another number. So for example there is no telephone number that starts with 911. This property has the consequence that the end of a code word can be recognized without further separator: If you dial 911, the telephone switch knows that the phone number is over, because no other phone number has the same beginning, and therefore connects you to the emergency services. The Morse alphabet is not a prefix code: The code of "E" is a prefix of the code of "A".

With the help of rooted trees, such prefix codes can be constructed. I want to generate binary codes, that is, codes whose code words consist only of the characters "0" and "1". For this purpose, we draw a binary regular rooted tree on which we write a 0 on the edges leading to the left and a 1 on the edges leading to the right.

We assign the characters to be encoded to the leaves. Such a character is then mapped to the sequence of 0s and 1s leading from the root to the corresponding leaf (Fig. 11.14).

No codeword is a prefix of another, otherwise the corresponding character would have to lie on the way to this other word. But we only coded leaves. Now a bitstream made up of these codewords can be decoded uniquely without using a pause character: For example,

01001100101110101011001000   can be decoded in   010 011 00 10 111 010 10 110 010 00.

Conversely, every prefix code can be represented by such a tree. The construction method is obvious: Start at the root and draw the walk for each codeword that leads to the left child for 0 and to the right child for 1. Because of the prefix property, the codewords are exactly the walks that end in the leaves of the resulting tree.

Our goal was to generate prefix codes so that frequently occurring characters are encoded short, rare characters have longer codes. The *Huffman algorithm* constructs

Fig. 11.14 Construction of a prefix code

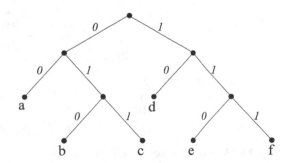

such a code, it can be proven that this represents an optimal prefix code (see Theorem 20.24).

Given is a source alphabet, from which the frequency of occurrence of the different characters is known. I would like to explain the algorithm using a concrete example. Our source alphabet consists of the characters a,b,c,d,e,f with the probabilities of occurrence:

$$a: 4\% \quad b: 15\% \quad c: 10\% \quad d: 15\% \quad e: 36\% \quad f: 20\%.$$

The algorithm proceeds as follows:

1. The characters of the alphabet are the leaves of a tree. We assign the frequency of the characters to the leaves.

    If you want to draw the tree, it is best to write the characters next to each other in the order of probability.

2. Find the two smallest nodes without a parent. If there are several, choose two arbitrary ones. Add a new node and connect it to these two. The new node should be the parent of the two initial nodes. Assign the sum of the two frequencies to this node.

    At the beginning, no node has a parent. When step 2 is carried out, the number of nodes without a parent is reduced by 1.

3. Repeat step 2 until only one node without a parent is left. This is the root of the tree.

In our example, this results in the code from Fig. 11.15.

The Huffman code is used, for example, in fax coding, where one knows the probabilities quite well for the number of consecutive white or black pixels. But it is still often used in modern compression methods, usually as part of multi-stage algorithms. For example as part of the jpeg algorithm for image compression.

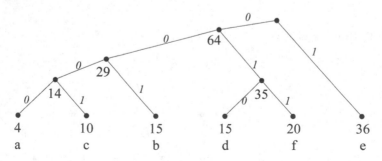

Fig. 11.15 Construction of the Huffman code

Prefix codes encode characters of constant length in differently long code words. An alternative compression approach searches for words of different lengths in the source text and translates them into code words of constant length. Such a method is, for example, realized by the widely used ZIP coding. During coding, a dictionary of frequently occurring character strings is created and—also using trees—these character strings are assigned a code word of fixed length. Most often the code words have a length of 9 bits, so that in addition to the 256 ASCII characters, 256 character strings can be provided with their own code.

11.3   Traversing Graphs

When we looked at search trees, we already saw that it is important to know methods that can find all the nodes of the tree. We now want to examine procedures that can reliably visit all the nodes of any graph, if possible exactly once. Most such visitation algorithms are based on the two prototypes *depth-first search* or *breadth-first search*. Let's first look at the depth-first search. We start at an arbitrary node x of the graph G and proceed according to the following recursive algorithm:

Traverse the graph starting at node x:
Mark x as visited.
For all unvisited nodes y that are adjacent to x:
  Go to y,
  Traverse the graph starting at node y:

Starting from x, follow a path until you reach a node to which no unvisited nodes are adjacent anymore. Then turn back to the next node that still has unvisited adjacent nodes. Visit such a node next and start from there again, as far as you can get, then turn back again and so on.

If you apply the depth-first search to a binary rooted tree, the depth-first search corresponds to the preorder traversal of the tree.

In Fig. 11.16 I have marked in which order the nodes are visited in a graph. When implementing, it still needs to be specified what exactly is meant by "for all unvisited adjacent nodes", a sequence must be specified. In the example, I started randomly from node 1 and then always took the path as far to the right as possible. Any other sequence would have been just as possible.

The number of unvisited nodes decreases by 1 at each step. Every node that is connected to x by a path is eventually caught by this search, and exactly once. If G is connected, this thus visits every node of G. If G has several connected components, we have to run the algorithm for each of these components.

Fig. 11.16 The depth-first
search

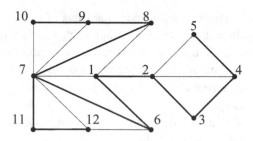

Each traversed edge leads to an unvisited node. If we process a connected graph with
n nodes with this algorithm, we therefore need $n - 1$ edges to visit the $n - 1$ nodes that
are different from the starting point. If we mark the used edges, we get a subgraph with
all n nodes and $n - 1$ edges. Such a graph is, according to Theorem 11.11, a tree. All
nodes of a connected graph can therefore be connected by the edges of a tree. Such a tree
is called *spanning tree* of the graph G. In Fig. 11.16 the edges that I used when traversing
the tree are shown in bold. You can see these edges, together with all the nodes, form a
tree, namely a spanning tree.

The depth-first search is suitable if you want to find the way out of a labyrinth. It is a
single-player algorithm: each path is first followed to the end.

> If you get lost in a corn maze the next time and use the depth-first search, you can find out
> what is meant by the runtime of an algorithm.

The breadth-first search is more suitable for teamwork: starting from a node, it swarms
in all directions, only gradually does it get further and further away from the starting
node. The breadth-first search algorithm for a connected graph with starting node x is
(for once not recursive):

1. Make the node x the current node and give it the number 1.
2. Visit all nodes adjacent to the current node and number them consecutively, starting
   with the next free number.
3. If not all nodes have been visited, make the node with the next number the current
   node and continue with step 2.

In Fig. 11.17 I have recorded the order of the visited nodes in a breadth-first search.
Again a spanning tree is created, but it looks completely different from the one in Fig.
11.16.

If we carry out a breadth-first search on a rooted tree, the levels will be successively
grazed until we reach the leaves. This is where the difference to a depth-first search, in
which all the branches are successively traversed to the leaves, is particularly noticeable.

Breadth-first search and depth-first search are methods that are applied to graphs
in this or a similar form over and over again in order to gain information about these
graphs. We have already seen that these algorithms can be used to determine whether a

Fig. 11.17 The breadth-first
search

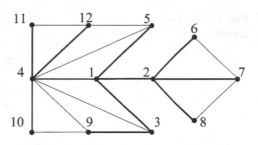

graph is connected. Depth-first search can also be used to determine whether a graph is connected multiple times, a property important, for example, in power or computer networks: Even after a line section has failed, these should still be connected.

Shortest Paths

Finding the shortest paths in networks is a task occuring very frequently in computer science. Graph traversal algorithms are also used for this. I would like to introduce you to the algorithm of Dijkstra, which he formulated in 1959. It is related to breadth-first search: Starting from one point, the shortest paths to all other points are determined. In doing so, one gradually moves further and further away from the starting point. The algorithm always finds the shortest path to the target point z if such a path exists at all. At the same time, all the shortest paths are found that lead to nodes that are closer to the starting point than z.

G is a network, the weight function is always positive. This condition states that paths from node to node always become longer. The shortest path from node x to node z is sought.

We divide the nodes of the graph into three sets: the set V of points that have already been visited and for which shortest paths are known, the set B of points that are adjacent to points from V, one could call B the boundary of V, and finally U, the unvisited points.

At the beginning, V only consists of the starting node x, to which the shortest path of length 0 is known. B consists of the nodes that are adjacent to x.

In each step of the algorithm, a node is selected from the boundary B and added to V, the points adjacent to this node are added to the boundary. The algorithm ends when the node z has been added to V or when no node is left in the boundary. This is the case when all nodes have been visited that are connected to the starting point x.

In Fig. 11.18 the iteration step is circled next to each node in which it is added to V, the sets V, B and U are drawn after four iterations of the algorithm.

Which node is added to V? Each node y of B is connected to at least one node of V by an edge. Since the shortest paths are known within V, we can use them to determine the shortest path to y, which only contains nodes of V. We assign this path length $l_B(y)$ to all points y of B. This does not necessarily give us the absolutely shortest path from x to y, because this could also contain other nodes from B or U. You can see this, for example, at

Fig. 11.18 The Dijkstra
algorithm

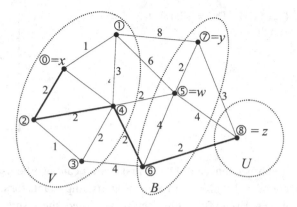

point y in Fig. 11.18: The shortest path from x to y has the length 8, the shortest path that only using from B has the length 9: $l_B(y) = 9$.

Now we are looking for a minimum of the distances in B. Let w be such a node for which $l_B(w)$ is minimal. The path from x to w cannot be further shortened, because every shorter path from x to w would have to touch another border point (for example y), but the path from x to this border point would already be at least as long as the one from x to w. The node w can therefore be removed from the border and added to V. At the same time, all nodes adjacent to w are added to the boundary. In the example, z is the last reached node, all shortest paths have to be calculated until the shortest path from x to z is found.

You can see the algorithm is time-consuming. The runtime increases quadratically with the number of nodes.

Nevertheless, this algorithm is still computable in a reasonable time, in contrast to the related problem of the traveling salesman: A sales representative wants to visit n cities and tries to find the shortest path that touches all n cities once and leads him back to his starting point. The task of the *traveling salesman* is probably one of the best-researched problems in computer science. There is no known algorithm solving it in polynomial time, that is, in a time is proportional to n^k when n represents the number of cities. It is believed no such algorithm exists. Already for about 1000 cities the problem is no longer computationally feasible in an acceptable time.

A mathematician could be content with to proof that there is no fast algorithm, but not the traveling salesman, who actually has to make his journey, and certainly not the computer scientist, who is given the task of finding a fast way. The typical approach of the computer scientist is to deviate from the pure doctrine and to look for solutions that are easily calculated, but also not too far from the optimal solution. He is looking for a compromise between the two goals "best possible way" and "lowest possible computing time". On the Internet you will find competitions in which good algorithms for the traveling salesman compete against each other.

11.4   Directed Graphs

In many applications, directions are assigned to the edges of a graph: one-way streets, flowcharts, project plans, or finite automata are examples of such graphs. In Fig. 11.19 you can see a finite state machine that can recognize comments in the source code of a program that begin with /* and end with */. The nodes are states, a transition from a state to another occurs when reading a certain character.

▶ **Definition 11.16** A *digraph* or *directed graph* G consists of a set $V = V(G)$, the nodes of G and a set $E = E(G)$ of ordered pairs $k = [x, y]$, with $x, y \in V$, the *directed edges* of G. The directed edges are also called *arrows* or *arcs*. The direction of the edge $k = [x, y]$ points from x to y. x is called the *tail* and y the *head* of the arc $[x, y]$.

You can see in Fig. 11.19 that there can now be two edges between different nodes, but in different directions. In the adjacency matrix of a digraph, the element a_{ij} has the value 1 if there is an arc from i to j. The adjacency matrix is therefore generally no longer symmetrical. For the above example it is:

$$\begin{pmatrix} 1 & 1 & 0 & 0 \\ 1 & 0 & 1 & 0 \\ 0 & 0 & 1 & 1 \\ 1 & 0 & 1 & 0 \end{pmatrix}.$$

A directed graph is nothing more than a relation on the set of nodes: it is xRy if and only if $[x, y] \in V$. See Definition 1.7 in Sect. 1.2. On the other hand, one can also say: every relation on a finite set represents a directed graph.

▶ **Definition 11.17** If x is a node of the directed graph G, then the number of arcs ending in x is called the *in-degree* of x, it is denoted by $d^-(x)$. The number of arcs starting in x is called the *out-degree* of x, it is denoted by $d^+(x)$.

Since each arc has exactly one head and one tail and each of these points contributes to the in-degree or out-degree of a node, it follows immediately:

▶ **Theorem 11.18** In every directed graph $G = (E, V)$ it is

$$\sum_{x \in V} d^+(x) = \sum_{x \in V} d^-(x) = |E|.$$

▶ **Definition 11.19** A node x of the directed graph G is called a *source*, if $d^-(x) = 0$, and a *sink*, if $d^+(x) = 0$.

In Definition 11.6 we have defined the terms walk, path, cycle and connectivity for undirected graphs. These transfer to directed graphs:

▶   **Definition 11.20** An *directed walk* from x_1 to x_n is a sequence of directed edges
        $[x_1, x_2][x_2, x_3], \ldots, [x_{n-1}, x_n]$, a *directed path* from x_1 to x_n is a directed walk
        in which all nodes are different, a *directed cycle* is a directed walk in which all
        nodes except for the start and end point are different. A directed graph G is called
        weakly connected, if the underlying graph, which is obtained when ignoring the
        directions, is a connected graph. G is called *strongly connected*, if every two nodes
        are connected by a directed path.

Particularly important for us will be directed graphs without cycles:

▶   **Definition 11.21** A directed graph in which there is no directed cycle is called a
        directed acyclic graph.

After many definitions finally a first, surprising result. Behind it is the fact that in such a
graph every path has a beginning and an end:

▶   **Theorem 11.22** In every directed acyclic graph there is at least one source and one
        sink.

Proof: Since the graph G is finite, there are only finitely many paths in it, among them
there is a longest path. Let x be the endpoint of this path. If $d^+(x) \neq 0$ is, then $[x, y]$
leads out of x. But since the path is maximal, it cannot be extended to y. The node y must
therefore already occur in the path. But then we would have found a cycle, a contradiction. So x is a sink. Similarly, the starting point of this longest path is a source.   □

The directed acyclic graphs have an interesting property that is particularly important
when processing the nodes. The order of processing can be such that when visiting the
node x, all the nodes have already been visited from which a path leads to x. So all the

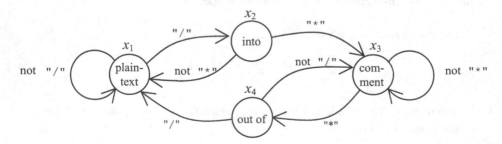

Fig. 11.19 A directed graph

"predecessors" of x are already processed. If you number the nodes of the graph according to this visit algorithm, you speak of a *topological sorting* of the graph.

▶   **Definition 11.23** The directed graph G with node set $\{x_1, x_2, \ldots, x_n\}$ is called *topologically sorted*, if for all nodes x_i it holds: If $[x_k, x_i]$ is an arc, then $k < i$.

▶   **Theorem 11.24** A directed graph G has a topological sorting if and only if it is acyclic.

Apparently, a graph cannot be topologically sorted if it contains a directed cycle. The nodes on this cycle cannot be numbered in ascending order, at some point the indices must become smaller again. The other direction of the theorem is more interesting. With mathematical induction, we construct a topological sort for every acyclic graph: If the graph only has one node, it is of course topologically sortable. If it has $n + 1$ nodes, we can find a source according to Theorem 11.22. We denote this by x_1. The source x_1 and all arcs emanating from it are removed from the graph. This results in an acyclic graph with n nodes that we can topologically sort. We denote the nodes of this sort by $\{x_2, x_3, \ldots, x_{n+1}\}$, so we have found the sort of G.   □

This induction also gives us a recursive algorithm for topological sorting:

Number the nodes of the directed acyclic graph G:
Find a source x of G, give it the next free number.
If x is not the only node of G:
   Remove x and the arcs starting from x, the result is the graph G:
   Number the nodes of the directed acyclic graph G:
Otherwise the numbering is finished.

This algorithm also represents a method for deciding whether a directed graph is acyclic: If you can no longer find a source although there are still unnumbered nodes, you have found a cycle.

As in the undirected graphs, the directed edges are often equipped with weights that represent, for example, transport capacities of pipelines, distances, duration or similar. In a network, the flow of material from a producer to a consumer can be recorded via a number of intermediaries. The producer is the source, the consumer the sink of the network, each arc is a transport route with a capacity (Fig. 11.20).

▶   **Definition 11.25** A directed graph is called *weighted directed graph* or *directed network* if each edge $[x, y]$ is assigned a weight $w(x, y) \in \mathbb{R}$.

A flow network is a directed, positively weighted network that has a source and a sink. The weight is also called capacity function. An important task of graph theory is to

Fig. 11.20 A directed
network

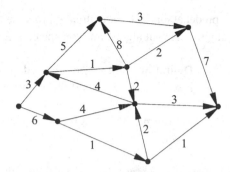

determine the maximum capacity of a flow network, that is, the maximum amount of the examined good that can be transported from the source to the sink. For no arc the capacity may be exceeded, and in each node, except for the source and the sink, as much must flow in as out. I do not want to deal with this question here, but at the end of the chapter I would like to discuss a problem from project planning.

The plan for the implementation of a project represents a directed graph: Starting from a source, for example the order placement, the project is divided into various modules that can be developed independently of each other. These are often divided into further sub-modules. There are dependencies between the individual modules: There are modules that can only be started when others are finished or have reached a defined status. In particular, during integration, everything comes together again: In different stages, the modules are assembled and tested until hopefully everything has grown together to form the big picture at the end of the handover.

A project plan defines durations for the processing of individual modules and milestones to which certain work must be done. Such a project plan can be regarded as a directed network. The arcs are the activities in the project, the weight of an arc represents the duration for the execution of this activity. The nodes denote the milestones of the project. A schedule assigns a time to each milestone at which it is to be achieved. How do you find an optimal schedule?

If there is a schedule at all, then the graph must be without cycles, otherwise something has gone completely wrong. The network in Fig. 11.20 is therefore not a project plan. We further assume that there is exactly one source and one sink in the graph. In Fig. 11.21 you can see an example of such a network plan. We assign the time 0 to the source, now we want to find the earliest time T at which the sink can be reached. Since each individual activity must be carried out, that is, really every path of the graph must be traversed, we are faced here with the problem of finding the *longest path* in the graph. For this I would like to introduce you to an algorithm.

Since the graph to be examined is acyclic, it can be topologically sorted. So we assume that the nodes $\{x_1, x_2, \ldots, x_n\}$ of the graph are numbered according to a topological sort.

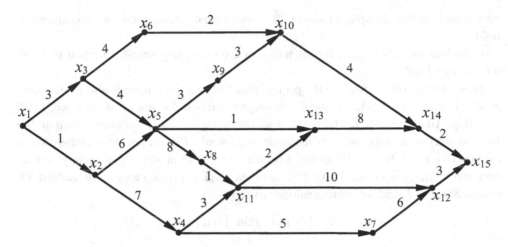

Fig. 11.21 The longest path in the project plan

For each node y we want to determine the number $L(y)$, which gives the longest length of a directed path from the source to y. First, we set $L(x_i) = 0$ for all i. Now the longest paths to all nodes y are determined one after the other. The value $L(y)$ can be increased during the process until all paths to y are considered:

1. Start with $i = 1$.
2. Visit the node x_i and examine all nodes y to which an arc $[x_i, y]$ exists. If $L(x_i) + w(x_i, y) > L(y)$, set $L(y) = L(x_i) + w(x_i, y)$, otherwise leave $L(y)$ unchanged.
3. As long as there are still nodes left: Increase i by 1 and go to step 2.

This algorithm works because when visiting the node x_i, the longest path $L(x_i)$ to it is already finally calculated: Because of the topological sort, no arc from later nodes leads back to x_i, so that $L(x_i)$ can no longer be changed. Thus $L(x_i) + w(x_i, y)$ is the longest path to y, which leads through x_i. Since in the course of time all predecessors of y are considered, at some point the longest path to y is also found.

It is not difficult to remember such a longest path from the source to the sink at the same time when calculating the $L(x_i)$.

The algorithm can be formulated very elegantly recursively:

$$L(x_1) = 0, \quad L(x_i) = \max_{\substack{k<i \\ [x_k, x_i] \in E}} \{L(x_k) + w(x_k, x_i)\}. \tag{11.1}$$

For the sink we thus obtain a value $L(x_n) = T$, where T is the desired minimum project duration. For each other node $L(x_i)$ denotes the earliest time at which this milestone can be reached. In the example, $L(x_{15}) = 29$, a longest path through the graph is shown in bold.

Calculate the values $L(x_i)$ for all nodes in the example yourself. Can you find any more longest paths?

How can one further analyze the project plan? At each node there is also a latest time at which it must be reached if the entire project duration T is not to be extended. I call this $S(x_i)$. To calculate it, the topological sorting helps us now in the other direction: We have to start at the sink and visit the nodes backwards. If I x_i examine, I already know all $S(x_k)$ for $k > i$. If $[x_i, x_k]$ is an arc, I have to subtract from $S(x_k)$ the necessary activity duration $w(x_i, x_k)$ to x_k and from all these times $S(x_k) - w(x_i, x_k)$ choose the earliest. Of course $S(x_n) = T$, and we obtain analogously to (11.1):

$$S(x_n) = T, \quad S(x_i) = \min_{\substack{k>i \\ [x_e, x_i] \in E}} \{S(x_k) - w(x_i, x_k)\}.$$

In the example $L(x_{14}) = 26$, while $S(x_{15}) = 29$. The activity $[x_{14}, x_{15}]$ is planned with 2 units, so that a margin of 1 unit remains.

In the meantime, we now know for each activity $[x, y]$ the earliest time $L(x)$ at which x can be reached, the time $S(y)$ at which y must occur at the latest, as well as the planned activity duration $w(x, y)$. The difference $S(y) - L(x) - w(x, y)$ is the time buffer: The activity $[x, y]$ can be extended by this period of time without endangering the entire project duration. If the buffer is 0, we speak of a *critical activity*. In every project there is at least one path that consists only of critical activities, for example the path that we found when determining $L(x_n)$. Such a path is called a *critical path*.

11.5   Comprehension Questions and Exercises

Comprehension Questions

1. A connected graph is a tree if it has one node more than edges. Why is it important in this theorem that the graph is connected?
2. If you declare another node to be the root in a rooted tree, you get a rooted tree again. Can it be that the two rooted trees have a different number of leaves?
3. Every walk k from x to y in a graph contains a path w from x to y. Of course, w is shorter or at most as long as k. Can there be a path from x to y that is longer than w?
4. Is a directed graph automatically turned into a graph if you remove the arrows from the edges?
5. Is a graph automatically turned into a directed graph if you assign a direction to each edge?

6. Can you represent a partial order on a finite set by a graph or by a directed graph?
7. In a directed graph, there can be two directed edges between the nodes x and y. Can there also be two directed edges between x and x?

Exercises

1. Write down the degree of each node in graph G from Fig. 11.22. How many even and odd nodes (= nodes of even and odd degree) does G contain?
   Set up the adjacency matrix for G.
2. Is it possible that there are nine people at a party, each of whom knows exactly five others?
3. G is a graph with n nodes and $n - 1$ edges. Show that G contains at least one endpoint or an isolated node (a node of degree 0).
4. Draw the syntax tree for the following expression:

$$c = (a+3)*b - 4*x + z*x/7;$$

   Visit the nodes of the tree using the three methods Inorder, Preorder, and Postorder.
5. Check whether the codes that consist of the following words are prefix codes, and if so, draw the corresponding binary tree:
   a) 11, 1010, 1011, 100, 01, 001, 000
   b) 00, 010, 011, 10, 110, 111, 0110, 0111
6. Construct a prefix code for the alphabet $\{a, b, c, d, e, f, g, h\}$ with the following frequency distribution $\{4, 6, 7, 8, 10, 15, 20, 30\}$ ($a = 4\%, b = 6\%, ..., h = 30\%$).
7. Dijkstra's algorithm for finding shortest paths in networks assumes the distance between two nodes is not negative. Why is that so? Give an example of a network in which this condition is not met and in which therefore the algorithm fails.
8. Given the following adjacency matrix of a network:

$$\begin{pmatrix} 0 & 0 & 0 & 4 & 1 & 2 & 0 & 0 & 0 \\ 0 & 0 & 4 & 2 & 6 & 0 & 4 & 2 & 0 \\ 0 & 4 & 0 & 2 & 0 & 0 & 2 & 0 & 4 \\ 4 & 2 & 2 & 0 & 3 & 2 & 0 & 0 & 2 \\ 1 & 6 & 0 & 3 & 0 & 0 & 0 & 8 & 0 \\ 2 & 0 & 0 & 2 & 0 & 0 & 0 & 0 & 1 \\ 0 & 4 & 2 & 0 & 0 & 0 & 0 & 3 & 0 \\ 0 & 2 & 0 & 0 & 8 & 0 & 3 & 0 & 0 \\ 0 & 0 & 4 & 2 & 0 & 1 & 0 & 0 & 0 \end{pmatrix}$$

Fig. 11.22 Exercise 1

The matrix entry a_{ij} just contains the length of the path from node i to node j. Try to draw a non-overlapping picture of the graph. Determine the shortest paths from the first node to the other nodes. Draw spanning trees in the graph that arise when starting from the first node when performing breadth-first search and depth-first search respectively.

9. The finite state machine from Fig. 11.19 is not quite perfect yet. If someone comes up with the idea of writing the following text:

    ```
    a//* this is a division */b
    ```

    then it fails. Complete the finite state machine so that it can also cope with this!
10. Show: An acyclic network that has exactly one source and one sink is weakly connected.

Part II
Analysis

The Real Numbers

12

Abstract

This chapter lays the foundation for investigations of convergent sequences and continuous functions. By the end of this chapter, you will

- know what characterizes the real numbers,
- know the properties of order and completeness on the real numbers,
- know the concept of a metric space and some examples of metric spaces,
- be able to handle neighborhoods in metric spaces,
- know open sets, closed sets, and know what the boundary of a set is.

12.1 The Axioms of Real Numbers

We have already used the real numbers in the first half quite often. Now we finally want to characterize them more precisely, as the properties of the real numbers are quite essential for the second half of the book. In Theorem 2.10 we have seen that there are numbers with which we would like to calculate, but which are not contained in $\mathbb{Q}$, for example $\sqrt{2}$. The rational numbers still have gaps. Just as $\mathbb{Z}$ from $\mathbb{N}$ and $\mathbb{Q}$ from $\mathbb{Z}$, one can also construct the real numbers $\mathbb{R}$ from $\mathbb{Q}$ formally by filling in all these gaps. However, it is not so easy to describe this filling of gaps mathematically precisely and I therefore will refrain from this construction. We will take another approach that has already

---

Supplementary Information The online version contains supplementary material available at
https://doi.org/10.1007/978-3-658-40423-9_12.

proved to be useful several times: We will collect the characteristic properties of $\mathbb{R}$, the axioms of the real numbers. The real numbers are then a set for us that satisfies these axioms. A little later we will find that the real numbers can be identified with the set of all decimal fractions. We can then also use the axioms in the following chapters to derive theorems about the real numbers.

The Order Axioms

First of all, the real numbers form a field. Please take a look at the field axioms from Definition 5.10 in Sect. 5.3 again. Another essential property of $\mathbb{R}$, which we have already used several times, is the ordering: Two elements can be compared with each other with respect to their size. $\mathbb{R}$ is a linearly ordered set with respect to the relation $\leq$ according to Definition 1.12.

An ordered field is linearly ordered as a set, the order additionally has some compatibility properties with the two operations. The properties of an ordered field can be described by the order axioms. There is no mention of the order at all at first, they state that there are positive and negative elements in ordered fields . The order and its properties can be derived from this.

▶   **Definition 12.1: The order axioms** A field K is called *ordered*, if there is a subset P (the prepositive cone) in it with the following properties:

(A1) $P, -P, \{0\}$ are disjoint $(-P := \{x \in K \mid -x \in P\})$.
(A2) $P \cup -P \cup \{0\} = K$.
(A3) $x, y \in P \Rightarrow x + y \in P, x \cdot y \in P$.

The elements from P are called *positive*, those from $-P$ are called *negative*. $P \cup \{0\}$ is called the *positive cone* of K.

Each element different from 0 is thus positive or negative. Sum and product of positive elements are positive again.

From these axioms, we can now immediately derive our order relations $<, \leq, >, \geq$ already known in $\mathbb{R}$:

▶   **Definition 12.2: The order relations in an ordered field** In an ordered field K with prepositive cone P, we define for all $x, y \in K$:

$$x < y \quad :\Leftrightarrow \quad y - x \in P,$$
$$x \leq y \quad :\Leftrightarrow \quad x < y \text{ or } x = y,$$
$$x > y \quad :\Leftrightarrow \quad y < x,$$
$$x \geq y \quad :\Leftrightarrow \quad y \leq x.$$

You can see that $y > 0$ is only true if $y \in P$. Thus, as expected, the prepositive cone P contains exactly the numbers greater than zero.

I would like to put together some basic properties of these relations now, which of course all follow from the order axioms. The properties a), b) and c) just say that $\leq$ is a linear order relation on the field K according to Definition 1.12.

▶   **Theorem 12.3** Let K be an ordered field and $x, y, z, w \in K$. Then hold:

a) $x \leq y$ and $y \leq x \Rightarrow x = y$ (the anti-symmetry).
b) $x \leq y, \; y \leq z \Rightarrow x \leq z$.   (the transitivity)
c) $x \leq y$ or $y \leq x$ (the linearity of the order)
d) $x < y, \; z \leq w \Rightarrow x + z < y + w$.   (the compatibility with „+“)
e) $x < y, \; z > 0 \Rightarrow xz < yz$
   $x < y, \; z < 0 \Rightarrow xz > yz$ (the compatibility with „·“)
f) $x > 0 \Rightarrow -x < 0$ and $x < y \Rightarrow -y < -x$
   $0 < x < y \Rightarrow 0 < y^{-1} < x^{-1}$.   (the transition to the inverse)

The calculations are elementary, although one sometimes has to think about it. I would like to derive the first and second property as examples:

to a): Let's first take $x \neq y$. Then $x \leq y$ means that $y - x \in P$ and $y \leq x$ has $x - y \in P$ as result, that is $y - x = -(x - y) \in -P$. Since P and $-P$ are disjoint, $y - x$ cannot be in P and $-P$ at the same time. The assumption $x \neq y$ must therefore be false.

to b): If $x = y$ or $y = z$, the assertion is clear. So let's take $x \neq y$ and $y \neq z$. Then $y - x \in P$ and $z - y \in P$, so also $(z - y) + (y - x) = z - x \in P$ and thus $x < z$.   □

The behavior with multiplications gives the most trouble when calculating with inequalities. If you multiply with negative numbers, the inequality sign turns around, otherwise not. This often leads to ugly case distinctions.

In an ordered field, according to the axiom (A3), $x \neq 0$ implies $x^2 > 0$. This is because if $x > 0$, then $x^2 > 0$ and for $x < 0$ it follows from $-x > 0$ that $(-x)^2 = x^2 > 0$ is again true. Therefore, $1 = 1^2 > 0$, so $-1 < 0$. It can therefore -1 never be the square of another number. You can see from this that complex numbers cannot carry the structure of an ordered field, no matter how much effort is put into the definition of the order.

Because of $0 < 1$, it follows from the rules $1 < 1 + 1 < 1 + 1 + 1 < \cdots < \underbrace{1 + 1 + \cdots + 1}_{n\text{-times}}$.

So these numbers are all different. If you remember the introduction of $\mathbb{N}$ in Sect. 3.1, then the set $\{1, 1 + 1, 1 + 1 + 1, \ldots\}$ just forms the natural numbers. The natural numbers $\mathbb{N}$ are therefore contained in any ordered field K. Since in a field the additive inverses and the quotients of two numbers must also be contained, K also contains the integers $\mathbb{Z}$ and the rational numbers $\mathbb{Q}$. $\mathbb{Q}$ itself is an ordered field and thus the smallest ordered field that exists. The prepositive cone of $\mathbb{Q}$ consists of the numbers $\mathbb{Q}_+ = \{\frac{m}{n} \mid m, n \in \mathbb{N}\}$.

We have met other fields in Sect. 5.3: the finite fields GF(p) and GF(p^n). But these fields cannot carry an order, since $\mathbb{N}$ is not contained in them.

The Completeness Axiom

We are on the way to characterizing the real numbers, and have seen that the order alone is not enough: the rational numbers are also ordered.

In contrast to the rational numbers, the real numbers are complete in a certain sense, there are no more gaps. We now want to formulate the term of completeness precisely.

▶ **Definition 12.4** Let K be an ordered field and $M \subset K$. An element $x \in M$ is called *maximum* of the set M, if for all $y \in M$ it holds $y \leq x$. Correspondingly, $x \in M$ is called *minimum* of M, if for all $y \in M$ it holds $y \geq x$.

If a set M has a maximum x, then this is uniquely determined: For if x' is another maximum, then from the Definition 12.4$x' \geq x$ and $x \geq x'$, it follows that $x = x'$. Similarly, the minimum is uniquely determined. We denote the maximum and minimum of M by $\max(M)$ and $\min(M)$, respectively.

Examples

Let $K = \mathbb{Q}$.

1. $M_1 = \{x \in \mathbb{Q} \mid x \leq 2\}$. It is $\max(M_1) = 2$, because $2 \in M_1$ and for all $y \in M_1$ it holds $y \leq 2$.
2. $M_2 = \{x \in \mathbb{Q} \mid x < 2\}$ has no maximum, because for every $x \in M_2$ there is a $y \in M_2$ with the property $x < y < 2$. You can, for example, choose $y = (x+2)/2$. Check that $x < (x+2)/2 < 2$!
3. $M_3 = \{x \in \mathbb{Q} \mid x^2 \leq 2\}$ has no maximum, because if x is the maximum, then $x^2 = 2$ or $x^2 < 2$. $x^2 = 2$ has no solution in $\mathbb{Q}$, so $x^2 < 2$. Then, similar to Example 2, there is always a number $y \in \mathbb{Q}$ with $x < y$ and $y^2 < 2$.
4. $M_4 = \mathbb{Z}$ has no maximum: For every integer z there is a larger integer, for example $z + 1$. ◀

▶ **Definition 12.5** Let K be an ordered field and $M \subset K$. M is called *bounded from above*, if there is a $x \in K$ with $x \geq y$ for all $y \in M$. The element x is then called *upper bound* of M. $s \in K$ is called *least upper bound* of M, if for every other upper bound x of M it holds $s \leq x$. The element s is then called *supremum* of M and is denoted by $\sup(M)$.

Similarly, the terms *bounded from below*, the *lower bound* and the *greatest lower bound* are defined. The greatest lower bound of a set is called the *infimum* of M and is denoted by $\inf(M)$.

The set M is called bounded if it is bounded from below and from above. Let's look at the examples from before again:

Examples of upper bounds

1. M_1 has, for example, 10^6, 4 and 2 as upper bounds. M_1 is not bounded from below. 2 is the smallest upper bound: $\sup(M_1) = 2$.
2. M_2 has the same upper bounds as M_1, but although M_2 has no maximum, it has a smallest upper bound, namely 2. Any smaller number belongs to M_2 and is therefore no longer an upper bound: $\sup(M_2) = 2$.
3. M_3 has, for example, 10^6, 4, 1.5, 1.42, 1.425 as upper bounds, but there is no smallest upper bound. We can always get closer to $\sqrt{2}$ with rational numbers, none of these numbers is the smallest rational number above $\sqrt{2}$. If $\sqrt{2}$ were a rational number, it would be the supremum. ◄

I would like to postpone the investigation of $M_4 = \mathbb{Z}$ for a moment.

If a set M has a maximum, this maximum is also the supremum at the same time. There are sets that have no maximum, but a supremum.

In the rational numbers there are sets that have upper bounds, but no supremum. With this we have found the tool with which we can grasp the existence of gaps in $\mathbb{Q}$: The gaps are the missing smallest upper bounds of sets that are bounded from above. There is nothing like this in $\mathbb{R}$, the real numbers are complete in this sense. The axiom applies:

▶ **Definition 12.6: The completeness axiom** An ordered field is called *complete*, if in it every non-empty set which is bounded from above has a supremum.

$\mathbb{Q}$ is not a complete field, as the example of the set M_3 shows.

With the axioms now assembled one can prove the great and important theorem:

▶ **Theorem 12.7** There is exactly one complete ordered field. This is called the *field of real numbers* $\mathbb{R}$.

In particular, to show the uniqueness of this field is a difficult mathematical task. But for us this is not so interesting. We only need the properties of $\mathbb{R}$, which can be concluded from the axioms.

For example, we see that $\sqrt{2}$ is a real number: for the supremum y of the set $\{x \in \mathbb{R} \mid x^2 \leq 2\}$ it is indeed $y^2 = 2$, thus $y = \sqrt{2}$.

Now let's look at the set of integers:

$\mathbb{Z}$ has no upper bound in $\mathbb{R}$, otherwise there would be a supremum $s \in \mathbb{R}$ of $\mathbb{Z}$. Then $s - 1$ is not an upper bound of $\mathbb{Z}$, so there is a $n \in \mathbb{Z}$ with $n > s - 1$. This means $n + 1 > s$, thus s is not an upper bound of $\mathbb{Z}$.

Similarly, it follows that every non-empty subset M of $\mathbb{Z}$ which is bounded from above has a maximum: Let $n_0 \in M$ be given. If M has no maximum, then with each number n also $n + 1$ is in M. By the induction principle it then follows $M = \{n \in \mathbb{Z} \mid n \geq n_0\}$, and this set is, just like $\mathbb{Z}$, not bounded from above.

These properties of $\mathbb{Z}$ are contained in the following theorem:

▶   **Theorem 12.8** For real numbers x, y it holds:

  a) to $x > 0$, $y > 0$ there is a $n \in \mathbb{N}$ with $n \cdot x > y$.
  b) to $x > 0$ there is a $n \in \mathbb{N}$ with $\frac{1}{n} < x$.
  c) to $x \in \mathbb{R}$ there is $m = \max\{z \in \mathbb{Z} \mid z \leq x\}$.

For a) we can choose a n with $n > \frac{y}{x}$. b) is a consequence: If $n \cdot x > 1$, then $\frac{1}{n} < x$ also applies.

To c): The set $\{z \in \mathbb{Z} \mid z \leq x\}$ is bounded from above and therefore as a subset of the integers has a maximum.   □

Above all, property b) will often be useful to us: In the next chapters we will use again and again that we can get as close to 0 as we like with the number sequence $\frac{1}{n}$.

▶   **Definition 12.9: The absolute value** For $x \in \mathbb{R}$

$$|x| := \begin{cases} x & \text{if } x \geq 0 \\ -x & \text{if } x < 0 \end{cases}$$

is called *absolute value* of x

▶   **Theorem 12.10: Properties of the absolute value** For $x, y \in \mathbb{R}$ it holds:

  a) $|x| \geq 0$, $|x| = 0 \Leftrightarrow x = 0$
  b) $|xy| = |x| \cdot |y|$.
  c) $|x + y| \leq |x| + |y|$.
  d) $\left| \dfrac{x}{y} \right| = \dfrac{|x|}{|y|}$.
  e) $\big| |x| - |y| \big| \leq |x - y|$.

I will leave it to you as an exercise to check the rules a) to c). Property d) follows from b) if one applies this rule to $\frac{x}{y}y$ and e) results from c) by replacing once x with $x - y$ and once y with $y - x$. $\qquad\square$

If you turn back to Theorem 5.16 in Sect. 5.3, you will find the same rules there for the absolute value of a complex number.

12.2 Topology

Topology is the "theory of places". It investigates questions such as whether points of a set are close together or far apart, what the boundary of a set is, whether sets have holes, whether they are connected or torn, or how a set can be deformed without tearing it apart. Topology is an important discipline in its own right in modern mathematics and can be applied to many classical areas of mathematics to gain new insights.

In the context of real and complex numbers, we will deal a lot with limits of sequences and with continuous functions. If a sequence converges, this means that the sequence elements move closer and closer to a limit. If a function is continuous, its graph does not make wild jumps. But what does "closer and closer" mean, what is a "jump"? In order to work with such concepts, we need the language of topology.

First we will define a distance function on the sets under investigation. For each two elements we can then say how far apart they are. The key concept for topological investigations lies in the concept of the neighborhoods of a point. To each point one can look at large and small neighborhoods and investigate which other points are contained in them.

▶ **Definition 12.11** Let X be a set and $d : X \times X \to \mathbb{R}$ a mapping. d is called a *metric* and (X, d) a *metric space*, if the following conditions are satisfied:

(M1) $d(x, y) \geq 0, d(x, y) = 0 \Leftrightarrow x = y$
(M2) For all $x, y \in X$ it holds $d(x, y) = d(y, x)$.
(M3) For all $x, y, z \in X$ it holds $d(x, y) \leq d(x, z) + d(z, y)$.

Think of $d(x, y)$ as the distance between the points x and y. (M3) is called the triangle inequality and in Fig. 12.1 you can see where the name comes from: The sum of the two

Fig. 12.1 The triangle inequality

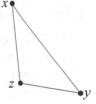

sides of a triangle is always greater than the third side. Equality can only hold in degenerate triangles, that is, if the three points are on one line.

1. The absolute value of real and complex numbers gives us a metric on $\mathbb{R}$ and $\mathbb{C}$: for $x, y \in \mathbb{R}$ respectively for $x, y \in \mathbb{C}$ let

$$d(x, y) := |x - y|.$$

(M1) and (M2) are immediately clear, (M3) applies, since $|x - y| = |x - z + z - y| \leq |x - z| + |z - y|$. In $\mathbb{R}$ and $\mathbb{C}$ we will always work with this metric.

2. In Definition 10.4 we have defined a norm on the real vector space $\mathbb{R}^n$. This norm $\|u\| := \sqrt{\langle u, u \rangle} = \sqrt{u_1^2 + u_2^2 + \cdots + u_n^2}$ has, according to Definition 10.2, the same properties as an absolute value and therefore

$$d(u, v) := \|u - v\|$$

is a metric on $\mathbb{R}^n$. We can interpret $d(u, v)$ in $\mathbb{R}^2$ and $\mathbb{R}^3$ as the distance between the two vectors u and v.

3. A central concept in coding theory is that of the *Hamming distance* between two code words: A code word is a n-tuple of 0 and 1, so the underlying set is $X = \{0, 1\}^n$. If $b = (b_1, b_2, \ldots, b_n)$ and $c = (c_1, c_2, \ldots, c_n)$ are two code words, then the Hamming distance between b and c is defined by

$$d(b, c) = \text{number of different digits of } b \text{ and } c.$$

$d((0, 0, 1, 0, 0, 1, 1, 0), (1, 1, 0, 0, 1, 1, 1, 0))$ is therefore 4. The Hamming distance is a metric on X. The properties (M1), (M2) are clear, the triangle inequality is a little more difficult to write down.

If a code is to detect transmission errors, not all elements of the set X can be valid code words. The valid code words then have a distance greater than 1 from each other. ◀

▶ **Definition 12.12** Let (X, d) be a metric space and $\varepsilon > 0$ a real number. Then for $x \in X$ the set

$$U_\varepsilon(x) := \{y \in X \mid d(x, y) < \varepsilon\}$$

is called ε-*neighborhood* of x.

1. In $\mathbb{R}$ is $U_\varepsilon(x) = \{y \mid |x - y| < \varepsilon\} = \{y \mid x - \varepsilon < y < x + \varepsilon\}$, which is the set of elements that differ from x by less than ε. In $\mathbb{C}$, the distance between y and x must be less than ε. This is the case for all points y that are located in a circle around x with radius ε.

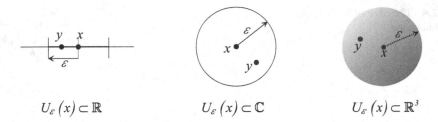

$$U_\varepsilon(x) \subset \mathbb{R} \qquad\qquad U_\varepsilon(x) \subset \mathbb{C} \qquad\qquad U_\varepsilon(x) \subset \mathbb{R}^3$$

Fig. 12.2 ε-neighborhoods

2. Accordingly, in $\mathbb{R}^3$, the ε-neighborhood of a point x represents the sphere around x with radius ε, in $\mathbb{R}^n$ the corresponding n-dimensional sphere. We can also regard the interval in $\mathbb{R}^1$ as a degenerated sphere.

3. In the Hamming metric $U_k(c)$ represents the set of codewords that differ from c in less than k places. With the help of these neighborhoods, error correction can be performed:

   A *t-error correcting code* is a code in which every transmission error can be detected and corrected up to t bits. If c' arrives at the receiver during the transmission of c and differs from c in at most t places, the original codeword can be recognized if c is the only allowed codeword in the neighborhood $U_{t+1}(c')$. When is this the case? Exactly when every two admissible words have a distance of at least $2t+1$ (Fig. 12.3). Then for each codeword b different from c it holds:
   $d(b,c') + d(c',c) \geq d(b,c)$, that is $d(b,c') \geq d(b,c) - d(c,c')$ and thus $d(b,c') \geq (2t+1) - t = t+1$, that is $b \notin U_{t+1}(c')$ (Fig. 12.3). ◄

The boundary of the sphere, that is the set of y with $d(x,y) = \varepsilon$, does not belong to the neighborhood. This has an important consequence: If $y \in U_\varepsilon(x)$, then there is always a $\delta > 0$ with $U_\delta(y) \subset U_\varepsilon(x)$: No matter how close we get to the boundary with y, there is always a whole neighborhood around y that is also contained in $U_\varepsilon(x)$ (Fig. 12.4). Sets with this property are called open sets in topology. The notions of closed sets, boundary points and the contact points of a set (Fig. 12.5) are closely related to this concept.

Abb. 12.3 Error correcting codes

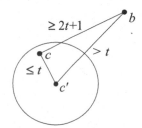

Fig. 12.4 The edge of a
ε-neighborhood

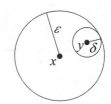

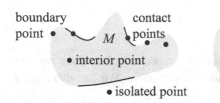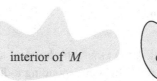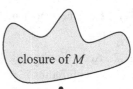

Fig. 12.5 Types of points

▶   **Definition 12.13** Let (X, d) be a metric space and $M \subset X$. M is called *open*, if for
   all $x \in M$ there is a $\varepsilon > 0$ with $U_\varepsilon(x) \subset M$.

   The element $x \in X$ is called *contact point* of M, if for every neighborhood $U_\varepsilon(x)$
   it holds: $U_\varepsilon(x) \cap M \neq \emptyset$.

   $x \in X$ is called *boundary point* of M, if x is a contact point of M and of $X \setminus M$.

   $x \in M$ is called *interior point* of M, if there is an neighborhood of x with
   $U_\varepsilon(x) \subset M$.

   $x \in M$ is called *isolated point* of M, if there is an neighborhood of x with
   $U_\varepsilon(x) \cap M = \{x\}$.

   The set M is called *closed*, if all contact points of M already belong to M.

   The *boundary R* is the set of boundary points of M. The set of contact points
   of M is called the *closure* of M and is denoted by $\overline{M}$, the set of interior points is
   called the *interior* of M.

The set X itself is both open and closed, as is the empty set. For each subset $M \subset X$, the
set $\overline{M}$ is closed, the interior of M is open.

   For the intervals in $\mathbb{R}$, I would like to introduce the following notations: Let $a, b \in \mathbb{R}$
and $a < b$. Then

$$[a, b] := \{x \in \mathbb{R} \mid a \leq x \leq b\},$$
$$]a, b[:= \{x \in \mathbb{R} \mid a < x < b\},$$
$$[a, b[:= \{x \in \mathbb{R} \mid a \leq x < b\},$$
$$]a, b] := \{x \in \mathbb{R} \mid a < x \leq b\},$$
$$[a, \infty[:= \{x \in \mathbb{R} \mid a \leq x\},$$
$$]-\infty, b] := \{x \in \mathbb{R} \mid x \leq b\},$$
$$]a, \infty[:= \{x \in \mathbb{R} \mid a < x\},$$
$$]-\infty, b[:= \{x \in \mathbb{R} \mid x < b\}.$$

With the last four notations, you should pay attention that ∞ is not a real number, but only a useful symbol. We also want to allow the limiting case $a = b$: $[a, a] = \{a\}$, $]a, a[=]a, a] = [a, a[= \emptyset$.

In the sense of the Definition 12.13, $[a, b]$ is closed and $]a, b[$ is open. The boundary of the first four sets is $\{a, b\}$. The closure of these sets is $[a, b]$ and their interior is $]a, b[$.

▶ **Theorem 12.14** In the metric space X, a subset $U \subset X$ is open if and only if its complement $X \setminus U$ is closed.

The proof of this theorem is typical for the ways of proofing in connection with neighborhoods. I would like to present it in detail here and you should try to carry it out. We have to show two directions. Let $V = X \setminus U$:

"$\Rightarrow$" Let U be open. Assume V is not closed. Then there is a contact point x of V that is not in V. Then $x \in U$, and the property of contact point says that for each neighborhood of x it holds: $U_\varepsilon(x) \cap V \neq \emptyset$. This is a contradiction to the fact that U is open.

"$\Leftarrow$" Let U be closed. Assume V is not open. Then there is a $x \in V$ so that for all neighborhoods of x it holds $U_\varepsilon(x) \cap U \neq \emptyset$. But this just means that x is a contact point of U. Since U is closed, then $x \in U$, a contradiction. □

Without proof, I would like to cite the following theorem:

▶ **Theorem 12.15** In the metric space X, any unions of open sets are open again, just as finite intersections of open sets are open. Similarly, any intersections and finite unions of closed sets are closed.

Above all, one must be careful that infinite intersections of open sets need not be open again. For example, the following intersection of open sets results in a closed interval:

$$\bigcap_{n \in \mathbb{N}} \left] a - \frac{1}{n}, b + \frac{1}{n} \right[= [a, b].$$

Find an example yourself that an infinite union of closed sets need not be closed anymore.

Now we want to use our topological methods for the first time to learn something new about the structure of the real and rational numbers:

▶ **Theorem 12.16** It is $\overline{\mathbb{Q}} = \mathbb{R}$, that is, every point of $\mathbb{R}$ is a contact point of $\mathbb{Q}$.

The proof uses Theorem 12.8: Let $x \in \mathbb{R}$ and $\varepsilon > 0$. Then there is a $n \in \mathbb{N}$ with the property $\frac{1}{n} < \varepsilon$ and a largest integer z below $n \cdot x$: It is $n \cdot x = z + r$ with $0 \le r < 1$. Then it is $x = \frac{z}{n} + \frac{r}{n}, \frac{r}{n} < \frac{1}{n} < \varepsilon$ and that means

$$\frac{z}{n} = x - \frac{r}{n} \in U_\varepsilon(x).$$ □

If $x \in \mathbb{R}$, then there is always a rational number in any neighborhood of x, no matter how small it may be. $\mathbb{Q}$ has gaps, as we have known for a long time, but these gaps are tiny, we can never find an interval in which no rational numbers are located. We also say of this property: "$\mathbb{Q}$ is dense in $\mathbb{R}$".

12.3   Comprehension Questions and Exercises

Comprehension Questions

1. Why can the field GF(p) not carry an order in which it becomes an ordered field?
2. Where does the term triangle inequality come from in a metric?
3. What is the difference between a boundary point and a contact point of a set?
4. What are the interior points, contact points and boundary points of the interval $[0, \pi[$ in $\mathbb{R}$?
5. What are the boundary points of the interval $[0, \pi[\cap \mathbb{Q}$ in $\mathbb{Q}$?
6. In a metric space, let $\overline{U_\varepsilon(x)}$ be the closure of a ε-neighborhood of x. Is there a neighborhood $U_\delta(y)$ of $y \in \overline{U_\varepsilon(x)}$ with $U_\delta(y) \subset \overline{U_\varepsilon(x)}$ for every $y \in \overline{U_\varepsilon(x)}$?
7. True or false: By specifying $x < y :\Leftrightarrow d(x,0) < d(y,0)$ for complex numbers x, y, an order is defined on the set $\mathbb{C}$ and $\mathbb{C}$ thus becomes an ordered field.

Exercises

1. Prove the statements c), d), e), f) from Theorem 12.3:
   c) $x \le y$ or $y \le x$.
   d) $x < yz \le w \Rightarrow x + z < y + w$
   e) $x < yz > 0 \Rightarrow xz < yz$
   f) $x > 0 \Rightarrow -x < 0$ and $x < y \Rightarrow -y < -x$

2. Let M, N be subsets of $\mathbb{R}$ that have a maximum. Let

$$M + N := \{x + y \mid x \in M, y \in N\}.$$

Show:

$$\max(M + N) = \max M + \max N.$$

3. Check for $x, y \in \mathbb{R}$ the triangle equation: $|x + y| \leq |x| + |y|$.

You can make many case distinctions here, or you use that $|x| = \max\{x, -x\}$ and apply Exercise 2.

4. a) For which $a \in \mathbb{R}$ does $-3a - 2 \leq 5 \leq -3a + 4$ hold?
   b) For which $a \in \mathbb{R}$ does $|a + 1| - |a - 1| = 1$ hold?
   c) Sketch the areas in $\mathbb{R}^2$, for which holds:
   $|x| + |y| = 2$,
   $|x| \leq 5$ and $|y| \leq 5$,
   $|x| \leq 5$ or $|y| \leq 5$.
5. Prove or disprove: The field $\mathbb{C}$ is an ordered field with the lexicographic order (see Example 3 after Definition 1.12).
6. Investigate whether the following mappings are metrics:
   a) Let X be a set and for $x, y \in X$ let

$$d(x, y) = \begin{cases} 1 & \text{if } x \neq y \\ 0 & \text{if } x = y. \end{cases}$$

   b) Let G be a network (see Definition 11.9). The weight $w(x, y)$ is always positive and $w(x, x) = 0$ for all x. Is w a metric on the set of nodes of G?
7. Show that for the Hamming distance of codewords the triangle inequality holds: $d(x, y) \leq d(x, z) + d(z, y)$.
8. Formulate with the help of the concept of the ε- neighborhood the statement" $\mathbb{Z}$ is dense in $\mathbb{R}$" and prove or disprove this statement.

Sequences and Series 13

Abstract

At the end of this chapter

- you will know what convergent sequences are, and be able to calculate limits of sequences,
- you will be able to deal with the ε of mathematicians,
- you will know the big O notation and be able to calculate and express the runtime of algorithms in the big O notation,
- you will know what series and convergent series are, and have learned important calculation rules to determine the limit of a convergent series,
- you will be able to represent real numbers in the decimal system and in other numeral systems and know the relationship of these representations to the theory of convergent series,
- you will have learned the number e and the exponential function as the first function defined by a convergent series.

Many classical mathematical functions are defined as limits of sequences of numbers or at least can be calculated as such. Thus, $\sin(x)$, $\log(x)$, a^x and many other functions are approximated by sequences of numbers when calculated on a computer. This poses the challenge of finding sequences that are easy to calculate and converge quickly to the desired limit in order to keep the calculation errors and the calculation time small.

Supplementary Information The online version contains supplementary material available at https://doi.org/10.1007/978-3-658-40423-9_13.

For computer scientists, non-convergent sequences are also interesting, namely those that describe the runtime of an algorithm in dependence on the input variables. Here, it is especially important to know how fast such a sequence grows.

Sequences of numbers and limits of such sequences are needed in many areas of mathematics. We will need them to investigate continuous functions and to introduce differential and integral calculus.

Series are nothing but special sequences of numbers. Function series, such as Taylor series and Fourier series play a special role here and are used in many technical applications and also in applications of computer science, for example in data compression.

13.1   Sequences of Numbers

▶   **Definition 13.1** Let M be a set. A *sequence* in M is a mapping

$$\varphi : \mathbb{N} \to M$$

$$n \mapsto a_n = \varphi(n).$$

We denote such a sequence with $(a_n)_{n \in \mathbb{N}}$, with $(a_n)_{n \geq 1}$ or simply with a_n.

Sequences do not always have to start with $n = 1$, sometimes they start with 0 or any other integer. We will mostly work with sequences of real or complex numbers, that is, with sequences in which $M = \mathbb{R}$ or $M = \mathbb{C}$.

Examples of number sequences

1. $a_n = a \in \mathbb{R}$ for all $n \in \mathbb{N}$ is the constant sequence.
2. For $i = \sqrt{-1} \in \mathbb{C}$, $(i^n)_{n \in \mathbb{N}}$ is the sequence: $i, -1, -i, 1, i, -1, -i, 1, \ldots$.
3. $(\frac{1}{n})_{n \in \mathbb{N}}$ is the sequence: $1, \frac{1}{2}, \frac{1}{3}, \frac{1}{4}, \frac{1}{5}, \ldots$.
4. $(\frac{n}{n+1})_{n \in \mathbb{N}}$ is the sequence: $\frac{1}{2}, \frac{2}{3}, \frac{3}{4}, \frac{4}{5}, \ldots, \frac{999}{1000}, \ldots$.
5. $(\frac{n}{2^n})_{n \in \mathbb{N}}$ is the sequence: $\frac{1}{2}, \frac{2}{4}, \frac{3}{8}, \frac{4}{16}, \ldots$.
6. $(a^n)_{n \in \mathbb{N}}$ with $a \in \mathbb{C}$ is the sequence $a, a^2, a^3, a^4, \ldots$ Example 2 is a special case of this sequence.
7. Let $b, c, d \in \mathbb{R}$ and $a_0 = b$, $a_n = a_{n-1} + c \cdot n + d$ for $n \geq 1$. Then $(a_n)_{n \in \mathbb{N}}$ represents the recursive definition of a sequence. If you turn back to Example 2 at the end of Sect. 3.3, you will see that this sequence describes the runtime of the Selection Sort.
8. $(\frac{1}{n}, \frac{1}{n^2})_{n \in \mathbb{N}}$ is a sequence in $\mathbb{R}^2$. Draw some sequence members! ◀

Convergent Sequences

If (X, d) is a metric space, a sequence in X converges to an element $a \in X$ if the sequence members get closer and closer to this limit. The language of topology allows us to capture this fuzzy formulation more precisely. The resulting expression is not quite simple, but the meaning is vivid:

▶   **Definition 13.2** Let (X, d) be a metric space, $(a_n)_{n\in\mathbb{N}}$ a sequence of elements from X. The sequence $(a_n)_{n\in\mathbb{N}}$ *converges to* a, we write for this $\lim\limits_{n\to\infty} a_n = a$, if it holds:

$$\forall \varepsilon > 0 \ \exists n_0 \in \mathbb{N} \ \forall n > n_0 : \ a_n \in U_\varepsilon(a).$$

In $\mathbb{R}$ and $\mathbb{C}$ is $U_\varepsilon(a) = \{x \mid |x - a| < \varepsilon\}$, there we can formulate the rule in this way:

$$\forall \varepsilon > 0 \ \exists n_0 \in \mathbb{N} \ \forall n > n_0 : \ |a_n - a| < \varepsilon.$$

The sequence is called *divergent*, if it does not converge.

What is behind it? In every neighborhood $U_\varepsilon(a)$, no matter how small it is, from a certain index n_0 on all further sequence elements lie. This means that almost all sequence elements, namely all except finitely many at the beginning of the sequence, are arbitrarily close to the limit. There are no more outliers: The whole tail of the sequence remains close to a. In Fig. 13.1 you can see how such sequences can look like in $\mathbb{R}$ or in $\mathbb{C}$.

There are unwritten naming conventions in mathematics that have almost the status of laws. For example, natural numbers are often denoted with m, n, real numbers with x, y, and similar. A particularly sacred cow for the mathematician is his ε. This can be $1, 1000, 10^6$ or 10^{-6} and the assertions about convergence remain correct. But every mathematician always has a small positive real number in his mind (whatever small may mean) and thinks of neighborhoods that can become arbitrarily small. And when a mathematician works with small neighborhoods, then these have a radius ε, not μ or q. You drive every corrector of an exercise or exam to despair when you answer the question about the definition of a convergent sequence, for example with:

$$\forall x < 0 \ \exists \varepsilon \in \mathbb{N} \ \forall \beta > \varepsilon : x < -|a_\beta - a|.$$

This is correct, but if you are unlucky, the line will simply be crossed out: To err is human. Such conventions of course make sense: They make waht is written easier to read because you can associate an unwritten meaning with a symbol. I stick to many such conventions in this book, even if I don't always mention them explicitly. If you read material in other textbooks on mathematics, you will find it helpful that similar terms are used there most of the time.

You also know such conventions from programming: programming guidelines containing, for example, formatting and naming rules, serve primarily not to restrict the freedom of the programmer, but to facilitate the reading of programs by other people than the developer.

Fig. 13.1 Convergent sequences

▶   **Theorem 13.3** If the sequence $(a_n)_{n \in \mathbb{N}}$ converges to a, then a is uniquely determined.

Proof: Assume a_n converges to a and to a', with $a \neq a'$ (Fig. 13.2). If you choose $\varepsilon < \frac{d(a,a')}{2}$, then there is a n_0 so that for $n > n_0$ holds $a_n \in U_\varepsilon(a)$, and similarly there is a n_1 with $a_n \in U_\varepsilon(a')$ for $n > n_1$. But $U_\varepsilon(a) \cap U_\varepsilon(a') = \emptyset$ (see Exercise 8 for this chapter).   □

Most of the sequences we will be dealing with are sequences of real numbers. The real numbers are ordered, unlike, for example, $\mathbb{C}$ or $\mathbb{R}^n$. In this case, we can formulate the following concepts:

▶   **Definition 13.4** Let $(a_n)_{n \in \mathbb{N}}$ be a sequence of real numbers. We say "a_n tends to ∞" and write $\lim\limits_{n \to \infty} a_n = \infty$ if, for all $r > 0$, there exists a $n_0 \in \mathbb{N}$ such that $a_n > r$ for all $n > n_0$. Similarly, we say "a_n tends to $-\infty$" ($\lim\limits_{n \to \infty} a_n = -\infty$) if, for all $r < 0$ and all $n > n_0$, it is $a_n < r$.

   The sequence $(a_n)_{n \in \mathbb{N}}$ is called *bounded* or *bounded from above* or *bounded from below* if the corresponding statement holds for the set $\{a_n \mid n \in \mathbb{N}\}$.

   The sequence $(a_n)_{n \in \mathbb{N}}$ is called *monotonically increasing (monotonically decreasing)* if, for all $n \in \mathbb{N}$, $a_{n+1} \geq a_n$ $(a_{n+1} \leq a_n)$ holds.

Remember that a sequence that tends to infinity is not "infinite" at some point, there is no number "infinite". The sequence is just not bounded from above.

However, when I talk about convergent sequences of real numbers below, I always mean that the limit a exists in the $\mathbb{R}$, that is, $a \neq \pm \infty$. If the case ∞ is allowed, I will mention this explicitly.

▶   **Theorem 13.5** Every convergent sequence of real numbers is bounded.

For if $a \in \mathbb{R}$ is the limit of the sequence, then from an index n_0 on, all sequence members are in $U_1(a)$. The set $\{a_n \mid n \in \mathbb{N}\}$ is therefore contained in the set $\{a_1, a_2, \ldots, a_{n_0}\} \cup U_1(a)$, and this set is bounded.   □

Let's look at the examples from above:

Fig. 13.2 The limit is unique

Examples

1. $a_n = a$ converges, $\lim_{n\to\infty} a_n = a$.

2. $(i^n)_{n\in\mathbb{N}}$ diverges! There are, for example, infinitely many sequence members in each neighborhood of 1, but the sequence also jumps out of each such neighborhood of 1 again and again. There is no index from which really all sequence members are close to 1. The same applies to the points -1, i, $-i$.

3. $(\frac{1}{n})_{n\in\mathbb{N}}$ converges to 0: Let $\varepsilon > 0$ be given. Then, according to Theorem 12.8, there is a n_0 with $\frac{1}{n_0} < \varepsilon$ and for $n > n_0$ it holds: $|\frac{1}{n} - 0| < \frac{1}{n_0} < \varepsilon$. We call $(\frac{1}{n})_{n\in\mathbb{N}}$ a null sequence.

4. $\lim_{n\to\infty} \frac{n}{n+1} = 1$. For $|\frac{n}{n+1} - 1| = |\frac{n-(n+1)}{n+1}| = \frac{1}{n+1}$. If $\frac{1}{n_0} < \varepsilon$ is as above, then for $n > n_0$ follows: $\frac{1}{n+1} < \frac{1}{n} < \varepsilon$.

5. For the sequence $(\frac{n}{2^n})_{n\in\mathbb{N}}$, the assumption is obvious that it converges to 0. If we use that for $n > 3$ always holds $2^n \geq n^2$, we get $|\frac{n}{2^n} - 0| < \frac{n}{n^2} = \frac{1}{n}$. We already know $1/n$ is less than ε if n is large enough.

6. In $(a^n)_{n\in\mathbb{N}}$ it all depends on the value of the number a: For $a = 1/2$ we get a convergent sequence, for $a = 2$ a divergent sequence. $a = 1$ results in convergence, $a = -1$ on the other hand again in divergence. In general:
   a) $\lim_{n\to\infty} a^n = 0$ for $|a| < 1$.
   b) a^n is divergent for $|a| \geq 1$, $a \neq 1$.
   c) $\lim_{n\to\infty} a^n = 1$ for $a = 1$.
   Let's calculate 6.a) for example: $|a| < 1 \Rightarrow \frac{1}{|a|} > 1$, that is $\frac{1}{|a|} = 1 + x$ or $\frac{1}{1+x} = |a|$ for a real number $x > 0$. In the exercises for mathematical induction in Chap. 3 you could calculate that in this case $(1 + x)^n \geq 1 + nx$ applies. This gives us:

$$|a^n| = |a|^n = \frac{1}{(1+x)^n} \leq \frac{1}{1+nx} < \varepsilon,$$

   if $1 + nx > \frac{1}{\varepsilon}$, that is, if $n > \frac{1}{x}(\frac{1}{\varepsilon} - 1)$. It doesn't matter at all that $\frac{1}{x}(\frac{1}{\varepsilon} - 1)$ is not an integer, we can simply choose the next larger integer for n_0. Please note that this calculation is also correct for complex numbers a!

7. The runtime of Selection Sort is not convergent, the sequence is monotonically increasing and not bounded from above: $\lim_{n\to\infty} a_n = \infty$.

8. In the sequence $(\frac{1}{n}, \frac{1}{n^2})_{n\in\mathbb{N}}$ both components converge to 0. Then $\|(\frac{1}{n}, \frac{1}{n^2}) - (0,0)\| = \sqrt{(\frac{1}{n})^2 + (\frac{1}{n^2})^2}$ also tends to 0, that is, $(\frac{1}{n}, \frac{1}{n^2})_{n\in\mathbb{N}}$ converges to $(0,0)$.
   In $\mathbb{R}^n$ it holds: A sequence converges to $(a_1, a_2, \ldots, a_n)$ if and only if for all i the i-th component of the sequence converges to a_i. ◄

You can see from these examples that it is sometimes not quite easy to calculate the limit, if it exists at all, even for simple sequences. Fortunately, there are a number of tricks with which we can reduce the convergence of sequences to those of already known sequences. We will hardly carry out any more such complex calculations as in Example 6. The following theorems provide us with some important tools. They use the order relation and are then only applicable in $\mathbb{R}$, but this is also by far the most important case.

▶   **Theorem 13.6: The comparison test for sequences** Let a_n, b_n, c_n be real number sequences and let $b_n \leq c_n \leq a_n$ for all n.

Further let $\lim\limits_{n\to\infty} a_n = \lim\limits_{n\to\infty} b_n = c$. Then $\lim\limits_{n\to\infty} c_n = c$ also applies.

Proof: Let $\varepsilon > 0$ be given. If n is sufficiently large, $c - \varepsilon < b_n \leq c_n \leq a_n < c + \varepsilon$ applies and thus $c_n \in U_\varepsilon(c)$.   □

With the phrase "for sufficiently large n" I mean: "there is a n_0 so that for all $n > n_0$ applies".

The limit of a sequence can therefore be determined if the sequence can be clamped between two other sequences, the common limit of which is already known. In Examples 4 and 5, we have already used this trick implicitly: We have clamped the sequences $\frac{n}{n+1} - 1$ and $\frac{n}{2^n}$ between the constant sequence 0 and the sequence $\frac{1}{n}$.

Closely related to this comparison test is the following theorem, the simple proof of which I would like to omit:

▶   **Theorem 13.7** Let a_n, b_n be real number sequences with $\lim\limits_{n\to\infty} a_n = a$ and $\lim\limits_{n\to\infty} b_n = b$. Let $a_n \leq b_n$ for all $n \in \mathbb{N}$. Then $a \leq b$ also applies.

But beware: From $a_n < b_n$ for all n it does not necessarily follow that $a < b$: In the limit, the "<"-sign can become a "≤"-sign. See Examples 1 and 5 for this.

In Theorems 13.6 and 13.7 you can replace the words "for all $n \in \mathbb{N}$" each time with "for all $n > n_0$" or with "for all sufficiently large n". In investigations of the convergence of sequences, it is always the ends of the sequences that matter. For small n, the sequences can do whatever they want, it has no influence on the limit.

▶   **Theorem 13.8: Calculation rules for sequences** Let a_n, b_n be real or complex number sequences with $\lim\limits_{n\to\infty} a_n = a$ and $\lim\limits_{n\to\infty} b_n = b$. Then holds:

a) $(a_n \pm b_n)_{n\in\mathbb{N}}$ is convergent with $\lim\limits_{n\to\infty} (a_n \pm b_n) = a \pm b$.

b) $(a_n \cdot b_n)_{n\in\mathbb{N}}$ is convergent with $\lim\limits_{n\to\infty} (a_n \cdot b_n) = a \cdot b$.

c) If $\lambda \in \mathbb{R}\ (\mathbb{C})$, then $(\lambda a_n)_{n\in\mathbb{N}}$ is convergent with $\lim\limits_{n\to\infty} \lambda a_n = \lambda a$.

d) If $b \neq 0$, then there is a n_0 so that for all $n > n_0$ it holds $b_n \neq 0$. Then

also $\left(\dfrac{1}{b_n}\right)_{n>n_0}$ and $\left(\dfrac{a_n}{b_n}\right)_{n>n_0}$ are convergent with $\lim\limits_{n\to\infty} \dfrac{1}{b_n} = \dfrac{1}{b}$ and

$\lim\limits_{n\to\infty} \dfrac{a_n}{b_n} = \dfrac{a}{b}$.

e) $(|a_n|)_{n\in\mathbb{N}}$ is convergent with $\lim\limits_{n\to\infty} |a_n| = |a|$.

f) Let $z_n = x_n + iy_n$ be a sequence of complex numbers. Then holds:
$$\lim_{n\to\infty} z_n = z = x + iy \Leftrightarrow \lim_{n\to\infty} x_n = x, \ \lim_{n\to\infty} y_n = y$$

I would like to prove part a) as an example. In doing so, a trick occurs that is used again and again in connection with the ε calculations: If a statement is true for all $\varepsilon > 0$, then you can replace a given ε with $\varepsilon/2, \varepsilon^2$ or other positive expressions depending on ε and the statement remains true:

So let $\varepsilon > 0$ be given. Then there is an n_0 so that for $n > n_0$ holds $|a_n - a| < \varepsilon/2$ and $|b_n - b| < \varepsilon/2$. It follows:

$$|(a_n \pm b_n) - (a \pm b)| = |(a_n - a) \pm (b_n - b)| \leq |(a_n - a)| + |(b_n - b)| < \frac{\varepsilon}{2} + \frac{\varepsilon}{2} = \varepsilon.$$

The other parts of the theorem are derived in principle similarly, although the individual conclusions are sometimes somewhat more complicated. $\qquad\square$

Examples of the application of this theorem

1. Let $a_n = 14/n$. Since $14/n = 14 \cdot (1/n)$, it holds: $\lim_{n\to\infty} 14/n = 14 \cdot \lim_{n\to\infty} 1/n = 14 \cdot 0 = 0$.

2. We can now also calculate the limit of the example 4 after Definition 13.1 as follows:

$$\lim_{n\to\infty} \frac{n}{n+1} = \lim_{n\to\infty} \frac{n}{n} \frac{1}{1+\frac{1}{n}} = \frac{\lim\limits_{n\to\infty} 1}{\lim\limits_{n\to\infty} 1 + \lim\limits_{n\to\infty} \frac{1}{n}} = \frac{1}{1+0} = 1.$$

3. Similarly, one can always calculate the limit of a quotient of 2 polynomials by factoring out the highest power of n in the numerator and denominator and by reducing:

$$\frac{6n^4 + 3n^2 + 2}{7n^4 + 12n^3 + n} = \frac{6 + \frac{3}{n^2} + \frac{2}{n^4}}{7 + \frac{12}{n} + \frac{1}{n^3}} \xrightarrow[n\to\infty]{} \frac{6}{7}. \quad \blacktriangleleft$$

The Big O Notation

In Sect. 3.3 we determined the runtime of some recursive algorithms. The runtime of an algorithm is, in addition to the performance of the machine, dependent on a number n, which is determined by the input. This number n, the size of the instance, can have very different meanings. It can be the size of a number, the number of elements of a list to be sorted, the length of a cryptographic key in bits, the dimension of a matrix, and much more. In each case, the runtime results as a mapping $f: \mathbb{N} \to \mathbb{R}^+$, where $f(n) = a_n$

denotes the   time complexity of the algorithm as a function of n. So this mapping is nothing other than a sequence. Such a sequence is usually not convergent, mostly it is monotonically increasing and unbounded, that is $\lim_{n\to\infty} a_n = \infty$. However, to assess the performance of an algorithm, it is quite essential to know how fast the sequence a_n grows and how to compare the growth of different sequences. In this context, the growth for large numbers n is important. Our goal is to find some meaningful prototypes of sequences and, for a calculated runtime a_n of an algorithm, to find the prototype b_n that best fits. We then say "The runtime of a_n is in the order of b_n". How can we make this term more precise? The following specification, the *big* O *notation*, has proven to be suitable:

▶   **Definition 13.9** For sequences a_n, b_n with positive real sequence members, one says a_n *is* O(b_n), if and only if the sequence a_n/b_n is bounded. That is

$$a_n \text{ is } O(b_n) :\Leftrightarrow \exists c \in \mathbb{R}^+ \forall n \in \mathbb{N}: a_n/b_n < c.$$

This is read as "a_n is of order b_n". For a_n is O(b_n) we also write $a_n = O(b_n)$.

In particular, according to Theorem 13.5 $a_n = O(b_n)$, if the sequence a_n/b_n is convergent. But this convergence does not always have to be the case.

If a_n and b_n only differ by a constant multiplicative factor c, then $a_n = O(b_n)$. This is useful for our purposes, because such a factor reflects the performance of the computer, which we do not want to take into account when assessing an algorithm.

With the help of this definition, the sequence a_n can first only be roughly estimated upwards, because of course for example for $a_n = an + b$ follows: a_n is $O(n^{1000})$. But we are not satisfied with that, we are looking for a few prototypes of sequences for b_n.

The comparability of sequences is facilitated by the following relation between the orders:

▶   **Definition 13.10** We say
a) $O(a_n) = O(b_n):\Leftrightarrow a_n$ is $O(b_n)$ and b_n is $O(a_n)$.
b) $O(a_n) < O(b_n):\Leftrightarrow a_n$ is $O(b_n)$ and b_n is not $O(a_n)$.

If you formally negate "a_n is O(b_n)", you get for "a_n is not O(b_n)" the rule:

$$\forall c \in \mathbb{R}^+ \exists n \in \mathbb{N}: a_n \geq c \cdot b_n.$$

Caution: Not all orders are comparable using this relation. For example, for the sequences

$$a_n = \begin{cases} 1 & n \text{ even} \\ n & n \text{ odd}, \end{cases} \qquad b_n = \begin{cases} n & n \text{ even} \\ 1 & n \text{ odd} \end{cases}$$

neither of the three relations $<, =$ or $>$ applies, because a_n is not O(b_n) and b_n is not O(a_n).

A simple criterion for determining whether $O(a_n) = O(b_n)$ is to examine the sequence a_n/b_n: If this sequence converges to a limit different from 0, it is bounded and according to Theorem 13.8 then b_n/a_n is convergent and thus bounded. So in this case $O(a_n) = O(b_n)$.

But note again: a_n/b_n does not always have to be convergent. Nevertheless, we can use this criterion to classify many algorithms.

For this, two simple

Examples

1. The runtime for the iterative calculation of the factorial of a number n we have already determined in Example 1 in Sect. 3.3: It was $a_n = n \cdot b + a$.

   With our rules for the calculation of limits, we get (look again at Example 3 after Theorem 13.8):

   $$\lim_{n \to \infty} \frac{nb + a}{n} = b,$$

   and thus $O(bn + a) = O(n)$.

2. When multiplying two $n \times n$ -matrices A and B, the element c_{ij} of the product matrix results as

   $$c_{ij} = \sum_{k=1}^{n} b_{ik} a_{kj}.$$

   These are n multiplications and $n - 1$ additions. In total, n^2 elements have to be calculated. If a is the time for a multiplication and b is the time for an addition, we get for the calculation of the product matrix the complexity $b_n = (a + b)n^3 - bn^2$. It holds:

   $$\lim_{n \to \infty} \frac{(a + b)n^3 - bn^2}{n^3} = a + b,$$

   thus $O((a + b)n^3 - bn^2) = O(n^3)$. ◀

The sequences $n, n^2, n^3, \ldots$ are prototypes of runtimes with which one likes to compare algorithms. Since for $k \in \mathbb{N}$ the sequence n^{k-1}/n^k is limited and n^k/n^{k-1} is unlimited, $O(n^{k-1}) < O(n^k)$ applies.

Similarly to the two examples, one shows:

▷   **Theorem 13.11: The   complexity of polynomial algorithms** For the sequence
    $a_n = \lambda_k n^k + \lambda_{k-1} n^{k-1} + \cdots + \lambda_1 n + \lambda_0$ with $\lambda_i \in \mathbb{R}^+$ it holds $O(a_n) = O(n^k)$.

In addition to the polynomial orders, there are some other important orders. It is:

$$O(1) < O(\log n) < O(n) < O(n \cdot \log n) < O(n^2) < O(n^3) < \ldots < O(n^k) < \ldots$$
$$< O(2^n) < O(3^n) < \ldots$$

An O(1)-algorithm has constant runtime, $O(\log n)$ is called *logarithmic*, $O(n)$ *linear* and
$O(n^2)$ *quadratic* runtime. Algorithms of the order $a^n, a > 1$ are called *exponential* algorithms. These are usually unusable for calculation purposes. In the following table I have
listed some values of such runtimes for you.

n	$\log n$	n	$n \cdot \log n$	n^2	n^3	2^n
10	3.32	10	33.22	100	1000	1024
100	6.64	100	66.44	10000	10^6	$1.27 \cdot 10^{30}$
1000	9.97	1000	9966	10^6	10^9	10^{301}
10 000	13.29	10 000	132 877	10^8	10^{12}	10^{3010}

In this table I have given the logarithm to the base 2. Later we will see that the base does not
play a role here. For different bases a, b is always $O(\log_a n) = O(\log_b n)$.

If you now go back to the examples at the end of Sect. 3.3 you will see that the Selection
Sort was of order n^2 , the Merge Sort of order $n \cdot \log n$. From the table you can see that
for larger numbers n this is a huge improvement.

Monotonic Sequences

Bounded sequences do not always have to be convergent, as the example $a_n = (-1)^n$
shows. But the important theorem is:

▷   **Theorem 13.12** Every monotonic and bounded sequence of real numbers is
    convergent.

The proof is essentially based on the completeness of the real numbers. In the rational
numbers, the   assertion is false. The limit of the monotonically increasing and bounded
sequence a_n is precisely the supremum a of the set $\{a_n \mid n \in \mathbb{N}\}$. We have to show that in
every neighborhood of a almost all, that is all except finitely many, sequence members lie:

Let $\varepsilon > 0$. Then for all $n \in \mathbb{N}$ $a_n \leq a < a + \varepsilon$, since a is an upper bound, and there is a n_0, so that $a - \varepsilon < a_{n_0}$, otherwise a would not be the least upper bound. But then because of the monotonicity for all $n > n_0$ holds: $a - \varepsilon < a_{n_0} \leq a_n \leq a < a + \varepsilon$, thus $a_n \in U_\varepsilon(a)$.

The conclusion for monotonically decreasing sequences proceeds analogously.   □

Let's look at some examples:

Examples

1. $a_n = n$ is monotonic, but not bounded.

2. $a_n = 1/n$ is monotonically decreasing and bounded, the limit is 0.

3. $a_n = 1 + \dfrac{1}{10} + \dfrac{1}{10^2} + \cdots + \dfrac{1}{10^n}$ is the sequence 1.1, 1.11, 1.111, ... This sequence is obviously monotonic and bounded, for example by 2. So it must be convergent. We will see later the limit is the rational number 10/9.

4. $a_n = 1 + \dfrac{1}{4} + \dfrac{1}{9} + \cdots + \dfrac{1}{n^2} = \sum\limits_{k=1}^{n} \dfrac{1}{k^2}$ is monotonically increasing. Is it bounded?

   We use a trick for this. For $k > 1, \dfrac{1}{k^2} < \dfrac{1}{k(k-1)} = \dfrac{1}{k-1} - \dfrac{1}{k}$ and thus we get:

   $$a_n \leq 1 + \underbrace{\left(1 - \dfrac{1}{2}\right)}_{k=2} + \underbrace{\left(\dfrac{1}{2} - \dfrac{1}{3}\right)}_{k=3} + \underbrace{\left(\dfrac{1}{3} - \dfrac{1}{4}\right)}_{k=4} + \cdots$$

   $$+ \underbrace{\left(\dfrac{1}{n-2} - \dfrac{1}{n-1}\right)}_{k=n-1} + \underbrace{\left(\dfrac{1}{n-1} - \dfrac{1}{n}\right)}_{k=n}$$

   $$= 2 - \dfrac{1}{n} < 2.$$

   The sequence is therefore bounded by 2 and thus convergent. Actually calculating the limit is a difficult task here. With means of integral calculus it can be shown $\lim_{n \to \infty} a_n = \dfrac{\pi^2}{6}$.

5. $a_n = 1 + \dfrac{1}{2} + \dfrac{1}{3} + \cdots + \dfrac{1}{n} = \sum\limits_{k=1}^{n} \dfrac{1}{k}$ is also monotonic. However, the following consideration shows that this sequence is not bounded:

   $$a_n = 1 + \dfrac{1}{2} + \underbrace{\dfrac{1}{3} + \dfrac{1}{4}}_{>2 \cdot \frac{1}{4} = \frac{1}{2}} + \underbrace{\dfrac{1}{5} + \dfrac{1}{6} + \dfrac{1}{7} + \dfrac{1}{8}}_{>4 \cdot \frac{1}{8} = \frac{1}{2}} + \underbrace{\dfrac{1}{9} + \cdots + \dfrac{1}{16}}_{>8 \cdot \frac{1}{16} = \frac{1}{2}}$$

   $$+ \underbrace{\dfrac{1}{17} + \cdots + \dfrac{1}{32}}_{>16 \cdot \frac{1}{32} = \frac{1}{2}} + \cdots + \dfrac{1}{n}.$$

If we group the members in this way, we always have

$$\frac{1}{2^n + 1} + \cdots + \frac{1}{2^{n+1}} > 2^n \frac{1}{2^{n+1}} = \frac{1}{2}$$

and thus $\sum_{k=1}^{2^n} \frac{1}{k} > 1 + \frac{n}{2}$. It takes a long time, but with this sequence we can exceed any given threshold. ◀

Try to determine the limits of some sequences with a computer. Calculate 10, 100, 1 000 000 sequence members or let the calculation run until the sequence members no longer change. You will notice that quite different behaviour is shown: Sometimes you get to the limit very quickly, sometimes very slowly, and even divergent sequences seem to converge on the computer occasionally. What all sequences have in common is that if we have determined a "limit" with the computer, we do not know how far it deviates from the actual limit. This can only be determined with mathematical methods.

13.2   Series

The last examples of sequences have one common property: Each sequence member is already a sum, and with each further sequence member something is added to this sum. Such sequences are called series. It would be a huge understatement to say that series play an important role in mathematics, many areas of mathematics are unthinkable without the theory of series, and we have to deal with it intensively.

▶   **Definition 13.13** Given a sequence $(a_n)_{n \in \mathbb{N}}$ of real or complex numbers.
   The *series* $\sum_{k=1}^{\infty} a_k$ is the sequence $(s_n)_{n \in \mathbb{N}} = (\sum_{k=1}^{n} a_k)_{n \in \mathbb{N}}$. If the sequence s_n converges to a, then we say the *series converges* and denote the limit a with $\sum_{k=1}^{\infty} a_k$. The *n*-th element of the sequence $\sum_{k=1}^{n} a_k$ is called the *n-th partial sum* of the series. The elements a_k are called the *elements* of the series.

One has to be careful with this designation: $\sum_{k=1}^{\infty} a_k$ is not a number at first! It is a series, and only if this series converges, the limit is denoted with it at the same time.

Just like sequences, series do not always have to begin at index $k = 1$, any integer, including negative numbers, is allowed as a starting index. In most cases, this beginning will be 0 or 1 for us.

Examples of series

1. We already know $\sum_{n=1}^{\infty} \frac{1}{n}$ and $\sum_{n=1}^{\infty} \frac{1}{n^2}$: The series $\sum_{n=1}^{\infty} \frac{1}{n}$ is divergent and $\sum_{n=1}^{\infty} \frac{1}{n^2} = \frac{\pi^2}{6}$.

2. An important series is the *geometric series*, which is defined for $q \in \mathbb{R}$ or $q \in \mathbb{C}$: $\sum_{k=0}^{\infty} q^k$. For all q with $|q| < 1$, $\sum_{k=0}^{\infty} q^k = \frac{1}{1-q}$.

   Proof: For the n-th partial sum, it holds (compare Theorem 3.3 in Sect. 3.2): $(1-q) \cdot \sum_{k=0}^{n} q^k = 1 - q^{n+1}$. From Example 6 after Theorem 13.5 we know that q^{n+1} is a null sequence and so we get:

$$(1-q) \cdot \sum_{k=0}^{\infty} q^k = (1-q) \cdot \lim_{n \to \infty} \sum_{k=0}^{n} q^k = \lim_{n \to \infty} (1-q) \cdot \sum_{k=0}^{n} q^k$$

$$= \lim_{n \to \infty} (1 - q^{n+1}) = 1 - \lim_{n \to \infty} q^{n+1} = 1. \qquad \square$$

3. The series $\sum_{n=0}^{\infty} \frac{1}{n!}$ is monotonically increasing. We try to find an upper bound again: For $k \geq 2$,

$$\frac{1}{k!} = \frac{1}{1 \cdot 2 \cdot 3 \cdot 4 \cdot \ldots \cdot k} \leq \frac{1}{1 \cdot \underbrace{2 \cdot 2 \cdot 2 \cdot \ldots \cdot 2}_{k-1 \text{ factors}}} = \frac{1}{2^{k-1}}$$

and therefore for all n:

$$s_n = 1 + \frac{1}{1!} + \frac{1}{2!} + \frac{1}{3!} + \cdots + \frac{1}{n!} \leq 1 + \frac{1}{2^0} + \frac{1}{2^1} + \frac{1}{2^2} + \cdots + \frac{1}{2^{n-1}}$$

$$\leq 1 + \sum_{k=0}^{\infty} \left(\frac{1}{2}\right)^k = 1 + \frac{1}{1 - \frac{1}{2}} = 3.$$

We used the geometric series from Example 2 immediately. The series is therefore bounded from above and thus convergent. ◄

The limit of this series is a famous number:

▶ **Definition 13.14** The limit $e := \sum_{n=0}^{\infty} \frac{1}{n!}$ is called *Euler's number*. The first digits in the decimal expansion are:

$$e = 2.718281828.$$

The number e is an infinite decimal fraction that never has a period. e is therefore not a rational number. e is, however, unlike $\sqrt{2}$, not a root of any polynomial with coefficients from $\mathbb{Q}$. Such numbers are called transcendental numbers.

Convergence Tests for Series

In general, it is difficult to see whether a given series is convergent or not. However, there is a fairly simple negative criterion:

▷   **Theorem 13.15** If the series $\sum_{n=0}^{\infty} a_n$ converges, then the sequence of the elements $(a_n)_{n\in\mathbb{N}}$ is a null sequence.

For if a is the limit of the series, then for each $\varepsilon > 0$ there is a $n_0 \in \mathbb{N}$ such that for all $n > n_0$ it holds:

$$|a_n| = |s_n - s_{n-1}| = |(s_n - a) - (s_{n-1} - a)| \le |s_n - a| + |s_{n-1} - a| < \varepsilon. \qquad \square$$

▷   **Definition 13.16** The series $\sum_{n=0}^{\infty} a_n$ is called *absolutely convergent*, if the series $\sum_{n=0}^{\infty} |a_n|$ converges.

▷   **Theorem 13.17** If the series $\sum_{n=0}^{\infty} a_n$ is absolutely convergent, then the series itself is convergent.

The proof of this theorem is profound. It can be reduced to the convergence of sequences, but for that one needs assertions about convergence of sequences, which I have not included in this book. Surprisingly, the theorem also holds for series of complex numbers: The convergence of a series of complex numbers, whose elements can jump around in the whole complex plane, already follows from the convergence of the absolute values of the series elements, that is, from the convergence of a real series.

▷   **Theorem 13.18: Calculation rules for series** If $\sum\limits_{n=0}^{\infty} a_n = a$ and $\sum\limits_{n=0}^{\infty} b_n = b$ are convergent series and c is a constant number, then:

a) $\sum\limits_{n=0}^{\infty} (a_n \pm b_n) = \sum\limits_{n=0}^{\infty} a_n \pm \sum\limits_{n=0}^{\infty} b_n = a \pm b,$

b) $\sum\limits_{n=0}^{\infty} c a_n = c \sum\limits_{n=0}^{\infty} a_n = ca,$

c) If $z_n = x_n + i y_n$ and $\sum\limits_{n=0}^{\infty} z_n$ is a series of complex numbers, then $\sum\limits_{n=0}^{\infty} z_n = z$ converges if and only if $\sum\limits_{n=0}^{\infty} x_n = x$ and $\sum\limits_{n=0}^{\infty} y_n = y$ are convergent. Then $z = x + iy.$

This theorem can be directly derived from the calculation rules for the limits of sequences (Theorem 13.8). Series are nothing but special sequences.

Some series can also be multiplied with each other. How does one carry that out? We try to catch all terms of the product of two series. We use a trick for that: We sort the terms according to the sum of the indices:

$$(a_0 + a_1 + a_2 + a_3 + \cdots + a_n + \cdots)(b_0 + b_1 + b_2 + b_3 + \cdots + b_n + \cdots)$$

$$\underset{\text{sum of indices } 0}{\underbrace{a_0 b_0}} + \underset{\text{sum of indices } 1}{\underbrace{(a_0 b_1 + a_1 b_0)}} + \underset{\text{sum of indices } 2}{\underbrace{(a_0 b_2 + a_1 b_1 + a_0 b_2)}} + \cdots + \sum_{k=0}^{n} \underset{\text{sum of indices } n}{\underbrace{a_k b_{n-k}}} + \cdots .$$

I wrote the equality sign in quotation marks here because of course one cannot actually calculate with infinite sums. There are no infinite sums after all, only series and their limits. Nevertheless, this "calculation" should serve as motivation for the following difficult theorem, which I cannot prove here. It does not always work, but one can multiply two convergent series if they are absolutely convergent:

▶ **Theorem 13.19** If $\sum_{n=0}^{\infty} a_n = a$ and $\sum_{n=0}^{\infty} b_n = b$ are absolutely convergent series, then the series $\sum_{n=0}^{\infty} (\sum_{k=0}^{n} a_k b_{n-k})$ is also absolutely convergent and it holds that

$$\sum_{n=0}^{\infty} \left(\sum_{k=0}^{n} a_k b_{n-k} \right) = \left(\sum_{n=0}^{\infty} a_n \right) \left(\sum_{n=0}^{\infty} b_n \right) = ab.$$

▶ **Theorem 13.20: The comparison test for series** If $\sum_{n=0}^{\infty} b_n$ is an absolutely convergent series and $|a_n| \leq |b_n|$ holds for all $n \in \mathbb{N}$, then the series $\sum_{n=0}^{\infty} a_n$ is also absolutely convergent and it holds that $\sum_{n=0}^{\infty} |a_n| \leq \sum_{n=0}^{\infty} |b_n|$.

For if $c \in \mathbb{R}^+$ is the limit of the series $\sum_{n=0}^{\infty} |b_n|$, then the sequence $s_n = \sum_{k=0}^{n} |a_k|$ is monotonic and bounded from above:

$$s_n = \sum_{k=0}^{n} |a_k| \leq \sum_{k=0}^{n} |b_k| \leq c$$

and thus, according to Theorem 13.12, convergent. This means nothing else than $\sum_{n=0}^{\infty} a_n$ is absolutely convergent.   □

If you calculate the series $\sum_{k=0}^{\infty} \frac{1}{k!}$ on your computer, you will find that you are already close to e after only a few terms. The series converges very quickly. But how can one determine that the series converges well if the limit is unknown? It is of course very important for numerical calculations to know when one can stop. Unfortunately, there are no off-the-shelf tests, but in this special case one can estimate the residual error. For this we use the calculation rule from Theorem 13.18b) and Theorem 13.20. Find out where! It is

$$e = \sum_{k=0}^{n} \frac{1}{k!} + r_{n+1}, \text{ with } r_{n+1} = \sum_{k=n+1}^{\infty} \frac{1}{k!}. \tag{13.1}$$

r_{n+1} is therefore the error we make if we stop the calculation after the n-th term. Now we can give an upper bound for r_{n+1} similar as before for the whole series:

$$
\begin{aligned}
r_{n+1} &= \frac{1}{(n+1)!}\left(1 + \frac{1}{n+2} + \frac{1}{(n+2)(n+3)} + \frac{1}{(n+2)(n+3)(n+4)} + \cdots\right) \\
&\leq \frac{1}{(n+1)!}\left(1 + \frac{1}{n+2} + \frac{1}{(n+2)^2} + \frac{1}{(n+2)^3} + \cdots\right) \\
&= \frac{1}{(n+1)!} \cdot \sum_{k=0}^{\infty}\left(\frac{1}{n+2}\right)^k = \frac{1}{(n+1)!} \cdot \frac{1}{1 - \frac{1}{n+2}} \\
&\leq \frac{1}{(n+1)!} \cdot 2 = \frac{2}{(n+1)!}.
\end{aligned}
$$

It is $2/13! \approx 3.21 \cdot 10^{-10}$, so the error occurs after 12 additions in the tenth place after the decimal point. Thus, the accuracy from Definition 13.14 is already given.

▶ **Theorem 13.21: The ratio test** Let the series $\sum_{n=0}^{\infty} a_n$ be given. If there is a number $0 < q < 1$ such that from an index n_0 on always holds $\left|\dfrac{a_{n+1}}{a_n}\right| \leq q$, then $\sum_{n=0}^{\infty} a_n$ is absolutely convergent. If from an index on always holds $\left|\dfrac{a_{n+1}}{a_n}\right| > 1$, then the series is divergent.

In particular, this theorem implies that $\sum_{n=0}^{\infty} a_n$ is convergent if $\lim_{n\to\infty} |\frac{a_{n+1}}{a_n}|$ exists and is less than 1. The ratio test is often applied in this form.

The second half is simple: If finally always $|\frac{a_{n+1}}{a_n}| > 1$, then a_n cannot be a null sequence, according to Theorem 13.15 then the series is divergent.

The first part of the criterion can be reduced to the comparison test using the geometric series: For all $n > n_0$ it holds:

$$|a_n| \leq |a_{n-1}|q \leq |a_{n-2}|q^2 \leq |a_{n-3}|q^3 \leq \cdots \leq |a_{n_0}|q^{n-n_0},$$

so $|a_n| \leq |a_{n_0}|q^{n-n_0}$. If $\sum_{n=n_0}^{\infty} |a_{n_0}|q^{n-n_0}$ were convergent, we could apply the comparison test and also $\sum_{n=n_0}^{\infty} a_n$ would be convergent, even absolutely convergent. So let's take a closer look at this series. Since we are allowed to move out the constant $|a_{n_0}|$ from the sum and because $q < 1$ it follows:

$$\sum_{n=n_0}^{\infty} |a_{n_0}| q^{n-n_0} = |a_{n_0}| \cdot \sum_{n=n_0}^{\infty} q^{n-n_0} = |a_{n_0}| \cdot \sum_{n=0}^{\infty} q^n = |a_{n_0}| \cdot \frac{1}{1-q}, \qquad (13.2)$$

there is really convergence. $\qquad \square$

I didn't quite argue correctly in the formula (13.2): In the first step I used Theorem 13.18. But this requires the convergence of the series. So we have to read the formula from right to left: Since the series converges, we are allowed to move in the constant $|a_{n_0}|$ and then the equation is correct. I will occasionally do similar calculations in the future. Always pay attention to the fact that such operations with series are only allowed if it turns out in the end that they are convergent and if we are then allowed to read the equations backwards.

Something else has to be considered about this test: Why is it not enough to demand $|a_{n+1}|/|a_n| < 1$? It could then be that the quotient gets closer and closer to 1 and that we can no longer bound it from above by an element $q < 1$. But then the whole argumentation with the geometric series doesn't work anymore. In fact, the assertion of the theorem can be false in such a case.

We want to apply the last theorem immediately to the investigation of a very important function, the exponential function:

▶ **Theorem 13.22** For all $z \in \mathbb{C}$ the series $\sum_{n=0}^{\infty} \frac{z^n}{n!}$ is absolutely convergent.
Because it is

$$\left| \frac{a_{n+1}}{a_n} \right| = \left| \frac{z^{n+1}/(n+1)!}{z^n/n!} \right| = \left| \frac{z}{n+1} \right| = \frac{|z|}{n+1}.$$

Since $\lim_{n \to \infty} |z|/(n+1) = 0$, from the ratio test follows the absolute convergence of the series. $\qquad \square$

To each $z \in \mathbb{C}$ we can assign the limit of a series, we get a function:

▶ **Definition 13.23: The complex exponential function** The function

$$\exp \colon \mathbb{C} \to \mathbb{C}, \quad z \mapsto \sum_{n=0}^{\infty} \frac{z^n}{n!}$$

is called *exponential function*.

▶ **Theorem 13.24** For all $z, w \in \mathbb{C}$ it holds:

a) $\exp(z + w) = \exp(z) \cdot \exp(w)$.
b) $\exp(\bar{z}) = \overline{\exp(z)}$.

For the proof of a) we need the product rule for series, Theorem 13.19:

$$\exp(z)\exp(w) = \sum_{n=0}^{\infty}\frac{z^n}{n!} \cdot \sum_{n=0}^{\infty}\frac{w^n}{n!} = \sum_{n=0}^{\infty}\left(\sum_{k=0}^{n}\frac{z^k}{k!}\frac{w^{n-k}}{(n-k)!}\right)$$

$$= \sum_{n=0}^{\infty}\frac{1}{n!}\left(\sum_{k=0}^{n}\frac{n!}{k!(n-k)!}z^k w^{n-k}\right)$$

$$\underset{\substack{\uparrow\\ \text{according to}\\ \text{binomial theorem}}}{=} \sum_{n=0}^{\infty}\frac{1}{n!}(z+w)^n = \exp(z+w).$$

The part b) can be reduced to Theorem 13.8f), which I do not want to carry out.   □

The exponential function can be found on any calculator, and the calculator does nothing but evaluate the first terms of this series. What about the speed of convergence? Similar to the series $\sum_{n=0}^{\infty}\frac{1}{n!}$ in (13.1), which is nothing but $\exp(1)$, we can carry out an error estimate. I will just give you the result:

▶   **Theorem 13.25** For $\exp(z) = \sum_{k=0}^{n}\frac{z^k}{k!} + r_{n+1}(z)$ it holds for the residual error after summation of the terms up to the index n:

$$|r_{n+1}(z)| \le \frac{2|z|^{n+1}}{(n+1)!}, \quad \text{if } |z| \le 1+\frac{n}{2}.$$

The residual error is here dependent from z. The unpleasant thing about this result is on the one hand that the error can always become larger, the greater the absolute value of z is, and on the other hand that the calculation is only correct for small z anyway. What do we do now if we want to calculate $\exp(100)$? To solve this problem, I have to put you off until we get to Sect. 14.2. There we will have a closer look to the properties of the exponential function.

13.3   Representation of Real Numbers in Numeral Systems

Natural numbers can be represented uniquely in different bases. We will now derive a similar procedure for the real numbers. As a result, we obtain that every positive real number can be written as a decimal fraction. The positive rational numbers are represented by finite or periodic decimal fractions. Such a decimal fraction is nothing else but a convergent series. But as with the representation of natural numbers, other numbers than 10 can also be used as base:

▶   **Definition 13.26** Let $b > 1$ be a fixed natural number. b is called the *base* and the numbers $0, 1, \ldots, b - 1$ are called the *digits of the base b system*. Let $z \in \mathbb{Z}$ and for all $n \geq z$ let $a_n \in \{0, 1, \ldots, b - 1\}$. For $n < 0$ let $1/b^n := b^{-n}$. A series of the form

$$\sum_{n=z}^{\infty} \frac{a_n}{b^n},$$

with $a_z \neq 0$ is called a *base b fraction*. We write

$$(a_z.a_{z+1}a_{z+2}a_{z+3} \ldots \mathrm{E}{-}z)_b.$$

If for $i > n$ it holds $a_i = 0$, we speak of a *finite base b fraction* and write $(a_z.a_{z+1}a_{z+2} \ldots a_n\mathrm{E}{-}z)_b$. If a series of digits repeats itself in a fraction over and over again, the fraction is called *periodic*. In this case, we define:

$$(a_z.a_{z+1}a_{z+2} \ldots a_n\overline{p_1 \ldots p_r}\mathrm{E}{-}z)_b$$
$$:= (a_z.a_{z+1}a_{z+2} \ldots a_np_1 \ldots p_rp_1 \ldots p_r \ldots \mathrm{E}{-}z)_b.$$

$p_1p_2 \ldots p_r$ is called the *period* of the fraction.

This is the first time we encounter series that can start with negative indices. For $z < 0$, the base b fraction has the form

$$\sum_{n=z}^{\infty} \frac{a_n}{b^n} = a_zb^{-z} + a_{z+1}b^{-z-1} + \cdots + a_{-1}b^1 + a_0b^0 + \frac{a_1}{b} + \frac{a_2}{b^2} + \frac{a_3}{b^3} + \cdots.$$

Often, the base b fraction is also written in the form $(a_za_{z+1} \ldots a_0.a_1a_2 \ldots)_b$ for $z \leq 0$ and $(0.\underbrace{0 \cdots 0}_{z-1 \text{ zeros}} a_za_{z+1} \ldots)_b$ for $z > 0$. The part before the dot corresponds exactly to the negative indices up to and including the index 0. This represents a natural number in the form we learned in Theorem 3.6 in Sect. 3.2. The part of the number after the dot starts with the index 1.

For $b = 10$, decimal fractions are usually written in one of these two forms, with the index designation omitted: The first notation is the exponential notation, which is how a computer usually outputs the numbers; the second notation is the familiar representation with the decimal point after the integer part.

Examples

1. $(1.1378\mathrm{E}2)_{10} = (113.78)_{10} = 1 \cdot 10^2 + 1 \cdot 10^1 + 3 \cdot 10^0 + \frac{7}{10} + \frac{8}{10^2} + \frac{0}{10^3} + \cdots = 113 + \frac{78}{100}$.

2. $(4.711\mathrm{E}{-}3)_{10} = (0.004711)_{10} = \frac{4}{10^3} + \frac{7}{10^4} + \frac{1}{10^5} + \frac{1}{10^6}$.

3. $(4.625\mathrm{E}1)_7 = (46.25)_7 = 4 \cdot 7^1 + 6 \cdot 7^0 + \frac{2}{7} + \frac{5}{49} = \frac{1685}{49}$.

4. $(3.3333\ldots)_{10} = 3 + \dfrac{3}{10} + \dfrac{3}{10^2} + \dfrac{3}{10^3} + \cdots = \displaystyle\sum_{n=0}^{\infty} \dfrac{3}{10^n} = 3 \cdot \sum_{n=0}^{\infty} \dfrac{1}{10^n} = 3 \cdot \dfrac{1}{1 - \frac{1}{10}} = 3 \cdot \dfrac{10}{9} = \dfrac{10}{3}.$

5. $x = (0.12\overline{34})_{10}$ also represents a rational number: To determine this, we use the following trick: $x \cdot 10\,000 - x \cdot 100 = 1234.\overline{34} - 12.\overline{34} = 1222.00$ and thus $x \cdot 9900 = 1222$, that is $x = 1222/9900$. ◀

These examples of base b fractions are convergent series. But does a base b fraction always converge? The answer is given to us by the next theorem:

▶ **Theorem 13.27** Every base b fraction is convergent. For the sum from index 1, the part after the dot of such a fraction, it holds $0 \leq \displaystyle\sum_{n=1}^{\infty} \dfrac{a_n}{b^n} \leq 1.$

We use the geometric series again and get:

$$\sum_{n=1}^{\infty} \frac{a_n}{b^n} \leq \sum_{n=1}^{\infty} \frac{b-1}{b^n} = (b-1) \sum_{n=1}^{\infty} \frac{1}{b^n}$$

$$= (b-1) \left(\sum_{n=0}^{\infty} \frac{1}{b^n} - 1 \right) = (b-1) \left(\frac{1}{1 - \frac{1}{b}} - 1 \right)$$

$$= (b-1) \left(\frac{1}{b-1} \right) = 1.$$

The part after the dot of the base b fraction is strictly monotonically increasing and bounded from above by 1. Therefore it is convergent. The complete base b fraction is then of course also convergent, because only a finite natural number is added.   □

Now we can deduce that the real numbers, which we have only described so far by their axioms, are exactly the base b fractions. As with the representation of the natural numbers, any base b greater than 1 is possible.

▶ **Theorem 13.28 The b-adic expansion** Let $b > 1$ be a natural number. Then holds:

a) Every base b fraction converges to a positive real number.
b) For every positive real number x there is a base b fraction that converges to x.
c) Every periodic base b fraction converges to a rational number.
d) For every positive rational number p/q there is a periodic or finite base b fraction that converges to p/q.

We have already proved part a) in Theorem 13.27.

To b): We can construct the base b fraction converging to x by induction. I only want to sketch the way here. To carry out the proof, you have to do the calculations on paper! In the base case, we first find for the number x a minimal index N so that $0 \le x < b^{N+1}$ is true and then also the first coefficient of the representation: This is the maximum $a_{-N} \in \{0, 1, \ldots, b-1\}$ with the property $a_{-N} b^N \le x$. If we have found the series $s_n = \sum_{k=-N}^{n} \frac{a_k}{b^k}$ up to the index n so that $s_n \le x < s_n + \frac{1}{b^n}$ is true, we look for the next term of the series: There must be a coefficient $a_{n+1} \in \{0, 1, \ldots, b-1\}$ with

$$s_n \le s_n + \frac{a_{n+1}}{b^{n+1}} \le x < s_n + \underbrace{\frac{a_{n+1}}{b^{n+1}} + \frac{1}{b^{n+1}}}_{\le \frac{1}{b^n}} \le s_n + \frac{1}{b^n}.$$

Then $s_{n+1} := \sum_{k=-N}^{n} \frac{a_k}{b^k} + \frac{a_{n+1}}{b^{n+1}}$ In this way, we can successively nest the number x. The series really converges to x.

Test the procedure with the number π and the base 10: It is $0 \le \pi < 10^1$, so $N = 0$ and $3 \le \pi$ is the first coefficient. If we have found the series up to 3.141 ($3.141 \le \pi < 3.141 + 0.001$), we look for the next digit so that we stay just below π, which is 5 here. Then $3.1415 \le \pi < 3.1415 + 0.0001$.

What is behind part c) of the theorem, we have just seen at the 5th example after Definition 13.26. In general terms: If x is a periodic base b fraction, whose period begins at the index n and has the period length k, then $b^{n+k} \cdot x - b^n \cdot x = m$ is a natural number and therefore $x = m/(b^{n+k} - b^n) \subset \mathbb{Q}$. Try this with more examples!

There is still the statement d) of the theorem missing, and for this I give an algorithm, the *division algorithm* for natural numbers p and q. If you look closely, then this procedure is exactly the same you have learned in school. But very likely you have not had the questions of the computer scientist in mind there: Why does the algorithm actually work? Does it have a termination condition that is always reached? This is what we want to deal with now. It is enough if we divide natural numbers p and q with $p/q < 1$. Because if $p/q \ge 1$, then we can represent p/q as $n + p'/q$, where $n, p' \in \mathbb{N}$ and $p'/q < 1$. We already know the representation of n as base b number. So our task is to find coefficients $a_1, a_2, a_3, \ldots$ with

$$\frac{p}{q} = \frac{a_1}{b} + \frac{a_2}{b^2} + \frac{a_3}{b^3} + \cdots = \sum_{n=1}^{\infty} \frac{a_n}{b^n}. \tag{13.3}$$

We carry out a continued division with remainder by q, as we know it from Theorem 4.11 in Sect. 4.2. We start with the division of $b \cdot p$, multiply the remainder each time by b and divide further:

$$b \cdot p = a_1 q + r_1, \qquad 0 \le r_1 < q,$$
$$b \cdot r_1 = a_2 q + r_2, \qquad 0 \le r_2 < q,$$
$$b \cdot r_2 = a_3 q + r_3, \qquad 0 \le r_3 < q, \tag{13.4}$$
$$\vdots$$
$$b \cdot r_n = a_{n+1} q + r_{n+1}, \quad 0 \le r_{n+1} < q.$$

The quotients a_i calculated in this way are all less than b, because from $b \le a_1$ and $p < q$ would follow $bp < bq \le a_1 q$ in contradiction to $bp = a_1 q + r_1 \ge a_1 q$. Since for all remainders holds $r_i < q$, the same argumentation applies to all a_i.

These a_i are exactly the coefficients from (13.3) we are looking for. Why? If we solve the lines of (13.4) successively for p, r_1, r_2 and so on and substitute, we get:

$$p = \frac{a_1}{b} \cdot q + \frac{r_1}{b}$$
$$r_1 = \frac{a_2}{b} \cdot q + \frac{r_2}{b}$$
$$\underset{\substack{\uparrow \\ \text{insert } r_1}}{\Rightarrow} \quad p = \frac{a_1}{b} \cdot q + \left(\frac{a_2}{b^2} \cdot q + \frac{r_2}{b^2} \right) = \left(\frac{a_1}{b} + \frac{a_2}{b^2} \right) \cdot q + \frac{r_2}{b^2}$$
$$r_2 = \frac{a_3}{b} \cdot q + \frac{r_3}{b}$$
$$\underset{\substack{\uparrow \\ \text{insert } r_2}}{\Rightarrow} \quad p = \frac{a_1}{b} \cdot q + \frac{a_2}{b^2} \cdot q + \left(\frac{a_3}{b^3} \cdot q + \frac{r_3}{b^3} \right) = \left(\frac{a_1}{b} + \frac{a_2}{b^2} + \frac{a_3}{b^3} \right) \cdot q + \frac{r_3}{b^3}$$
$$\vdots$$
$$r_n = \frac{a_n}{b} \cdot q + \frac{r_{n+1}}{b}$$
$$\Rightarrow \quad p = \left(\sum_{k=1}^{n+1} \frac{a_k}{b^k} \right) \cdot q + \frac{r_{n+1}}{b^{n+1}},$$

in total, therefore,

$$\frac{p}{q} = \left(\sum_{k=1}^{n+1} \frac{a_k}{b^k} \right) + \frac{r_{n+1}}{b^{n+1} q}.$$

This series converges to p/q, because the difference $\frac{r_{n+1}}{b^{n+1} q}$ tends to 0.

But when does the algorithm end? If the remainder r_n is once equal to 0, then $p/q = (0.a_1 a_2 \ldots a_n)_b$, that is, a finite fraction. If the remainder remains unequal to 0, then at some point a remainder must repeat itself, because there are only finitely many remainders smaller than q. Then the fraction is periodic, because from this point on all divisions repeat themselves. If, for example, $r_n = r_s$ is for a $n > s$, then $a_{s+1} = a_{n+1}$ and therefore: $p/q = (0.a_1 \ldots a_s \overline{a_{s+1} \ldots a_n})_b$.

Let us calculate the number $5/6$ in different bases:

1. $b = 10$: $5 \cdot 10 = 8 \cdot 6 + 2$
   $$2 \cdot 10 = 3 \cdot 6 + 2$$
   $$5/6 = (0.8\overline{3})_{10}.$$

2. $b = 2$: $5 \cdot 2 = 1 \cdot 6 + 4$
   $$4 \cdot 2 = 1 \cdot 6 + 2$$
   $$2 \cdot 2 = 0 \cdot 6 + 4$$
   $$5/6 = (0.1\overline{10})_2.$$

3. $b = 12$: $5 \cdot 12 = 10 \cdot 6 + 0$ ◀
   $$5/6 = (0.A)_{12}, \text{ where } A \text{ is the digit } 10.$$

You can see that a number can be infinitely periodic in some number systems and finite in others.

Fractions are also stored in the computer in exactly the same form as presented in Definition 13.26: Such a number is determined by the sign, the exponent and the mantissa. The mantissa is the sequence of a_n, which of course is always finite in the computer. The whole thing happens, as you can imagine, with the base 2. This is the *floating point representation* of the number. A float in Java, for example, includes 32 bits: 1 bit for the sign, 8 bits for the exponent and 23 bits for the mantissa. The mantissa is actually 24 bits long: The first bit is always 1 and therefore does not have to be stored. For the exponent, the numbers from -126 to $+127$ are possible. This is a total of 254 values, so there is still space for two more values in the 8 bits of the exponent: This also allows you to encode 0 and number overflows. So a typical float looks like $-1.\underbrace{0010 \cdots 101}_{23\,\text{bit}} \cdot 2^{98}$ in the computer.

When outputting to the screen, the number is then converted into a decimal number and output approximately as $-3.56527E+29$. Here E+29 is to be read as "$\cdot 10^{29}$".

Rounding errors can therefore already occur when inputting and outputting, for example when a decimal number entered cannot be converted into a finite binary fraction. Further errors can occur through mathematical operations. In Sect. 18.1 we will deal with problems in numerical calculations in more detail.

13.4 Comprehension Questions and Exercises

1. Can a sequence have multiple limits?
2. Are there bounded sequences of real numbers that do not converge?
3. Which is greater: $O(n^{1000})$ or $O(2^n)$?

4. A sequence of complex numbers $(z_n)_{n \in \mathbb{N}}$ is to be called bounded if the sequence of absolute values $(|z_n|)_{n \in \mathbb{N}}$ is bounded. Is every convergent sequence of complex numbers bounded?

5. A series is nothing else but a sequence. Why?

6. The formula $\sum_{n=1}^{\infty} a_n$ has two different meanings in the theory of infinite series. Which are they?

7. Is there a difference between $0.\overline{9}$ and 1 in the decimal representation?

8. What property does the decimal expansion have for an irrational number?

Exercises

1. Give an example for sequences $(a_n)_{n \in \mathbb{N}}$ and $(b_n)_{n \in \mathbb{N}}$ with $a_n < b_n$ for all $n \in \mathbb{N}$ and $\lim_{n \to \infty} a_n = \lim_{n \to \infty} b_n$.

2. Find the limits of the sequences (if they exist):

   a) $\dfrac{3n^2 + 2n + 1}{n^3 - 3n^2 - 1}$

   b) $\dfrac{n^2 + 5}{n^2 - 5}$

   c) $\left(1 + \dfrac{3}{n!}\right)^7$

3. Prove or disprove the assertions:

   a) If a_n and b_n diverge, then $a_n + b_n$ diverges and $a_n \cdot b_n$ diverges.

   b) If a_n is bounded and b_n is a null sequence, then $a_n \cdot b_n$ is a null sequence.

4. You probably know the story of Achilles and the tortoise: In a race, the tortoise gets a head start on Achilles. Achilles can never catch up to the tortoise, because he always has to reach the point where the tortoise was just a moment ago. But by then, the tortoise has already moved on. Can you help Achilles?

5. Check the following series for convergence:

   a) $\displaystyle\sum_{n=0}^{\infty} \dfrac{n^2}{2^n}$

   b) $\displaystyle\sum_{n=1}^{\infty} \dfrac{n!}{1 \cdot 3 \cdot 5 \cdots (2n-1)}$

   c) $\displaystyle\sum_{n=0}^{\infty} \dfrac{2^{n+1}}{n!}$

6. Determine the order of the algorithm

$$x^n = \begin{cases} 1 & \text{if } n = 0 \\ x^{\frac{n}{2}} \cdot x^{\frac{n}{2}} & \text{if } n \text{ even} \\ x^{n-1} \cdot x & \text{else} \end{cases}$$

Flip back to Exercise 9 in Chap. 3.

7. Give natural numbers p, q for which holds:

a) $\dfrac{p}{q} = 0.4711\overline{4711}$

b) $\dfrac{p}{q} = 0.1230\overline{434}$

c) $\dfrac{p}{q} = \sqrt{2}$

8. Show: If $\varepsilon < d(a,b)/2$ in a metric space, then

$$U_\varepsilon(a) \cap U_\varepsilon(b) = \emptyset.$$

9. Expand the number $1/5!$ into a decimal fraction, a binary fraction, and a hexadecimal fraction.

10. Find finite decimal numbers that cannot be converted into finite binary fractions.

Continuous Functions

14

Abstract

The study of continuous functions is at the center of analysis. If you have worked through this chapter

- you know what a continuous real or complex function is and can check if a given function is continuous,
- you have learned about functions of several variables and the concept of continuity for such functions,
- you master important properties of continuous functions,
- you know many elementary continuous functions: power function and root, exponential function, the logarithm and trigonometric functions,
- and have learned a whole new definition of the number π.

In this chapter we deal with mappings between subsets of real or complex numbers: It is $K = \mathbb{R}$ or $\mathbb{C}$, $D \subset K$ and $f : D \to K$. Depending on whether the underlying field is $\mathbb{R}$ or $\mathbb{C}$, we speak of real or complex functions. In a short section we will also deal with functions of several variables, that is, with mappings whose domain is a subset of the $\mathbb{R}^n$.

Supplementary Information The online version contains supplementary material available at https://doi.org/10.1007/978-3-658-40423-9_14.

14.1   Continuity

In a first approach, continuous real functions without gaps in the domain are those functions whose graph can be drawn with a pencil without taking it off. The essential property of continuous functions is that the graph must not make any jumps. In Fig. 14.1 you can see the images of a continuous and discontinuous function in this sense. This description is not the whole truth, only the good continuous functions behave like this. Most of the time, however, we will have to deal with such. How can this property be formulated mathematically precisely?

The concept of convergence of sequences helps us to describe jumps: Let's look at the point of discontinuity a in Fig. 14.1 and argue first intuitively: If we approach the x-axis to the point a with a sequence whose sequence elements sometimes lie to the left and sometimes to the right of a, then the function values of the sequence elements diverge. The sequence of function values is not convergent. This property is characteristic for points of discontinuity. At the point b we have a similar situation: If a sequence converges to b and there are infinitely many sequence elements different from b, then the sequence of function values converges, but not to the function value $f(b)$.

First, I would like to formulate what the limit of a function at a point is. The definition includes not only limits for points in the domain of the function, but also for all boundary points of the set (see Definition 12.13). This formulation will allow us later to extend the domain of a function sensibly. If we add the boundary points to the domain D of the function, we just get the set of contact points of D. For these points we explain the concept of the limit of a function.

▶   **Definition 14.1** Let $K = \mathbb{R}$ or $\mathbb{C}$, $D \subset K$ and $f : D \to K$ be a function. Let x_0 be a contact point of D. If for all sequences $(x_n)_{n \in \mathbb{N}}$ with $x_n \in D$, it holds:

$$\lim_{n \to \infty} x_n = x_0 \Rightarrow \lim_{n \to \infty} f(x_n) = y_0,$$

then one says: *f has in x_0 the limit* y_0 and writes for this

$$\lim_{x \to x_0} f(x) = y_0.$$

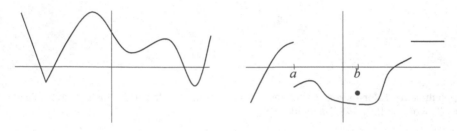

Fig. 14.1   Continuity

If $K = \mathbb{R}$ and the domain is not bounded from above (from below), then we say $\lim_{x \to \infty} f(x) = y_0$ ($\lim_{x \to -\infty} f(x) = y_0$), if for all sequences $(x_n)_{n \in \mathbb{N}}$ with $\lim_{n \to \infty} x_n = \infty$ ($-\infty$) it holds $\lim_{n \to \infty} f(x_n) = y_0$. For y_0, the value $\pm\infty$ is also allowed in this definition.

If x_0 is a point of the domain D, then there exists the function value $f(x_0)$. If the limit $\lim_{x \to x_0} f(x)$ exists, then it is equal to $f(x_0)$. Because the condition in Definition 14.1 must hold for all sequences, in particular for example for the constant sequence, which for all n has the value $x_n = x_0$. Then, of course, $\lim_{n \to \infty} f(x_n) = f(x_0)$ for this sequence and thus the limit is fixed. You can see in Fig. 14.1 neither in the point a nor in the point b the limits exist. For this a few

Examples

1. Let $f: \mathbb{R} \to \mathbb{R}$, $x \mapsto ax + b$. In every point $x_0 \in \mathbb{R}$, the limit of f exists: If $(x_n)_{n \in \mathbb{N}}$ is a sequence with $\lim_{n \to \infty} x_n = x_0$, then $f(x_n) = ax_n + b$ and according to the calculation rules for convergent sequences from Theorem 13.8 it follows $\lim_{n \to \infty} f(x_n) = ax_0 + b = f(x_0)$.
2. The floor function: For all $x \in \mathbb{R}$, let $\lfloor x \rfloor := \text{Max}\{z \in \mathbb{Z} \mid z \le x\}$ be the largest integer less than or equal to x. You can see the graph of the function $f(x) = \lfloor x \rfloor$ in Fig. 14.2.
   The limit $\lim_{x \to 1} f(x)$ does not exist: To do this, we choose two test sequences that converge to 1: $x_n = 1 - 1/n$, $y_n = 1 + 1/n$. Then $\lim_{n \to \infty} f(x_n) = 0$ and $\lim_{n \to \infty} f(y_n) = 1$.
3. Let $f: \mathbb{R} \setminus \{0\} \to \mathbb{R}$, $x \mapsto 1$. This function has a gap in the domain at 0, see Fig. 14.3. However, the number 0 is a contact point of the domain and the limit of the function exists at this point: It is $\lim_{x \to 0} f(x) = 1$.
4. $\lim_{x \to \infty} \frac{1}{x} = 0$, $\lim_{x \to \infty} x^2 = \infty$. The limit $\lim_{x \to 0} \frac{1}{x}$ does not exist, however. Why? ◄

From the examples you can see that it is often easier to disprove the existence of the limit than to prove it: To disprove it, you only have to find two sequences whose function values have different limits, or even one sequence whose function values do not converge. In Example 2, this would have been the sequence $1 + (-1)^n/n$. If you want to

Fig. 14.2 The floor function

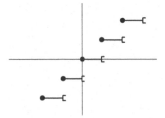

prove the existence of a limit, you have to check the convergence of every conceivable sequence. Fortunately, the computational rules for convergent sequences from Theorem 13.8 make our work easier: They can be transferred to limits of functions almost literally:

▶ **Theorem 14.2** Let $f, g \colon D \to K$ be real or complex functions, a a contact point of D and $\lim_{x \to a} f(x) = b$, $\lim_{x \to a} g(x) = c$. Then holds:

a) $\lim_{x \to a} (f(x) \pm g(x)) = b \pm c$.

b) $\lim_{x \to a} (f(x) \cdot g(x)) = b \cdot c$.

c) For $\lambda \in \mathbb{R} \, (\mathbb{C})$ it is $\lim_{x \to a} (\lambda f(x)) = \lambda b$.

d) For $c \neq 0$ it is $\lim_{x \to a} (f(x)/g(x)) = b/c$.

e) $\lim_{x \to a} |f(x)| = |b|$.

f) If $K = \mathbb{C}$, then

$$\lim_{z \to a} f(z) = b = u + iv \quad \Leftrightarrow \quad \lim_{z \to a} \mathrm{Re}(f(z)) = u, \quad \lim_{z \to a} \mathrm{Im}(f(z)) = v.$$

Here, $\mathrm{Re} f$ and $\mathrm{Im} f$ are the functions from D to $\mathbb{R}$ that map the element z with $f(z) = u + iv$ to the real part u respectively to the imaginary part v.

▶ **Theorem 14.3** Let $f \colon D \to K$ and $g \colon E \to K$ be real or complex functions with $f(D) \subset E$. Then it holds for the composition $g \circ f \colon D \to K$, $x \mapsto g(f(x))$:

$$\lim_{x \to x_0} f(x) = y_0 \quad \text{and} \quad \lim_{y \to y_0} g(y) = z_0 \quad \Rightarrow \quad \lim_{x \to x_0} g \circ f(x) = z_0.$$

For $\lim_{n \to \infty} x_n = x_0$ implies $\lim_{n \to \infty} f(x_n) = f(x_0)$ and from this

$$\lim_{n \to \infty} g(f(x_n)) = g(f(x_0)). \qquad \qquad \Box$$

With the help of this concept of limit, we can now formulate what we understand by a continuous function:

Fig. 14.3 A gap in the domain

▶ **Definition 14.4** Let $K = \mathbb{R}$ or $\mathbb{C}$, $D \subset K$ and $f : D \to K$ be a function.

a) f is called continuous in $x_0 \in D$, if it holds: $\lim\limits_{x \to x_0} f(x) = f(x_0)$.

b) f is called continuous, if f is continuous in all points $x_0 \in D$.

Let us examine the examples after Definition 14.1 for continuity:

Examples

1. $f : \mathbb{R} \to \mathbb{R}$, $x \mapsto ax + b$ is continuous in every point $x \in \mathbb{R}$, thus f is continuous.
2. $f(x) = \lfloor x \rfloor$ is continuous for all $x \in \mathbb{R} \setminus \mathbb{Z}$. In $x \in \mathbb{Z}$ the function is not continuous!
3. $f : \mathbb{R} \setminus \{0\} \to \mathbb{R}$, $x \mapsto 1$ is continuous on the whole domain, even though it cannot be drawn in one line. This function has another special property: 0 is a contact point of the domain and the limit of the function at this point exists and is equal to 1. If we define $f(0) := 1$ at this point, then the resulting function is continuous on $\mathbb{R}$. ◀

▶ **Theorem 14.5: Continuos extension** Let $f : D \to K$ be a continuous real or complex function. Let x_0 be a contact point of D and $x_0 \notin D$. If the limit $\lim_{x \to x_0} f(x) = c$ exists, then the function

$$\tilde{f} : D \cup \{x_0\}, \quad x \mapsto \begin{cases} f(x) & \text{for } x \in D \\ c & \text{for } x = x_0 \end{cases}$$

is continuous. We say "f is continuously extendable to x_0."

Theorems 14.2 and 14.3 can now be directly transferred to continuous functions:

▶ **Theorem 14.6** Let $f, g : D \to K$ be real or complex functions, f and g be continuous in $x_0 \in D$. Then holds:

a) $f \pm g$ is continuous in x_0.
b) $f \cdot g$ is continuous in x_0.
c) For $\lambda \in K$, $\lambda \cdot f$ is continuous in x_0.
d) If $g(x_0) \neq 0$, then f/g is continuous in x_0.
e) $|f|$ is continuous in x_0.
f) If $K = \mathbb{C}$, then f is continuous in z if and only if $\text{Re} f$ and $\text{Im} f$ are continuous in z.

▶ **Theorem 14.7** Let $f : D \to K$ and $g : E \to K$ be real or complex functions with $f(D) \subset E$. Then for the composition $g \circ f : D \to K$, $x \mapsto g(f(x))$ holds: If f is continuous in x_0 and g is continuous in $f(x_0)$, then $g \circ f$ is continuous in x_0.

With these theorems, we can reduce the continuity of composite functions to the individual parts. Thus, for example, all polynomials $f \in \mathbb{R}[X]$ or $f \in \mathbb{C}[X]$ are continuous on the entire domain. Let us look at the polynomial $x \mapsto ax^2 + b$, for example:

First, the mapping $g\colon x \mapsto x$ is continuous. Then $g \cdot g\colon x \mapsto x \cdot x$ is continuous and so is $x \mapsto ax^2$, $x \mapsto bx$, and finally the sum $x \mapsto ax^2 + b$. For general polynomials, induction is once again required.

Functions of several variables

I would now like to go into functions $f\colon U \to \mathbb{R}$, where U is a subset of $\mathbb{R}^n$, that is, real-valued functions of several variables. Examples of this are

$$f\colon \mathbb{R}^2 \to \mathbb{R}, \quad (x, y) \mapsto x^2 + y^2,$$

$$g\colon \mathbb{R}^3 \setminus \{(x, y, z) | x = \pm y\} \to \mathbb{R}, \quad (x, y, z) \mapsto \frac{z}{x^2 - y^2}.$$

These mappings represent an important special case of the functions $f\colon U \to \mathbb{R}^m$, $U \subset \mathbb{R}^n$. Each such "vector-valued" function has the form

$$f\colon \mathbb{R}^n \supset U \to \mathbb{R}^m,$$

$$(x_1, \ldots, x_n) \mapsto (f_1(x_1, \ldots, x_n), f_2(x_1, \ldots, x_n), \ldots, f_m(x_1, \ldots, x_n)),$$

that is, it consists of a m-tuple of real-valued functions of several variables. We will restrict ourselves to examining these component functions $f\colon \mathbb{R}^n \supset U \to \mathbb{R}$.

You have already learned that mathematicians like to rely on theories they already know when they encounter new problems, and here too we will try to use as much of our knowledge about functions of one variable as possible. A very important concept in this context is the notion of the partial function. If you keep all the variables fixed except for one, you get a function of one variable again:

▶ **Definition**   14.8 Let $U \subset \mathbb{R}^n$, $f\colon U \to \mathbb{R}$ be   a   function   and $(a_1, \ldots, \hat{a_i}, \ldots, a_n) \in \mathbb{R}^{n-1}$.
Let $D = \{x \in \mathbb{R} \mid (a_1, \ldots a_{i-1}, x, a_{i+1}, \ldots, a_n) \in U\} \subset \mathbb{R}$. Then for $i = 1, \ldots, n$

$$f_i\colon \quad D \to \mathbb{R}, \quad x \mapsto f(a_1, \ldots, a_{i-1}, x, a_{i+1}, \ldots, a_n)$$

is a real function. It is called the *partial function* of f.

A function with n variables not only has n partial functions, but infinitely many: The variables that are different from the i-th component are all kept fixed, but each value is allowed for that! If, for example, in the function $x^2 + y^2$ the second variable is kept fixed, one obtains the partial functions x^2, $x^2 + 1$, $x^2 + 2$, $x^2 + 3$ and so on.

Can functions of several variables be drawn?

If $U \subset \mathbb{R}^n$ and $f : U \to \mathbb{R}$ is a function, then the graph of f is the set $\{(x, f(x)) \mid x \in U\} \subset \mathbb{R}^n \times \mathbb{R} = \mathbb{R}^{n+1}$.

On paper or on the screen we can draw the $\mathbb{R}^2$, at most projections of the $\mathbb{R}^3$. Functions $f : \mathbb{R}^2 \to \mathbb{R}$ can thus still be visualized if one conceives the domain as a subset of the x_1-x_2-plane and adds the function value as the x_3-component to each argument (Fig. 14.4 left).

Now, for example, you can choose a series of partial functions f_1 and f_2 and project the plot into the $\mathbb{R}^2$; this creates a spatial impression. In the middle of Fig. 14.4 you see the graph of the function $f : \mathbb{R}^2 \to \mathbb{R}$, $(x, y) \mapsto x^2 + y^2$, a paraboloid, drawn in this form. Another common method of graphical representation are *contour lines*, which you may know from maps or from the isobars of weather maps: For a series of x_3 values, all pre-images are marked in the x_1-x_2-plane. With a little practice, you can get along well with a map, but if the contour lines are not marked with the corresponding function values (the heights), it may happen that you think you are climbing to a peak and actually landing in a depression. The right part of Fig. 14.4 shows contour lines of the paraboloid. The closer the contour lines run, the steeper the slope. For functions of more than two variables, there are no satisfactory representation options anymore.

The paraboloid has no holes, cracks and jumps, it represents the graph of a continuous function. We can literally transfer the term of continuity to functions of several variables, because the $\mathbb{R}^n$ is a metric space with the distance $\| \cdot \|$ and in such spaces there are convergent sequences according to Definition 13.2:

▶ **Definition 14.9** Let $U \subset \mathbb{R}^n$ and $f : U \to \mathbb{R}$ be a function. Let x_0 be a contact point of U. If for all sequences $(x_k)_{k \in \mathbb{N}}$ with $x_k \in U$ it holds:

$$\lim_{k \to \infty} x_k = x_0 \quad \Rightarrow \quad \lim_{k \to \infty} f(x_k) = y_0,$$

then we say: "*f has in x_0 the limit y_0*" and write for it

$$\lim_{x \to x_0} f(x) = y_0.$$

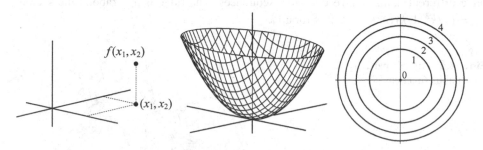

Fig. 14.4 Functions $f : \mathbb{R}^2 \to \mathbb{R}$

▶ **Definition 14.10** Let $U \subset \mathbb{R}^n$ and $f : U \to \mathbb{R}$ be a function.

a) f is called *continuous in* $x_0 \in U$, if it holds: $\lim_{x \to x_0} f(x) = f(x_0)$.

b) f is called *continuous*, if f is continuous in all points $x_0 \in U$.

Many questions concerning functions of several variables can be reduced to the investigation of partial functions. How is it with continuity? First of all it holds:

▶ **Theorem 14.11** If $U \subset \mathbb{R}^n$ and $f : U \to \mathbb{R}$ are continuous, then all partial functions of f are continuous.

This is easy to see: If $f_i : D \to \mathbb{R}$, $x \mapsto f(a_1, \ldots, a_{i-1}, x, a_{i+1}, \ldots, a_n)$ is a partial function and the sequence x_k of real numbers converges in D to x_0, then $(a_1, \ldots, a_{i-1}, x_k, a_{i+1}, \ldots, a_n)_{k \in \mathbb{N}}$ converges to $(a_1, \ldots, a_{i-1}, x_0, a_{i+1}, \ldots, a_n)$ and because of the continuity of f also $f_i(x_k) = f(a_1, \ldots, a_{i-1}, x_k, a_{i+1}, \ldots, a_n)$ converges to $f(a_1, \ldots, a_{i-1}, x_0, a_{i+1}, \ldots, a_n) = f_i(x_0)$.

But one should not expect too much from the partial functions, the inverse of this theorem is false! A counter-example is sufficient: Let

$$f : \mathbb{R}^2 \to \mathbb{R}, \quad (x, y) \mapsto \begin{cases} \dfrac{xy^2}{x^2 + y^4} & (x, y) \neq (0, 0) \\ 0 & (x, y) = (0, 0) \end{cases}$$

All partial functions are continuous: For $a_2 \neq 0$ $f(x, a_2) = \frac{x a_2^2}{x^2 + a_2^4}$ is continuous, for $a_1 \neq 0$ $f(a_1, y) = \frac{a_1 y}{a_1^2 + y^4}$ is continuous. There only remain the partial functions $f(x, 0)$ and $f(0, y)$ to be investigated. But these are everywhere equal to 0, thus also continuous.

The function f itself is not continuous at the point $(0, 0)$. See Fig. 14.5 for this. My drawing program naturally has difficulty drawing the function near the point of discontinuity, but you can see: If you walk along the ridge towards the point $(0, 0)$, you always stay at the same height. But if you approach this point on the x-axis, you always stay at height 0. So we can find sequences that converge to $(0, 0)$, but whose function values have different limits. I give two such sequences: The ridge is a parabola, the sequence $x_k = (1/k^2, 1/k)$ runs on it. It is for all k

Fig. 14.5 An only partially
continuous function

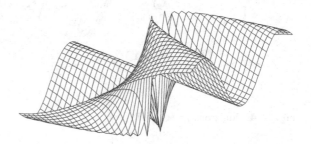

$$f(x_k) = \frac{(1/k^2) \cdot (1/k)^2}{(1/k^2)^2 + (1/k)^4} = \frac{1/k^4}{2/k^4} = 0.5,$$

and thus also $\lim_{x_k \to (0,0)} f(x_k) = 0.5$. But for the sequence $y_k = (1/k, 0)$ it is always $f(y_k) = 0$, so $\lim_{y_k \to (0,0)} f(y_k) = 0$.

14.2   Elementary Functions

Before we investigate further properties of continuous functions, I would like to introduce you to some important functions.

First of all, some basic terms:

▶ **Definition 14.12** Let $f : \mathbb{R} \to \mathbb{R}$ be a function. Then f is called

even	⇔ for all $x \in \mathbb{R}$ holds $f(x) = f(-x)$,
odd	⇔ for all $x \in \mathbb{R}$ holds $f(x) = -f(-x)$,
monotonically increasing	⇔ for all $x, y \in \mathbb{R}$ with $x < y$ is $f(x) \leq f(y)$,
strictly monotonically increasing	⇔ for all $x, y \subset \mathbb{R}$ with $x < y$ is $f(x) < f(y)$,
monotonically decreasing	⇔ for all $x, y \in \mathbb{R}$ with $x < y$ is $f(x) \geq f(y)$,
strictly monotonically decreasing	⇔ for all $x, y \in \mathbb{R}$ with $x < y$ is $f(x) > f(y)$,
periodic with period T	⇔ for all $x \in \mathbb{R}$ is $f(x + T) = f(x)$.

In Fig. 14.6 you can see the graphs of an even, an odd, a monotonic, a strictly monotonic and a periodic function. You can see that an even function is symmetrical to the y-axis, an odd function is point-symmetrical to the origin.

The terms monotonically increasing and decreasing can also be formulated for subsets of $\mathbb{R}$, as well as the terms even and odd, provided that with x always $-x$ lies in the domain. For complex functions, the terms do not make sense, because in complex numbers there is no order. Note that a strictly monotonic function is always injective, because from $x \neq y$ it follows of course in this case always $f(x) \neq f(y)$.

Fig. 14.6 Types of functions

Examples of functions

1. $f: \mathbb{R} \to \mathbb{R}$, $x \mapsto ax + b$ is strictly monotonically increasing for $a > 0$, strictly monotonically decreasing for $a < 0$. For $a = 0$, f is monotonically increasing and monotonically decreasing. The zeros of a function are the numbers that are mapped to 0. $f(x)$ has exactly one zero, unless $a = 0$. The graph of this function is a line.

2. The function $f: \mathbb{R} \to \mathbb{R}$, $x \mapsto ax^2 + bx + c$, $a \neq 0$ represents a parabola. The function is not monotonic, for $a = 0$ it is a linear function. In Fig. 14.7 you can see the graphs of some parabolas, the form depends on the values a, b, c.
   We know that the zeros of this function are given by the formula

   $$x_{1/2} = \frac{-b \pm \sqrt{b^2 - 4ac}}{2a}.$$

   Depending on whether the value under the square root is greater, smaller or equal to 0, there are two, no or one real zero, i.e. intersections of the graph with the x-axis.
   The complex function $f: \mathbb{C} \to \mathbb{C}$, $x \mapsto ax^2 + bx + c$, $a \neq 0$ always has zeros. Unfortunately, complex functions can no longer be painted: The graph $\Gamma_f = \{(x, f(x)) \mid x \in \mathbb{C}\}$ would be a subset of $\mathbb{R}^4$. We just manage to use projections to draw subsets of $\mathbb{R}^3$.

3. The power function $f: \mathbb{C} \to \mathbb{C}$, $x \mapsto x^n$, $n \in \mathbb{N}_0$ has already been encountered several times. Now we want to put together a few calculation rules for it. First we still define for all $x \in \mathbb{C}$: $x^0 := 1$. For $x, y \in \mathbb{C}$ then

   $$(x \cdot y)^n = x^n \cdot y^n,$$

   applies due to the commutative law of multiplication.

   $$x^n \cdot x^m = x^{n+m},$$

   because $x^n x^m = \underbrace{(x \cdots x)}_{n\text{-times}} \cdot \underbrace{(x \cdots x)}_{m\text{-times}} = \underbrace{x \cdot x \cdots x}_{n+m\text{-times}}.$

   $$(x^n)^m = x^{n \cdot m} = (x^m)^n,$$

Fig. 14.7 Parabolas

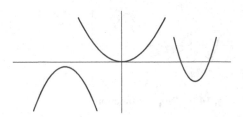

because $(x^n)^m = \underbrace{\underbrace{(x \cdots \cdots x)}_{n\text{-times}} \cdot \underbrace{(x \cdots \cdots x)}_{n\text{-times}} \cdots \underbrace{(x \cdots \cdots x)}_{n\text{-times}}}_{m\text{-times}} = \underbrace{x \cdot x \cdots \cdots x}_{m \cdot n\text{-times}}.$

The fixation $x^0 := 1$ seemed to fall from the sky at first. Now it becomes clear that this is the only sensible possibility: If we want the above rules to also apply to $n = 0$, then $x^0 \cdot x^1 = x^{0+1} = x^1$ must be. This is only true for $x^0 = 1$.

All of these functions are polynomial functions and therefore continuous. With the power function, we now look at the restriction to $\mathbb{R}_0^+$, the positive real numbers including 0: $f: \mathbb{R}_0^+ \to \mathbb{R}_0^+, x \mapsto x^n$. This function is strictly monotonically increasing for $n > 0$, because from $0 \le x < y$ it follows from the order laws of the real numbers $0 \le x^n < y^n$. This makes the function injective. Is it also surjective? Since $\mathbb{R}$ is complete, it is likely that every positive real number has a nth root. We can only prove this at the end of this chapter. But I want to use it already now: The function $f: \mathbb{R}_0^+ \to \mathbb{R}_0^+, x \mapsto x^n$ is bijective and therefore has an inverse function, the *root function*. This is our next example of an elementary function:

4. $\sqrt[n]{\ }: \mathbb{R}_0^+ \to \mathbb{R}_0^+, x \mapsto \sqrt[n]{x}, n > 0$ is the inverse function to $f: \mathbb{R}_0^+ \to \mathbb{R}_0^+, x \mapsto x^n$. Therefore it is $\sqrt[n]{x^n} = (\sqrt[n]{x})^n = x$.

The restriction of this function to non-negative real numbers is essential: $x \mapsto x^n$ is not always injective on the real or complex numbers and therefore not invertible as a function. Inverse functions can only be defined on bijective parts of a function. So is for example $f: \mathbb{R}_0^- \to \mathbb{R}_0^+, x \mapsto x^2$ a bijective mapping, but not $f: \mathbb{R} \to \mathbb{R}_0^+, x \mapsto x^2$. ◀

▷ **Definition 14.13** For $x \in \mathbb{R}^+$ and $n, m \in \mathbb{N}$ we define:

$$x^{\frac{1}{n}} := \sqrt[n]{x}, \quad x^{\frac{n}{m}} := \sqrt[m]{x^n}, \quad x^{-\frac{n}{m}} := \frac{1}{x^{\frac{n}{m}}}.$$

Why these abbreviations? With these names, the same rules of calculation apply to $x^{\frac{1}{n}}$ and $x^{\frac{n}{m}}$ as for the power function from Example 3 above:

▷ **Theorem 14.14** For all $x, y \in \mathbb{R}^+$ and for all rational numbers $p, q \in \mathbb{Q}$ it holds:

a) $(xy)^p = x^p \cdot y^p$,

b) $x^p \cdot x^q = x^{p+q}$,

c) $(x^p)^q = x^{p \cdot q}$.

Using the bijectivity of the root, these rules can be reduced to the corresponding rules of the power function. Let us calculate rule b) first, assuming positive exponents: It is

$$(x^{n/m} \cdot x^{p/q})^{qm} = (x^{n/m})^{qm} \cdot (x^{p/q})^{qm} = (\sqrt[m]{x^n})^{qm} \cdot (\sqrt[q]{x^p})^{qm}$$
$$= ((\sqrt[m]{x^n})^m)^q \cdot ((\sqrt[q]{x^p})^q)^m = (x^n)^q \cdot (x^p)^m = x^{nq+pm}.$$

From this, we take the $(q \cdot m)$th root and obtain

$$x^{\frac{n}{m}} \cdot x^{\frac{\ell}{q}} = (x^{nq+pm})^{\frac{1}{qm}} = x^{\frac{nq+pm}{qm}} = x^{\frac{n}{m}+\frac{\ell}{q}}.$$

Calculate for yourself that the relation also holds for negative exponents. For this you have to use that x^{-n} according to Definition 14.13 is the multiplicative inverse of x^n. □

We can now define new functions:

More examples of functions

5. Let $a \in \mathbb{R}^+$. $f_a \colon \mathbb{Q} \to \mathbb{R}^+$, $q \mapsto a^q$ is called the *exponential function to the base a*. This function is declared according to Definition 14.13 for all rational numbers and the calculation rules from Theorem 14.14 apply. In particular, therefore, for all rational numbers p and q it is true:

$$f_a(p) \cdot f_a(q) = f_a(p+q). \tag{14.1}$$

Please note: For numbers $x \in \mathbb{R} \setminus \mathbb{Q}$, a^x is not yet defined. We don't know yet what $3^{\sqrt{2}}$ or e^{π} is!

6. In Chap. 13 we encountered an exponential function for the first time. In Definition 13.23 we defined it as the limit of a series. In contrast to the function from Example 5, it is defined for all elements $z \in \mathbb{C}$:

$$\exp \colon \mathbb{C} \to \mathbb{C}, \quad z \mapsto \sum_{n=0}^{\infty} \frac{z^n}{n!}.$$

◀

There is currently some confusion of language because I have denoted the functions from Examples 5 and 6 with the same name. We will soon know the reason for this. For the function exp we have already calculated in Theorem 13.24 that

$$\exp(z) \cdot \exp(w) = \exp(z+w) \tag{14.2}$$

is true for all $z, w \in \mathbb{C}$, i.e. the same relation as in (14.1). Furthermore,

$$\exp(0) = \sum_{n=0}^{\infty} \frac{0^n}{n!} = \underbrace{\frac{0^0}{0!}}_{=\frac{1}{1}} + \underbrace{\frac{0^1}{1!}}_{=0} + \cdots = 1, \quad \exp(1) = e = 2.718281\ldots.$$

We will use this to show that the exponential function "exp" is the same as the exponential function to the base e:

▶   **Theorem 14.15** For $q \in \mathbb{Q}$, holds $\exp(q) = e^q$.

Proof: First let $n \in \mathbb{N}$. Then

$$\exp(n) = \exp(\underbrace{1 + 1 + \cdots + 1}_{n\text{-times}}) = \exp(1)^n = e^n.$$

If $n, m \in \mathbb{N}$, then

$$\exp\left(\frac{n}{m}\right)^m = \exp\left(\underbrace{\frac{n}{m} + \frac{n}{m} + \cdots + \frac{n}{m}}_{m\text{-times}}\right) = \exp(n) = e^n,$$

From which, by taking the mth root, it follows:

$$\exp\left(\frac{n}{m}\right) = \sqrt[m]{e^n} = e^{\frac{n}{m}}.$$

Lastly, we must investigate $q \in \mathbb{Q}$, $q < 0$: Then $\exp(q)\exp(-q) = \exp(0) = 1$ and thus again

$$\exp(q) = \frac{1}{\exp(-q)} = \frac{1}{e^{-q}} = e^q.$$ □

Therefore, the following definition is sensible:

▶   **Definition 14.16** For $z \in \mathbb{C}$, let $e^z := \exp(z)$.

This extends the exponential function f_e from Example 5 from $\mathbb{Q}$ to $\mathbb{C}$ and eliminates the confusion. Now we can calculate e^π, but still not $3^{\sqrt{2}}$. Other bases than e are currently forbidden for non-rational exponents.

As an exercise, I leave it to you to check the following assertion, it is similar to the proof of Theorem 14.15:

$$\text{for } q \in \mathbb{Q} \text{ and } x \in \mathbb{R} \text{ is } e^{xq} = (e^x)^q. \tag{14.3}$$

▶   **Theorem 14.17** The exponential function $\exp\colon \mathbb{C} \to \mathbb{C}$, $z \mapsto \exp(z)$ is continuous.

In Fig. 14.8 you can see the graph of the real exponential function e^x: The function increases very slowly to the origin and then grows almost explosively. Remember? Algorithms with exponential growth are considered incalculable.

To prove continuity: We first show that $\exp$ is continuous at 0. Let z_n be a null sequence in $\mathbb{C}$, we have to show that then $\lim_{n\to\infty} \exp(z_n) = \exp(0) = 1$. We use

Fig. 14.8 The exponential
function

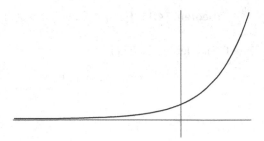

Theorem 13.25, in which we estimated the residual error after summation up to the
index m:

$$|r_{m+1}(z)| \leq \frac{2|z|^{m+1}}{(m+1)!}, \quad \text{if } |z| \leq 1 + \frac{m}{2}. \tag{14.4}$$

Here it is enough for us that $\exp(z_n) = 1 + r_1(z_n)$. The sequence $|z_n|$ will be less than 1 at
some point, and then

$$|\exp(z_n) - 1| = |r_1(z_n)| \leq \frac{2|z_n|}{1!} = 2|z_n|.$$

$2 \cdot |z_n|$ converges to 0 and so $\lim_{n\to\infty} \exp(z_n) = 1$. If now $z \neq 0$, any sequence that converges to z can be written in the form $z + h_n$, where h_n is a null sequence. Because of
$\exp(z + h_n) = \exp(z) \exp(h_n)$ it follows

$$\lim_{n\to\infty} \exp(z_n) = \lim_{n\to\infty} \exp(z) \cdot \exp(h_n) = \exp(z) \cdot \lim_{n\to\infty} \exp(h_n) = \exp(z) \cdot 1. \qquad \Box$$

How is e^x calculated for real exponents on the computer? We first assume $x > 0$. In Theorem 13.25 we have seen that the exponential series $\sum_{n=0}^{\infty} \frac{x^n}{n!}$ converges very quickly for
small x, for example for $x < 1$. The error can be estimated according to formula (14.4).
If now x is large, we look for the largest natural number $n \leq x$. Then $x = n + h$ where
$h < 1$ and we can calculate:

$$e^x = e^{n+h} = e^n \cdot e^h.$$

To calculate e^n we already know a fast recursive algorithm (see (3.4) in Sect. 3.3), and
we can determine e^h with the help of the quickly converging series. For negative exponents x we simply calculate $1/e^{-x}$.

How the exponential function behaves for complex exponents, we will see from the
next theorems, which reveal a surprising connection to the trigonometric functions sine
and cosine:

▶   **Theorem 14.18: Properties of the exponential function** Let $z = x + iy \in \mathbb{C}$,
      $x, y \in \mathbb{R}$. Then holds:

a) $e^z = e^x \cdot e^{iy}$.

b) $e^x > 0$.

c) $|e^{iy}| = 1$.

d) $|e^z| = e^x$.

Proof: a) follows directly from (14.2).

To b): If $x > 0$, then $e^x = \sum_{n=0}^{\infty} \frac{x^n}{n!} > 0$. For $x < 0$ we then get $e^x = 1/e^{-x} > 0$.

For c) we need Theorem 13.24b) and the properties of the absolute value from Theorem 5.16: This gives us

$$|e^{iy}|^2 = e^{iy} \cdot \overline{e^{iy}} = e^{iy} \cdot e^{\overline{iy}} = e^{iy} \cdot e^{-iy} = e^{iy-iy} = e^0 = 1.$$

d) follows from the first three properties:

$$|e^z| = |e^x e^{iy}| = |e^x| \cdot |e^{iy}| = e^x \cdot 1 = e^x. \qquad \square$$

The most important conclusion we can draw at the moment is from property c): If y is a real number, then e^{iy} always has absolute value 1, that is, it lies on the unit circle.

Flip back briefly to the introduction of sine and cosine in Theorem 7.2: The coordinates of a point (a, b) on the unit circle are given by sine and cosine of the angle between (a, b) and the x-axis. On the other hand, these coordinates are just the real part and the imaginary part of the complex number $z = a + ib$. See Figure 14.9 for this.

This establishes a relation between the exponent y and the angle α. Each exponent has an associated angle. You probably already know that mathematicians are so distinguished that they do not measure angles like ordinary people from 0° to 360°, but in so-called *radians*. Behind it is exactly the discovery we have just made: we simply denote the angle that belongs to the point e^{iy} with y (Fig. 14.10).

We still do not know that there is really such a y for each angle. The angle 0° belongs to $y = 0$, because $e^{i0} = 1 + i \cdot 0 = (1, 0)$. We will see that with increasing angle, the value of y also increases continuously. With the help of integral calculus (in the exercises of Chap. 16), we will soon be able to calculate that the length of the circular arc from $(1, 0)$ to the point $(\cos y, \sin y) = e^{iy}$ is exactly y. The conversion between degrees

Fig. 14.9 Radian 1

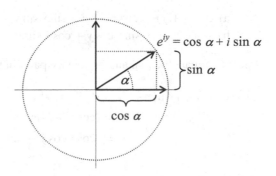

Fig. 14.10 Radian 2

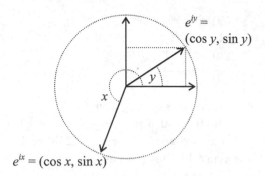

$$e^{iy} = (\cos y, \sin y)$$

$$e^{ix} = (\cos x, \sin x)$$

and radians is just as easy as between euros and dollars (but much more stable). For all calculations with the functions cosine and sine, radians have proven to be much simpler and more practical than degrees. It could be called a natural angle measurement. So don't be afraid of it!

▶   **Definition and Theorem 14.19: The functions cosine and sine** Let $y \in \mathbb{R}$. Then

$$\cos(y) := \mathrm{Re}(e^{iy}), \quad \sin(y) := \mathrm{Im}(e^{iy}).$$

The number y is called the *angle* between the x-axis and the direction of e^{iy}. The functions cosine and sine are continuous functions with domain $\mathbb{R}$.

The continuity follows from Theorem 14.7f).   □

▶   **Theorem 14.20** For all $y \in \mathbb{R}$ it holds that $(\cos y)^2 + (\sin y)^2 = 1$.

Proof: According to Theorem 14.18c), $|e^{iy}| = 1$ and thus $|e^{iy}|^2 = (\cos y)^2 + (\sin y)^2 = 1$. □

You could object that this is nothing really new, but we have not yet proven it.

In the future, I will write $\cos^2 x$ and $\sin^2 x$ for $(\cos x)^2$ and $(\sin x)^2$, respectively.

▶   **Theorem 14.21: Addition formulas for cosine and sine**

a) $\cos(x + y) = \cos x \cos y - \sin x \sin y$.
b) $\sin(x + y) = \sin x \cos y + \cos x \sin y$.

Proof: Compare the real and imaginary parts in the following calculation:

$$\cos(x+y) + i\sin(x+y) = e^{i(x+y)} = e^{ix}e^{iy}$$
$$= (\cos x + i \sin x)(\cos y + i \sin y)$$
$$= (\cos x \cos y - \sin x \sin y) + i(\sin x \cos y + \cos x \sin y). \quad □$$

Do you remember the rotations in $\mathbb{R}^2$? They were described by rotation matrices. The composition of two rotations means the multiplication of the two rotation matrices. If we first rotate by x and then by y, we get the rotation by $x + y$:

$$\begin{pmatrix} \cos(x+y) & -\sin(x+y) \\ \sin(x+y) & \cos(x+y) \end{pmatrix} = \begin{pmatrix} \cos y & -\sin y \\ \sin y & \cos y \end{pmatrix} \begin{pmatrix} \cos x & -\sin x \\ \sin x & \cos x \end{pmatrix}$$

$$= \begin{pmatrix} \cos y \cos x - \sin y \sin x & -\cos y \sin x - \sin y \cos x \\ \sin y \cos x + \cos y \sin x & -\sin y \sin x + \cos y \cos x \end{pmatrix}.$$

Compare the entries with Theorem 14.21: Isn't that a beautiful piece of mathematical magic? Two completely different theories produce the same result.

The following theorem is interesting again when it comes to concrete calculations of cosine and sine. For a long time, the values could only be looked up in tables, in the computer age we calculate the values using quickly convergent series:

▶   **Theorem 14.22: The series expansions of sine and cosine** It is

$$\cos x = 1 - \frac{x^2}{2!} + \frac{x^4}{4!} - \frac{x^6}{6!} + \frac{x^8}{8!} - \cdots = \sum_{n=0}^{\infty} \frac{(-1)^n}{(2n)!} \cdot x^{2n},$$

$$\sin x = x - \frac{x^3}{3!} + \frac{x^5}{5!} - \frac{x^7}{7!} + \frac{x^9}{9!} - \cdots = \sum_{n=0}^{\infty} \frac{(-1)^n}{(2n+1)!} \cdot x^{2n+1}.$$

For according to Theorem 13.18c) the series $\sum_{n=0}^{\infty} \frac{(ix)^n}{n!} = e^{ix}$ as well as $\sum_{n=0}^{\infty} \mathrm{Re}\left(\frac{(ix)^n}{n!}\right) = \mathrm{Re}(e^{ix})$ and $\sum_{n=0}^{\infty} \mathrm{Im}\left(\frac{(ix)^n}{n!}\right) = \mathrm{Im}(e^{ix})$ are convergent and it holds:

$$e^{ix} = \mathrm{Re}(e^{ix}) + i\,\mathrm{Im}(e^{ix})$$

$$= 1 + \frac{ix}{1!} + \frac{\overbrace{i^2 x^2}^{-1x^2}}{2!} + \frac{\overbrace{i^3 x^3}^{-ix^3}}{3!} + \frac{\overbrace{i^4 x^4}^{1x^4}}{4!} + \frac{\overbrace{i^5 x^5}^{ix^5}}{5!} + \frac{\overbrace{i^6 x^6}^{-1x^6}}{6!} + \frac{\overbrace{i^7 x^7}^{-ix^7}}{7!} + \cdots$$

$$\mathrm{Re}(e^{ix}) = 1 \quad\quad - \frac{x^2}{2!} \quad\quad + \frac{x^4}{4!} \quad\quad - \frac{x^6}{6!} \quad\quad + \cdots$$

$$\mathrm{Im}(e^{ix}) = \quad \frac{x}{1!} \quad\quad - \frac{x^3}{3!} \quad\quad + \frac{x^5}{5!} \quad\quad - \frac{x^7}{7!} + \cdots$$

14.3   Properties of Continuous Functions

Theorems About Continuous Functions

The question of zeros of functions occurs again and again in all technical and economic applications of mathematics, and so the efficient calculation of the zeros of continuous functions is one of the standard problems of numerical analysis. The following important theorem provides us with a first method for calculating zeros of continuous functions.

▶ **Theorem 14.23: Bolzano's theorem** Let $f: [a, b] \to \mathbb{R}$ be continuous, let $f(a)$ and $f(b)$ have different signs. Then there is a x in $[a, b]$ with $f(x) = 0$.

With the idea that continuous functions can be drawn without detaching, this theorem is clear: If f is smaller than 0 at a, then the graph must cross the x-axis at some point on the way to b, whether it wants to or not. Of course, we must not rely on the idea. We carry out a proof that simultaneously provides us with a constructive calculation method, the interval nesting, which is also called *bisection*. We assume that $f(a) < 0$ and $f(b) > 0$. If $f(a) > 0$ and $f(b) < 0$ applies, we only have to exchange the inequality signs in point a).

We inductively construct a sequence $[a_n, b_n]$ of intervals with the properties

a) $f(a_n) \leq 0$, $f(b_n) \geq 0$,
b) $a \leq a_{n-1} \leq a_n < b_n \leq b_{n-1} \leq b$, that is $[a_n, b_n] \subset [a_{n-1}, b_{n-1}]$ for $n > 0$,
c) $b_n - a_n = \frac{1}{2^n}(b - a)$.

We start with $a_0 = a$, $b_0 = b$ (Fig. 14.11). If the sequence of intervals is constructed up to the index n, we halve the interval and look to see in which half the zero is.

Let $c = \dfrac{a_n + b_n}{2}$. If $f(c) < 0$, we set $a_{n+1} = c$, $b_{n+1} = b_n$, otherwise $a_{n+1} = a_n$ and $b_{n+1} = c$.

c is between a_n and b_n and $b_{n+1} - a_{n+1} = \dfrac{1}{2}(b_n - a_n) = \dfrac{1}{2}\dfrac{1}{2^n}(b - a)$, so a), b) and c) are fulfilled. According to b) a_n is monotonically increasing and bounded, b_n monotonically decreasing and bounded. Both sequences are therefore convergent.

$\lim_{n\to\infty} (b_n - a_n) = \lim_{n\to\infty} 1/2^n \cdot (b - a) = 0$ implies $\lim_{n\to\infty} a_n = \lim_{n\to\infty} b_n =: x$. This common limit x is a zero of the function. Because f is continuous, $\lim_{n\to\infty} f(a_n) = \lim_{n\to\infty} f(b_n) = f(x)$ applies. Further, for all $n \in \mathbb{N}$ $f(a_n) \leq 0$, 0 is therefore an upper bound for the sequence and thus it must be $\lim_{n\to\infty} f(a_n) \leq 0$. Similarly it follows $\lim_{n\to\infty} f(b_n) \geq 0$. This can only be the case if $f(x) = 0$.   □

Fig. 14.11 Bolzano's theorem

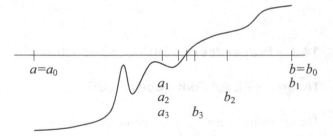

$a = a_0$ $\qquad\qquad\qquad\qquad\qquad\qquad\qquad\qquad\qquad\qquad$ $b = b_0$
$\qquad\qquad\qquad\qquad\qquad\qquad a_1 \qquad\qquad\qquad\qquad\qquad b_1$
$\qquad\qquad\qquad\qquad\qquad\qquad a_2 \qquad\qquad\qquad b_2$
$\qquad\qquad\qquad\qquad\qquad\qquad a_3 \qquad b_3$

The interval nesting can be implemented to numerically find zeros of functions: If you have found two points a, b whose function values have different signs, you can localize a zero with an error of $|b - a|/2^n$ in n steps. This method always works. There are indeed faster methods that we will also get to know, but these can sometimes go wrong.

Caution: Even if a "zero" x_0 has been found that only deviates by a value ε from the real zero x, one cannot make any statement about how far the function value $f(x_0)$ deviates from 0: If the function is very steep at this point, $f(x_0)$ can be far from 0. So you need to know more about the function to assess the quality of the found zero.

Bolzano's theorem has a whole range of interesting consequences. When calculating eigenvalues of matrices in Chap. 9 we have already used the following assertion:

▶ **Theorem 14.24** Every real polynomial of odd degree has at least one zero.

The sign of the polynomial $a_n x^n + a_{n-1} x^{n-1} + \cdots + a_1 x + a_0$ is determined for large values of $|x|$ solely by the first term, since this grows the strongest. If n is odd, then for $x < 0$ it is also $x^n < 0$ and $x > 0$ implies $x^n > 0$. So you will always find two values for x, so that the function values have different signs and thus Bolzano's theorem is applicable. □

Another immediate consequence of Bolzano's theorem is the intermediate value theorem:

▶ **Theorem 14.25: The intermediate value theorem** Let $f : [a, b] \rightarrow \mathbb{R}$ be continuous, $x, y \in [a, b]$ with $f(x) = m$ and $f(y) = M$ and $m \leq M$. Then for every $d \in [m, M]$ there is a c between x and y with $f(c) = d$.

Expressed a little less cryptically, this means: Every number that lies between two function values m and M of a continuous function has a preimage. The image has no gaps between m and M (Fig. 14.12).

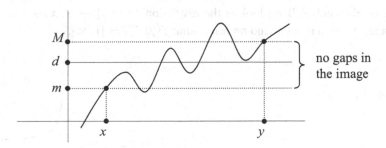

Fig. 14.12 The intermediate value theorem

Fig. 14.13 The image of a
function

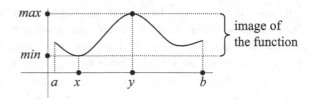

image of
the function

Even though it looks like more: the intermediate value theorem is only a reformula-
tion of Bolzano's theorem: Apply Bolzano's theorem to the function $g(x) := f(x) - d$.
Then a zero of g is a "d" point of f.   □

You can even describe the image of such a real function more precisely: The image is
a closed interval, the function values have a minimum and a maximum. The proof of this
is much more elaborate, I do not want to carry it out (Fig. 14.13).

▶   **Theorem 14.26** Let $f: [a, b] \to \mathbb{R}$ be a continuous function. Then the image set
      $f([a, b])$ of f is bounded and has a minimum and a maximum, that is, there are
      $x, y \in [a, b]$ so that $f(x) = \min f([a, b])$ and $f(y) = \max f([a, b])$.

I would like to add one last important result to the theorems about continuous functions,
also without proof:

▶   **Theorem 14.27** Let I be an interval and $f: I \to M$ a bijective continuous real
      function. Then the inverse function $f^{-1}: M \to I$ is also continuous.

The interval I in this theorem can be open, closed, or half-open, and ∞ is also allowed as
an interval boundary.

In all the theorems 14.23 to 14.27, the requirements placed on the domain are very
essential. Let's take the function $f: \mathbb{R} \setminus \{0\} \to \mathbb{R}, \ x \mapsto 1/x$ as an example (Figure
14.14): It is $f(-1) = -1$, $f(1) = 1$, but it has no zero and between $f(-1)$ and $f(1)$, not
every value is taken either. If we look at the restriction $f:]0, 1[\to \mathbb{R}, \ x \mapsto 1/x$, we see
that the image has no maximum and no minimum: $f(]0, 1[) =]1, \infty[$.

Figure 14.14 The function
$x \mapsto 1/x$

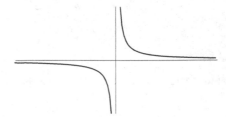

We can now further investigate the elementary functions that we learned in Sect. 14.2.

In Example 4 after Definition 14.12 I introduced the root. There I have already used that the power $f\colon \mathbb{R}_0^+ \to \mathbb{R}_0^+$, $x \mapsto x^n$ is surjective. Now we check that with the help of the intermediate value theorem. To $y \in \mathbb{R}_0^+$ we therefore have to find a x with $x^n = y$. Since the intermediate value theorem is based on a closed interval, we first tinker with the function f: We are looking for a $b \in \mathbb{R}_0^+$ so that $f(b) = b^n > y$. There is always one: If $y < 1$, then we choose $b = 1$, and for $y \geq 1$ we take $b := y^n$, because then $y^n \geq y$. Now we apply the intermediate value theorem to the function $\tilde{f}\colon [0, b] \to \mathbb{R}_0^+$, $x \mapsto x^n$: Every number between 0^n and b^n has a preimage, in particular there is a x with $x^n = y$. $f\colon \mathbb{R}_0^+ \to \mathbb{R}_0^+$, $x \mapsto x^n$ is therefore a bijective function.

According to Theorem 14.27, the inverse function to the power function is also continuous:

▶　　**Theorem 14.28** The nth root $\sqrt[n]{\ }\colon \mathbb{R}_0^+ \to \mathbb{R}_0^+$, $x \mapsto \sqrt[n]{x}$ is bijective and continuous.

Logarithm and General Exponential Function

In the exercises for this chapter you can compute that the real exponential function

$$\exp\colon \mathbb{R} \to \mathbb{R}^+, \quad x \mapsto \sum_{n=0}^{\infty} \frac{x^n}{n!}$$

is an injective function (this does not apply to the complex exponential function!). The same argument as for the power function shows us that exp is also surjective. This means that there must be a continuous inverse function (Fig. 14.15).

▶　　**Theorem and Definition 14.29: Natural logarithm** The inverse function of the exponential function exp is called the *natural logarithm* and is denoted by $\log_e$ or ln:

$$\ln\colon \mathbb{R}^+ \to \mathbb{R}, \quad x \mapsto \ln x.$$

Fig. 14.15 The logarithm

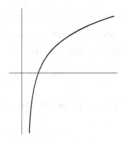

Here, $e^{\ln x} = x$ and $\ln e^y = y$. The natural logarithm is continuous and it holds:

a) $\ln(a \cdot b) = \ln a + \ln b$
b) $\ln(a^p) = p \cdot \ln a$ for all $p \in \mathbb{Q}$.

The rules a) and b) can be traced back to the corresponding rules of the exponential function: If $a = e^x$, $b = e^y$, then we get:

$$\ln(a \cdot b) = \ln(e^x \cdot e^y) = \ln(e^{x+y}) = x + y = \ln a + \ln b,$$

and further from

$$a^p = (e^{\ln a})^p = e^{p \cdot \ln a} \tag{14.5}$$

by applying the logarithm to both sides, the rule b). (Note that in (14.5) the assertion (14.3) has been used!)   □

The rule a) is based on the slide rule principle: The multiplication of two numbers can be traced back to the addition of the logarithms. If you put numbers on a ruler not in a linear scale, but logarithms of numbers, you get by putting together the distances that correspond to $\ln a$ and $\ln b$ the distance $\ln a \cdot b$, so you can read off the product of a and b.

Remember that we could not yet say what $a \in \mathbb{R}^+$ and for non-rational exponents x is? The equation (14.5) now gives us a way to extend the exponentiation sensibly from $\mathbb{Q}$ to $\mathbb{R}$:

▶   **Theorem and Definition 14.30: The General Exponential Function** Let $a \in \mathbb{R}^+$. Then the function

$$f_a : \mathbb{R} \to \mathbb{R}^+, \quad x \mapsto e^{x \cdot \ln a} =: a^x.$$

is called the *exponential function to the base a*. It is bijective and continuous and it holds:

$$a^{x+y} = a^x a^y. \tag{14.6}$$

f_a is the composition of the two bijective and continuous functions

$$g : \mathbb{R} \to \mathbb{R}, \, x \mapsto x \cdot \ln a, \quad h : \mathbb{R} \to \mathbb{R}^+, \, x \mapsto e^x,$$

and thus itself also bijective and continuous. On the rational numbers, f_a coincides with our previous definition $a^{\frac{m}{n}} = \sqrt[n]{a^m}$ according to (14.5). The equation (14.6) follows from the corresponding rule for the exponential function with base e:

$$a^{x+y} = e^{(x+y)\ln a} = e^{x \ln a} \cdot e^{y \ln a} = a^x \cdot a^y.$$ □

The exponential function is the only continuous function that has the property (14.6):

▶   **Theorem 14.31** Let $f : \mathbb{R} \to \mathbb{R}$ be a continuous function in which for all $x, y \in \mathbb{R}$ it holds: $f(x + y) = f(x)f(y)$. Then $f(x) = a^x$ for $a := f(1)$.

Just as in Theorem 14.15 one can first calculate that for all rational numbers $q \in \mathbb{Q}$ it holds: $f(q) = a^q$. Now let r be any real number. Then there is a sequence of rational numbers $(q_n)_{n \in \mathbb{N}}$ that converges to r. For example, one can use the sequence of base b fractions that converges to r (see Theorem 13.28b). Since both $f(x)$ and a^x are continuous functions, then it follows:

$$f(r) = f(\lim_{n \to \infty} q_n) = \lim_{n \to \infty} f(q_n) = \lim_{n \to \infty} a^{q_n} = a^{\lim_{n \to \infty} q_n} = a^r. \qquad \square$$

The function a^x has a continuous inverse:

▶   **Definition 14.32: Logarithm to base a** The inverse function to the exponential function $x \mapsto a^x$, $a > 0$ is called the *logarithm* to base a:

$$\log_a : \mathbb{R}^+ \to \mathbb{R}, \quad x \mapsto \log_a x.$$

So it is $a^{\log_a x} = x$, $\log_a(a^y) = y$ and

$$\log_a(x \cdot y) = \log_a(x) + \log_a(y), \quad \log_a(x^y) = y \cdot \log_a(x). \tag{14.7}$$

(14.7) can be derived from (14.6) just as for the natural logarithm. As you know, the logarithm to base 2 is particularly important for computer scientists.

If $a, b \in \mathbb{R}^+$ then $\log_b x = \log_b(a^{\log_a x}) = \log_a x \cdot \log_b a$. The logarithms to different bases a and b therefore only differ by the constant factor $\log_b a$. In particular, we obtain from this for the big O notation of the logarithms:

$$O(\log_a n) = O(\log_b n) \quad \text{for all } a, b > 0.$$

The Trigonometric Functions

In Definition 14.19 we defined the continuous functions cosine and sine as the real part or imaginary part of the function e^{ix}:

$$\cos(x) := \operatorname{Re}(e^{ix}), \quad \sin(x) := \operatorname{Im}(e^{ix}).$$

This resulted in the series expansions

$$\cos x = 1 - \frac{x^2}{2!} + \frac{x^4}{4!} - \frac{x^6}{6!} + \frac{x^8}{8!} - \cdots, \quad \sin x = x - \frac{x^3}{3!} + \frac{x^5}{5!} - \frac{x^7}{7!} + \frac{x^9}{9!} - \cdots.$$

By inserting we get $\cos 0 = 1$. From the series expansion we can see quickly that $\cos 2 < 0$ (with the help of an error estimate we can also prove this). Bolzano's theorem now says that there is at least one zero between 0 and 2. With a little more effort, one can see from the series expansion that the cosine is strictly monotonically decreasing in the interval between 0 and 2. The zero between 0 and 2 is therefore unique and can be calculated arbitrarily accurately, for example by means of an bisection. The first decimal digits of the zero are 1.570796327.

▶ **Definition 14.33** The number $\pi = 3.141592654\ldots$ is defined as twice the first positive zero of the cosine.

You know π, for example, from the formulas $r^2\pi$ for the area and $2r\pi$ for the circumference of a circle. Of course, our π from Definition 14.33 is exactly the same number. To check this, we just have to be a little patient. Like Euler's number e, the number π is an infinite non-periodic decimal number, so in particular irrational.

Because of $\cos(\pi/2) = 0$ and $\cos^2 x + \sin^2 x = 1$ we have $\sin(\pi/2) = +1$ or -1. From the series expansion of the sine we can see that only $+1$ is possible. This gives us:

$$e^{i(\pi/2)} = \cos(\pi/2) + i\sin(\pi/2) = i$$

and

$$e^{i\pi} = e^{i(\pi/2)+i(\pi/2)} = e^{i(\pi/2)} \cdot e^{i(\pi/2)} = i^2 = -1,$$

so $\cos\pi = -1$ and $\sin\pi = 0$. The same argumentation gives $e^{i2\pi} = e^{i\pi} \cdot e^{i\pi} = 1 = e^0$. From this we get for all $x \in \mathbb{R}$:

$$e^{i(x+2\pi)} = e^{ix} \cdot e^{i2\pi} = e^{ix}.$$

The functions e^{ix}, $\cos x$, $\sin x$ are therefore periodic with a period of 2π.

With similar considerations, a whole range of other calculation rules and relationships between cosine and sine can be derived. The formularies are full of them. Just one more example:

$$\cos(x + \pi/2) + i\sin(x + \pi/2) = e^{i(x+\pi/2)} = e^{ix}e^{i(\pi/2)} = ie^{ix}$$
$$ie^{ix} = i(\cos x + i\sin x) = -\sin x + i\cos x,$$

and the comparison between the real part and the imaginary part results in

$$\sin x = -\cos(\pi/2 + x), \quad \cos x = \sin(\pi/2 + x).$$

You can see from this that cosine and sine are quite similar, they are only shifted a little against each other. Now we can finally plot the graphs of the functions, see Fig. 14.16.

Sine and cosine are not injective and therefore cannot be globally inverted. However, if we look at restrictions on intervals, the bijectivity is given (Fig. 14.17).

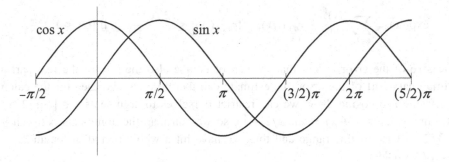

Fig. 14.16 cosine and sine

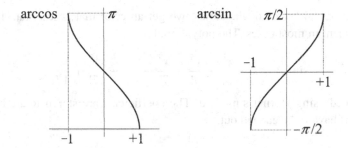

Fig. 14.17 arccosine and arcsine

▶ **Theorem and Definition 14.34** $\cos\colon [0, \pi] \to [-1, 1]$ and $\sin\colon [-\pi/2, \pi/2] \to [-1, 1]$ are bijective and continuous. The continuous inverse functions are called *arccosine* and *arcsine*:

$$\arccos\colon [-1, 1] \to [0, \pi], \quad \arcsin\colon [-1, 1] \to [-\pi/2, \pi/2].$$

Other important trigonometric functions, which I do not want to go into in more detail, are

$$\tan x = \frac{\sin x}{\cos x}, \quad \cot x = \frac{\cos x}{\sin x},$$

whose domains have gaps at the zeros of cosine or sine. They have the inverse functions arctan and arccot.

Numerical Calculation of Trigonometric Functions

The power series expansion of sine and cosine can be used to approximate function values very efficiently. How far do you have to add? From Theorem 13.25 we know that for the residual error of the exponential function it holds:

$$\exp(ix) = \sum_{k=0}^{n} \frac{(ix)^k}{k!} + r_{n+1}(ix), \quad |r_{n+1}(ix)| \leq \frac{2|ix|^{n+1}}{(n+1)!}, \quad \text{if } |ix| \leq 1 + \frac{n}{2}.$$

If the size of the error is less than ε, then of course also the size of the real part and the imaginary part of the error, the estimate can therefore also be used for the calculation of sine and cosine. Now we can restrict ourselves to arguments $x \in [-\pi/2, \pi/2]$. It is namely $\sin(\pi/2 + x) = \sin(\pi/2 - x)$, so we can map the arguments between $\pi/2$ and $3/2 \cdot \pi$ back to this range and thus we have hit a whole period of length 2π. For $|x| \leq \pi/2$ it holds:

$$r_{n+1}(ix) \leq \frac{2(\pi/2)^{n+1}}{(n+1)!}.$$

If we add the first 17 terms of the series, we get an error that is less than $1.06 \cdot 10^{-12}$, which is sufficient in most cases. The polynomial

$$\sin(x) \approx x - \frac{x^3}{3!} + \frac{x^5}{5!} - \frac{x^7}{7!} + \cdots - \frac{x^{15}}{15!} + \frac{x^{17}}{17!}$$

can be evaluated using Horner's method. The coefficients are stored in a table, only 16 multiplications have to be carried out.

Radian and polar coordinates

Please take another look at Fig. 14.10 after Theorem 14.18. We want to investigate how the point $e^{iy} = (\cos y, \sin z)$ on the unit circle moves when y changes. Starting at 0, $\cos y$ first decreases and $\sin y$ increases; the point moves up on the unit circle. At the point $\pi/2$, cosine is equal to 0 and sine is equal to 1, this corresponds to the angle 90°. Then sine becomes smaller again and cosine negative, e^{iy} moves further around to the left on the circle. The value $y = \pi$ corresponds to 180° and at $y = 2\pi$ the circle is finally closed again: $e^{i2\pi} = (1, 0)$. The intermediate value theorem guarantees that for each point w on the unit circle there is really a value $y \in [0, 2\pi[$ with $w = (\cos y, \sin y)$. This value is unique, it is the radian between w and the x-axis.

With this knowledge, we can now identify each complex number unequal to 0 (that is, each element of $\mathbb{R}^2$ different from 0) by *polar coordinates*: Let $z \in \mathbb{C}$, $z \neq 0$ and $r := \|z\|$. Then z/r has length 1, so it is a point on the unit circle, and there is exactly one angle $\varphi \in [0, 2\pi[$ with $e^{i\varphi} = z/r$, that is, with $z = r \cdot e^{i\varphi}$, where r and φ are uniquely determined:

▶ **Theorem and Definition 14.35** Each complex number $z \neq 0$ has a unique representation $z = r \cdot e^{i\varphi}$ with $r \in \mathbb{R}^+$, $\varphi \in [0, 2\pi[$.

For each vector $(x, y) \in \mathbb{R}^2 \setminus \{(0, 0)\}$, there are uniquely determined numbers $r \in \mathbb{R}^+$ and $\varphi \in [0, 2\pi[$ with $(x, y) = (r \cdot \cos \varphi, r \cdot \sin \varphi)$.

Figure 14.18 Polar
coordinates

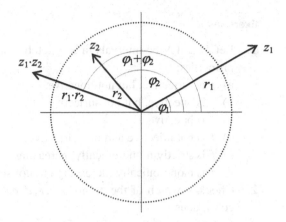

(r, φ) are called the *polar coordinates* of z or of (x, y). r is the distance of the element from the origin, φ the angle between the element and the x-axis.

The representation of complex numbers by polar coordinates allows for a simple geometric interpretation of the multiplication of two complex numbers: Let $z_1, z_2 \in \mathbb{C}$. Then

$$z_1 \cdot z_2 = r_1 \cdot e^{i\varphi_1} \cdot r_2 \cdot e^{i\varphi_2} = r_1 r_2 \cdot e^{i(\varphi_1 + \varphi_2)}.$$

That is, the absolute values of the two complex numbers multiply and the angles they form with the x-axis add up, see Figure 14.18.

14.4   Comprehension Questions and Exercises

Comprehension Questions

1. What does the continuity of the function f at the point x mean?
2. Why is it often easier to show that a function is discontinuous at a point than to show that it is continuous there?
3. There are also functions on $\mathbb{Z}/p\mathbb{Z}$. Why does the concept of continuity not make sense for such functions?
4. Let $f \colon [a, b] \to \mathbb{R}$ be a continuous mapping. Is there a $x_0 \in [a, b]$ with the property $f(x_0) = \frac{f(a) + f(b)}{2}$?
5. In Bolzano's theorem, an interval is required as the domain of the continuous function f. Why? Sketch a continuous function without this requirement, in which Bolzano's theorem does not hold.
6. How does the calculator calculate cos and sin?
7. sin and cos are surjective functions from $\mathbb{R}$ to $[-1, +1]$. Why does the inverse function not have the image $\mathbb{R}$?
8. Does every point in $\mathbb{R}^2$ have a unique representation in polar coordinates?

Exercises

1. Let M and N be intervals in $\mathbb{R}$. Sketch graphs of functions from $M \to N$ with the properties:
    a) f is surjective, but not injective.
    b) f is injective, but not surjective.
    c) f is bijective.
    d) f is not injective and not surjective.
    e) f is strictly monotonically increasing.
    f) f is monotonically increasing but not strictly monotonically increasing.
2. Check at which of the points a, b, c, d, e, f the function sketched in Fig. 14.19 is continuous.
3. Is the function $f : \mathbb{R} \setminus \{0\} \to \mathbb{R}$, $x \mapsto 2 - |x|/x$ continuous at the point 0?
4. Show that the function $f : \mathbb{R} \setminus \{1\} \to \mathbb{R}$, $x \mapsto (2x^4 - 6x^3 + x^2 + 3)/(x - 1)$ can be extended continuously at the point $x = 1$.
5. Show: For $q \in \mathbb{Q}$ and $x \in \mathbb{R}$, $\exp(xq) = \exp(x)^q$. Use the idea of the proof of Theorem 14.14.
6. Show that the real exponential function $\exp : \mathbb{R} \to \mathbb{R}$, $x \mapsto \sum_{n=0}^{\infty} \frac{x^n}{n!}$ is injective. Hint: First show that for $x \neq 0$ it holds: $\exp(x) \neq \exp(0) = 1$.
7. Calculate the number π to 8 digits using nested intervals.
8. Calculate $\lim_{x \to 0} \frac{\sin x}{x}$ and $\lim_{x \to 0}(x \cdot \cos x)$. Use the power series expansion of $\sin x$ for the first limit, and the fact that $|\cos x| \leq 1$ for all x for the second limit.
9. A famous theorem in number theory states that for large n, the number of prime numbers in the interval $[1, n]$ is approximately $n / \ln(n)$. How large is the proportion of prime numbers to all natural numbers of length 512 bits? The first bit should be 1.
    These prime numbers are needed to generate keys in the RSA algorithm.

10. What do you say to the following proof: $\frac{1}{-1} = \frac{-1}{1} \Rightarrow \frac{\sqrt{1}}{\sqrt{-1}} = \frac{\sqrt{-1}}{\sqrt{1}} \Rightarrow \frac{1}{i} = \frac{i}{1} \Rightarrow 1 = i^2$, in contradiction to $i^2 = -1$?

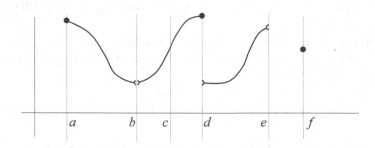

Fig. 14.19 Exercise 2

Differential Calculus

15

Abstract

A long chapter lies before you. If you manage it

- you know the definition of the derivative for functions of one or more variables,
- you can apply rules for the calculation of derivatives,
- you have calculated the derivatives of many elementary functions,
- you can determine extrema and inflection points of real functions,
- you know power series and the radius of convergence of power series,
- you can calculate Taylor polynomials and Taylor series for differentiable functions and estimate approximation errors,
- you have learned the basics of differential calculus of functions of several variables.

15.1 Differentiable Functions

If you want to connect a series of points in a drawing with a line, you would like to have a "nice" curve through the points. There are certainly different ideas about what is nice. The graphic editor with which I drew most of the pictures in this book, for example, does not connect 5 given points with straight line segments, but as shown in the right half of Fig. 15.1. Both pictures show graphs of continuous functions, but the right curve is not only continuous, it also has no corners, it is smooth.

Supplementary Information The online version contains supplementary material available at https://doi.org/10.1007/978-3-658-40423-9_15.

349

Fig. 15.1 Corners in functions

Differential calculus provides us with the mathematical means to describe and display such smooth curves. It investigates whether functions have corners, how steep they are or how curved they are, and also extreme values and changes in curvature behavior can be analyzed in this way.

With few exceptions, we will work with real functions in this chapter. But occasionally we will also investigate complex-valued functions, in particular the function e^{ix}, which we learned in the last chapter. The following theorems also apply to these and therefore I will formulate them accordingly.

So let $D \subset \mathbb{R}$, $K = \mathbb{R}$ or $\mathbb{C}$, $x_0 \in D$ and let $f : D \to K$ be a function. We investigate the mapping to the *difference quotient*

$$D \setminus \{x_0\} \to K, \quad x \mapsto \frac{f(x) - f(x_0)}{x - x_0}.$$

In $\mathbb{R}$ this function has a simple interpretation: $\frac{f(x)-f(x_0)}{x-x_0}$ specifies the slope of the line through the points $(x, f(x))$ and $(x_0, f(x_0))$: If the quotient is 0, the line is horizontal, the greater it is, the steeper it runs. If it is positive, the line rises to the right, if it is negative, it falls to the right.

The slope of a line of the form $g(x) = mx + c$ is just the factor m: For two points on the line $\dfrac{g(x) - g(x_0)}{x - x_0} = \dfrac{m(x - x_0)}{x - x_0} = m$ applies.

Now what happens if the point x gets closer and closer to x_0 ? In Fig. 15.2 you can see that the line then approaches the tangent at the point x_0 more and more. Is there a limit for $x \to x_0$? If so, this limit denotes the slope of the tangent at the point x_0.

▶   **Definition 15.1** Let $D \subset \mathbb{R}$, $K = \mathbb{R}$ or $\mathbb{C}$, $x_0 \in D$ and let $f : D \to K$ be a function. If the limit exists, then

$$f'(x_0) := \lim_{x \to x_0} \frac{f(x) - f(x_0)}{x - x_0}$$

Fig. 15.2 Slope of a function

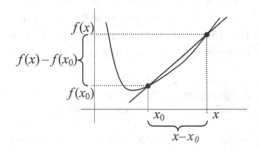

is called the *derivative* of f at x_0, f is called differentiable at x_0. If f is differentiable at every point $x \in D$, then f is called *differentiable*. In this case,

$$f' : D \to K, \quad x \mapsto f'(x)$$

is a function, the *derivative* of f.

Often the derivative is also denoted by $\frac{df}{dx}(x_0)$. This reminds one of the origin as limit of a fraction. I don't very much like this notation, because it can easily lead one to interpret the derivative as the quotient of two "numbers" df and dx. But this is wrong.

Some remarks on this definition:

1. The statement "the limit exists" means in particular that there must be sequences x_n in $D \setminus \{x_0\}$ that converge to x_0. So it makes no sense to investigate the differentiability of functions in isolated points of the domain, that is, in points around which there is an entire neighborhood that does not belong to D.
2. If the derivative of a function f exists at the point x_0, then the tangent at x_0 has the slope $f'(x_0)$ and is therefore a line of the form $f'(x_0) \cdot x + c$. This can be written as $f'(x_0)(x - x_0) + f(x_0)$.
3. For $h := x - x_0$, it is $\displaystyle \lim_{x \to x_0} \frac{f(x) - f(x_0)}{x - x_0} = \lim_{h \to 0} \frac{f(x_0 + h) - f(x_0)}{h}$. In this form, the quotient is often easier to calculate.

   The use of the letter h is based on a similar convention as for the ε. Under h you can always imagine a small number, but in contrast to ε it does not have to be greater than 0, $x_0 + h$ can approach x_0 from the left and from the right or jump back and forth wildly.

4. An important physical interpretation of the derivative: If $p(t)$ is the position of a body at time t, then $\frac{p(t) - p(t_0)}{t - t_0}$ is the average velocity in the time interval $[t_0, t]$ and $p'(t)$ is the instantaneous velocity. The change in velocity over time, $\frac{p'(t) - p'(t_0)}{t - t_0}$, is the average acceleration, and $p''(t)$, the derivative of the derivative, represents the instantaneous acceleration. For time derivatives, physicists often write $\dot{p}(t)$ or $\ddot{p}(t)$ instead of $p'(t)$ or $p''(t)$.
5. Derivatives also play a major role in mathematical economics. An example: If $f(x)$ is a cost curve, that is, a function that describes the cost of producing x parts of a product, then the derivative $f'(x)$ is the marginal cost. It represents the cost of producing the next part.

▶   **Theorem 15.2** Let $D \subset \mathbb{R}$, $K = \mathbb{R}$ or $\mathbb{C}$, $x_0 \in D$ and let $f : D \to K$ be a function. Then the following properties are equivalent:

a) f is differentiable with $f'(x_0) = a$.
b) There is a function $\varphi : D \to K$ with $\lim_{x \to x_0} \varphi(x) = 0$ and $a \in K$ with

$$f(x) = f(x_0) + a \cdot (x - x_0) + \varphi(x) \cdot (x - x_0). \tag{15.1}$$

If f is differentiable with $f'(x_0) = a$, then we have to find the function φ. To do this, we define

$$\varphi(x) = \begin{cases} \dfrac{f(x) - f(x_0)}{x - x_0} - f'(x_0) & \text{for } x \neq x_0, \\ 0 & \text{for } x = x_0. \end{cases}$$

This function satisfies the equation (15.1), and since

$$\lim_{x \to x_0} \left(\frac{f(x) - f(x_0)}{x - x_0} - f'(x_0) \right) = \lim_{x \to x_0} (f'(x_0) - f'(x_0)) = 0$$

$\lim\limits_{x \to x_0} \varphi(x) = 0$ also applies.

Conversely, if the equation (15.1) is given, then for $x \neq x_0$:

$$\frac{f(x) - f(x_0)}{x - x_0} = a + \varphi(x)$$

and thus $\lim\limits_{x \to x_0} \dfrac{f(x) - f(x_0)}{x - x_0} = a$, that is $a = f'(x_0)$.   □

We therefore have two descriptions of the derivatives. The definition as the limit of the difference quotient gives us a concrete calculation possibility at hand. From the second representation $f(x) = f(x_0) + f'(x_0) \cdot (x - x_0) + \varphi(x) \cdot (x - x_0)$ we can read an intuitive geometric interpretation:

The closer x approaches x_0, the better $f(x)$ agrees with $f(x_0) + f'(x_0) \cdot (x - x_0)$. The remainder $\varphi(x) \cdot (x - x_0)$ goes to 0.

The line $g(x) := f(x_0) + f'(x_0) \cdot (x - x_0)$ is the tangent to f at x_0. This tangent represents an approximation of the function f near x_0, while the error $f(x) - g(x) = \varphi(x)(x - x_0)$ goes to 0 faster than $x - x_0$ (Fig. 15.3).

As an immediate consequence of Theorem 15.2, it follows that every differentiable function is continuous:

Fig. 15.3 The tangent

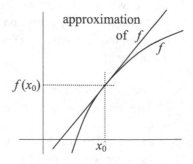

Fig. 15.4 A non-differentiable function

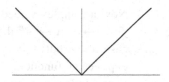

▶ **Theorem 15.3** If the function f is differentiable in x_0, then it is also continuous in x_0.

We check the definition of continuity. It holds

$$\lim_{x \to x_0} f(x) = \lim_{x \to x_0} [f(x_0) + f'(x_0) \cdot (x - x_0) + \varphi(x) \cdot (x - x_0)] = f(x_0). \qquad \square$$

Conversely, the differentiability of a function does not follow from its continuity. For example, the absolute value $f: \mathbb{R} \to \mathbb{R}$, $x \mapsto |x|$ (Fig. 15.4) is not differentiable at 0, because

$$\frac{f(x) - f(0)}{x - 0} = \begin{cases} +1 & \text{for } x > 0 \\ -1 & \text{for } x < 0 \end{cases}$$

Examples of differentiable functions and their derivatives

1. $f(x) = c$: $\displaystyle\lim_{x \to x_0} \frac{f(x) - f(x_0)}{x - x_0} = \lim_{x \to x_0} \frac{c - c}{x - x_0} = \lim_{x \to x_0} 0 = 0.$

   If you use Theorem 15.2b), then $c = f(x) = f(x_0) + 0 \cdot (x - x_0) + 0 \cdot (x - x_0).$

2. $f(x) = a \cdot x$: $\displaystyle\lim_{x \to x_0} \frac{ax - ax_0}{x - x_0} = \lim_{x \to x_0} \frac{a(x - x_0)}{x - x_0} = \lim_{x \to x_0} a = a.$

   Again, you can easily check Theorem 15.2b).

3. $f(x) = x^n$: It is $f'(x) = n \cdot x^{n-1}$, because

$$\lim_{h \to 0} \frac{(x + h)^n - x^n}{h}$$

$$= \lim_{h \to 0} \frac{\cancel{x^n} + \binom{n}{1} h x^{n-1} + \binom{n}{2} h^2 x^{n-2} + \binom{n}{3} h^3 x^{n-3} + \cdots + h^n - \cancel{x^n}}{h}$$

$$= \lim_{h \to 0} \frac{h \left(\binom{n}{1} x^{n-1} + \binom{n}{2} h x^{n-2} + \binom{n}{3} h^2 x^{n-3} + \cdots + h^{n-1} \right)}{h}$$

$$= \lim_{h \to 0} \binom{n}{1} x^{n-1} + \binom{n}{2} h x^{n-2} + \binom{n}{3} h^2 x^{n-3} + \cdots + h^{n-1}$$

$$= \lim_{h \to 0} \binom{n}{1} x^{n-1} = n x^{n-1}.$$

4. Now a complex-valued function comes into play for the first time. Let $a \in \mathbb{C}$ and $f: \mathbb{R} \to \mathbb{C}, x \mapsto e^{ax}$ the exponential function. Of particular importance to us is the case $a = i$, but of course a real value of a is also possible, then we get the real exponential function.

First a little preparation: Remember that for small values of $|ax|$ it holds (see Theorem 13.25):

$$e^{ax} = 1 + ax + r_2(ax) \quad \text{with} \quad |r_2(ax)| \le \frac{2|ax|^2}{2!} = |ax|^2.$$

So $|\frac{e^{ah}-1}{h} - a| = |\frac{e^{ah}-1-ah}{h}| = |\frac{r_2(ah)}{h}| \le |a|^2 |h|$. Since for $h \to 0$ also $|a|^2|h|$ goes to 0, we get $\lim_{h \to 0} |\frac{e^{ah}-1}{h} - a| = 0$, that is $\lim_{h \to 0} \frac{e^{ah}-1}{h} = a$. With this intermediate result we can calculate the derivative of e^{ax}:

$$(e^{ax})' = \lim_{h \to 0} \frac{e^{a(x+h)} - e^{ax}}{h} = \lim_{h \to 0} \frac{e^{ax} e^{ah} - e^{ax}}{h}$$

$$= \lim_{h \to 0} \frac{e^{ah} - 1}{h} \cdot e^{ax} = a e^{ax}.$$

In particular, $(e^x)' = e^x$, that is, the exponential function reproduces itself in the derivative. This is the reason why the exponential function occurs so often in nature: There are many natural functions whose changes, for example changes over time, are proportional to the function itself. Think, for example, of the growth of a family of mice or the cooling of a hot cup of tea. Such processes can be described using the exponential function. ◀

Differentiation Rules

▶   **Theorem 15.4** Let $D \subset \mathbb{R}$, $K = \mathbb{R}$ or $\mathbb{C}$, $x_0 \in D$ and let $f, g: D \to K$ be functions that are differentiable in $x_0 \in D$. Then for $\lambda \in \mathbb{R}$ or $\lambda \in \mathbb{C}$, the functions $\lambda \cdot f$, $f \cdot g$ and $f + g$ are also differentiable in x_0 and it holds:

a) $(f + g)'(x_0) = f'(x_0) + g'(x_0)$,
b) $(\lambda f)'(x_0) = \lambda \cdot f'(x_0)$,
c) $(f \cdot g)'(x_0) = f'(x_0)g(x_0) + f(x_0)g'(x_0)$ (the *product rule*).
d) If $g(x) \ne 0$ is in a neighborhood of x_0, then $\frac{f}{g}$ is also differentiable at the point x_0 and it holds $\left(\frac{f}{g}\right)'(x_0) = \dfrac{f'(x_0)g(x_0) - f(x_0)g'(x_0)}{g(x_0)^2}$ (the *quotient rule*).

As an example, I would like to derive the product rule, the other rules can be shown similarly, partly with more, partly with less effort:

$$(fg)'(x_0) = \lim_{h \to 0} \frac{f(x_0 + h)g(x_0 + h) - f(x_0)g(x_0)}{h}$$

$$= \lim_{h \to 0} \frac{1}{h} \big(f(x_0 + h)g(x_0 + h) - f(x_0)g(x_0)$$

$$\underbrace{- f(x_0 + h)g(x_0) + f(x_0 + h)g(x_0)}_{=0} \big)$$

$$= \lim_{h \to 0} \frac{f(x_0 + h)(g(x_0 + h) - g(x_0)) + g(x_0)(f(x_0 + h) - f(x_0))}{h}$$

$$= \lim_{h \to 0} \left(f(x_0 + h) \frac{g(x_0 + h) - g(x_0)}{h} + g(x_0) \frac{f(x_0 + h) - f(x_0)}{h} \right)$$

$$= f(x_0)g'(x_0) + g(x_0)f'(x_0). \qquad \square$$

Now we can calculate a whole range of other derivatives.

More examples

5. First, to derive a polynomial: $a_n x^n + a_{n-1} x^{n-1} + \ldots + a_1 x + a_0$ has the derivative $na_n x^{n-1} + (n-1)a_{n-1} x^{n-2} + \ldots + a_1$. This follows immediately from Theorem 15.4a) and b).

6. It is $(e^{ix})' = ie^{ix} = i(\cos x + i \sin x) = -\sin x + i \cos x$. On the other hand, according to Theorem 15.4a) $(e^{ix})' = (\cos x + i \sin x)' = \cos' x + i \sin' x$. From this we obtain by comparison of the real and imaginary parts the derivatives of cosine and sine:

$$\cos' x = -\sin x, \qquad \sin' x = \cos x.$$

7. For $x \neq 0$, the derivative of $(\frac{1}{x^n})'$ can be determined using the quotient rule. For $f = 1, f' = 0, g = x^n, g' = nx^{n-1}$, we have:

$$\left(\frac{1}{x^n} \right)' = \left(\frac{0 - nx^{n-1}}{x^{2n}} \right) = -nx^{(n-1)-2n} = -nx^{-n-1} = -n \frac{1}{x^{n+1}}.$$

We can also write this as: $(x^{-n})' = -n \cdot x^{-n-1}$. So now for all $z \in \mathbb{Z}$ the rule applies: $(x^z)' = zx^{z-1}$, by the way also for $z = 0$!

8. In the domain of the tangent, it is

$$\tan' x = \left(\frac{\sin x}{\cos x} \right)' = \frac{\sin' x \cos x - \sin x \cos' x}{\cos^2 x} = \frac{\cos^2 x + \sin^2 x}{\cos^2 x} = \frac{1}{\cos^2 x}.$$

You can deduce the derivative of the cotangent in the same way: $\cot' x = -1/\sin^2 x$. ◀

▶   **Theorem 15.5: The chain rule** Let D, $E \subset \mathbb{R}$, $K = \mathbb{R}$ or $\mathbb{C}$, $x \in D$ and
let $f: D \to \mathbb{R}$ and $g: E \to K$ be functions with $f(D) \subset E$. So it is possible to build the composition $g \circ f$. Let f be differentiable in $x_0 \in D$ and g in $f(x_0)$. Then $g \circ f$ is also differentiable in x_0 and it is

$$(g \circ f)'(x_0) = g'(f(x_0)) \cdot f'(x_0).$$

First, g is derived at the point $f(x_0)$, this derivative must be multiplied by the derivative of the "inner function" at the point x_0.

This time I would like to use the description from Theorem 15.2b) to prove differentiability: Let $y = f(x)$, $y_0 = f(x_0)$. So there are functions φ, ψ with $\lim_{x \to x_0} \varphi(x) = 0$, $\lim_{y \to y_0} \psi(y) = 0$ and

$$y - y_0 = f(x) - f(x_0) = (f'(x_0) + \varphi(x)) \cdot (x - x_0),$$
$$g(y) = g(y_0) + (g'(y_0) + \psi(y)) \cdot (y - y_0).$$

Then we get by inserting $(y - y_0)$ and multiplying out

$$g(y) = g(y_0) + (g'(y_0) + \psi(y)) \cdot (f'(x_0) + \varphi(x)) \cdot (x - x_0)$$
$$= g(y_0) + g'(y_0)f'(x_0)(x - x_0)$$
$$+ \underbrace{[g'(y_0)\varphi(x) + \psi(f(x))f'(x_0) + \psi(f(x))\varphi(x)]}_{=: \, \delta(x)}(x - x_0),$$

$$= g(y_0) + g'(y_0)f'(x_0) \cdot (x - x_0) + \delta(x) \cdot (x - x_0),$$

thus
$$g(f(x)) = g(f(x_0)) + g'(f(x_0))f'(x_0) \cdot (x - x_0) + \delta(x) \cdot (x - x_0).$$

$x \to x_0$ implies $y = f(x) \to f(x_0) = y_0$, $\delta(x)$ is a function with $\lim_{x \to x_0} \delta(x) = 0$ and according to Theorem 15.2b) that just means $g(f(x_0))' = g'(f(x_0)) \cdot f'(x_0)$.   □

Two more examples

9. $\sin^n x$ is the composition of the mappings $f: x \mapsto \sin x$ and $g: y \mapsto y^n$. It is $f'(x) = \cos x$, $g'(y) = ny^{n-1}$ and therefore

$$(\sin^n x)' = n \cdot (\sin x)^{n-1} \cdot \cos x.$$

10.   For $a > 0$, $a^x := e^{x \ln a}$. The inner function is $f: x \mapsto x \ln a$, the outer function $g: y \mapsto e^y$:

$$(a^x)' = \underbrace{e^{x \ln a}}_{g'(f(x))} \cdot \underbrace{(\ln a)}_{f'(x)} = a^x \cdot \ln a. \quad \blacktriangleleft$$

The scheme is always the same:

- Identify the inner and outer function g and f,
- Differentiate g at the point $f(x)$, that is, form $g'(y)$ and set $f(x)$ for y,
- Multiply with the derivative $f'(x)$.

More examples follow after the next theorem:

▶   **Theorem 15.6** The inverse function rule: Let I be an interval in $\mathbb{R}$, $f : I \to \mathbb{R}$ an injective differentiable function and $f'(x) \neq 0$ for all $x \in I$. Then the inverse function $g : f(I) \to \mathbb{R}$ is also differentiable and it holds

$$g'(x) = \frac{1}{f'(g(x))}.$$

We show the differentiability in the point $y_0 = f(x_0)$. It is $y = f(x)$, thus $x = g(y)$. According to (15.1) follows from the differentiability of f:

$$f(g(y)) = f(g(y_0)) + f'(g(y_0))(g(y) - g(y_0)) + \varphi(g(y))(g(y) - g(y_0))$$

with $\lim_{x \to x_0} \varphi(x) = 0$. From this we get

$$y - y_0 = f(g(y)) - f(g(y_0)) = [f'(g(y_0)) + \varphi(g(y))](g(y) - g(y_0)),$$

and so for the differential quotient of g at the point y_0:

$$\frac{g(y) - g(y_0)}{y - y_0} = \frac{1}{f'(g(y_0)) + \varphi(g(y))}.$$

g is continuous as the inverse function of a continuous function and therefore $y \to y_0$ implies $g(y) \to g(y_0) = x_0$. So for the limit holds:

$$\lim_{y \to y_0} \frac{g(y) - g(y_0)}{y - y_0} = \frac{1}{f'(g(y_0))}. \qquad \square$$

This theorem has an intuitive meaning (see Fig. 15.5): we get the graph of the inverse function g to the function f by reflecting f at the angle bisector of x- and y-axis. If t is the tangent to f at x_0 with slope a, then the tangent s to g at the point y_0 is exactly the reflection of t, i.e., the inverse function of t. If t has the linear equation $y = ax + b$, then s has the equation $x = (1/a) \cdot (y - b)$. The slope of g in y_0 is thus $1/a$.

Now we can calculate the derivatives of all elementary functions that we have learned in Sec. 14.2. We continue our series of examples:

Fig. 15.5 The derivative of
the inverse function

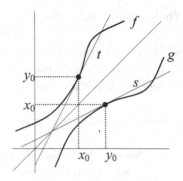

Examples

11. $\ln'(x)$. The logarithm is the inverse function of the exponential function $\exp(x)$
    and therefore it holds:

$$\ln'(x) = \frac{1}{\exp'(\ln(x))} = \frac{1}{\exp(\ln(x))} = \frac{1}{x}.$$

Since the logarithm is only defined for positive arguments, there can be no problems with the zero of the denominator.

12. $\sqrt[n]{x}\,' = (x^{1/n})'$. The nth root is the inverse image of x^n. With $y = x^{1/n}$ we get:

$$(x^{1/n})' = \frac{1}{(y^n)'} = \frac{1}{ny^{n-1}} = \frac{1}{n(x^{1/n})^{n-1}} = \frac{1}{n \cdot x^{1-1/n}} = \frac{1}{n}x^{(1/n)-1}.$$

13. Now you can calculate the derivative of $x^{n/m} = (x^n)^{1/m}$ yourself using the chain
    rule. Here is the result:

$$(x^{m/n})' = \frac{m}{n}x^{(m/n)-1}.$$

14. For all real exponents α the derivative rule applies, which we derived for natural
    numbers in Example 3 after Theorem 15.3: $(x^\alpha)' = \alpha \cdot x^{\alpha-1}$. Because for $\alpha \in \mathbb{R}$ it
    is $x^\alpha := e^{\alpha \ln x}$ and we can apply the chain rule to the inner function $f \colon x \mapsto \alpha \ln x$
    and the outer function $g \colon y \mapsto e^y$:

$$(x^\alpha)' = \underbrace{e^{\alpha \ln x}}_{g'(f(x))} \cdot \underbrace{(\alpha/x)}_{f'(x)} = \alpha \cdot x^\alpha /x = \alpha x^{\alpha-1}.$$

Please note the difference to the calculation of $(a^x)'$. We determined this derivative earlier.

15. For $x \in \,]{-}1, 1[$ it is $\arcsin'(x) = \frac{1}{\sin'(\arcsin x)} = \frac{1}{\cos(\arcsin x)}$. The function values of
    arcsin lie in the interval $]{-}\pi/2, \pi/2[$.
    In this range, the cosine is always positive and from $(\sin y)^2 + (\cos y)^2 = 1$ we
    get $\cos y = \sqrt{1 - (\sin y)^2}$. We insert this above and it results in:

$$\arcsin'(x) = \frac{1}{\sqrt{1 - (\sin(\arcsin x))^2}} = \frac{1}{\sqrt{1 - x^2}}.$$

In the points ± 1 the derivative does not exist. Can you see why? ◀

The derivative of a differentiable function is again a function and therefore possibly differentiable again. In this way, one can form higher-order derivatives, which are needed in many applications of differential calculus. I give you a recursive definition of the notion of the n-th derivative:

▶ **Definition 15.7: Higher-order derivatives** Let I be an interval, $f : I \to \mathbb{R}$ function and $x_0 \in I$.

$n = 1$: f is called in x_0 *1 times differentiable*, if f is differentiable in x_0. It is $f^{(1)}(x_0) := f'(x_0)$. If f is differentiable in a neighborhood U of x_0, then the function $f^{(1)} = f' : U \to \mathbb{R}$ is called the *1st derivative* of f.

$n > 1$: f is called in x_0 *n times differentiable*, if f is differentiable at least $(n - 1)$ times in a neighborhood U of x_0 and $f^{(n-1)}$ is 1 times differentiable in x_0. It is $f^{(n)}(x_0) := f^{(n-1)'}(x_0)$. If f is n times differentiable in a neighborhood U of x_0, then the function $f^{(n)} = f^{(n-1)'} : U \to \mathbb{R}$ is called the *nth derivative* of f.

$n = \infty$: f is called in x_0 *differentiable infinitely often*, if f is n times differentiable for all $n \in \mathbb{N}$.

Derivatives of lower order are often denoted with primes like the first derivative: $f''(x), f'''(x)$.

In the definition of the nth derivative, it is not sufficient to only demand that f is $(n - 1)$ times differentiable at the point x_0. After all, $f^{(n-1)}$ must be a function that is to be differentiated in x_0. To determine the limit of the differential quotient, there must be sequences in the domain of $f^{(n-1)}$ that converge to x_0, but are different from x_0. For simplicity, we require that $f^{(n-1)}$ must exist in a whole neighborhood of x_0.

▶ **Definition 15.8** The function $f : I \to \mathbb{R}$ is called n times continuously differentiable in I, if f is n times differentiable in I and $f^{(n)}$ is continuous.

Examples

1. e^x, $\sin x$, $\cos x$ as well as all polynomials are infinitely often differentiable on $\mathbb{R}$.

2. ln is infinitely often differentiable on $\mathbb{R}^+$.

3. $f: \mathbb{R} \to \mathbb{R}$, $x \mapsto x \cdot |x|$ is continuous and differentiable: For $x < 0$, $f(x) = -x^2$ and has the derivative $-2x$. For $x > 0$, $f(x) = x^2$ with derivative $2x$. At point 0 we have to look at the difference quotient:

$$\lim_{x \to 0} \frac{f(x) - f(0)}{x - 0} = \lim_{x \to 0} \frac{x \cdot |x|}{x} = \lim_{x \to 0} |x| = 0.$$

The limit exists, so f is also differentiable in 0. The derivative is:

$$f': \mathbb{R} \to \mathbb{R}, \quad x \mapsto 2|x|.$$

This function is, like the absolute value function, continuous on $\mathbb{R}$, but not differentiable at point 0, it has a kink there. So f is once continuously differentiable on $\mathbb{R}$.

> In this calculation, I applied a very typical approach to the investigation of functions: Continuity and differentiability are local properties of the function. If we investigate differentiability at a point x_0, only a very small neighborhood of x_0 plays a role. In this small neighborhood we can also represent the given function by another formula if it helps us further: In the example, we replaced $x \cdot |x|$ by $x \cdot x$ or by $x \cdot (-x)$, whose derivatives we already know. No matter how close we get to 0 with x_0, this replacement always works for a whole neighborhood of x_0. Only in 0 itself do we have to look more closely.

4. Even if it is hard to imagine: There are functions that are differentiable, but whose derivative is not continuous anymore. This seems to contradict common sense, which says that the slope of the tangent of a differentiable function cannot make jumps. Here we see that imagination and mathematical reality unfortunately do not always match.

   However, the examples for such functions are a bit wild, at least they are not the kind that you can draw with a pencil. I would like to give you such a function (Fig. 15.6):

Fig. 15.6 A non-continuously differentiable function

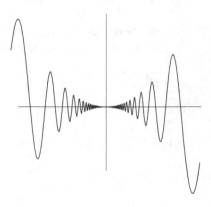

$$f(x) := \begin{cases} x^2 \sin(1/x) & \text{if } x \neq 0 \\ 0 & \text{if } x = 0 \end{cases}$$

For $x \neq 0$ we can calculate the derivative with the product and chain rule. Check that for $x \neq 0$ it holds:

$$f'(x) = 2x \sin(1/x) - \cos(1/x).$$

Now we look at the point 0. There the function is differentiable, the limit of the difference quotient is

$$\lim_{x \to 0} \frac{x^2 \sin(1/x) - 0}{x - 0} = \lim_{x \to 0} x \cdot \sin(1/x) = 0,$$

because $\sin(1/x)$ is bounded by ± 1. The function $f'(x)$ is therefore defined on all of $\mathbb{R}$, but it is not continuous in point 0, because the limit $\lim_{x \to 0} f'(x)$ does not exist. For the sequence $x_k = 1/(\pi k)$, for example, it holds:

$$f'(x_k) = \frac{2}{\pi k} \sin(\pi k) - \cos(\pi k) = \begin{cases} +1 & \text{if } k \text{ is odd} \\ -1 & \text{if } k \text{ is even.} \end{cases}$$

The sequence of function values does not converge. The derivative jumps near 0 constantly between $+1$ and -1 back and forth. The drawing in Fig. 15.6 can no longer represent the graph in the near the origin. ◀

Calculation of Extrema

▶ **Definition 15.9** Let I be an interval, $x_0 \in I$, $f: I \to \mathbb{R}$. We say "f has in x_0 a *local maximum (a local minimum)*", if there is an neighborhood $U_\varepsilon(x_0)$ such that $f(x) < f(x_0)$ (respectively $f(x) > f(x_0)$) for all $x \in (U_\varepsilon(x_0) \cap I) \setminus \{x_0\}$ holds. Local maxima and minima are called local *extrema* of f (Fig. 15.7).

▶ **Theorem 15.10** Let $f: I \to \mathbb{R}$ be differentiable in a neighborhood U_ε of $x_0 \in I$. If f has a local extremum in x_0, then $f'(x_0) = 0$.

Intuitively, this means: In a local extremum, the tangent to the function is horizontal.

Proof: We assume that the extremum in x_0 is a maximum. Then for all $x \in]x_0 - \varepsilon, x_0[$ the difference quotient $\frac{f(x) - f(x_0)}{x - x_0} > 0$, since the numerator and denominator are less

Fig. 15.7 A local maximum

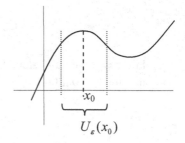

$U_\varepsilon(x_0)$

than 0. Similarly, for all $x \in \,]x_0, x_0 + \varepsilon[$ the quotient $\frac{f(x) - f(x_0)}{x - x_0} < 0$. Because of the differentiability, the limit of the difference quotient exists for $x \to x_0$. This limit, which is just $f'(x_0)$, can only be 0.   □

You can see from this proof that the theorem is really only valid for points in the interior of the domain: From the right and from the left, one must be able to approach x_0. If one wants to determine all the extrema of a function $f : I \to \mathbb{R}$, one must consider the following candidates:

1. The boundary points of the interval (if they exist),
2. The points with $f'(x) = 0$,
3. The points in which f is not differentiable.

In Fig. 15.8 you can see that extrema may or may not occur at points of every type. In particular, it also does not follow from $f'(x) = 0$ that x is an extremum!

   For further characterization of the extrema, we need an important theorem, the mean value theorem of differential calculus. The following theorem serves as preparation:

▶   **Theorem 15.11: Rolle's theorem** Let $f : [a, b] \to \mathbb{R}$ be continuous and $f : \,]a, b[\,\to \mathbb{R}$ differentiable. Let $f(a) = f(b)$. Then there is at least one $x_0 \in \,]a, b[$ with the property $f'(x_0) = 0$.

If f is constant, then the statement is clear. Otherwise, the image of f has a minimum and a maximum according to Theorem 14.26. f must therefore have an extremum in the open interval $]a, b[$ (Fig. 15.9), and for this, according to Theorem 15.10, $f'(x) = 0$ holds.   □

Fig. 15.8 Candidates for extrema

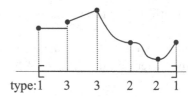

type: 1   3   3   2   2   1

Fig. 15.9 Rolle's Theorem

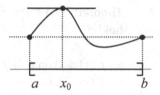

Now the announced central theorem of the theory:

▶ **Theorem 15.12: The mean value theorem of differential calculus** Let $f : [a, b] \to \mathbb{R}$ be a continuous function and $f :]a, b[\to \mathbb{R}$ differentiable. Then there is a point $x_0 \in]a, b[$ with the property

$$f'(x_0) = \frac{f(b) - f(a)}{b - a}.$$

$\frac{f(b)-f(a)}{b-a}$ is exactly the slope of the connecting line between $f(a)$ and $f(b)$. The mean value theorem states that there is a point on the graph at which the slope of the function takes on the "average slope" between a and b.

The trick in the proof consists in distorting the function f so that the conditions of Theorem 15.11 are fulfilled (Fig. 15.10): We subtract from f the connecting line s between the endpoints of the graph. For the function

$$g(x) = f(x) - \frac{f(b) - f(a)}{b - a}(x - a)$$

we have $g(a) = g(b)$ and therefore, according to Theorem 15.12, there is a x_0 with $g'(x_0) = 0$. The derivative of g is $g'(x) = f'(x) - \frac{f(b)-f(a)}{b-a}$, so that $f'(x_0) = \frac{f(b)-f(a)}{b-a}$ is really true.   □

The mean value theorem makes the local shape of a function accessible to us. The interval I, from which the following result speaks, can of course only be a small part of the actual domain, but in this part we can now describe its shape:

Fig. 15.10 The mean value theorem of differential calculus

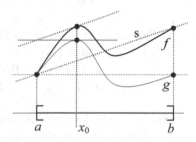

▶   **Theorem 15.13** For a function f, which is differentiable on the open interval I, holds:

a) $f'(x) > 0$ on $I \Rightarrow f$ is strictly monotonically increasing on I.
b) $f'(x) < 0$ on $I \Rightarrow f$ is strictly monotonically decreasing on I.
c) $f'(x) \geq 0$ on $I \Rightarrow f$ is monotonically increasing on I.
d) $f'(x) \leq 0$ on $I \Rightarrow f$ is monotonically decreasing on I.
e) $f'(x) = 0$ on $I \Rightarrow f$ is constant on I.

I only show part a), the other points can be derived similarly: Let $x_1, x_2 \in I$ and $x_1 < x_2$. Then there is a x_0 between them with $\frac{f(x_2)-f(x_1)}{x_2-x_1} = f'(x_0) > 0$. That can only be if $f(x_2) > f(x_1)$. So f is strictly monotonically increasing.   □

As an immediate consequence of e) we get that two functions with the same derivative only differ by a constant:

▶   **Theorem 15.14** If $f: I \to \mathbb{R}$ and $g: I \to \mathbb{R}$ are differentiable functions on the interval I and $f'(x) = g'(x)$ holds for all $x \in I$, then there is a $c \in \mathbb{R}$ with $f(x) = g(x) + c$ for all $x \in I$.

For $f - g$ has the derivative 0, so it is constant.   □

We come to the famous curve sketching, which you have probably been plagued with extensively in school. In essence, it is based on Theorem 15.13. But we will now also apply the assertions of this theorem to derivatives of higher order.

▶   **Theorem 15.15: Extremum test**
   Part 1: Let $f:]a, b[\to \mathbb{R}$ be differentiable and $x_0 \in]a, b[$ with $f'(x_0) = 0$.
               Then f has a local maximum in x_0 if the derivative $f'(x)$ is positive directly to the left of x_0 (that is, in an interval $]x_0 - \varepsilon, x_0[$) and negative directly to the right of x_0. f has a local minimum in x_0 if the derivative $f'(x)$ is negative directly to the left of x_0 and positive directly to the right of x_0.

   Part 2: Let $f:]a, b[\to \mathbb{R}$ be twice continuously differentiable and $x_0 \in]a, b[$ with $f'(x_0) = 0$.
               a) If $f''(x_0) < 0$, then f has a local maximum in x_0.
               b) If $f''(x_0) > 0$, then f has a local minimum in x_0.

For part 1: If the derivative is positive to the left, f grows there, if it is negative to the right it falls again to the right of x_0, and at the point x_0 itself there is a maximum. For the minimum, one proceeds analogously.

For part 2: Since f'' is still continuous, $f''(x) < 0$ applies in a whole interval around x_0. According to Theorem 15.13 f' is therefore strictly monotonically decreasing in this area. Because of $f'(x_0) = 0$, a sign change from $+$ to $-$ takes place at the point x_0. The first part of the theorem then says that a maximum is present at the point x_0. Assertion b) follows again analogously.   □

If the first and second derivatives are 0 at the point x_0, no statement can be made about the point x_0: It can be an extremum, or a so-called *terrace point*, a point with a horizontal tangent, but not an extremum. A simple example of this situation is the function $x \mapsto x^3$ at the point 0.

A differentiable function f is called *concave* in the interval I if the derivative (i.e. the slope) decreases in the interval I and *convex* if the derivative increases. An *inflection point* is a point at which the curvature changes, i.e. a point at which the derivative has an extreme value.

In Fig. 15.11 I have sketched a function f with its first and second derivative. You can see from this that the extrema of the first derivative are zeros of the second derivative.

Without proof, I would like to cite the following theorem, which describes the curvature of a function. There are no new ideas behind it, you just go down one derivative:

▶   **Theorem 15.16** Let $f \colon]a, b[\ \to \mathbb{R}$ be twice differentiable.

a) If $f''(x) < 0$ for all $x \in]a, b[$, then f is concave.
b) If $f''(x) > 0$ for all $x \in]a, b[$, then f is convex.
c) If f is three times continuously differentiable in x_0, $f''(x_0) = 0$ and $f'''(x_0) \neq 0$, then f has an inflection point in x_0.

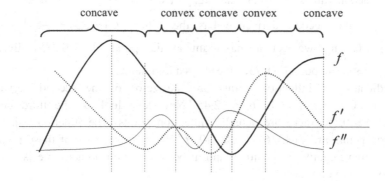

Fig. 15.11   The curvature

Please note that Theorems 15.15 and 15.16 only apply inside the domain of differentiable functions. Boundary points, discontinuities and non-differentiable points of the function must be examined separately.

In curve sketching one checks from a given function:

- the domain,
- the zeros,
- the extrema,
- the inflection points,
- the curvature.

With this knowledge, a useful sketch of the function can usually be made. Let's go through an example:

Example

Let $f(x) = \frac{x^2-1}{x^3}$. The domain is the set $\{x \in \mathbb{R} \mid x \neq 0\}$. The zeros are the zeros of the numerator: $x^2 - 1 = 0$, that is $x_{1/2} = \pm 1$.

We calculate the first three derivatives using the quotient rule:

$$f'(x) = \left(\frac{x^2-1}{x^3}\right)' = \frac{2x \cdot x^3 - (x^2-1) \cdot 3x^2}{x^6} = \frac{-x^4+3x^2}{x^6} = \frac{-x^2+3}{x^4},$$

$$f''(x) = \left(\frac{-x^2+3}{x^4}\right)' = \frac{2x^2-12}{x^5},$$

$$f'''(x) = \left(\frac{2x^2-12}{x^5}\right)' = \frac{-6x^2+60}{x^6}.$$

For the zeros of f' we set $-x^2 + 3 = 0$ and get $x_{3/4} = \pm\sqrt{3} \approx 1.73$.

The function values at these points are: $f(\pm\sqrt{3}) = \frac{3-1}{3\cdot(\pm\sqrt{3})} \approx \pm 0.385$.

At $x_{3/4}$, the second derivative has the values $f''(\pm\sqrt{3}) = \frac{2\cdot 3-12}{9\cdot(\pm\sqrt{3})}$, so it is $f''(\sqrt{3}) < 0$, and we get a maximum at the point $(1.73, 0.385)$. Because of $f''(-\sqrt{3}) > 0$, the point $(-1.73, -0.385)$ is a minimum.

Candidates for inflection points are the zeros of the second derivative: It is $f''(x) = 0$ for $x_{5/6} = \pm\sqrt{6} \approx \pm 2.45$. Now let's look at the third derivative: $f'''(\pm\sqrt{6}) > 0$, so there are actually inflection points. The function values of the inflection points are $f(\pm\sqrt{6}) \approx \pm 0.340$. f'' has positive slope at these points, so it changes from negative to positive, and at both inflection points there is a transition from concave to convex.

How does $f(x)$ behave near the definition gap? For $x < 0$ the function is positive and gets bigger and bigger, for $x > 0$ it is $f(x) < 0$ and gets smaller and smaller as x approaches 0. Now we can sketch the graph of the function quite well, see Fig. 15.12. ◀

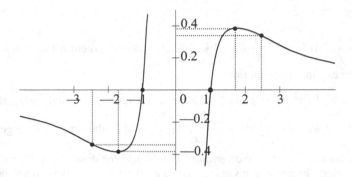

Fig. 15.12 A curve sketching

15.2 Power Series

The problem of approximating a given function by other, usually simpler to handle functions, occurs again and again in mathematics. Polynomials are often chosen for this purpose. Calculus provides us with a powerful tool for constructing such approximation polynomials. To prepare, we first examine power series. These can be thought of as polynomials of "infinite degree". Of course, this does not exist, it is an infinite series of functions. When evaluating such a polynomial, one will break off after a finite number of summands.

▶ **Definition 15.17** Let $(a_k)_{k\in\mathbb{N}}$ be a sequence of real numbers. Then

$$\sum_{k=0}^{\infty} a_k x^k \tag{15.2}$$

is called a real power series.

So far, this formula represents neither a function nor a series of numbers. We only know: If you insert a real number x into (15.2), you get a series, which is either convergent or not. But we can make a function out of it:

$$f: \left\{x \in \mathbb{R} \,\middle|\, \sum_{k=0}^{\infty} a_k x^k \text{ is convergent}\right\} \to \mathbb{R}, \ x \mapsto \sum_{k=0}^{\infty} a_k x^k.$$

This is the function I always mean when I talk about a power series. The starting index of a power series does not have to be 0, but at least a natural number greater than or equal to 0.

We have already met some of these series:

---

1. $\sum\limits_{k=0}^{\infty} \left(\dfrac{1}{k!}\right) x^k$. It is $a_k = \dfrac{1}{k!}$ for all k, the series is convergent for all $x \in \mathbb{R}$ and represents the exponential function.

2. $\sum\limits_{k=0}^{\infty} x^k = \sum\limits_{k=0}^{\infty} 1 \cdot x^k$. Here $a_k = 1$ is for all k. We have calculated earlier that for all x
   with $|x| < 1$ we have $\sum\limits_{k=0}^{\infty} x^k = \dfrac{1}{1-x}$. If $|x| \geq 1$, then the series is divergent.

   Even for $x = -1$ we have divergence! Flip back to the definition of the convergence
   of a series: The partial sums of the series are $1, 0, 1, 0, 1, \ldots$, so they do not form a convergent sequence.

3. $\sum\limits_{k=1}^{\infty} \dfrac{1}{k} x^k$ is also convergent for $|x| < 1$: For this we use the ratio test from Theorem
   13.19: For the quotient of two consecutive series elements we have

   $$\lim_{k\to\infty} \left| \frac{(1/(k+1))x^{k+1}}{(1/k)x^k} \right| = \lim_{k\to\infty} \left| \frac{k}{k+1} x \right| = |x| < 1$$

   and thus the series is convergent. For $x = 1$ the series is divergent, then it just rep-
   resents the harmonic series, see Example 5 at the end of Sec. 13.1. For $|x| > 1$ this
   quotient will become greater than 1 at some point, so there is divergence. The case
   $x = -1$ is still interesting: In this case the series is convergent, but I do not want to
   prove it here.

4. The trigonometric series $\sum\limits_{k=0}^{\infty} \dfrac{(-1)^k}{(2k)!} x^{2k}$ and $\sum\limits_{k=0}^{\infty} \dfrac{(-1)^k}{(2k+1)!} x^{2k+1}$ for cosine and sine
   also represent power series, which are convergent for all real numbers.

5. A polynomial $b_0 + b_1 x + b_2 x^2 + \cdots + b_n x^n$ can also be interpreted as a power
   series that is always convergent, where:

   $$a_k = \begin{cases} b_k & \text{for } k \leq n \\ 0 & \text{for } k > n. \end{cases} \quad \blacktriangleleft$$

▶ **Definition 15.18** Let $\sum\limits_{k=0}^{\infty} a_k x^k$ be a power series. Then

$$R := \begin{cases} \sup\left\{ x \in \mathbb{R} \;\middle|\; \sum\limits_{k=0}^{\infty} a_k x^k \text{ is convergent} \right\}, & \text{if the supremum exists,} \\ \infty & \text{else} \end{cases}$$

is called the *radius of convergence* of the power series $\sum\limits_{k=0}^{\infty} a_k x^k$.

The radius of convergence is ∞ or it is $R \geq 0$, because for $x = 0$ there is always convergence.

▶   **Theorem 15.19** Let $\sum\limits_{k=0}^{\infty} a_k x^k$ be a power series with radius of convergence $R > 0$.
Then the series converges absolutely for all for all $x \in \mathbb{R}$ with $|x| < R$.

We can reduce this theorem to the comparison test in Theorem 13.20: Let x be given
with $|x| < R$. Since R is the supremum of the elements for which the series converges,
there must be a x_0 with $|x| < x_0 < R$, so that $\sum_{k=0}^{\infty} a_k x_0^k$ converges. Then, according to
Theorem 13.15, the sequence of numbers $a_k x_0^k$ is a null sequence and therefore also
bounded.
For example, let $|a_k x_0^k| < M$. It follows:

$$|a_k x^k| = |a_k x_0^k| \frac{|x^k|}{|x_0^k|} \leq M \cdot \left| \frac{x}{x_0} \right|^k = M \cdot q^k, \quad q < 1.$$

The series $\sum_{k=0}^{\infty} M q^k = M \sum_{k=0}^{\infty} q^k$ converges as a geometric series, so its elements are
always larger then the elements of the investigated series, which is therefore absolutely
convergent.   □

By the way, one can also prove that for all x with $|x| > R$ divergence occurs. For the
boundary points of the interval $[-R, R]$ no statement is possible: In the points $x = \pm R$
both convergence and divergence can occur, as you could see in Example 3 before. For
the range of convergence of a power series with radius of convergence $R > 0$, all of the
following sets are possible: $[-R, R],]-R, R[, [-R, R[,]-R, R]$.
Examples 1, 4 and 5 have an infinite radius of convergence, examples 2 and 3 have a
radius of convergence of 1. Sometimes the radius of convergence of a power series can
be easily calculated using the ratio test:

▶   **Theorem 15.20** Let $\sum_{k=0}^{\infty} a_k x^k$ be a power series and let at least from a certain
index on all a_k be different from 0. If the following limit exists in $\mathbb{R} \cup \{\infty\}$, then
the radius of convergence is:

$$R = \lim_{k \to \infty} \left| \frac{a_k}{a_{k+1}} \right|.$$

Let's calculate the case $R < \infty$: Then

$$\lim_{k \to \infty} \left| \frac{a_{k+1} x^{k+1}}{a_k x^k} \right| = \lim_{k \to \infty} \left| \frac{a_{k+1}}{a_k} \right| |x| = \frac{1}{R} |x|.$$

By the ratio test, convergence is at least then when $|x|/R < 1$, thus when $|x| < R$ is, and
divergence for $|x| > R$. Please carry out the case $R = \infty$ yourself!   □

You can try out this rule with examples 1, 2 and 3, but the procedure is not directly applicable to trigonometric series. Can you see why?

If R is the radius of convergence of a power series, then

$$f: \;]-R, R[\;\rightarrow\; \mathbb{R}, \quad x \mapsto \sum_{k=0}^{\infty} a_k x^k$$

represents a real function. This has surprisingly good properties:

▶   **Theorem 15.21** If $\sum_{k=0}^{\infty} a_k x^k$ is a power series with radius of convergence $R > 0$,

then the function $f(x) = \sum_{k=0}^{\infty} a_k x^k$ is continuous and differentiable in the interval

$]-R, R[$. The power series $\sum_{k=1}^{\infty} k a_k x^{k-1}$ also has the radius of convergence R and in

the open interval $]-R, R[$ represents the derivative of the function f.

This means that you can differentiate a power series term by term, just like a polynomial.

The proof of this theorem is difficult, it uses properties of series and functions that I have not introduced to you. I can therefore only quote the theorem. Fortunately, it is easy to apply. I will show you two

---

Examples

1. The first result we already know, it would be bad if something else came out now:

$$(e^x)' = \sum_{k=0}^{\infty} \frac{x^k}{k!} = \sum_{k=1}^{\infty} \frac{k x^{k-1}}{k!} = \sum_{k=1}^{\infty} \frac{x^{k-1}}{(k-1)!} = \sum_{k=0}^{\infty} \frac{x^k}{k!} = e^x.$$

2. The second example is tricky, but here we get something really new:
    Let $|x| < 1$. Then

$$\left(\sum_{k=1}^{\infty} \frac{1}{k} x^k \right)' = \sum_{k=1}^{\infty} \frac{k}{k} x^{k-1} = \sum_{k=1}^{\infty} x^{k-1} = \sum_{k=0}^{\infty} x^k = \frac{1}{1-x}.$$

On the other hand, we can calculate $(\ln(1-x))' = -\frac{1}{1-x}$ using the chain rule. So we see that $\sum_{k=1}^{\infty} \frac{1}{k} x^k$ and $-\ln(1-x)$ have the same derivative. This means that the functions only differ by a constant c: $\sum_{k=1}^{\infty} \frac{1}{k} x^k = -\ln(1-x) + c$. If we insert 0 on the left and right for x, we get $0 = c$ and now we have a series representation for the logarithm:

$$\ln(1-x) = \sum_{k=1}^{\infty} \frac{-1}{k} x^k \quad \text{for } |x| < 1. \; \blacktriangleleft$$

We have seen earlier that power series are important in order to calculate function values concretely. Unfortunately, the series found now for the logarithm is very badly convergent and unusable for efficient calculations. But we will improve the representation soon, see Example 2 after Theorem 15.24.

15.3 Taylor Series

The goal of this section is to approximate as many functions as possible by power series, as we have just succeeded for the logarithm. We only examine the functions at one point x_0, so we only want to find a local approximation.

Remember that the tangent of a function at the point x_0 gave an approximation to the function (compare Theorem 15.2):

$$f(x) = f(x_0) + f'(x_0)(x - x_0) + F_1.$$

The error F_1 goes to 0 if x goes to x_0, even $F_1/(x - x_0)$ goes to 0. There is hope that with polynomials of higher degree we can achieve an even better approximation near x_0 (Fig. 15.13):

$$f(x) = f(x_0) + f'(x_0)(x - x_0) + A_2(x - x_0)^2 + F_2,$$
$$f(x) = f(x_0) + f'(x_0)(x - x_0) + A_2(x - x_0)^2 + A_3(x - x_0)^3 + F_3.$$

We wish that F_2 goes to 0 even faster than F_1, F_3 faster than F_2 and so on. The question is: Can we determine such A_i and F_i?

Let's first assume that $x_0 = 0$, that f is infinitely often differentiable in a neighborhood of 0, and that there is a power series that represents f in a neighborhood of 0:

$$f(x) = a_0 + a_1 x + a_2 x^2 + \cdots = \sum_{k=0}^{\infty} a_k x^k.$$

Fig. 15.13 Approximations

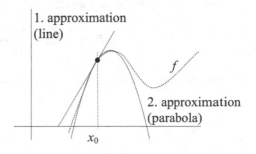

1. approximation (line)

f

2. approximation (parabola)

x_0

In this case, we have already given our desired approximation. As a power series, f can be differentiated an arbitrary number of times in this neighborhood, and we can determine the derivatives of f by differentiating the summands:

$$f'(x) = a_1 + 2a_2 x + 3a_3 x^2 + 4a_4 x^3 + \cdots \qquad \Rightarrow \qquad f'(0) = a_1,$$
$$f''(x) = 2 \cdot a_2 + 2 \cdot 3 \cdot a_3 x + 3 \cdot 4 \cdot a_4 x^2 + \cdots \qquad \Rightarrow \qquad f''(0) = 2a_2,$$
$$f'''(x) = 2 \cdot 3 \cdot a_3 + 2 \cdot 3 \cdot 4 \cdot a_4 x + \cdots \qquad \Rightarrow \qquad f'''(0) = 2 \cdot 3 \cdot a_3,$$
$$f^{(4)}(x) = 2 \cdot 3 \cdot 4 \cdot a_4 + \cdots \qquad \Rightarrow \qquad f^{(4)}(0) = 2 \cdot 3 \cdot 4 \cdot a_4.$$

If we continue this calculation, we get for the coefficients a_k:

$$a_k = \frac{f^{(k)}(0)}{k!}.$$

Under certain conditions, a function f can therefore be "expanded" into a power series:

$$f(x) = \sum_{k=0}^{\infty} \frac{f^{(k)}(0)}{k!} x^k = f(0) + f'(0)x + \frac{f''(0)}{2!} x^2 + \cdots .$$

We have to solve the following problems in this context:

- What are the conditions exactly?
- If the expansion is not exact, can the function at least be approximated?
- What can we say about functions that are not infinitely differentiable?

▶   **Theorem 15.22** Let the function f be $(n+1)$ times differentiable in the interval I, $x, x_0 \in I$. Then:

$$f(x) = f(x_0) + \frac{f'(x_0)}{1!}(x - x_0) + \frac{f''(x_0)}{2!}(x - x_0)^2 + \cdots$$
$$+ \frac{f^{(n)}(x_0)}{n!}(x - x_0)^n + R_n(x - x_0).$$

Here, for a ϑ between x and x_0:

$$R_n(x - x_0) = \frac{f^{(n+1)}(\vartheta)}{(n+1)!}(x - x_0)^{n+1}. \qquad (15.3)$$

R_n depends on $x - x_0$, so we read "R_n of $(x - x_0)$", not "R_n times $(x - x_0)$".

▶ **Definition 15.23** In Theorem 15.22,

$$j_{x_0}^n(f)(x - x_0) := f(x_0) + \frac{f'(x_0)}{1!}(x - x_0) + \frac{f''(x_0)}{2!}(x - x_0)^2 + \cdots$$
$$+ \frac{f^{(n)}(x_0)}{n!}(x - x_0)^n$$

is called the *nth degree Taylor polynomial* or *nth Taylor polynomial* or *n-jet* of f in x_0. The term $R_n(x - x_0)$ is called the *nth remainder* of the Taylor polynomial.

The Taylor polynomial and the remainder are dependent on the *point of expansion* x_0: If you shift this point, you get a different polynomial. The remainder $R_n(x - x_0)$ just indicates the error that arises when you replace $f(x)$ by $j_{x_0}^n(f)(x - x_0)$, because it is $f(x) = j_{x_0}^n(f)(x - x_0) + R_n(x - x_0)$.

We obtain an equivalent, alternative notation in which a n-jet looks a little more like a polynomial, if we set, as we have sometimes done before, $x = x_0 + h$, that is $h = x - x_0$. Then

$$f(x_0 + h) = f(x_0) + \frac{f'(x_0)}{1!}h + \frac{f''(x_0)}{2!}h^2 + \cdots + \frac{f^{(n)}(x_0)}{n!}h^n + R_n(h)$$
$$= j_{x_0}^n(f)(h) + R_n(h).$$

For small values of h, $f(x_0 + h)$ is therefore approximated by $j_{x_0}^n(h)$.

But now to the proof of Theorem 15.22: I only carry it out for $n = 1$. But the complete idea is contained in it, the general case only requires a little more writing.

We therefore calculate $R_1 = R_1(x - x_0)$. First let $r = \frac{R_1}{(x-x_0)^2}$. Then:

$$f(x) = f(x_0) + f'(x_0)(x - x_0) + r \cdot (x - x_0)^2. \tag{15.4}$$

From this we construct the following function h, which is defined and differentiable between x and x_0:

$$h(y) := f(x) - f(y) - f'(y)(x - y) - r \cdot (x - y)^2.$$

(15.4) implies $h(x_0) = 0$. Calculate $h(x)$: Everything dissolves into nothing, also $h(x) = 0$. Why the whole thing? According to the mean value theorem 15.12 there is now an element ϑ between x and x_0, with $h'(\vartheta) = 0$. We calculate the derivative of h at the point ϑ:

$$0 = h'(\vartheta)$$
$$= -f'(\vartheta) - f''(\vartheta)(x - \vartheta) - f'(\vartheta)(-1) - r \cdot 2(x - \vartheta)(-1)$$
$$= (-f''(\vartheta) + r \cdot 2)(x - \vartheta)$$

and thus $r = \frac{f''(\vartheta)}{2!}$, so that $R_1 = r \cdot (x - x_0)^2 = \frac{f''(\vartheta)}{2!}(x - x_0)^2$ results. □

It is particularly nice that the remainder converges very well to 0, provided that the $(n+1)$th derivative is also continuous:

▷ **Theorem 15.24** Let the function f be $(n+1)$ times continuously differentiable in the interval I, $x, x_0 \in I$. Then in the Taylor expansion of f for the nth remainder holds:

$$R_n(x - x_0) = \varphi_n(x - x_0) \cdot (x - x_0)^n \quad \text{mit} \ \lim_{x \to x_0} \varphi_n(x - x_0) = 0.$$

As a proof: According to (15.3)

$$|\varphi_n(x - x_0)| = \left| \frac{f^{(n+1)}(\vartheta)}{(n+1)!} (x - x_0) \right|.$$

The $(n+1)$th derivative $f^{(n+1)}$ is continuous and therefore takes a minimum and maximum between x and x_0, the absolute value is bounded from above by a number C and obtain

$$|\varphi_n(x - x_0)| \le \left| \frac{C}{(n+1)!} (x - x_0) \right|,$$

and thus $\lim_{x \to x_0} \varphi_n(x - x_0) = 0$.   □

If x goes to x_0, then the remainder R_n converges to 0 more strongly than the nth power of $(x - x_0)$. This is also called say "f and its Taylor polynomial agree at the point x_0 in nth".

For infinitely often differentiable functions, we can not only set up a Taylor polynomial, but also write down a whole series formally:

▷ **Definition 15.25** If f is infinitely often differentiable in the interval I and $x, x_0 \in I$, then the series

$$j_{x_0}(f)(x - x_0) := \sum_{k=0}^{\infty} \frac{f^{(k)}(x_0)}{k!} (x - x_0)^k$$

is called the *Taylor series* of f at the point x_0.

Replace $x - x_0$ again by h, then you will see that the Taylor series is nothing but a power series.

▷ **Theorem 15.26** The Taylor series of the function f converges exactly at the point $(x - x_0)$ to $f(x)$ if the sequence of remainders satisfies: $\lim_{n \to \infty} R_n(x - x_0) = 0$. In this case

$$f(x) = \sum_{k=0}^{\infty} \frac{f^{(k)}(x_0)}{k!} (x - x_0)^k.$$

This result is not very profound according to our current knowledge. Of course, one must investigate how the remainder behaves for ever larger n in order to assess the convergence of the series. Unfortunately, now all conceivable cases can occur here, for example:

- The Taylor series of f diverges everywhere except for $x = x_0$.
- The Taylor series converges, but not to the function f.
- The Taylor series converges to f.

Note the subtle but essential difference between the two expressions

$$\lim_{x \to x_0} R_n(x - x_0) \quad \text{and} \quad \lim_{n \to \infty} R_n(x - x_0). \tag{15.5}$$

The first expression in (15.5) always converges to 0 according to Theorem 15.24. However, if one keeps the number $x \neq x_0$ fixed, then the sequence R_n of remainders does not have to go to 0 if n becomes larger. In such cases, the Taylor series of the function f does not represent the function f.

But finally to applications of the theory.

Examples

1. The Taylor polynomial for the logarithm near 1 is

$$\ln(1 + h) = \sum_{k=0}^{n} \frac{\ln^{(k)}(1)}{k!} h^k + R_n(h).$$

The derivatives are:

$$\ln'(x) = x^{-1},$$
$$\ln''(x) = -x^{-2},$$
$$\ln^{(3)}(x) = 2x^{-3},$$
$$\ln^{(4)}(x) = -2 \cdot 3 \cdot x^{-4},$$
$$\vdots$$
$$\ln^{(k)}(x) = (-1)^{k+1}(k - 1)! x^{-k},$$

so that for $k > 0$: $\ln^{(k)}(1) = (-1)^{k+1}(k - 1)!$. Because of $\ln(1) = 0$ we finally get:

$$\ln(1 + h) = \sum_{k=1}^{n} \frac{(-1)^{k+1}}{k} h^k + R_n(h)$$

$$= h - \frac{1}{2}h^2 + \frac{1}{3}h^3 - \frac{1}{4}h^4 + \cdots \pm \frac{1}{n}h^n + R_n(h).$$

For the remainder, according to Theorem 15.22 holds:

$$|R_n(h)| = \left| \frac{\ln^{(n+1)}(\vartheta)}{(n+1)!} h^{n+1} \right| = \left| \frac{(-1)^n n! \vartheta^{-(n+1)}}{(n+1)!} h^{n+1} \right| = \frac{1}{n+1} \cdot \left| \frac{1}{\vartheta^{n+1}} \right| \cdot |h^{n+1}|$$

for an element ϑ between 1 and $1+h$. At least for $0 < h < 1$ one can see that $1/\vartheta^{n+1}$ and h^{n+1} are always less than 1 and therefore the sequence $R_n(h)$ converges to 0, although very slowly. With a little more effort, one can show that convergence also exists for $-1 < h \le 0$. So the Taylor series converges to the logarithm and we have found a new power series representation: For $|h| < 1$ we have

$$\ln(1+h) = \sum_{k=1}^{\infty} \frac{(-1)^{k+1}}{k} h^k.$$

In Example 2 after Theorem 15.21 we had found a very similar power series: For $|h| < 1$ it held:

$$\ln(1-h) = \sum_{k=1}^{\infty} \frac{-1}{k} h^k.$$

If you replace h by $-h$, you will see that it is actually the same series expansion we obtained there in a completely different way.

To calculate the logarithm practically, we conjure up from these two poorly convergent series a better convergent one, with the help of which we can also determine the logarithm for all $x > 0$: For all h with $|h| < 1$ is

$$\ln\left(\frac{1+h}{1-h}\right) = \ln(1+h) - \ln(1-h) = \sum_{k=1}^{\infty} \frac{(-1)^{k+1}}{k} h^k - \sum_{k=1}^{\infty} \frac{-1}{k} h^k$$

$$= 2\left(h + \frac{h^3}{3} + \frac{h^5}{5} + \cdots\right) = 2 \sum_{k=0}^{\infty} \frac{h^{2k+1}}{2k+1}.$$

However, for $|h|$ near 1 the convergence rate decreases again. Why can we determine the logarithm for all $x > 0$ with this series? This follows from the fact that the mapping

$$]-1, 1[\to \mathbb{R}^+, \quad h \mapsto \frac{1+h}{1-h} = x$$

is bijective. You can calculate this yourself.

2. Let us consider the function $f(x) = x^\alpha$ for $x > 0$ and $\alpha \in \mathbb{R}$. The derivatives are

$$f'(x) = \alpha x^{\alpha-1},$$
$$f''(x) = \alpha(\alpha - 1)x^{\alpha-2},$$
$$\vdots$$
$$f^{(k)}(x) = \alpha(\alpha - 1) \cdots (\alpha - k + 1)x^{\alpha-k}.$$

Now we determine the Taylor polynomial at the point of expansion $x_0 = 1$. We obtain:

$$(1 + h)^\alpha = f(1)h^0 + \sum_{k=1}^{n} \frac{\alpha(\alpha - 1) \cdots (\alpha - k + 1)}{k!} h^k + R_n(h).$$

For the fraction behind the summation sign we introduce the abbreviation

$$\binom{\alpha}{k} := \frac{\alpha(\alpha - 1) \cdots (\alpha - k + 1)}{k!}.$$

We call this expression the generalized binomial coefficient. If we set $\binom{\alpha}{0} := 1$, then

$$(1 + h)^\alpha = \sum_{k=0}^{n} \binom{\alpha}{k} h^k + R_n(h).$$

It can be shown with some effort that for $|h| < 1$ the sequence of the remainders converges to 0, for these h is therefore

$$(1 + h)^\alpha = \sum_{k=0}^{\infty} \binom{\alpha}{k} h^k. \tag{15.6}$$

Does this sound familiar to you? For $\alpha \in \mathbb{N}$ you get exactly the binomial theorem 4.8, which we derived in Sec. 4.1: For $k > n$ the binomial coefficient then always becomes 0 and (15.6) becomes a finite sum that ends with the $k = n$. The series from (15.6) is therefore called the *binomial series*. It can be used, for example, to approximate square roots:

$$\sqrt{1 + h} = \sum_{k=0}^{2} \binom{\frac{1}{2}}{k} h^k + R_2(h) = 1 + \frac{h}{2} - \frac{h^2}{8} + \binom{\frac{1}{2}}{3} \vartheta^{1/2-3} h^3.$$

for a ϑ between 1 and $1 + h$. Let's say $|h| \le 1/2$. Then the error is maximal when $\vartheta = 1/2$ (as small as possible) and $|h| = 1/2$ (as large as possible). $\binom{\frac{1}{2}}{3}$

has the value $1/16$, and thus $|R_2(h)| < (1/16) \cdot 5.66 \cdot (1/8) \approx 0.044$. The sum $1 + h/2 - h^2/8$ therefore represents a quite reasonable approximation to $\sqrt{1+h}$ in this area.

3. You might know the following theorem from physics: The position of a particle that is uniformly accelerated is fixed for all times, if at a certain time t_0 the position, velocity and acceleration of the particle are known. Why is that so? Let's denote the position at time t with $p(t)$, then the velocity is $v(t) = p'(t)$ and the acceleration is $a(t) = v'(t) = p''(t)$. "Uniformly accelerated" means that all higher derivatives vanish: The acceleration does not change. If we now set up the Taylor polynomial for the function s, we get:

$$p(t_0 + h) = p(t_0) + v(t_0)h + \frac{1}{2}a(t_0)h^2 + 0.$$

The Taylor polynomial ends after the third term and represents the function p for all $h \in \mathbb{R}$. ◀

If you look at this example closely, you will notice that p, v and a are not real-valued functions at all; after all, the position of a particle consists of three spatial components, and v and a have certain directions: The functions have the $\mathbb{R}^3$ as their codomain. But that doesn't matter, the x-,y- and z-components of the functions are real-valued functions, to which we can apply all our theorems.

15.4   Differential Calculus of Functions of Several Variables

We will now investigate real-valued functions of several variables, i.e. functions $f: U \to \mathbb{R}$, where U is a subset of the $\mathbb{R}^n$. The theory of functions of several variables is a vast and important discipline of mathematics, which I can only touch upon here. I would like to introduce some basic results of the differential calculus of such functions below, most of them without proof.

We can use our knowledge of the differential calculus of functions of one variable when we investigate the partial functions: At a point $(a_1, a_2, \ldots, a_n)$ inside the domain, there are n partial functions that go through this point. If we differentiate these partial functions, we get the slope of the function at this point in the different coordinate directions. In order to ensure that one can approach a point a from the domain U of the function f from all coordinate directions, we will, for simplicity, always take the set U as an open subset of the $\mathbb{R}^n$ from now on.

▶   **Definition 15.27** Let $U \subset \mathbb{R}^n$ be an open set, $f: U \to \mathbb{R}$ a function and $a = (a_1, \ldots, a_n) \in U$. If the derivative of the partial function

$$f_i: x_i \mapsto f(a_1, \ldots, a_{i-1}, x_i, a_{i+1}, \ldots, a_n)$$

exists at the point $x_i = a_i$, then $f'_i(x_i)$ is called the *partial derivative* of f with respect to x_i at the point a and is denoted by

$$\frac{\partial f}{\partial x_i}(a) \quad \text{or} \quad \frac{\partial}{\partial x_i} f(a).$$

f is called *partially differentiable* if the partial derivatives of f exist for all $a \in U$, and *continuously partially differentiable* if these partial derivatives are all continuous.

If the function f is partially differentiable with respect to the variable x_i in each point $x \in U$, then the derivative $\frac{\partial f}{\partial x_i} : x \mapsto \frac{\partial f}{\partial x_i}(x)$ is itself a function of U to $\mathbb{R}$, and can therefore be partially differentiated again, if necessary, even with respect to other variables than x_i. The second partial derivatives of a function f of n variables are denoted by

$$\frac{\partial^2 f}{\partial x_i \partial x_j}(x) := \frac{\partial}{\partial x_i}\left(\frac{\partial f}{\partial x_j}\right)(x), \quad \frac{\partial^2 f}{\partial x_i^2}(x) := \frac{\partial}{\partial x_i}\left(\frac{\partial f}{\partial x_i}\right)(x),$$

and the higher derivatives by

$$\frac{\partial^k}{\partial x_{i_1} \partial x_{i_2} \ldots \partial x_{i_k}} f(x) := \frac{\partial}{\partial x_{i_1}} \frac{\partial}{\partial x_{i_2}} \cdots \frac{\partial}{\partial x_{i_k}} f(x).$$

The order of the indices is not quite uniform in the literature: With me, $\frac{\partial^2 f}{\partial x_i \partial x_j}$ means that first j is derived, then i. Sometimes it is interpreted the other way around. In a moment you will see that you usually don't have to worry about it.

Examples

1. Let $f(x, y) = x^2 + y^2$. Then $\frac{\partial f}{\partial x}(x, y) = 2x$, $\frac{\partial f}{\partial y}(x, y) = 2y$. The second partial derivatives are $\frac{\partial^2 f}{\partial x^2} = 2 = \frac{\partial^2 f}{\partial y^2}$, $\frac{\partial^2 f}{\partial x \partial y} = 0 = \frac{\partial^2 f}{\partial y \partial x}$.

2. Let $f(x, y) = x^2 y^3 + y \ln x$. Then

$$\frac{\partial f}{\partial x} = 2xy^3 + \frac{y}{x}, \qquad \frac{\partial f}{\partial y} = 3x^2 y^2 + \ln x,$$

$$\frac{\partial^2 f}{\partial x^2} = 2y^3 - \frac{y}{x^2}, \quad \frac{\partial^2 f}{\partial x \partial y} = 6xy^2 + \frac{1}{x} = \frac{\partial^2 f}{\partial y \partial x}, \quad \frac{\partial^2 f}{\partial y^2} = 6x^2 y. \qquad \blacktriangleleft$$

The mixed second partial derivatives agree in each case. If you calculate a few examples, you will find that this is no coincidence. There is the amazing statement:

▶ **Theorem 15.28** For every twice continuously differentiable function $f \colon U \to \mathbb{R}$
   it holds for all $i, j = 1, \ldots, n$:

$$\frac{\partial^2 f}{\partial x_i \partial x_j} = \frac{\partial^2 f}{\partial x_j \partial x_i}.$$

The proof for this consists of a clever multiple application of the mean value theorem of
differential calculus.

In contrast to functions of one variable, the continuity of the function itself does not
follow from the partial differentiability of a function. For this, the partial derivatives
must also be continuous:

▶ **Theorem 15.29** If $f \colon U \to \mathbb{R}$ is continuously partially differentiable, then f is
   also continuous.

▶ **Definition 15.30** If f is in a partially differentiable, then the vector

$$\operatorname{grad} f(a) = \left(\frac{\partial f}{\partial x_1}(a), \frac{\partial f}{\partial x_2}(a), \ldots, \frac{\partial f}{\partial x_n}(a) \right)$$

is called the *gradient* of f at the point a.

If f is in an open set U partially differentiable, then grad is a function from U to $\mathbb{R}^n$.
   Here is an example: If $f(x, y, z) = e^{x+2y} + 2x \sin z + z^2 xy$, then

$$\operatorname{grad} f(x, y, z) = (e^{x+2y} + 2 \sin z + z^2 y, 2e^{x+2y} + z^2 x, 2x \cos z + 2zxy).$$

Extrema

With the help of partial derivatives, local extrema can be determined. First the definition,
which can be formulated analogously to the one-dimensional case:

▶ **Definition 15.31** Let $U \subset \mathbb{R}^n$ be open and $f \colon U \to \mathbb{R}$ a function. $a \in U$ is called
   a *local maximum* (or *minimum*) of f, if there is an neighborhood $U_\varepsilon(a)$, so that
   for all $x \in U_\varepsilon(a) \setminus \{a\}$ it holds $f(x) < f(a)$ (or $f(x) > f(a)$). A local minimum or
   maximum is called a *local extremum*.

▶ **Theorem 15.32** Let $U \subset \mathbb{R}^n$ be open and $f \colon U \to \mathbb{R}$ partially differentiable.
   Then for every local extremum $a \in U$ holds $\operatorname{grad} f(a) = 0$.

For the partial functions $f_i(a_1, a_2, \ldots, x_i, \ldots, a_n)$ obviously have a local extremum at the
point $x_i = a_i$ and therefore $\frac{\partial f}{\partial x_i}(a_i) = 0$ must be true for all i.   □

Fig. 15.14 A saddle point

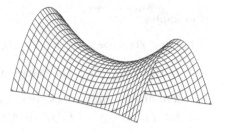

A point $a \in \mathbb{R}^n$ is called a *stationary point*, if $\operatorname{grad} f(a) = 0$. Such a point does not necessarily have to be a maximum or minimum, as you can see in Fig. 15.14, which shows the function $x^2 - y^2$. In this function, the partial functions through (0,0) have a maximum in x-direction and a minimum in y-direction. Therefore, both partial derivatives are equal to 0. Such a point is called a *saddle point*.

In order to decide whether a minimum or a maximum is present, the second derivatives have to be consulted again. But now there are more than one of them:

▶   **Definition 15.33** If $f: U \to \mathbb{R}$ is twice continuously partially differentiatiable in $a \in U$, then the second partial derivatives form a matrix

$$H_f(a) := \begin{pmatrix} \frac{\partial^2 f}{\partial x_1 \partial x_1}(a) & \cdots & \frac{\partial^2 f}{\partial x_1 \partial x_n}(a) \\ \vdots & \ddots & \vdots \\ \frac{\partial^2 f}{\partial x_n \partial x_1}(a) & \cdots & \frac{\partial^2 f}{\partial x_n \partial x_n}(a) \end{pmatrix}.$$

This matrix is called the *Hessian* matrix of f at the point a.

In order to analyze extrema, the Hessian matrix has to be examined. In general, this is difficult, but for the case of two variables one can still write down a simple criterion:

▶   **Theorem 15.34** Let $a = (a_1, a_2)$ be a stationary point of the function $f: U \to \mathbb{R}$, $U \subset \mathbb{R}^2$, that is $\operatorname{grad} f(a) = 0$. Then holds:

a) If $\det(H_f(a)) > 0$ and $\frac{\partial^2 f}{\partial x_1{}^2}(a) < 0$, then a is a maximum.

b) If $\det(H_f(a)) > 0$ and $\frac{\partial^2 f}{\partial x_1{}^2}(a) > 0$, then a is a minimum.

c) If $\det(H_f(a)) < 0$, then a is not an extremum.

d) If $\det(H_f(a)) = 0$, then no decision is possible.

If the function f has more than two variables, one must also investigate subdeterminants of the Hessian matrix.

Examples

1. Let $f(x, y) = x^2 + y^2$ be the paraboloid. It is $\operatorname{grad} f(0,0) = (0,0)$ and $H_f(0,0) = \begin{pmatrix} 2 & 0 \\ 0 & 2 \end{pmatrix}$. The determinant is greater than 0 and $\dfrac{\partial^2 f}{\partial x^2} = 2 > 0$, so $(0,0)$ is a minimum, which does not surprise us very much.

2. Let $f(x, y) = y^3 - 3x^2 y$. Again, $\operatorname{grad} f(0,0) = (0,0)$, and the calculation of the second derivatives yields $H_f(x, y) = \begin{pmatrix} -6y & -6x \\ -6x & 6y \end{pmatrix}$, so $\det(H_f(0,0)) = 0$, we cannot make a decision about the type of stationary point in this case. ◀

The Regression Line

As an application of the determination of extrema, I would like to calculate the *regression line* through a number of points. Often, functional technical, physical or economic relationships between different quantities are empirically obtained, for example through experiments or through samples. In many cases, one tries to approximate the measured values by a line. A simple example: In a spring, the extension is proportional to the force acting. If you hang different weights on the spring and measure the extension each time, the measurement points must be connectable by a line, the slope of this line results exactly in the proportionality factor, the spring constant. Due to measurement errors, the points do not lie exactly on a line, the line that can be best placed through the points is sought.

The criterion for this is: The sum of the squared deviations of the points from the line should be minimal. The squares are chosen to obtain positive values for all deviations. Strictly speaking, only the deviation of the y-component of the points (x_i, y_i) from the line is minimized. The line is found by calculating the extremum of a function of two variables:

Given the n data points (x_i, y_i), the line has the form $y = ax + b$, the deviation of the i-th data point from the line is $ax_i + b - y_i$, see Fig. 15.15. The line parameters a, b are unknown, they must be determined so that the function

Fig. 15.15 The regression line

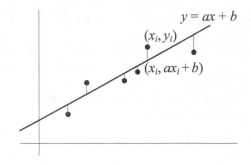

$$F(a, b) = \sum_{i=1}^{n} (ax_i + b - y_i)^2$$

is minimal. F is a function of two variables, so we can apply Theorem 15.34. Something unusual is that the x_i and y_i are constant and a and b are variable, but that should not bother us. First we look for stationary points. Necessary for this is $\frac{\partial F}{\partial a} = 0, \frac{\partial F}{\partial b} = 0$. It is

$$\frac{\partial F}{\partial a} = \sum_{i=1}^{n} 2(ax_i + b - y_i)x_i = 0,$$

$$\frac{\partial F}{\partial b} = \sum_{i=1}^{n} 2(ax_i + b - y_i)1 = 0.$$

Sorted by a and b and divided by 2 we get the equations

$$\left(\sum_{i=1}^{n} x_i^2\right)a + \left(\sum_{i=1}^{n} x_i\right)b = \sum_{i=1}^{n} y_i x_i,$$

$$\left(\sum_{i=1}^{n} x_i\right)a + nb = \sum_{i=1}^{n} y_i.$$

These are two linear equations for a and b, which usually have exactly one solution. The Hessian matrix is

$$\begin{pmatrix} \sum_{i=1}^{n} 2x_i^2 & \sum_{i=1}^{n} 2x_i \\ \sum_{i=1}^{n} 2x_i & 2n \end{pmatrix},$$

one can calculate that $\det(H_f(a)) > 0$ is always true. Furthermore, $\frac{\partial^2 F}{\partial a^2} = \sum_{i=2}^{n} 2x_i^2 > 0$, so there is actually a minimum. To calculate the regression line, the sums of the x_i, the y_i, the x_i^2 and the $x_i y_i$ must be calculated, with these coefficients a two-dimensional system of linear equations is solved.

15.5 Comprehension Questions and Exercises

Comprehension Questions

1. If $f: D \to \mathbb{R}$ and $x_0 \in D$, then the difference quotient is a mapping of $D \setminus \{x_0\} \to \mathbb{R}$. Why does x_0 have to be removed from the domain here?
2. Let $f: [a, b] \to \mathbb{R}$ be a continuous mapping. Is f then differentiable in the open interval $]a, b[$?
3. Let $f: [a, b] \to \mathbb{R}$ be continuous and differentiable and for a $x_0 \in]a, b[$ let $f'(x_0) = 0$. Does f then have a local extremum at x_0?
4. Which points in the domain of a continuous function are potential candidates for extrema?

5. What condition must a function f fulfill if one wants to set up a Taylor series for it?
6. Let $R_n(x - x_0)$ be the remainder when expanding a function f into a Taylor series. Which of the two following limits is always 0?

$$\lim_{n \to \infty} R_n(x - x_0) \quad \text{oder} \quad \lim_{x \to x_0} R_n(x - x_0).$$

Exercises

1. Calculate the derivatives of $\sqrt{2x - 1}$, $\cot x$, $\cos^n(x/n)$.
2. Determine the derivative of the following functions twice each:
    a) $(3x + 5x^2 - 1)^2$ by product rule and chain rule,
    b) $10/(x^3 - 2x + 5)$ by quotient rule and chain rule.
3. At which points is the function $x \mapsto x|\sin x|$ differentiable? Explain why the function is not differentiable everywhere.
4. Is the function $x \mapsto x \sin |x|$ differentiable at 0?
5. Let $f:]-1, +1[\to \mathbb{R}^+, x \mapsto (1 + x)/(1 - x)$.
    a) Show that f is bijective.
    b) Calculate the first and second derivative of the function f.
6. Show that the function $f: \mathbb{R} \setminus \{0\} \to \mathbb{R}, x \mapsto \ln |x|$ is differentiable throughout its domain and calculate the derivative. (Use that for $x > 0$ holds $\ln(x)' = 1/x$.)
7. Calculate the derivative of $\log_a(x)$ using the statement that $\log_a(x)$ is the inverse function of a^x.
8. Confirm the result from Exercise 7 using the statement $\log_a(x) = \ln(x)/\ln(a)$.
9. Carry out a curve sketching with the two following functions and sketch the functions:
    a) $x^3 + 4x^2 - 3x$.
    b) $x^2 \cdot e^{-2x}$.
10. Sketch the graph of a non-differentiable function for which the mean value theorem of differential calculus does not apply.
11. Let $f: [0, 4[\to \mathbb{R}, x \mapsto \begin{cases} -x + 2 & \text{for } 0 \le x < 1 \\ -x^2 + 4x - 2 & \text{for } 1 \le x < 4 \end{cases}$
    Examine in which points the function is continuous and differentiable, and calculate all maxima and minima of the function.
12. Calculate the radii of convergence of the following power series:
    a) $\displaystyle\sum_{n=0}^{\infty} \frac{x^n}{2^n}$
    b) $\displaystyle\sum_{n=0}^{\infty} \frac{n^2 x^n}{2n + 1}$
    c) $\displaystyle\sum_{n=0}^{\infty} \frac{x^n}{(2n + 1)!}$

13. Determine for the function $\sqrt{x}$ the Taylor polynomial of 3rd degree at the point $x = 1$. Use this to calculate for $h = 0, 0.5, 1$ an approximation for $\sqrt{x + h}$. Determine the resulting error thus arising in comparison to the value from the calculator.

14. Write a program to determine zeros with bisection. Take the derivative of the function $f : x \mapsto (1/x)\sin(x) \, (x \neq 0)$ and determine some zeros of the derivative numerically. Then sketch the graph of the function f.

15. Determine all partial derivatives of first order of the functions

   a) $3x_1^4 x_2^3 x_3^2 x_4^1 + 5x_1 x_2 + 8x_3 x_4 + 1$.

   b) $xe^{yz} + \dfrac{\sqrt{xz}}{\ln y}, x, y, z > 0$.

16. Determine all stationary points of the following function:

$$x^3 - 3x^2 y + 3xy^2 + y^3 - 3x - 21y.$$

17. A rectangular box (length x, width y, height z), which is open at the top, should have a capacity of 32 liters. Determine x, y and z so that the material consumption for the box is minimal.

Integral Calculus

16

Abstract

At the end of this chapter you will have

- learned the integral as the area under a piecewise continuous function,
- understood the connection between differential and integral calculus,
- derived calculation rules for determining integrals and used these rules to calculate many integrals,
- calculated volumes of shapes and arc lengths using integration,
- calculated integrals with integration limits at the boundary of the domain (improper integrals),
- learned how to represent piecewise continuous functions as Fourier series,
- and you know the discrete Fourier transform.

Determining the areas of shapes that are not bounded by line segments is an old mathematical problem, think for example of determining the area of a circle. With the help of integral calculus we can tackle this task. The integral of a real function determines the area enclosed between the graph of the function and the x-axis. We obtain this area with the help of a limit: it is approximated better and better by a sequence of rectangles.

Integration will prove to be useful for many other tasks as well: we can for example also calculate arc lengths of curves or the volume of a shape. An important and surpris-

Supplementary Information The online version contains supplementary material available at https://doi.org/10.1007/978-3-658-40423-9_16.

387

ing result of the theory establishes a connection between differential calculus and integral calculus: differentiation and integration are inverse operations to each other. This opens up new areas of application for integral calculus.

16.1   The Integral of Piecewise Continuous Functions

Fig. 16.1 shows some areas that we want to calculate with the help of the integral. You will see that we not only determine areas that are bounded all around by the graph and the x-axis, like the left area. For a function that is defined between a and b, we bound the areas on the left and right with the help of the vertical lines through a and b. We can also determine areas under the graph in this way for functions with discontinuities.

Let us restrict the set of functions that we want to integrate more precisely:

▶   **Definition 16.1** A function $f \colon [a, b] \to \mathbb{R}$ is called *piecewise continuous* if f has at most finitely many discontinuities and if at each such discontinuity s the two limit values $\lim\limits_{x \to s, x < s} f(x)$ and $\lim\limits_{x \to s, x > s} f(x)$ exist.

Such a function can therefore be composed of finitely many continuous pieces. The second condition of the definition states that f does not vanish into infinity in the discontinuities, like for example $1/x$ in the point 0. The function only has jump discontinuities, the image of the function is bounded.

Now we want to approximate the area by rectangles. We take bars that become narrower and narrower and whose upper boundary just cuts the graph, see Fig. 16.2.

▶   **Definition 16.2** A partition Z of the interval $I = [a, b]$ is defined by $k + 1$ numbers $x_0, x_1, \ldots, x_k$ with $a = x_0 < x_1 < x_2 < \cdots < x_k = b$. Here

$$\Delta x_i := x_i - x_{i-1} \qquad \text{the length of the } i - \text{th subinterval,}$$
$$\Delta Z := \text{Max}\{\Delta x_i\} \qquad \text{the fineness of the partition.}$$

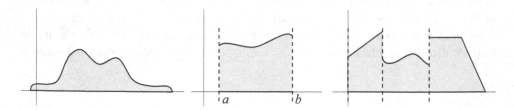

Fig. 16.1 Integrable functions

Fig. 16.2 Approximation of
the area

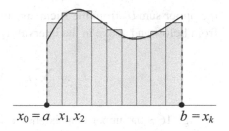

$$x_0 = a \quad x_1 \; x_2 \qquad\qquad\qquad b = x_k$$

Let $f: [a,b] \rightarrow \mathbb{R}$ be a piecewise continuous function. For $i = 1, \ldots, k$ let $y_i \in [x_{i-1}, x_i]$. Then

$$S := \sum_{i=1}^{k} f(y_i) \cdot \Delta x_i \tag{16.1}$$

is called a *Riemann sum* of f.

There are various concepts of integration that are applicable to different classes of functions. The definition of the integral in the form presented by me is called the *Riemann integral* after the German mathematician Bernhard Riemann (1826–1866). For piecewise continuous functions, different definitions of the integral yield the same results.

Note that according to this definition, areas below the x-axis are calculated negative.

▶   **Theorem and Definition 16.3** Let $f: [a,b] \rightarrow \mathbb{R}$ be piecewise continuous. $(Z_n)_{n \in \mathbb{N}}$ be a sequence of partitions of $[a,b]$ with the property $\lim_{n \to \infty} \Delta Z_n = 0$ and S_n a Riemann sum to the partition Z_n. Then the limit $\lim_{n \to \infty} S_n$ exists and is not dependent on the specific sequence of partitions.

This limit is called the *definite integral* of f over $[a, b]$ and is denoted by

$$\int_a^b f(x)dx.$$

The function f is called *integrable*, the numbers a, b are called *limits of integration*, the function f is called *integrand* and x is called *variable of integration*.

I only want to sketch the procedure for the proof of the theorem: f can be decomposed into continuous parts, we can therefore assume the function to be continuous. The sequence Z_n of decompositions arises by subdividing the interval $[a, b]$ over and over again. For the decomposition Z_n it follows from Theorem 14.26 that f in the interval $[x_{i-1}, x_i]$ has a minimum m_i and a maximum M_i. Now we can form the lower sum L_n and the upper sum U_n from the Riemann sum S_n. These are just the bars that bound the graph from below and above in the interval $[x_{i-1}, x_i]$:

$$L_n := \sum_{i=1}^{k} m_i \cdot \Delta x_i, \quad U_n := \sum_{i=1}^{k} M_i \cdot \Delta x_i.$$

In Fig. 16.3 the upper sum is shown in grey and the lower sum is dotted. The actual Riemann sum S_n meanders somewhere in between, for all n it holds:

$$L_n \le S_n \le U_n.$$

What happens if we switch to a finer decomposition, for example Z_{n+1}? Then the upper sum can only become smaller and the lower sum can only become larger. You can see this in Fig. 16.4 for the lower sum. The sequence of upper sums is therefore monotonically decreasing, that of the lower sums monotonically increasing.

The lower sums are bounded from above, for example by U_1, the upper sums are bounded from below, for example by L_1. According to Theorem 13.12 monotonic and bounded sequences are always convergent. There are therefore $L := \lim_{n \to \infty} L_n$ and $U := \lim_{n \to \infty} U_n$, and of course $L \le U$. In fact, lower sums and upper sums converge to the same limit. Although this appears intuitive, there is still a lot of work involved here, which requires more knowledge of continuous functions than I have presented. So let's just believe $L = U$. Since S_n is trapped between L_n and U_n, the sequence S_n has no choice but to also converge to the common limit.

Fig. 16.3 A Rieman sum

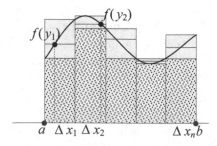

Fig. 16.4 The lower sums

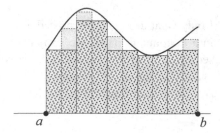

It remains to be shown that this limit is independent of the specific partitioning of the interval. This requires a little index shifting, but is elementary to check. The idea is to always find common refinements for different partitions. But I don't want to do that either. □

The integral sign represents a stylized S, finally the integral is the limit of a sum. Behind the integral sign is not a product " $f(x)$ times a small piece of dx ", but rather a notation that is intended to remind us of the origin from the sum (16.1). The definite integral itself is no longer a function, but a real number.

The naming of the integration variable is arbitrary, it is

$$\int_a^b f(x)dx = \int_a^b f(y)dy = \int_a^b f(t)dt = \int_a^b f(b)db.$$

Usually the integration variable is denoted with a different character than the limits of integration.

An computer scientist who is used to working with local and global variables should have no problems with this.

So far in our integrals $a < b$ must always be true. In order to avoid case distinctions and to include boundary cases, we define:

▶ **Definition 16.4** If $a < b$ and f is integrable between a and b, then

$$\int_b^a f(x)dx := -\int_a^b f(x)dx, \quad \int_a^a f(x)dx := 0.$$

These specifications are also compatible with the following theorem, which assembles simple calculation rules for integrable functions. They easily follow from the definition of the integral and the known arithmetic rules for limits. I omit the proofs:

▶ **Theorem 16.5** Let f, g be integrable functions a, b, c in the domain and $\alpha \in \mathbb{R}$. Then the following holds:

a) $\int_a^b f(x)dx = \int_a^c f(x)dx + \int_c^b f(x)dx.$

b) $\int_a^b (f(x) + g(x))dx = \int_a^b f(x)dx + \int_a^b g(x)dx.$

c) $\int_a^b \alpha f(x)dx = \alpha \int_a^b f(x)dx.$

d) If $f(x) \le g(x)$ for all $x \in [a, b]$, then $\int_a^b f(x)dx \le \int_a^b g(x)dx.$

Integration examples

1. $\int_a^b c\,dx = c(b - a)$. This is simply the area of the rectangle determined by the constant function c between a and b (Fig. 16.5).
2. $\int_0^{2\pi} \sin x\,dx = 0$. Since the areas above and below the x-axis cancel each other out, the integral is 0 (Fig. 16.6). What would be interesting, of course, is the calculation of $\int_0^{\pi} \sin x\,dx$, but we will have to move that a bit.
3. $\int_a^b x\,dx$: For this, we can use the formula for the area of a triangle "$\frac{1}{2}$· base line · height": We get $\int_a^b x\,dx = \frac{1}{2}b^2 - \frac{1}{2}a^2$ (Fig. 16.7).
4. If our entire mathematics is correct, then the area of a circle with radius 1 must be exactly π. A circle is not a function, but the semicircle over the x-axis (Fig. 16.8):

Fig. 16.5 Rectangle

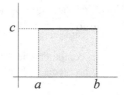

Fig. 16.6 Sine

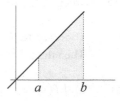

Fig. 16.7 Trapezoid

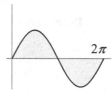

Fig. 16.8 Semicircle

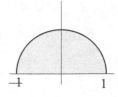

Since for the points of the circle $x^2 + y^2 = 1$, the function is $y = \sqrt{1 - x^2}$. Is really $\int_{-1}^{1} \sqrt{1 - x^2} dx = \pi/2$? To check this, we need to learn more about the calculation of integrals. ◄

▶ **Theorem 16.6: The mean value theorem for integrals** Let f be continuous on $[a, b]$. Then there is an element $y \in [a, b]$ with the property

$$\int_a^b f(x)dx = (b - a)f(y).$$

As a continuous function on a closed interval, f has a minimum $m = f(\alpha)$ and a maximum $M = f(\beta)$. We define $\mu := \dfrac{1}{b - a} \cdot \int_a^b f(x)dx$. Since for all x holds $m \le f(x) \le M$, it follows from Theorem 16.5d):

$$\int_a^b m dx \le \int_a^b f(x)dx \le \int_a^b M dx \quad \Rightarrow \quad m(b - a) \le \underbrace{\int_a^b f(x)dx}_{=\mu(b-a)} \le M(b - a)$$

$$\Rightarrow \quad m \le \mu \le M.$$

According to the Intermediate Value Theorem 14.25, there is now a y between α and β with the property $f(y) = \mu$, that is

$$\int_a^b f(x)dx = (b - a)f(y). \qquad \qquad \Box$$

The following is the rescue ring for calculating integrals: The fundamental theorem of calculus establishes the connection between differentiation and integration and allows us to use the knowledge we have collected in the last chapter to our advantage again.

▶ **Definition 16.7** Let I be an interval. A differentiable function $F: I \to \mathbb{R}$ is called the *antiderivative* or *primitive* of $f: I \to \mathbb{R}$ if for all $x \in I$ it holds: $F'(x) = f(x)$.

▶ **Theorem 16.8** If F, G are antiderivatives of f, then $F(x) - G(x) = c \in \mathbb{R}$.

This is because, according to Theorem 15.14, functions with the same derivative only differ by a constant.

▶ **Theorem 16.9: The fundamental theorem of calculus** Let I be an interval, $a \in I$ and $f: I \to \mathbb{R}$ a continuous function. Then the function $F_a: I \to \mathbb{R}$, which is defined by

$$F_a(x) := \int_a^x f(t)dt \tag{16.2}$$

is an antiderivative of f, that is:

$$F_a'(x) = \frac{d}{dx} \int_a^x f(t)dt = f(x).$$

Any other antiderivative of f has the form $F(x) = F_a(x) + c$ for a $c \in \mathbb{R}$.

Look closely at the integral (16.2): To make the number a function again, we have now made the upper limit of integration variable. The derivative of this integral function is just the function itself.

The proof is surprisingly not difficult anymore, we have already done the preliminary work. We simply calculate the derivative of F_a, that is the limit of the difference quotient. Let us first look at the numerator and apply the mean value Theorem 16.6:

$$F_a(x+h) - F_a(x) = \int_a^{x+h} f(t)dt - \int_a^x f(t)dt = \int_x^{x+h} f(t)dt = f(y_h) \cdot h.$$

For each h there is, according to the mean value theorem, an element y_h between x and $x + h$. Now we get for the derivative:

$$F_a'(x) = \lim_{h \to 0} \frac{F_a(x+h) - F_a(x)}{h} = \lim_{h \to 0} \frac{1}{h} \cdot f(y_h) \cdot h = \lim_{h \to 0} f(y_h),$$

And since $h \to 0$ also $y_h \to x$. Then $\lim_{h \to 0} f(y_h) = f(x)$, thus $F_a'(x) = f(x)$. The second part of the theorem follows from Theorem 16.8.   □

Now we can give a specific recipe for calculating integrals:

▶   **Theorem 16.10: Computation of integrals** If F is any antiderivative of the function f, then:

$$\int_a^b f(x)dx = F(b) - F(a) =: F(x)\Big|_a^b.$$

For there is a $c \in \mathbb{R}$ with $F(x) = F_a(x) + c$. Then $F(b) = F_a(b) + c$ and $F_a(a) = \int_a^a f(t)dt = 0$ implies $F(a) = F_a(a) + c = c$. This gives us:

$$\int_a^b f(t)dt = F_a(b) = F(b) - c = F(b) - F(a).$$ □

To compute a specific integral, proceed as follows:

• Find an antiderivative F of f. As a test, we can always take the derivative.
• Form $F(b) - F(a)$.

In the last chapter we have determined the derivatives of many functions. If we now read all these results from right to left, we suddenly get the antiderivatives of many functions. Here is a small table:

$f(x)$	$F(x)$	$f(x)$	$F(x)$
$x^z \; (z \in \mathbb{Z} \setminus \{-1\})$	$\dfrac{1}{z+1} x^{z+1}$	$\dfrac{1}{\cos^2 x}$	$\tan x$
$\dfrac{1}{x} \; (x \neq 0)$	$\ln|x|$	$\dfrac{1}{\sin^2 x}$	$-\cot x$
$\sin x$	$-\cos x$	$e^{\alpha x}$	$\dfrac{1}{\alpha} e^{\alpha x}$
$\cos x$	$\sin x$	$\dfrac{1}{\sqrt{1-x^2}}$	$\arcsin x$

Examples for the calculation of definite integrals

1. $\displaystyle\int_a^b c\,dx.$

   The constant c is the derivative of $c \cdot x$, so $\int_a^b c\,dx = c \cdot x \big|_a^b = cb - ca.$

2. $\displaystyle\int_a^b x\,dx = \frac{1}{2} x^2 \Big|_a^b = \frac{1}{2} b^2 - \frac{1}{2} a^2.$

   Fortunately, these are the same results as in the first examples according to Theorem 16.5. But now we can also calculate new things, for example:

3. $\displaystyle\int_0^\pi \sin x\,dx = -\cos x \big|_0^\pi = (-\cos \pi) - (-\cos 0) = 1 - (-1) = 2.$ ◄

Integration Rules

We don't know the antiderivative for the logarithm yet. Also, the antiderivative of $\sqrt{1-x^2}$, which we need to calculate the area of a circle, is unfortunately not included here. We need additional methods of integration, which we will derive from the corresponding differentiation rules.

First, a common notation:

▶ **Definition 16.11** An arbitrary antiderivative of the integrable function f is denoted
by $\int f(x)dx$ and is called the *indefinite integral* of f.

This notation is not quite clean, since $\int f(x)dx$ is only determined up to a constant. In
general, this does not cause any problems, since when forming a definite integral this
constant disappears again.

In the following theorem, I would like to summarize the most important integration rules.

▶ **Theorem 16.12: Integration rules**
a) The *integral for piecewise continuous functions* :
   If $f : [a, b] \rightarrow \mathbb{R}$ is piecewise continuous with discontinuities $a_1, a_2, \ldots, a_n$,
   then with $a_0 = a$, $a_{n+1} = b$:

$$\int_a^b f(x)dx = \int_{a_0}^{a_1} f(x)dx + \int_{a_1}^{a_2} f(x)dx + \cdots + \int_{a_n}^{a_{n+1}} f(x)dx = \sum_{k=0}^{n} \int_{a_k}^{a_{k+1}} f(x)dx.$$

b) The *linearity*:
   For integrable functions f, g and $\alpha, \beta \in \mathbb{R}$ it holds:

$$\int \alpha f(x) + \beta g(x)dx = \alpha \int f(x)dx + \beta \int g(x)dx.$$

c) The *partial integration*:
   If f and g are continuously differentiable on the interval I, then for $a, b \in I$:

$$\int_a^b f'(x)g(x)dx = f(x)g(x)\Big|_a^b - \int_a^b f(x)g'(x)dx, \tag{16.3}$$

or for the indefinite integral:

$$\int f'(x)g(x)dx = f(x)g(x) - \int f(x)g'(x)dx. \tag{16.4}$$

d) The *substitution rule*:
   Let f be continuous on the interval I and g be continuously differentiable on an
   interval J. Let $g(J) \subset I$, so that the composition $f \circ g$ is possible. If F is an antiderivative of f, then it holds:

$$\int_a^b f(g(x))g'(x)dx = \int_{g(a)}^{g(b)} f(y)dy = F(y)\Big|_{g(a)}^{g(b)}, \tag{16.5}$$

or:

$$\int f(g(x))g'(x)dx = F(g(x)). \tag{16.6}$$

The rules a) and b) are included here for completeness, they follow from the definition of piecewise continuity or from Theorem 16.5. New are partial integration and substitution. Partial integration can be directly traced back to the product rule: It is

$$(f(x)g(x))' = f'(x)g(x) + f(x)g'(x).$$

An antiderivative of $(f(x)g(x))'$ is according to the fundamental theorem of calculus $f(x)g(x)$, on the right side we get the antiderivative

$$\int f'(x)g(x)dx + \int f(x)g'(x)dx,$$

and thus

$$f(x)g(x) = \int f'(x)g(x)dx + \int f(x)g'(x)dx,$$

from which we obtain by rearranging (16.4). Inserting the limits of integrationresults in (16.3).

The substitution rule is the counterpart to the chain rule of differential calculus. It states (compare Theorem 15.5): $f(g(x))' = f'(g(x))g'(x)$.

If we replace f in it with its antiderivative F and f' correspondingly with f, we get $F(g(x))' = f(g(x))g'(x)$. So $F(g(x))$ is the antiderivative of $f(g(x))g'(x)$, which is precisely the assertion (16.6). If we calculate the definite integral, we get (16.5):

$$\int_a^b f(g(x))g'(x)dx = F(g(x))\big|_a^b = F(y)\big|_{g(a)}^{g(b)} = \int_{g(a)}^{g(b)} f(y)dy. \quad \square$$

Examples

First of all, for the application of partial integration. With this rule, an integral of a product $\int u(x)v(x)dx$ can be computed by setting $u(x) = f'(x)$ and $v(x) = g(x)$. This results in the integral $\int f(x)g'(x)dx$, which is hopefully easier to solve.

1.
$$\int \underbrace{x}_{g}\,\underbrace{e^x}_{f'}\,dx = \underbrace{x}_{g} \cdot \underbrace{e^x}_{f} - \int \underbrace{1}_{g'} \cdot \underbrace{e^x}_{f}\,dx$$
$$= x \cdot e^x - e^x = (x-1)e^x.$$

As a test, you should always take the derivative:
$$((x-1)e^x)' = 1 \cdot e^x + (x-1)e^x = xe^x.$$

2.
$$\int \underbrace{x}_{g}\,\underbrace{\sin x}_{f'}\,dx = \underbrace{x}_{g} \cdot \underbrace{(-\cos x)}_{f} - \int \underbrace{1}_{g'}\,\underbrace{(-\cos x)}_{f}\,dx$$
$$= -x \cdot \cos x + \sin x.$$

Try it yourself!

As you can see, you can eliminate the factor x in a product by partial integration. If the derivative of a function u is known, but not the integral, the approach $g = u$, $f' = 1$ sometimes helps:

3.
$$\int_a^b \ln x\, dx = \int_a^b \underbrace{1}_{f'} \cdot \underbrace{\ln x}_{g}\, dx = x \cdot \ln x\Big|_a^b - \int_a^b \underbrace{x}_{f} \cdot \underbrace{\frac{1}{x}}_{g'}\, dx$$

$$= x \cdot \ln x\Big|_a^b - x\Big|_a^b = (x \ln x - x)\Big|_a^b.$$

Here I used the rule in the form (16.3), the antiderivative of $\ln x$ is $x \cdot \ln x - x$.

4.
$$\int \underbrace{\cos^2 x}_{f' \cdot g}\, dx = \underbrace{\sin x \cos x}_{f \cdot g} - \int \underbrace{-\sin^2 x}_{f \cdot g'}\, dx \quad \text{and}$$

$$\int \sin^2 x\, dx = \int 1 - \cos^2 x\, dx = \int 1\, dx - \int \cos^2 x\, dx.$$

This reduced the integration of $\cos^2 x$ to the integration of $\cos^2 x$. Are we chasing our own tail? No, because if we insert, we get:

$$\int \cos^2 x\, dx = \sin x \cos x + \underbrace{\int 1\, dx}_{=x} - \int \cos^2 x\, dx$$

$$\Rightarrow \quad 2\int \cos^2 x\, dx = \sin x \cos x + x,$$

so $\frac{1}{2}(\cos x \sin x + x)$ is the antiderivative of $\cos^2 x$. Try it out!

Now for the substitution. The rule "$y = g(x)$, $dy = g'(x)dx$" is helpful when performing the integration. First of all, the functions f and g must always be identified:

5.
$$\int_a^b e^{\sin x} \cos x\, dx.$$

It is $f(y) = e^y$, $g(x) = \sin x$. This gives us:

$$\int_a^b e^{\sin x} \cos x\, dx \underset{\substack{\uparrow \\ y = \sin x \\ dy = \cos x\, dx}}{=} \int_{\sin a}^{\sin b} e^y\, dy = e^y\Big|_{\sin a}^{\sin b} = e^{\sin x}\Big|_a^b.$$

At the last equality sign you see that the antiderivative is $e^{\sin x}$.

6. For any continuously differentiable function g that has no zeros in the interval under consideration, the integral $\int \frac{g'(x)}{g(x)}\, dx = \int \frac{1}{g(x)} g'(x)\, dx$ can be calculated. It is $f(y) = 1/y$, the antiderivative F is $F(y) = \ln |y|$. With (16.6) we obtain:

$$\int \frac{g'(x)}{g(x)} dx = \ln|g(x)|.$$

7. We determine $\int_0^{2\pi} \cos kx\, dx$ for $k \in \mathbb{N}$. We set $f(y) = \cos y$ and, $g(x) = kx$.
   Then $g'(x) = k$, $g(0) = 0$, $g(2\pi) = 2\pi k$. Our function does not quite fit yet, but we
   can always insert a constant factor:

$$\int_0^{2\pi} \cos kx\, dx = \frac{1}{k} \int_0^{2\pi} \underbrace{\cos kx}_{f(g(x))} \cdot \underbrace{k}_{g'(x)} dx \underset{\substack{\uparrow \\ y=kx,dy=kdx}}{=} \frac{1}{k} \int_0^{2\pi k} \cos y\, dy$$

$$= \frac{1}{k} \sin y \Big|_0^{2\pi k} = 0 - 0 = 0.$$

In the last three examples, we interpreted the integrand as $f(g(x))g'(x)$ by careful observation or small rearrangements. Unfortunately, this rarely works out except for nicely constructed exercise examples. I would like to introduce you to a second type of use of the substitution rule. We now read the equation (16.5) from right to left:

$$\int_{g(a)}^{g(b)} f(y)dy = \int_a^b f(g(x))g'(x)dx.$$

If we cannot determine an antiderivative for f, we replace y with a cleverly chosen function $g(x)$. Afterwards we have to multiply $f(g(x))$ with the afterburner $g'(x)$ and hope that we can calculate the integral on the right side more easily now. If we choose an invertible function for g, we can write the substitution rule in the form:

$$\int_a^b f(y)dy = \int_{g^{-1}(a)}^{g^{-1}(b)} f(g(x))g'(x)dx.$$

8. Now we are finally able to calculate the area of a circle (see Example 4 after
   Theorem 16.5): $\int_{-1}^1 \sqrt{1-y^2}dy$ is the area of the semicircle with radius 1. We
   choose as substitution function $g(x) = \sin x$. The sine between $-\pi/2$ and $\pi/2$ is
   bijective, that is, invertible, and therefore:

$$\int_{-1}^1 \sqrt{1-y^2}dy \underset{\substack{\uparrow \\ y=\sin x \\ dy=\cos x dx}}{=} \int_{\sin^{-1}(-1)}^{\sin^{-1}(1)} \sqrt{1-(\sin x)^2} \cos x dx$$

$$= \int_{-\pi/2}^{\pi/2} \sqrt{1-(\sin x)^2} \cos x dx.$$

Between $-\pi/2$ and $\pi/2$, it holds $\sqrt{1-(\sin x)^2} = \cos x$. We have already calculated the antiderivative of $\cos^2 x$ in Example 4, and so we get:

$$\int_{-1}^{1} \sqrt{1 - y^2}\, dy = \int_{-\pi/2}^{\pi/2} \cos^2 x\, dx = \frac{1}{2}(\cos x \sin x + x)\Big|_{-\pi/2}^{\pi/2}$$

$$= \frac{1}{2}\left(\cos\left(\frac{\pi}{2}\right)\sin\left(\frac{\pi}{2}\right) + \left(\frac{\pi}{2}\right)\right)$$

$$-\frac{1}{2}\left(\cos\left(-\frac{\pi}{2}\right)\sin\left(-\frac{\pi}{2}\right) + \left(-\frac{\pi}{2}\right)\right)$$

$$= \frac{\pi}{2}.$$

Now we finally know that our number π, defined as abstractly as "twice the first positive zero of the power series of the cosine", really corresponds to the well-known mathematical constant π. ◀

For the set of *rational functions*, that is, functions that can be represented as a quotient of polynomials, one can always specify an antiderivative:

▶   **Theorem 16.13** If $f(x), g(x)$ are real polynomials, then a rational function is given by $h(x) = f(x)/g(x)$, which is defined outside the zeros of the denominator. There is always an antiderivative to $h(x)$.

With the help of the rather laborious method of partial fraction decomposition, one can decompose $h(x)$ into a sum of fractions of the form

$$\frac{a}{(x-b)^k} \quad \text{or} \quad \frac{a+bx}{(x^2 - cx + d)^k}$$

with $a, b, c, d \in \mathbb{R}$ and $k \in \mathbb{N}$. The integrals of these fractions can be found in the integral table or with a computer algebra system. I don't want to go into that any further.

16.2   Applications of the Integral

Let us recall the beginning of this chapter: If $f: [a, b] \to \mathbb{R}$ is piecewise continuous, $a = x_0 < x_1 < x_2 < \cdots < x_k = b$ a partition of the interval, $\Delta x_i := x_i - x_{i-1}$ and $y_i \in [x_{i-1}, x_i]$, then the Riemann sums converge to the integral of the function, regardless of the partition, as long as the partitions refine to 0.

$$S_k := \sum_{i=1}^{k} f(y_i) \cdot \Delta x_i, \quad \lim_{k \to \infty} S_k = \int_a^b f(x)\, dx.$$

We can make use of this statement for further tasks:

Volumes of Shapes

Let a shape be given in the three-dimensional space $\mathbb{R}^3$. As a function of x, let the cross-sectional area $F(x)$ be known (Fig. 16.9).

1. The cylinder:

$$F(x) = \begin{cases} 0 & x < a, x > b \\ r^2\pi & a \leq x \leq b. \end{cases}$$

2. The sphere: For $|x| < R$ the radius at the point x is: $r(x) = \sqrt{R^2 - x^2}$ and thus

$$F(x) = \begin{cases} 0 & x < -R, x > R \\ (R^2 - x^2)\pi & -R \leq x \leq R. \end{cases}$$

3. The cone:

$$F(x) = \begin{cases} 0 & x < 0, x > H \\ \dfrac{R^2\pi}{H^2}x^2 & 0 \leq x \leq H. \end{cases} \quad \blacktriangleleft$$

Now if we want to calculate a volume, we put slices of thickness Δx_i next to each other. If $y_i \in [x_{i-1}, x_i]$, then the volume of the i-th slice is $V_i = F(y_i)\Delta x_i$ and for the volume of the entire shape we get the approximation:

$$V = \sum_{i=1}^{k} F(y_i)\Delta x_i.$$

But that is exactly the Riemann sum of the function F. Therefore,

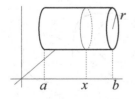
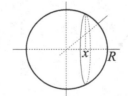
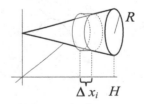

Fig. 16.9 Volume of shapes

▶ **Theorem 16.14** If in a three-dimensional shape the cross-sectional area in the x direction is given by an integrable function $F\colon [a,b] \to \mathbb{R}$, then for the volume V of this shape holds

$$V = \int_a^b F(x)dx.$$

Example

The volume of a cone of height H with base radius R is:

$$\int_0^H \frac{R^2\pi}{H^2}x^2dx = \frac{R^2\pi}{H^2}\frac{x^3}{3}\bigg|_0^H = \frac{R^2\pi H^3}{H^2 3} = \frac{1}{3}R^2\pi H = \frac{1}{3}\cdot \text{base area}\cdot\text{height.} \quad ◀$$

I leave the calculation of the volume of a sphere as an exercise. Compare your result with the value from the formulary!

The Arc of a Curve

The graph of a function $f\colon [a,b] \to \mathbb{R}$ is an example of a two-dimensional curve: $\Gamma_f = \{(x,f(x))\mid x \in [a,b]\}$. But not every curve can be interpreted as a function graph, see Fig. 16.10.

Think of a plane curve as the path an ant travels over time. If we describe the x- and y-component of the path as a function of time t, we get the parametrization of the curve and have reduced the problem to real functions again:

▶ **Definition 16.15** Let $x,y\colon [a,b] \to \mathbb{R}$ be continuously differentiable functions. Then the mapping

$$s\colon [a,b] \to \mathbb{R}^2, \quad t \mapsto (x(t),y(t))$$

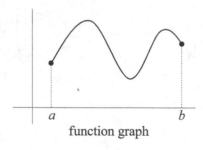
function graph

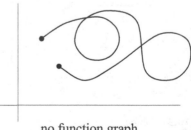
no function graph

Fig. 16.10 Curves

is called the *parametrizationof a curve s.* The variable t is the *parameter,* $[a,b]$ is the *parameter interval* and $\{s(t) \mid t \in [a,b]\} \subset \mathbb{R}^2$ is the set of points of the curve.

1. If $f \colon [a,b] \to \mathbb{R}$ is a continuously differentiable function, then

$$s \colon [a,b] \to \mathbb{R}^2, \quad t \mapsto (t, f(t))$$

   is a parametrization of the graph of the function.
2. $s_1 \colon [-1, 1] \to \mathbb{R}^2, \; t \mapsto (t, \sqrt{1 - t^2})$ is a parametrization of the semicircle with radius 1 around $(0,0)$, see Example 4 after Theorem 16.5. The entire circle cannot be given in this form; $s_2 \colon [0, 2\pi] \to \mathbb{R}^2, t \mapsto (r \cos t, r \sin t)$ is however a parametrization of the full circle with radius r around $(0,0)$. ◀

You see that the parametrization of a curve is by no means unique, which corresponds to the fact that paths can be traversed at different speeds. Curves can also intersect, in the example of the circle the initial and final point are the same, that is also possible. Of course, one can also investigate curves in $\mathbb{R}^3$ and in $\mathbb{R}^n$, which describe the flight of a fly or a hyperspaceship.

With the help of integral calculus, we can determine the arc length of such a curve. I restrict myself to the two-dimensional case. We want to reduce the task to Riemann sums again and divide the curve into ever shorter pieces, which we approximate by line segments.

We investigate the curve $s \colon [a,b] \to \mathbb{R}^2, \; t \mapsto (x(t), y(t))$ with the partition $a = t_0 < t_1 < \cdots < t_n = b$ and use the notation as in Fig. 16.11:

$$\Delta t_i = t_i - t_{i-1},$$
$$\Delta x_i = x(t_i) - x(t_{i-1}),$$
$$\Delta y_i = y(t_i) - y(t_{i-1}),$$
$$\Delta s_i = \sqrt{\Delta x_i^2 + \Delta y_i^2}.$$

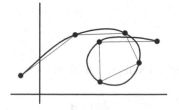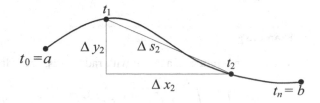

Fig. 16.11 The arc length

We refine the partition so that the Δt_i go to 0. For the approximate arc length L_n we then have:

$$L_n = \sum_{i=1}^{n} \Delta s_i = \sum_{i=1}^{n} \sqrt{\Delta x_i^2 + \Delta y_i^2}. \tag{16.7}$$

From the mean value theorem of differential calculus 15.12, applied to $x(t)$ and $y(t)$, we learn that there is a $u_i \in \Delta t_i$ and a $v_i \in \Delta t_i$ with the property

$$x'(u_i) = \frac{x(t_i) - x(t_{i-1})}{t_i - t_{i-1}} = \frac{\Delta x_i}{\Delta t_i}, y'(v_i) = \frac{y(t_i) - y(t_{i-1})}{t_i - t_{i-1}} = \frac{\Delta y_i}{\Delta t_i},$$

i.e. $\Delta x_i = x'(u_i)\Delta t_i$ and $\Delta y_i = y'(v_i)\Delta t_i$. Inserted into (16.7) we get:

$$L_n = \sum_{i=1}^{n} \sqrt{x'(u_i)^2 + y'(v_i)^2} \cdot \Delta t_i, \quad u_i, v_i \in \Delta t_i.$$

This is not quite, but almost a Riemann sum of the function $\sqrt{x'(t)^2 + y'(t)^2}$. The only problem is that the elements u_i, v_i in the interval Δt_i can be different.

At this point I would like to abbreviate the derivation. It can be shown that for finer and finer partitions, the arc length L can really be calculated as the limit of Riemann sums of this function, it is

▶   **Theorem 16.16** Let $s \colon [a,b] \to \mathbb{R}^2$, $t \mapsto (x(t), y(t))$ be the parametrization of a curve with continuously differentiable functions x and y. Then the arc length L of the curve is equal to

$$L = \int_a^b \sqrt{x'(t)^2 + y'(t)^2}\, dt.$$

And since $(t, f(t))$ is the parametrization of the function f, we immediately get as a consequence:

▶   **Theorem 16.17** The graph of a continuously differentiable function $f \colon [a,b] \to \mathbb{R}$ has the arc length $\int_a^b \sqrt{1 + f'(t)^2}\, dt$.

Example

The circumference of a circle with radius r is, according to Example 2 from before:

$$L = \int_0^{2\pi} \sqrt{r^2 \sin^2 t + r^2 \cos^2 t}\, dt = \int_0^{2\pi} r\, dt = rt \big|_0^{2\pi} = 2\pi r. \blacktriangleleft$$

Improper Integrals

The integral $\int_a^b \frac{1}{x} dx$ exists for all $a, b > 0$ and for all $a, b < 0$. But what happens if a approaches 0, or b approaches infinity (Fig. 16.12)? Does there exist for $a < 0$ and $b > 0$ a finite area between a and b? To answer such questions, we need to carry out limit considerations for the integrals.

▶ **Definition 16.18** Let the function f be defined on $[a, b[$ and let the integral $\int_a^c f(x)dx$ exist for all c between a and b. Then

$$\int_a^b f(x)dx := \lim_{c \to b} \int_a^c f(x)dx$$

is called the *improper integral* of f if this limit exists. For b, the value ∞ is also allowed, for the limit, the values $\pm\infty$ are allowed.

Examples

1. $\int_0^1 \frac{1}{x} dx = \lim_{c \to 0} \int_c^1 \frac{1}{x} dx - \lim_{c \to 0} (\ln x \big|_c^1) = \lim_{c \to 0} (-\ln c) = \infty.$

2. $\int_1^\infty \frac{1}{x} dx = \lim_{c \to \infty} \int_1^c \frac{1}{x} dx = \lim_{c \to \infty} (\ln x \big|_1^c) = \lim_{c \to \infty} (\ln c) = \infty.$

   The function $1/x$ therefore has no finite area. It looks different with $1/x^2$: The area between 1 and ∞ remains finite:

3. $\int_1^\infty \frac{1}{x^2} dx = \lim_{c \to \infty} \int_1^c \frac{1}{x^2} dx = \lim_{c \to \infty} \left(-\frac{1}{x}\big|_1^c\right) = \lim_{c \to \infty} \left(-\frac{1}{c} + \frac{1}{1}\right) = 1.$

4. $\int_{-1}^1 \frac{1}{\sqrt{|x|}} dx = \int_{-1}^0 \frac{1}{\sqrt{|x|}} dx + \int_0^1 \frac{1}{\sqrt{|x|}} dx$ (Fig. 16.13):

   The antiderivative of $1/\sqrt{x}$ is $2\sqrt{x}$. It follows that

   $$\int_0^1 \frac{1}{\sqrt{|x|}} dx = \lim_{c \to 0} 2\sqrt{x}\big|_c^1 = 2.$$

Fig. 16.12 Improper integrals 1

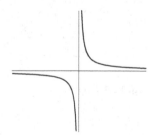

Fig. 16.13 Improper
integrals 2

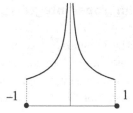

It is clear that the left half is just as big, but we can also check it again using the
substitution rule:

$$\int_{-1}^{0} \frac{1}{\sqrt{|x|}}dx = \int_{-1}^{0} \frac{1}{\sqrt{-x}}dx \underset{\underset{y=-x, dy=-dx}{\uparrow}}{=} -\int_{+1}^{0} \frac{1}{\sqrt{y}}dy = \int_{0}^{1} \frac{1}{\sqrt{y}}dy = 2. \qquad \blacktriangleleft$$

16.3 Fourier Series

We have already often approximated functions by sequences of functions in the last
chapters, for example by polynomials for graphical representation, by power series for
numerical calculations or by step functions or trapezoidsl for integration. Now we learn
a new important approximation method, the Fourier series. There are many applications
for this and for related representations in technology and also in computer science. Given
functions are approximated by sequences of trigonometric functions, by $\cos(nx)$ and
$\sin(nx)$, $n \in \mathbb{N}$ (Fig. 16.14).

All these functions are periodic with the period 2π. For $n > 1$ the frequency increases,
that is the number of oscillations in a fixed interval. These functions also have shorter
periods than 2π, but that should not interest us at the moment. $\cos(nx)$ and $\sin(nx)$ are of
the same shape, only shifted against each other.

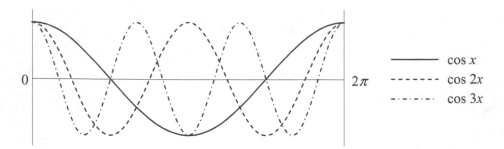

Fig. 16.14 $\cos(nx)$

In this section we will only investigate periodic functions. If f is periodic with period 2π, then with the help of the transformation $x \mapsto x \cdot (2\pi/T)$ one can easily create a function with any period T: $f(x \cdot (2\pi/T))$ is then periodic with period T, because

$$f\left((x+T) \cdot \frac{2\pi}{T}\right) = f\left(x \cdot \frac{2\pi}{T} + \frac{T2\pi}{T}\right) = f\left(x \cdot \frac{2\pi}{T} + 2\pi\right) = f\left(x \cdot \frac{2\pi}{T}\right).$$

We restrict ourselves in our following investigations to the period 2π, all statements can be transferred to functions of arbitrary period T by replacing everywhere x by $x \cdot (2\pi/T)$.

The idea behind the Fourier expansion is that many functions can be composed of such individual oscillations: The sound of a violin consists of a fundamental frequency with some overtones, which are also oscillations. Even the most complex piece of music is only made up of an overlay of such oscillations. For example, the running noise of a turbine, the data transmitted over a channel, or the color information of an image that is scanned point by point. Which class of functions can be decomposed in this way?

The following relations between cosine and sine are an essential basis for the Fourier expansion of a function:

▶   **Theorem 16.19** For all $n, m \in \mathbb{N}$ the orthogonality relations apply:

$$\frac{1}{\pi} \int_0^{2\pi} \cos(mx)\cos(nx)dx = \begin{cases} 0 & m \neq n \\ 1 & m = n \end{cases}$$

$$\frac{1}{\pi} \int_0^{2\pi} \sin(mx)\sin(nx)dx = \begin{cases} 0 & m \neq n \\ 1 & m = n \end{cases} \tag{16.8}$$

$$\int_0^{2\pi} \cos(mx)\sin(nx)dx = 0$$

There is a very close relationship to linear algebra: The set of functions that are integrable on $[0, 2\pi]$ is a real vector space. The integral $\frac{1}{\pi} \int_0^{2\pi} f(x)g(x)dx$ represents a kind of scalar product (see Exercise 9 of this chapter). The formulas (16.8) state that the functions $\cos nx$ and $\sin nx$ form an orthonormal system with respect to this scalar product: They are mutually perpendicular and have length 1. This orthonormal system is almost a basis: As we will see in Theorem 16.20, we can approximate all vectors of the space by linear combinations of $\cos nx$ and $\sin nx$.

I calculate the first of these relations as an example: In the formulary we find the rule $\cos\alpha \cos\beta = 1/2(\cos(\alpha - \beta) + \cos(\alpha + \beta))$. If $m \neq n$, then we get from it, according to Example 7 at the end of Sect. 16.1,

$$\int_0^{2\pi} \cos(mx)\cos(nx)dx = \frac{1}{2} \int_0^{2\pi} \cos((m-n)x) - \cos((m+n)x)dx = 0,$$

and if $n = m$, then the integral

$$\int_0^{2\pi} \cos^2(mx)dx = \frac{1}{m}\int_0^{2\pi} \cos^2(mx)mdx$$

$$\underset{\substack{\uparrow \\ y=mx \\ dy=mdx}}{=} \frac{1}{m}\int_0^{2\pi m} \cos^2 ydy = \frac{1}{m}\left[\frac{1}{2}(\sin(y)\cos(y) + y)\right]\Big|_0^{2\pi m} = \pi.$$

The antiderivative of $\cos^2 x$ we had already calculated in Example 4 after Theorem 16.12.   □

Now let real numbers a_n, $n \in \mathbb{N}_0$ and b_n, $n \in \mathbb{N}$ be given, and first of all assume that the following series converges for all $x \in \mathbb{R}$ and thus represents a function f:

$$f(x) = \frac{1}{2}a_0 + \sum_{n=1}^{\infty}(a_n \cos(nx) + b_n \sin(nx)). \qquad (16.9)$$

This series is then called the *Fourier series* of f. Since all the individual functions involved are periodic, the function f is also periodic with period 2π.

Let's try to calculate the coefficients a_k, b_k for a function f which can be represented in the form (16.9). We use Theorem 16.19: The trick is to multiply from the right with $\cos(kx)$ or $\sin(kx)$ and then to integrate from 0 to 2π:

$$\int_0^{2\pi} f(x)\cos(kx)dx = \int_0^{2\pi}\left(\frac{1}{2}a_0 + \sum_{n=1}^{\infty}(a_n\cos(nx) + b_n\sin(nx))\right)\cos(kx)dx.$$

Under certain conditions, which I will not go into here, summation and integration can be interchanged (for example, if $\sum_{n=0}^{\infty} a_n$ and $\sum_{n=0}^{\infty} b_n$ are absolutely convergent). Since the trigonometric functions are orthogonal to each other, almost everything cancels out in this rearrangement:

$$\int_0^{2\pi} f(x)\cos kxdx = \underbrace{\int_0^{2\pi}\frac{1}{2}a_0\cos kxdx}_{=0} + \sum_{n=1}^{\infty}a_n\underbrace{\int_0^{2\pi}\cos(nx)\cos(kx)dx}_{=0 \text{ for } n\neq k}$$

$$+ \sum_{n=1}^{\infty}b_n\underbrace{\int_0^{2\pi}\sin(nx)\cos(kx)dx}_{=0}$$

$$= a_k\int_0^{2\pi}\cos(kx)\cos(kx)dx = a_k\pi.$$

This is how we get the coefficients a_k. If we multiply (16.9) by $\sin(kx)$ and integrate, we get the b_k, and finally an integration without any multiplication gives us the coefficient a_0:

$$a_k = \frac{1}{\pi} \int_0^{2\pi} f(x) \cos kx\, dx, \quad b_k = \frac{1}{\pi} \int_0^{2\pi} f(x) \sin kx\, dx,$$
$$a_0 = \frac{1}{\pi} \int_0^{2\pi} f(x)\, dx. \tag{16.10}$$

There is also linear algebra behind this calculation: The i-th coordinate of the vector v in an orthonormal basis is obtained by scalar multiplication of the vector with the basis vector b_i: $v_i = \langle v, b_i \rangle$. Try it out! This is exactly the scalar product we have carried out here.

The first coefficient $a_0/2$ can be interpreted quite simply: It is $2\pi \cdot \frac{a_0}{2} = \int_0^{2\pi} f(x)\, dx$, so $a_0/2$ is the "average" function value between 0 and 2π: The area of the rectangle with the height $a_0/2$ is equal to the area under f.

If a function f has a Fourier series representation and summation and integration can be interchanged, then the coefficients have the form calculated in (16.10). More interesting is the question for which functions f such a Fourier series actually exists. Information about this gives the next theorem. It is very difficult to prove, but the derivation above should provide some motivation for its content:

▶ **Theorem 16.20** Let f be piecewise continuous and periodic with period 2π. Let the coefficients a_k, b_k be determined as in (16.10). Further let

$$S_n(x) := \frac{1}{2}a_0 + \sum_{k=1}^{n} (a_k \cos(kx) + b_k \sin(kx)).$$

Then

$$\lim_{n \to \infty} \int_0^{2\pi} (S_n(x) - f(x))^2 dx = 0. \tag{16.11}$$

What does (16.11) mean? $\int_0^{2\pi} (S_n(x) - f(x))^2 dx$ measures the difference between the areas of $S_n(x)$ and f, with all surface areas being positive because of the square. This means that the area between f and S_n always gets smaller, it goes to 0. The relation (16.11) is called: S_n *converges in mean square to* f. The coefficients a_k, b_k form null sequences, and if f is good, then the convergence also goes quite quickly. However, for each individual point x it does not have to be the case that $\lim_{n \to \infty} S_n(x) = f(x)$, in general really only the area between the functions goes to 0.

The formula (16.11) can be expressed again in the language of linear algebra: $\frac{1}{\pi}\int_0^{2\pi}(S_n(x)-f(x))^2dx$ is nothing other than the scalar product $\langle S_n(x)-f(x), S_n(x)-f(x)\rangle$. In the norm that belongs to this scalar product, it therefore applies that $\lim_{n\to\infty}\|S_n(x)-f(x)\| = 0$. This means that with respect to this norm S_n converges to f.

A particularly interesting property of the Fourier approximation is the ability to approximate discontinuous functions by continuous functions. S_n is continuous everywhere. We will do this right away.

Example of a Fourier expansion

We examine the function

$$f(x) = \begin{cases} 1 & 0 \le x < \pi \\ 0 & \pi \le x < 2\pi, \end{cases}$$

which is continued periodically on $\mathbb{R}$. Imagine a data line in which the bits 1 and 0 are continuously sent. Now

$$a_0 = \frac{1}{\pi}\int_0^{2\pi}f(x)dx = \frac{1}{\pi}\int_0^{\pi}1dx = \frac{1}{\pi}\pi = 1,$$

$$a_k = \frac{1}{\pi}\int_0^{\pi}1\cos kx\,dx = \frac{1}{k\pi}\int_0^{\pi}k\cos kx\,dx = \frac{1}{k\pi}\int_0^{k\pi}\cos y\,dy$$

$$= \frac{1}{k\pi}\sin y\Big|_0^{k\pi} = 0,$$

$$b_k = \frac{1}{\pi}\int_0^{\pi}1\sin kx\,dx = \cdots = \frac{1}{k\pi}\int_0^{k\pi}\sin y\,dy$$

$$= -\frac{1}{k\pi}\cos y\Big|_0^{k\pi} = -\frac{1}{k\pi}\cdot\begin{cases} 1-1 & k\text{ even} \\ -1-1 & k\text{ odd,} \end{cases}$$

so

$$b_k = \begin{cases} 0 & k\text{ even} \\ \frac{2}{k\pi} & k\text{ odd} \end{cases},$$

and thus we obtain the Fourier expansion of $f(x)$:

$$f(x) = \frac{1}{2} + \frac{2}{\pi}\left(\sin(x) + \frac{\sin(3x)}{3} + \frac{\sin(5x)}{5} + \frac{\sin(7x)}{7} + \cdots\right). \blacktriangleleft$$

In Fig. 16.15 you can see very well how the difference area develops in the approximation. The overshooters at the discontinuities are typical: These remain even with higher

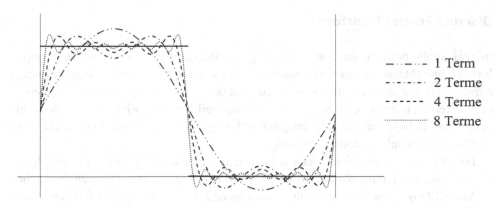

Fig. 16.15 A Fourier expansion

iterations, so the approximation always differs a whole piece from the given function at individual points.

> However, in this particular example, it holds for every fixed x in which the function is continuous $\lim_{n \to \infty} S_n(x) = f(x)$. Try to visualize this with a drawing!

Every data channel has attenuation properties that limit its capacity. This attenuation is frequency-dependent: Different frequencies are filtered out to different degrees for technical reasons. If you know these parameters, you can adjust the corresponding frequencies in the above Fourier expansion and see how the data that you initially fed into the line as nice rectangles comes out at the other end. Can you still identify the 0s and 1s there? In this way, you can determine the capacity of a data line.

Similarly, for example, by analyzing the running noise of a turbine, you can identify changes that are very likely to indicate a defect.

The human ear, by the way, works on the same principle, it performs a frequency analysis: Different frequencies stimulate different parts of the cochlea in the inner ear.

> Fourier series were developed by Joseph Fourier (1768–1830). Fourier was a contemporary and confidant of Napoleon and is considered one of the fathers of mathematical physics. He came across his series while analyzing heat conduction problems. His results were certainly recognized by the mathematical establishment of his time, but he was severely criticized for the poor mathematical representation and the incomplete proof. Fourier was not deterred by this criticism, nor was the wide ranging applicability of his theory affected by it. In fact, some of his theorems were not fully proven until the 20th century. In 2006, the mathematician Lennart Carleson received the Abel Prize (see the note after Definition 5.2) mainly for a work from 1966 in which he showed that every continuous function can be represented as the sum of its Fourier series. But Fourier just had the right nose. The dispute between the nagging mathematicians who cling to the pure doctrine and the users who recklessly calculate is still present today. In defense of the mathematicians, it should be said that the toolbox should already be in order, even if the user sometimes grabs the wrong wrench.

Discrete Fourier Transform

In real applications, a function is rarely given analytically as in the above example. Instead, one obtains the function through individual measured values, as with the turbine noise mentioned above, with a sound recording or when scanning an image. So there is only a certain number of points of the function available from which the Fourier coefficients are to be calculated. The integration that has to be carried out to calculate these coefficients can only be done numerically.

For the sake of simplicity, let's stay with our period of 2π and start with $N+1$ evenly distributed interpolation points (Fig. 16.16). The distance between two points is $\Delta x_i = (2\pi)/N$, the point i has the x-coordinate $x_i = i \cdot (2\pi)/N$. The function values $f(x_i)$ at these points are known. For integration, we simply form the Riemann sums. For the coefficients a_k, we set the approximation:

$$a_k = \frac{1}{\pi} \int_0^{2\pi} f(x) \cos(kx) dx = \frac{1}{\pi} \sum_{i=0}^{N-1} f(x_i) \cos(kx_i) \cdot \frac{2\pi}{N} = \frac{2}{N} \cdot \sum_{i=0}^{N-1} f(x_i) \cos(kx_i),$$

just as $b_k = \frac{2}{N} \sum_{i=0}^{N-1} f(x_i) \sin(kx_i)$ and $a_0 = \frac{2}{N} \cdot \sum_{i=0}^{N-1} f(x_i)$. The coefficient a_0 is again twice the average of the N function values.

Of course, this numerical integration makes errors that become larger the larger k becomes, because then the function wobbles back and forth more and more wildly. With our few interpolation points, we can hardly hope to get a reasonably accurate integral. One first effect of this error is that the sequences a_k, b_k are no longer null sequences. In fact, they are periodic sequences, because from

$$a_{k+N} = \frac{2}{N} \cdot \sum_{i=0}^{N-1} f(x_i) \cos((k+N)x_i),$$

$$\cos((k+N)x_i) = \cos\left((k+N) \cdot i\frac{2\pi}{N}\right) = \cos\left(k \cdot i\frac{2\pi}{N} + i2\pi\right)$$

$$= \cos\left(k \cdot i\frac{2\pi}{N}\right) = \cos(kx_i)$$

it follows that $a_{k+N} = a_k$ for all $k > 0$. Similarly, $b_k = b_{k+N}$.

Fig. 16.16 Interpolation points of a function

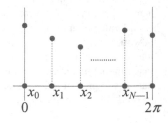

x_0 x_1 x_2 x_{N-1}

0   2π

More regularities can be discovered in the coefficients: Because of $\cos x = \cos(2\pi - x)$ and $\sin x = -\sin(2\pi - x)$, a similar calculation shows that $a_{N-k} = a_k$ and $b_{N-k} = -b_k$ holds. The coefficients therefore only have to be calculated up to half of N. I use the floor function below: $\lfloor N/2 \rfloor$ has the value $N/2$ for even N and $(N-1)/2$ for odd N.

▶ **Definition 16.21** The mapping which assigns the following N coefficients to the N values $f(x_i)$ of the function f at the points $x_i = i \cdot (2\pi)/N$, where $i = 0, 1, \ldots, N-1$ and $k = 1, \ldots \lfloor N/2 \rfloor$:

$$a_0 = \frac{2}{N} \cdot \sum_{i=0}^{N-1} f(x_i), \quad a_k = \frac{2}{N} \sum_{i=0}^{N-1} f(x_i) \cos(kx_i),$$

$$b_k = \frac{2}{N} \sum_{i=0}^{N-1} f(x_i) \sin(kx_i),$$

is called *discrete Fourier transform (DFT)*.

If N is even, then $b_{N/2} = 0$, so that in any case only N "real" coefficients have to be calculated. From the N given points of the function $f(x)$, exactly N Fourier coefficients are determined.

The absolutely amazing thing about the deliberately error-prone construction is that the function can still be represented as a "Fourier series" using these coefficients. Even a finite sum using the N calculated coefficients is sufficient, and it is not just an approximation, the representation is exact:

▶ **Theorem 16.22** The function f can be calculated at the points $x_i = i \cdot (2\pi)/N$ from the coefficients of the discrete Fourier transform. For odd N:

$$f(x_i) = \frac{a_0}{2} + \sum_{k=1}^{\lfloor N/2 \rfloor} (a_k \cos(kx_i) + b_k \sin(kx_i))$$

and for even N:

$$f(x_i) = \frac{a_0}{2} + \sum_{k=1}^{N/2-1} (a_k \cos(kx_i) + b_k \sin(kx_i)) + \frac{a_{N/2}}{2} \cos((N/2) \cdot x_i).$$

I do not want to prove this theorem, but it is interesting that the proof does not require any analytical tools. It is a purely algebraic calculation in which, using elementary properties of sine and cosine, a linear system of equations for the N coefficients is solved.

The runtime for calculating the Fourier coefficients is of the order $O(N^2)$ in both directions. In the 1960s, implementations were developed that reduce this effort to $O(N \cdot \log N)$, a tremendous improvement. Such implementations are summarized under the name *Fast Fourier Transform (FFT)*. The possible uses for the Fourier transform increased enormously. For example, real-time transforms are now also possible, as they are, for example, required for the turbine already mentioned. I would like to sketch one application example for you: the use of the DFT for image compression:

Let's start with a black and white image. Already when the image is read in by a scanner, data loss occurs: on the one hand through the size of the pixels, on the other hand through the continuous course of the gray values being pressed into a finite scale, typically 256 gray levels, which occupy one byte of memory space per pixel. Nevertheless, the memory requirements for such naked image data are enormous.

If the image is further compressed, another data loss usually results. But this is designed so that no visible deterioration of the image quality takes place for the eye. For example, another coarsening of the gray levels can be carried out, because 256 gray values are difficult to distinguish for the human eye.

How do you proceed? In a first approach, the gray values could be rounded even further. This would have to be done in the same way for all values, since all gray levels are of equal importance. No account can be taken of the content of the image. So for soft, low-contrast images, perhaps a completely different type of rounding would produce good results than for high-contrast images.

At this point, the Fourier transformation can be used: First we decompose the image into manageable parts, for example into sub-images with a size of 8×8 pixels. The function values $f(x_i)$ are the gray levels of the pixels.

Now we carry out the DFT line by line with these points, we get the coefficients a_k and b_k. For the function values $f(x_i)$ it then applies:

$$f(x_i) = \frac{a_0}{2} + \sum_{k=1}^{15}(a_k \cos(kx_i) + b_k \sin(kx_i)) + \frac{a_{16}}{2} \cos(32x_i).$$

The data that we store compressed are no longer the function values themselves, but the Fourier coefficients. It turns out that these play a completely different role for the image information. $a_0/2$, for example, represents the mean gray value of the image, an important information that should be stored very precisely. Various experiments and experiences show that the higher-frequency components (larger k) contribute much less to the image information than the low-frequency components (small k). Therefore, the coefficients with a larger index can be rounded much coarser or even completely thrown away without causing a visible image loss. With the same amount of data, this results in much less information loss than without the Fourier transformation.

In the well-known jpeg compression, images are treated with the discrete cosine transformation (DCT), which is a close relative of the Fourier transformation. This is carried out on sub-images of the size 8×8, but not line by line, but in a two-dimensional form.

The resulting coefficients are then rounded according to fixed rules laid down in tables. These rounded coefficients are then Huffman-coded. The Huffman coding is a very effective supplement to the DCT, as the probability of occurrence of different coefficients can vary greatly.

The approach of approximating functions by linear combinations of other, linearly independent functions and then calculating with the coefficients of these approximations is fruitful in many areas of mathematics over and over again. The transformation of functions using wavelets is very current: Starting from a wavelet prototype, a family of orthogonal functions is formed, with the help of which it is possible to approximate the original function well in different scales, on a large and small scale. There are also discrete, fast variants of the wavelet transformation, which can be used, for example, in data compression with even better results than the Fourier transform. The latest jpeg standard also includes the use of wavelets.

16.4   Comprehension Questions and Exercises

Comprehension questions

1. Let $f: [a, b] \to \mathbb{R}$ be continuous. Does the definite integral $\int_a^b f(x)dx$ then exist?
2. Let $f: [-a, a] \to \mathbb{R}$ be an integrable, even function. What is $\int_{-a}^a f(x)dx$?
3. Explain the fundamental theorem of calculus.
4. If the antiderivative $F(x)$ of the function $f(x)$ exists, is $F: [a, b] \to \mathbb{R}$ then continuous and differentiable?
5. $f: [a, b] \to \mathbb{R}$ is a non-periodic continuous function. Can you still set up a Fourier series for f?
6. In the note to Theorem 16.19 I said that the functions $\cos(nx)$, $\sin(nx)$ on the vector space of functions integrable on $[0, 2\pi]$ almost form a basis. Why only almost? Take a look again at the definitions 6.16 and 6.5.
7. Can you explain the term "convergence in the mean square"?

Exercises

1. Calculate the following integrals:
   a) $\int_0^{2\pi} \sin(x) \cos(x)dx$,
   b) $\int x \ln x dx$,
   c) $\int_a^b x^2 e^x dx$,
   d) $\int_0^1 (3x - 2)^2 dx$,
   e) $\int_{-2}^2 \frac{1}{2x-8}dx$.

2. Calculate $\int_0^{2\pi} \cos^2(nx)dx$, $n \in \mathbb{N}$. Use Example 4 after Theorem 16.12.

3. Calculate an antiderivative for $\tan(x)$ in the range $-\pi/2 < x < \pi/2$. Use Example 6 after Theorem 16.12.

4. One of the two following integrals can be calculated (as adefinite integral), one cannot. Calculate the integral or explain why it does not work:

   a) $\int_{-3}^{3} \frac{7}{3x+4} dx$,

   b) $\int_{-1}^{5} \frac{2}{6x+9} dx$.

5. Calculate the arc length of the unit circle between $(0, 0)$ and $(\cos\alpha, \sin\alpha)$.

   As a result, you find that to the angle α in radians belongs exactly the circular arc of length α.

6. Calculate the volume of a sphere with radius R.

7. Compute the Fourier series for the periodic "sawtooth wave" $f(x) = x$ for $0 < x \leq 2\pi$. Use the formulary or a computer algebra system to determine the necessary integrals.

8. Implement the discrete Fourier transform in both directions and confirm by tests Theorem 16.22.

9. Check whether a scalar product is given on the set of functions integrable in the interval $[0, 2\pi]$ by the relation $\langle f, g \rangle = \frac{1}{\pi} \int_{0}^{2\pi} f(x)g(x)dx$. See Definition 10.1 for this. One of the four conditions is not fulfilled. Which one?

Differential Equations

17

Abstract

If you have worked through this chapter

- you understand the importance of differential equations,
- you can recognize important types of differential equations,
- you know solution methods for separable differential equations and for first order linear differential equations,
- and you can solve linear homogeneous differential equations with constant coefficients completely.

17.1 What are Differential Equations?

Late risers have the following serious problem to solve if they do not want to come to the lecture too late: The coffee from the machine is too hot to drink, it has to be brought to drinking temperature as quickly as possible. Does the coffee cool down faster if you add the sugar immediately and then wait, or is it smarter to wait for a while and then add the sugar?

Let's do a proper problem analysis: We are given $T_K(t)$, the coffee temperature at time t, the temperature of the surrounding air T_L and the maximum drinking temperature T_m. Let's first neglect the sugar. We assume that the cooling takes place more quickly the

Supplementary Information The online version contains supplementary material available at https://doi.org/10.1007/978-3-658-40423-9_17.

greater the temperature difference between coffee and air is, the cooling is proportional to this difference.

This cooling is nothing other than the change in temperature over time, that is, its derivative $T_K'(t)$. We get the equation:

$$T_K'(t) = c(T_K(t) - T_L) \tag{17.1}$$

with a proportionality factor $c < 0$: Since the temperature decreases, the derivative is negative.

We are looking for a function $T_K(t)$ which satisfies (17.1). Let's guess first. The desired function looks similar to its derivative, and so the exponential function is a good choice. We try

$$T_K(t) = \alpha e^{\beta t} + \gamma.$$

We have to take a few variables into the approach, because T_K is not exactly the exponential function. Let's take the derivative: It is $T_k'(t) = \alpha \beta e^{\beta t}$ and inserted in (17.1) we get

$$\beta \alpha e^{\beta t} = c(\alpha e^{\beta t} + \gamma - T_L).$$

For $\gamma = T_L$ and $\beta = c$ this equation is fulfilled, the function

$$T_K(t) = \alpha e^{ct} + T_L \tag{17.2}$$

is a solution of the equation (17.1) for all α. We can also specify α: At time 0, the time of pouring the coffee, the coffee has a fixed temperature T_0, and since e^0 is equal to 1, it follows that $T_0 = \alpha + T_L$, that is, $\alpha = T_0 - T_L$. Fig. 17.1 shows the undisturbed cooling curve.

Now the sugar comes into play: The dissolution of the sugar represents a chemical process that consumes a certain amount of energy, i.e. heat. This leads to a "sudden" cooling by d degrees. Before and after, the exponential cooling curve applies again. In Fig. 17.2 you can see how the temperature develops in this case: In the dashed curve, the sugar is first dissolved and then waiting takes place. The exponential cooling starts with the initial value $T_0 - d$, which in the function (17.2) results in a different initial value α'.

There is a number δ so that $\alpha' = \alpha e^\delta$. For this δ therefore $\alpha' e^{ct} + T_L = \alpha e^{ct+\delta} + T_L$ applies. This means, the dashed function $\alpha e^{ct+\delta} + T_L$ is only shifted parallel to the t-axis by a piece compared to $\alpha e^{ct} + T_L$ from Fig. 17.1.

Fig. 17.1 The cooling

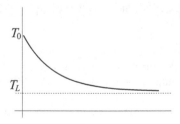

Fig. 17.2 Cooling with sugar

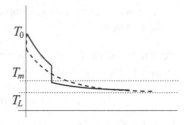

If you wait for a while and then dissolve the sugar, the cooling takes place faster. You can see that in the solid curve.

Unfortunately, the computational determination of the optimal sugar injection time takes longer than breakfast.

An equation of the form $T_K'(t) = c(T_K(t) - T_L)$ is called a *differential equation*. In differential equations, in addition to an unknown function, derivatives of this function also occur. In our first example, we guessed a solution. However, this was only completely determined by another condition: the *initial condition* $T_K(0) = T_0$.

In the next example, we examine a pendulum. In order to have to calculate only in one dimension, we use a spring pendulum: A weight hangs on a spring and can move in y-direction, in the rest position the pendulum should be at the point $y = 0$. We set the pendulum in oscillation and want to determine the displacement $y(t)$ as a function of time t.

Hooke's law is known from physics: With the displacement y, the restoring force $F(t)$ is proportional to y. At any time t therefore $F(t) = a \cdot y(t)$ applies. Here $a < 0$ is the spring constant.

Newton's law states that the acceleration that the weight experiences is proportional to the force acting. The acceleration is the second derivative of the function $y(t)$, and so we get a second equation $F(t) = b \cdot y''(t)$, where $b > 0$. Combined, this results in a differential equation for y:

$$y''(t) = (a/b) \cdot y(t) = c \cdot y(t). \tag{17.3}$$

$c < 0$ is a system constant that is determined by the spring and the weight hanging on it. The second derivative looks similar to the function itself, maybe the exponential function will work again:

$$y(t) = \alpha e^{\beta t} + \gamma, \quad y''(t) = \beta^2 \alpha e^{\beta t}.$$

Substituting into (17.3) we get $\beta^2 \alpha e^{\beta t} = c(\alpha e^{\beta t} + \gamma)$, so

$$(\beta^2 - c)\alpha e^{\beta t} = c\gamma. \tag{17.4}$$

A pretty boring solution would be $\alpha = \gamma = 0$. Also $\beta = 0$ is not interesting. Let's ignore these solutions, so (17.4) must hold for different values of t. This can only be the case if $\beta^2 - c = \gamma = 0$, so $\beta^2 = c$. Because of $c < 0$ there is no real solution for β, the first attempt has failed.

Later we will see that the complex roots can also help us find solutions to the real differential equation.

We know more functions whose second derivative resembles the function: sine and cosine. Let's try it: Let

$$y_1(t) = \alpha \sin(\beta t) + \gamma, \quad y_1''(t) = -\beta^2 \alpha \sin(\beta t).$$

We set $-\beta^2 \alpha \sin(\beta t) = c(\alpha \sin(\beta t) + \gamma)$ in (17.3) again. Then for $\gamma = 0$ and $\beta = \sqrt{-c}$ the function $y_1(t) = \alpha \cdot \sin(\sqrt{-c} \cdot t)$ is a solution of the differential equation (17.3).

Similarly, we find that $y_2(t) = \alpha \cdot \cos(\sqrt{-c} \cdot t)$ represents a solution. Is one of the two solutions the right one? Are there any other solutions?

Initial conditions must still be taken into account: We set the pendulum in oscillation and then start a stopwatch. When the stopwatch starts, the displacement should be 1 meter. This initial condition is not fulfilled by the function $y_1(t)$, because $y_1(0)$ is 0, but by $y_2(t)$: We get the solution $y_2(t) = 1 \cdot \cos(\sqrt{-c} \cdot t)$.

There are other useful initial conditions. The speed of the weight is very important in the moment when the stopwatch starts. We could measure, for example, that the initial speed $y'(0) = v$ is. But a different initial speed than 0 is never possible with the function $y_2(t) = \cos(\sqrt{-c} \cdot t)$, because $y_2'(0) = 0$. Is there an analytical solution at all, which fulfills both initial conditions? And are there perhaps other initial conditions that are necessary to uniquely determine a solution?

In nature, for example in physics, chemistry and biology, but also in technology and economy, problems occur again and again in which there is a relationship between the investigated variables and their temporal or spatial derivatives. These relationships can be described by differential equations. The theory of the solution of such equations is an important field of applied mathematics. In general, the analytical solution of differential equations is very difficult, often even impossible. We will look at some not too complex types of differential equations in this chapter. For these we will carry out the following program:

- Find all possible solutions to a given differential equation.
- Formulate precise initial conditions that describe the system.
- Determine a specific solution of the differential equation that fits these initial conditions.

▶   **Definition 17.1** An equation in which, in addition to the independent variable x and a sought function $y = y(x)$, also their derivatives up to the order n occur, is called *nth order ordinary differential equation* If x_0 is in the domain of y, values $y(x_0), y'(x_0), \ldots, y^{(n-1)}(x_0)$ are called *initial conditions*, the differential equation together with its initial conditions is called *initial value problem*.

If there are ordinary differential equations, then there must of course also be extraordinary ones: These are differential equations that contain more than one variable, for example location and time, or even three-dimensional spatial coordinates. Such equations are called *partial differential equations*. We cannot deal with these here.

An (ordinary) differential equation can therefore be written in the form

$$F(x, y, y', y'', \ldots, y^{(n)}) = 0 \quad (\text{detailed: } F(x, y(x), y'(x), y''(x), \ldots, y^{(n)}(x)) = 0).$$

Examples

1. The coffee problem $T_K'(t) = c(T_K(t) - T_L)$ can be written with $y = T_K(x)$ as $F(x, y, y') = y' - cy - cT_L = 0$.
2. In the pendulum example we get $F(x, y, y', y'') = y'' - cy = 0$. Here x and y' don't appear explicitly in the equation, but that shouldn't bother us.
3. $F(x, y, y') = y' - ay + by^2 = 0$ and $F(x, y, y') = y' + 2xy - 2x = 0$ are further differential equations, for which we cannot make any progress with guessing. For these equations we will develop solution methods. Can you find them? ◀

As we have seen from our first examples, the solutions of differential equations can contain parameters: For example, $y'' = 0$ has the solutions $y = ax + b$ with the independent parameters $a, b \in \mathbb{R}$. The equation $y = 3x + 7$ is a particular solution of $y'' = 0$.

▶ **Definition 17.2** A solution of a *n*th order differential equation is called *general* if it contains n independent parameters. The solution is called *particular* if it contains no parameters.

Example

Solve $y'' = f(x)$. The function y' is an antiderivative of $f(x)$, so

$$y'(x) = \int_{x_0}^x f(t)dt + c.$$

Furthermore y must be an antiderivative of y' that is $y(x) = \int_{x_0}^x y'(s)ds + d$, and therefore:

$$y(x) = \int_{x_0}^x \left(\int_{x_0}^s f(t)dt + c \right) ds + d = \int_{x_0}^x \left(\int_{x_0}^s f(t)dt \right) ds + c(x - x_0) + d,$$

is the general solution of the differential equation $y'' = f(x)$ with the parameters c and d. ◀

The determination of the solution of a differential equation is often referred to as integration.

17.2   First Order Differential Equations

Separable Differential Equations

▷   **Definition 17.3** A first order differential equation $F(x, y, y') = 0$ is called *separable* if it can be written in the form

$$y' = f(x)g(y) \tag{17.5}$$

where $f : I \to \mathbb{R}$, $g : J \to \mathbb{R}$ are continuous functions on intervals I, J.

If for $x_0 \in I$, $y_0 \in J$ the initial condition $y(x_0) = y_0$ is given, and if $g(y) \neq 0$ in the interval J, then (17.5) is solvable. So that I don't have to work with too many different names, I will use the same letter for the integration variable as for the upper limit of integration. Let

$$G(y) := \int_{y_0}^{y} \frac{1}{g(y)} dy, \quad F(x) = \int_{x_0}^{x} f(x) dx$$

be the antiderivatives of $1/g(y)$ respectively $f(x)$. On J is $G'(y) = 1/g(y) \neq 0$, therefore G is strictly monotonic and has an inverse function G^{-1}. Then

$$y(x) := G^{-1}(F(x)) \tag{17.6}$$

is the solution of the initial value problem $y' = f(x)g(y)$, $y(x_0) = y_0$.

We can just try it: It is $G(y(x)) = F(x)$. If we differentiate this equation to the right and left (to the left using the chain rule), we get $G'(y(x)) \cdot y'(x) = F'(x)$, so $1/g(y(x)) \cdot y'(x) = f(x)$, that is $y'(x) = f(x)g(y)$.

It remains to check the initial condition: Because of $G(y_0) = 0$ and $F(x_0) = 0$ we get: $y(x_0) = G^{-1}(F(x_0)) = G^{-1}(0) = y_0$. I summarize the result:

▷   **Theorem 17.4** The initial value problem $y' = f(x)g(y)$, with functions $f : I \to \mathbb{R}$, $g : J \to \mathbb{R}$, and the initial value $y(x_0) = y_0 \in J$ and $g \neq 0$ on J, has the unique solution y, which is obtained by solving the following equation for y.

$$\int_{y_0}^{y} \frac{1}{g(y)} dy = \int_{x_0}^{x} f(x) dx. \tag{17.7}$$

There is a mnemonic which sends a shiver down the spine of the true mathematician, but it works: Write the differential equation in the form $dy/dx = f(x)g(y)$ and then pretend that dy/dx is a regular fraction. Bring all the y in the equation to the left and all the x to the right and you get $(1/g(y))dy = f(x)dx$. Now put integral hooks in front of it and you have (17.7).

1. Let's take another look at the coffee problem $T_K'(x) = c(T_K(x) - T_L)$, $T_K(0) = T_0$. We write it now in the form:

$$y' = \underbrace{c}_{f(x)} \underbrace{(y - T_L)}_{g(y)}$$

and get:

$$\int_{T_0}^{y} \frac{1}{y - T_L} dy = \int_0^x c\, dt \quad \Rightarrow \quad \ln(y - T_L)\big|_{T_0}^{y} = cx\big|_0^x$$

$$\Rightarrow \quad \ln(y - T_L) - \ln(T_0 - T_L) = cx - 0$$

$$\Rightarrow \quad \ln\left(\frac{y - T_L}{T_0 - T_L}\right) = cx.$$

We apply the inverse function of the logarithm:

$$\frac{y - T_L}{T_0 - T_L} = e^{cx} \quad \Rightarrow \quad y = (T_0 - T_L)e^{cx} + T_L,$$

fortunately the same result that we worked out at the beginning of the chapter by guessing.

2. The undisturbed reproduction of a happy rabbit family can be described by the differential equation $y'(x) = c \cdot y(x)$, where $y(x)$ is the size of the population, and $y'(x)$ is the growth at time x. We get exponential growth $y(x) = e^{cx}$ as the solution. Even in Australia, there was a point where the rabbits had reached the end of the line: resources became scarce, growth slowed. This slowed growth can be described by the following differential equation:

$$y'(x) = \underbrace{\alpha}_{f(x)} \cdot \underbrace{y(x)(1 - \beta y(x))}_{g(y)}. \tag{17.8}$$

The term $1 - \beta y(x)$ represents the braking factor: as soon as $y(x)$ comes close to $1/\beta$, growth becomes smaller. For $y(x) = 1/\beta$, we would get zero growth. Let's try the solution with the initial condition $y(0) = y_0$:

$$\int_{y_0}^{y} \frac{1}{y - \beta y^2} dy = \int_0^x \alpha\, dx = \alpha x.$$

You will find the left integral in the formulary: $\int \frac{1}{y - \beta y^2} dy = \ln(\frac{\beta y}{1 - \beta y})$. Try it! This gives:

$$\ln\left(\frac{\beta y}{1 - \beta y}\right) - c = \alpha x \quad \left(c = \ln\left(\frac{\beta y_0}{1 - \beta y_0}\right)\right)$$

and from there by applying the exponential function $\dfrac{\beta y}{1 - \beta y} = e^{\alpha x + c}$.

If we solve this equation for y, we get $y = \dfrac{e^{\alpha x + c}}{\beta(e^{\alpha x + c} + 1)}$.

Fig. 17.3 shows the graph of the function, the rabbit curve. You can see that $\lim_{x \to \infty} y(x) = 1/\beta$ is the limit of growth. If the initial value y_0 is greater than $1/\beta$, we get no growth, but a shrinkage to $1/\beta$.

3. Of course, differential equations can also be solved in which the function $f(x)$ is not constant. Let's take

$$y' = \frac{x}{y - 1}.$$

Since x and y must be defined in an interval, it must be $y > 1$ or $y < 1$. By the initial value $y(1/2) = 1/2$, it is already $y < 1$ predetermined. Now we can integrate:

$$\int_{1/2}^{y} (y - 1)\,dy = \int_{1/2}^{x} x\,dx \quad \Rightarrow \quad \frac{y^2}{2} - y\,\bigg|_{1/2}^{y} = \frac{x^2}{2}\,\bigg|_{1/2}^{x}$$

$$\Rightarrow \quad \frac{y^2}{2} - y - \frac{1}{8} + \frac{1}{2} = \frac{x^2}{2} - \frac{1}{8}$$

$$\Rightarrow \quad \frac{y^2}{2} - y + \frac{1}{2} - \frac{x^2}{2} = 0$$

$$\Rightarrow \quad y^2 - 2y + 1 - x^2 = 0$$

$$\Rightarrow \quad y = 1 \pm \sqrt{1 - 1 + x^2} = 1 \pm x.$$

$y = 1 \pm x$? Are there two solutions? No, from the initial condition it follows that only $y = 1 - x$ can be a solution. ◀

First Order Linear Differential Equations

Differential equations in which the function y and its derivatives only occur linearly are called *linear differential equations*. In the first order, they have the form

$$y' + a(x)y = f(x).$$

Fig. 17.3 The rabbit curve

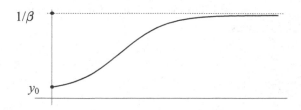

If the function $f(x)$ is equal to 0 on the right side, the equation is called *homogeneous*, otherwise *inhomogeneous*. For homogeneous first order linear equations, the general solution can be given:

▶ **Theorem 17.5** If $a(x)$ is continuous on the interval I, the general solution of the differential equation $y' + a(x)y = 0$ is:

$$y(x) = c \cdot e^{-A(x)},$$

where $c \in \mathbb{R}$ and $A(x)$ is an antiderivative of $a(x)$.   □

By trying it out, one finds that this is the solution.

Now we tackle the inhomogeneous equation. The solution principle of the *variation of constants* helps us here, which is sometimes useful for other differential equations as well. We start from the solution $y_h(x) = c \cdot e^{-A(x)}$ of the homogeneous equation. But now we replace the constant $c \in \mathbb{R}$ by a function $c(x)$ and try to determine it so that $y_s(x) = c(x) \cdot e^{-A(x)}$ becomes a solution of the inhomogeneous equation. We simply insert y_s in the differential equation:

$$\underbrace{c'(x)e^{-A(x)} - a(x)c(x)e^{-A(x)}}_{y'_s} + a(x)\underbrace{c(x)e^{-A(x)}}_{y_s} = f(x).$$

This results in $c'(x)e^{-A(x)} = f(x)$ or $c'(x) = f(x)e^{A(x)}$, i.e. a differential equation for $c(x)$. Any antiderivative

$$c(x) = \int f(x)e^{A(x)}dx$$

is a solution. This gives us the

▶ **Theorem 17.6** The differential equation $y' + a(x)y = f(x)$, $f, a : I \to \mathbb{R}$ continuous, $x_0 \in I$, has the general solution

$$y = \left[\int_{x_0}^{x} f(t)e^{A(t)}dt + c \right] e^{-A(x)},$$

where $A(x)$ is an antiderivative of $a(x)$ and $c \in \mathbb{R}$.

Example

Determine the general solution of the differential equation $y' + 2xy = 2x$.

Here $a(x) = 2x$, an antiderivative is $A(x) = x^2$. The general solution of the equation according to Theorem 17.6 is

$$y(x) = \left[\int_0^x 2t \cdot e^{t^2} dt + c \right] e^{-x^2} \underset{\substack{\uparrow \\ y=t^2, dy=2tdt}}{=} \left[\int_0^{x^2} e^y dy + c \right] e^{-x^2}$$

$$= (e^{x^2} \underbrace{- 1 + c}_{=d})e^{-x^2} = 1 + de^{-x^2}$$

with $d \in \mathbb{R}$. The check gives: $-2xde^{-x^2} + 2x(1 + de^{-x^2}) = 2x$. ◄

17.3 nth Order Linear Differential Equations

▶ **Definition 17.7** A differential equation of the form

$$y^{(n)} + a_1(x)y^{(n-1)} + \cdots + a_{n-1}(x)y' + a_n(x)y = f(x) \qquad (17.9)$$

is called a *nth order linear differential equation* The functions $a_i, f : I \to \mathbb{R}$ are assumed to be continuous on the interval I. If $f = 0$, the equation is called *homogeneous*, otherwise *inhomogeneous*.

The following existence and uniqueness theorem holds for these differential equations, which I would like to quote without proof:

▶ **Theorem 17.8** Let $y^{(n)} + a_1(x)y^{(n-1)} + \cdots + a_{n-1}(x)y' + a_n(x)y = f(x)$ be a nth order linear differential equation, $a_i, f : I \to \mathbb{R}$ and $x_0 \in I$. Then there is a unique solution y of this initial value problem for the initial values $y(x_0) = b_0, y'(x_0) = b_1$, $\ldots, y^{(n-1)}(x_0) = b_{n-1}$. This solution exists on the whole interval I.

A nice result, but once again a typical mathematician's theorem: It doesn't help us at all to find solutions, it can only serve to calm us down when we get stuck on a specific task. More interesting are the two following theorems, which give us information about the structure of the solutions and help us to sort the solutions and check them for completeness. And to prove Theorem 17.10 we need Theorem 17.8, so that one is not completely superfluous either.

▶ **Theorem 17.9** The set H of solutions $y : I \to \mathbb{R}$ of the homogeneous linear differential equation $y^{(n)} + a_1(x)y^{(n-1)} + \cdots + a_{n-1}(x)y' + a_n(x)y = 0$, $a_i : I \to \mathbb{R}$ is a $\mathbb{R}$-vector space. Each solution y of the inhomogeneous equation $y^{(n)} + a_1(x)y^{(n-1)} + \cdots + a_n(x)y = f(x)$, $a_i, f : I \to \mathbb{R}$ has the form $y_s + y_h$, where $y_h \in H$ and y_s is a particular solution of the inhomogeneous differential equation.

Similarly to the solution of systems of linear equations, we therefore only have to determine one solution of the inhomogeneous equation in addition to all solutions of the homogeneous equation to solve the system completely.

The set H is a subset of the vector space of all real functions on I. We therefore only have to check the subspace criterion (see Theorem 6.4 in Section 6.2). According to Theorem 17.8, H is not empty, and are $y_1, y_2 \in H$, $\lambda \in \mathbb{R}$, then:

$$(y_1 + y_2)^{(n)} + a_1(x)(y_1 + y_2)^{(n-1)} + \cdots + a_n(x)(y_1 + y_2)$$
$$= (y_1^{(n)} + a_1(x)y_1^{(n-1)} + \cdots + a_n(x)y_1)$$
$$+ (y_2^{(n)} + a_1(x)y_2^{(n-1)} + \cdots + a_n(x)y_2) = 0,$$
$$(\lambda y_1)^{(n)} + a_1(x)(\lambda y_1)^{(n-1)} + \cdots + a_n(x)(\lambda y_1)$$
$$= \lambda(y_1^{(n)} + a_1(x)y_1^{(n-1)} + \cdots + a_n(x)y_1) = 0,$$

so H is a vector space.

Similarly, one shows: If y_t is in addition to y_s another solution of the inhomogeneous equation, then $y_t - y_s \in H$, that is $y_t = y_s + y_h$ for a $y_h \in H$.   □

▶   **Theorem and Definition 17.10** The solution space of a homogeneous nth order linear differential equation has dimension n. A basis of this solution space is called a *fundamental system.*

According to Theorem 17.8, we can construct solutions $y_0, y_1, \ldots, y_{n-1}$ for a $x_0 \in I$ which satisfy the initial conditions

$$\begin{pmatrix} y_0(x_0) \\ y_0'(x_0) \\ \vdots \\ y_0^{(n-1)}(x_0) \end{pmatrix} = \begin{pmatrix} 1 \\ 0 \\ \vdots \\ 0 \end{pmatrix}, \quad \begin{pmatrix} y_1(x_0) \\ y_1'(x_0) \\ \vdots \\ y_1^{(n-1)}(x_0) \end{pmatrix} = \begin{pmatrix} 0 \\ 1 \\ \vdots \\ 0 \end{pmatrix}, \quad \cdots, \quad \begin{pmatrix} y_{n-1}(x_0) \\ y_{n-1}'(x_0) \\ \vdots \\ y_{n-1}^{(n-1)}(x_0) \end{pmatrix} = \begin{pmatrix} 0 \\ 0 \\ \vdots \\ 1 \end{pmatrix}$$

$$(17.10)$$

These solutions form a basis, because they are linearly independent and generate the whole space:

First, I show that the set is generating: Let y be any solution of the differential equation. Then there are real numbers $b_0, b_1, \ldots, b_{n-1}$ with the property $y(x_0) = b_0$, $y'(x_0) = b_1$, ..., $y^{(n-1)}(x_0) = b_{n-1}$. As we have constructed our solutions, this initial condition also applies to the linear combination $z = b_0 y_0 + b_1 y_1 + \cdots + b_{n-1} y_{n-1}$. Since the solution with this initial condition is unique, it follows $z = y$.

To show linear independence, we still have to elaborate a little. I include the following assertion about the Wronskian:

▶ **Definition 17.11** Let $y_0, y_1, \ldots, y_{n-1}: I \to \mathbb{R}$ $(n-1)$-times differentiable. Then the function

$$W(x) := W[y_0, y_1, \ldots, y_{n-1}](x) := \det \begin{pmatrix} y_0(x) & y_1(x) & \cdots & y_{n-1}(x) \\ y_0'(x) & y_1'(x) & \cdots & y_{n-1}'(x) \\ \vdots & \vdots & \ddots & \vdots \\ y_0^{(n-1)}(x) & y_1^{(n-1)}(x) & \cdots & y_{n-1}^{(n-1)}(x) \end{pmatrix}$$

is called *Wronskian* of $y_0, y_1, \ldots, y_{n-1}$.

▶ **Theorem 17.12** If the $(n-1)$-times differentiable functions $y_0, y_1, \ldots, y_{n-1}: I \to \mathbb{R}$ are linearly dependent, then $W[y_0, y_1, \ldots, y_{n-1}](x) = 0$ for all $x \in I$.

What does linear independence of functions mean? If a linear combination $\lambda_0 y_0 + \lambda_1 y_1 + \cdots + \lambda_{n-1} y_{n-1} = \vec{0}$ yields the zero vector, then each coefficient must be 0. If the functions are linearly dependent, there is such a linear combination in which not all $\lambda_i = 0$. The zero vector $\vec{0}$ represents the zero function, which has the value 0 everywhere. By differentiation, we successively obtain

$$\lambda_0 y_0 + \lambda_1 y_1 + \cdots + \lambda_{n-1} y_{n-1} = \vec{0},$$
$$\lambda_0 y_0' + \lambda_1 y_1' + \cdots + \lambda_{n-1} y_{n-1}' = \vec{0},$$

$$\vdots$$

$$\lambda_0 y_0^{(n-1)} + \lambda_1 y_1^{(n-1)} + \cdots + \lambda_{n-1} y_{n-1}^{(n-1)} = \vec{0}.$$

This means nothing else than for all $x \in I$ the vector $(\lambda_0, \lambda_1, \ldots, \lambda_{n-1})$, which is different from $\vec{0}$, is a solution to the homogeneous linear equation system

$$\begin{pmatrix} y_0(x) & y_1(x) & \cdots & y_{n-1}(x) \\ y_0'(x) & y_1'(x) & \cdots & y_{n-1}'(x) \\ \vdots & \vdots & \ddots & \vdots \\ y_0^{(n-1)}(x) & y_1^{(n-1)}(x) & \cdots & y_{n-1}^{(n-1)}(x) \end{pmatrix} \begin{pmatrix} x_0 \\ x_1 \\ \vdots \\ x_{n-1} \end{pmatrix} = \begin{pmatrix} 0 \\ 0 \\ \vdots \\ 0 \end{pmatrix}$$

As we have learned in Theorem 9.4 in Sect. 9.1, this is only possible if the determinant of the coefficient matrix is 0.   □

Now back to the proof of linear independence of the solution functions (17.10): By contraposition to Theorem 17.11 it holds: If the Wronskian is anywhere unequal 0, then the functions contained therein are linearly independent. At the point x_0 the found solutions however yield as Wronskian just the determinant of the identity matrix. Thus Theorem 17.10 is proven.   □

Let's look at the example of the pendulum at the beginning of the chapter again: We had guessed for the linear differential equation $y''(t) + 0 \cdot y'(t) - c \cdot y(t) = 0$ the two solutions:

$$y_1(t) = \cos \beta t, \quad y_2(t) = \sin \beta t,$$

with $\beta = \sqrt{-c}$. Are these solutions linearly independent? We calculate $W(x)$:

$$W(x) = \det \begin{pmatrix} \cos(\beta x) & \sin(\beta x) \\ -\beta \sin(\beta x) & \beta \cos(\beta x) \end{pmatrix} = \beta \cos^2(\beta x) + \beta \sin^2(\beta x) = \beta \neq 0.$$

This means that y_1 and y_2 form a basis. Now we also know that the initial conditions $y(0) = 1$, $y'(0) = v$, which we had set in this example, were reasonable and complete. The unique solution of the initial value problem is $y(t) = 1 \cdot \cos \beta t + (v/\beta) \cdot \sin \beta t$.

Linear Differential Equations with Constant Coefficients

For differential equations in the form (17.9) there is no general solution method. If the coefficient functions however are constant real numbers, one can specify a basis of the solution space. We first solve the homogeneous system

$$y^{(n)} + a_1 y^{(n-1)} + \cdots + a_{n-1} y' + a_n y = 0. \tag{17.11}$$

We try the exponential function $e^{\lambda x}$. Then

$$y(x) = e^{\lambda x}, \quad y'(x) = \lambda e^{\lambda x}, \quad y''(x) = \lambda^2 e^{\lambda x}, \quad \ldots, \quad y^{(n)}(x) = \lambda^n e^{\lambda x}.$$

Inserted in (17.11) we obtain for all x:

$$\lambda^n e^{\lambda x} + a_1 \lambda^{n-1} e^{\lambda x} + \cdots + a_{n-1} \lambda^1 e^{\lambda x} + a_n e^{\lambda x} = (\lambda^n + a_1 \lambda^{n-1} + \cdots + a_{n-1} \lambda + a_n) e^{\lambda x}$$
$$= 0.$$

If λ is a root of $\lambda^n + a_1 \lambda^{n-1} + \cdots + a_{n-1} \lambda + a_n$, then $e^{\lambda x}$ is a solution of (17.11). So we have to investigate this polynomial:

▶ **Definition 17.13** The polynomial $p(\lambda) := \lambda^n + a_1 \lambda^{n-1} + \cdots + a_{n-1} \lambda + a_n$ is called *characteristic polynomial* of the differential equation $y^{(n)} + a_1 y^{(n-1)} + \cdots + a_{n-1} y' + a_n y = 0$.

From the roots of the characteristic polynomial we can construct a fundamental system for (17.11). We have to distinguish the following cases:

Case 1: λ is a simple real root. Then $e^{\lambda x}$ is a solution of the differential equation.

Case 2: $\lambda = \alpha + i\beta$ is a simple complex root. Then $e^{\lambda x}$ provides a complex solution to the differential equation: $y\colon \mathbb{R} \to \mathbb{C}$, $x \mapsto e^{\lambda x}$ satisfies (17.11). It is easy to check that then the real part and the imaginary part of $e^{\lambda x}$ are solutions. Now

$$e^{\lambda x} = e^{(\alpha+i\beta)x} = e^{\alpha x}e^{i\beta x} = e^{\alpha x}(\cos \beta x + i\sin \beta x).$$

So $e^{\alpha x} \cos \beta x$ and $e^{\alpha x} \sin \beta x$ are solution functions.

Now one root has given us two solutions. Since there can only be n linearly independent solutions in total, it looks as if we could get too many solutions. But that's not the case: With $\alpha + i\beta$, also $\alpha - i\beta$ is a root of $p(\lambda)$ (see Theorem 5.41 in Section 5.6), and from the conjugate root we get the solutions $e^{\alpha x} \cos(-\beta x) = e^{\alpha x} \cos \beta x$ and $e^{\alpha x} \sin(-\beta x) = -e^{\alpha x} \sin \beta x$, so no new linearly independent functions.

Case 3: Multiple roots occur. I would like to give the corresponding solutions without further calculation: If λ is a k-fold real root, then the k functions $x^i e^{\lambda x}, i = 0, \ldots, k-1$ are linearly independent solutions and $\lambda = \alpha + i\beta$ (and thus also $\alpha - i\beta$) is a k-fold complex root, then the $2k$ functions $x^i e^{\alpha x} \cos \beta x$, $x^i e^{\alpha x} \sin \beta x$, $i = 0, \ldots, k-1$ are the corresponding solutions.

In each case, we have therefore found exactly n solutions from the polynomial $p(\lambda)$. With the help of the Wronskian, one can determine that they are all linearly independent. They form a fundamental system of the differential equation (17.11).

Examples

1. $y'' - 2y' + y = 0$ has the characteristic polynomial $\lambda^2 - 2\lambda + 1$ with the roots $\lambda_{1/2} = 1 \pm \sqrt{1-1} = 1$, so a double root. A fundamental system is e^x, $x \cdot e^x$, the general solution is $y(x) = ae^x + bxe^x$. Let's try the two solutions:

$$y = e^x: \qquad\qquad\qquad e^x - 2e^x + e^x = 0,$$
$$y = xe^x: \quad \underbrace{(xe^x + e^x + e^x)}_{y''} - 2\underbrace{(xe^x + e^x)}_{y'} + xe^x = 0.$$

2. $y'' - 2y' + 5y = 0$ has the characteristic polynomial $\lambda^2 - 2\lambda + 5$ with the roots $\lambda_{1/2} = 1 \pm \sqrt{1-5} = 1 \pm 2i$. A fundamental system consists of $e^x \cos 2x$, $e^x \sin 2x$. Let's test the first solution: It is

$$y' = e^x \cos 2x - 2e^x \sin 2x,$$
$$y'' = e^x \cos 2x - 2e^x \sin 2x - 2e^x \sin 2x - 4e^x \cos 2x$$

and therefore

$$\underbrace{e^x \cos 2x - 4e^x \sin 2x - 4e^x \cos 2x}_{y''} \underbrace{-2e^x \cos 2x + 4e^x \sin 2x}_{-2y'} + 5e^x \cos 2x = 0.$$

3. We have already examined the differential equation of the pendulum. Usually it is written in the form $y'' + \omega_0^2 y = 0$, it has the solutions $\sin \omega_0 t$ and $\cos \omega_0 t$. The oscillation period is $T = 2\pi/\omega_0$. The number $\omega_0 = 2\pi/T$ represents the oscillation frequency of the system, that is, the number of oscillations per unit of time, up to the factor 2π. The differential equation arose from the assumption that the driving force $F = my''$ is proportional to the displacement y. With the calculated solutions, the pendulum continues to swing infinitely. Of course, this is not true in reality, there is a deceleration due to friction. How can one integrate this damping into the differential equation? The restoring force is reduced by friction. Experimentally, it is found that the friction force is proportional to the velocity of the pendulum, that is, $F_R = \gamma y'$. This gives us the differential equation that describes a damped oscillation (spring constant, weight and friction are summarized in the parameters $\alpha > 0$ and ω_0, the names are convention):

$$y'' + 2\alpha y' + \omega_0^2 y = 0.$$

The characteristic polynomial is $\lambda^2 + 2\alpha\lambda + \omega_0^2 = 0$ and has the roots

$$\lambda_{1/2} = -\alpha \pm \sqrt{\alpha^2 - \omega_0^2} = -\alpha \pm \beta.$$

Here two cases have to be distinguished: Is $\alpha^2 - \omega_0^2 > 0$, then we get two real roots, both of which are less than 0. Here α is relatively large, there is a strong damping. Imagine that the pendulum is dipped in honey. The general solution is

$$y = c_1 e^{\lambda_1 x} + c_2 e^{\lambda_2 x}.$$

In this case, no proper oscillation takes place at all, the function has at most one maximum and at most one root. Fig. 17.4 shows some possible shapes of the curve. In the second case $\alpha^2 - \omega_0^2 < 0$ we get two complex roots $-\alpha \pm i\omega_1$ with $\omega_1 = \sqrt{\omega_0^2 - \alpha^2}$. Here α is relatively small, so there is a weak damping, for example, the pendulum swings in the air. The general solution here is

$$y = c_1 e^{-\alpha x} \sin \omega_1 x + c_2 e^{-\alpha x} \cos \omega_1 x.$$

Fig. 17.4 The strong damping

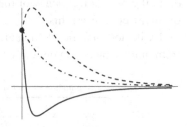

Fig. 17.5 The weak damping

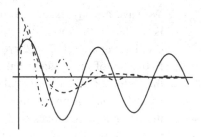

The curve represents an oscillation with the frequency $\omega_1 < \omega_0$ whose amplitude decreases. The oscillation therefore proceeds somewhat slower than undamped. Fig. 17.5 shows examples of this solution.

The limiting case $\alpha^2 = \omega_0^2$ results in the solution

$$y = (c_1 + c_2 x)e^{-\alpha x}.$$

The course of the curve corresponds here to that of the strong damping. ◀

Inhomogeneous Linear Differential Equations

The determination of a particular solution in the inhomogeneous case is generally very difficult. There are many approaches to this, one of which consists again of the variation of the constants: If $y_1, y_2, \ldots, y_n$ is a fundamental system of the homogeneous equation, one tries to determine functions $c_i(x)$ so that $c_1(x)y_1 + c_2(x)y_2 + \cdots + c_n(x)y_n$ is a solution of the inhomogeneous system.

This solution can still be written down—with many integral hooks—but often it can no longer be calculated analytically.

As an example, we look at the pendulum equation one last time, this time it is driven by an external force: We give the pendulum a push at regular intervals. For simplicity, we assume that this additional force is periodically applied in the form of a cosine function. We obtain the inhomogeneous differential equation

$$y'' + 2\alpha y' + \omega_0^2 y = a \cos \omega t = f(t). \tag{17.12}$$

ω_0 is the natural frequency of the pendulum, the frequency ω of the driving force can be quite different from this.

I will not carry out the integration, but I would like to tell you the result, which has a tremendous impact on our daily lives. The solution is still calculable analytically, it reads:

$$y(t) = \underbrace{\frac{a}{\sqrt{(\omega_0^2 - \omega^2)^2 + 4\alpha^2\omega^2}}}_{\text{amplitude of oscillation}} \cdot \cos(\underbrace{\omega t}_{\substack{\text{oscillation} \\ \text{with frequency} \\ \text{of driving force}}} - \underbrace{\varphi}_{\substack{\text{phase} \\ \text{is shifted}}}), \quad \varphi = \arctan\frac{2\alpha\omega}{\omega_0^2 - \omega^2}.$$

If ω_0 is close to ω, and the damping (that is, α) is small, the denominator of the amplitude will be very small, and the amplitude can be very large. The further ω is from ω_0, the smaller the amplitude of the oscillation will be.

The differential equation (17.12) not only describes the spring pendulum, but in principle any oscillating system. The negative effect of the presented solution is the resonance desaster: The opera singer can make glasses explode, storms can be able to set bridges into oscillation so that they tear. Look on the internet for the Tacoma Narrows Bridge! Positive effects are experienced in the rocking chair and when listening to the radio or making a phone call: The electrical oscillating circuit in the receiver consists of a capacitor and an inductor. The natural frequency of the system can be adjusted by changing the capacity of the capacitor to the transmission frequency so that the oscillating circuit is exactly driven by this and no other frequency.

17.4 Comprehension Questions and Exercises

Comprehension Questions

1. What is a separable differential equation?
2. Let a nth order linear differential equation be given. Can one then specify not only the initial values $y(x_0), y'(x_0), \ldots, y^{(n-1)}(x_0)$ but also $y^{(n)}(x_0)$?
3. If f and g are solutions of a linear differential equation, is then $f \cdot g$ also a solution of the differential equation?
4. If f and g are solutions of a linear differential equation, then $f + g$ is also a solution. Does this statement also hold for nonlinear differential equations?
5. Are $\sin(x)$ and $\cos(x + \pi/2)$ linearly independent functions?

Exercises

1. Solve the initial value problems:
   a) $y' = \frac{1+x}{y}$, $y(1) = 1$

   b) $y' = xy + 2y$, $y(1) = 2$
2. Find the general solutions of the differential equations:
   a) $y' + \frac{1}{x}y = \sin x$
   b) $y' + (2x - 1)y = xe^x$
3. Determine a fundamental system of solutions of the differential equations:
   a) $y''' - 3y'' + 4y' - 2y = 0$
   b) $y'' - 6y' + 9y = 0$
4. Show that the solutions of the differential equation $y' + a(x)y = f(x)$ given in Theorem 17.5 and Theorem 17.6 include all possible solutions. Use Theorem 17.10 for this.
5. Show that the following functions are linearly independent:
   a) $e^{\alpha x}, e^{\beta x}$ $(\alpha \neq \beta)$
   b) $e^{\alpha x}, xe^{\alpha x}$

Numerical Methods 18

Abstract

The final chapter from the second part of the book familiarizes you with applications of theoretical mathematics to specific computational tasks. At the end of this chapter

- you know that calculation errors are inevitable and can estimate their size and their propagation during calculation operations,
- you can calculate zeros and fixed points of nonlinear equations using different methods,
- you can determine smooth interpolation curves between given points in the $\mathbb{R}^2$,
- you can solve integrals numerically,
- and determine solutions to first order differential equations.

18.1 Problems with Numerical Calculations

I don't need to explain to you the importance of computers for mathematical calculations. The largest civilian concentrations of computers are at weather services and large film studios in America. Everywhere, equations are solved, roots are determined, functions and differential equations are integrated and curves, shapes and shadows are calculated like mad.

Supplementary Information The online version contains supplementary material available at https://doi.org/10.1007/978-3-658-40423-9_18.

Leaving aside the ever more powerful computer algebra systems, however, computers do not calculate with formulas, but with numbers, and in doing so they make errors in calculation; not because they would miscalculate, but because they simply cannot represent real numbers accurately enough. It can happen that such errors amplify during long calculations and lead to unusable results. Before we therefore deal with some important numerical algorithms, I would like to explain the problem of uncertainty of calculation to you in more detail. I will do this using the example of solving systems of linear equations. In Chap. 8.8 we dealt extensively with the Gaussian algorithm. This also represents the first important numerical method.

Real Numbers in the Computer

If you carry out a calculation on the computer and have the result displayed on the screen, it usually appears in the form

$$\pm a_1.a_2a_3 \ldots a_n\mathrm{E}\pm m. \tag{18.1}$$

This is the *floating point representation* of the number. The sequence of digits $a_1a_2a_3 \ldots a_n$ is called the *mantissa* of the number, the coefficient a_1 is always unequal to 0. $\mathrm{E}\pm m$ stands for "$\cdot 10^{\pm m}$".

This is exactly how real numbers are stored in the computer, but not in decimal form, but to the base 2 (see also the end of Sec. 13.3).

To store a real number, just like for an integer, only a finite memory space is available, which has to store the sign, mantissa and exponent. In Java, for example, this is 8 bytes for a double, i.e. 64 bits. Numbers like $\sqrt{2}$ or π cannot be represented precisely in this way and, of course, only finitely many numbers can be stored in total.

Every real number that we enter into the computer, or that arises during a calculation, is rounded to the next number representable in this form. This results in errors. We distinguish between the absolute error a_x and the relative error r_x. If x' is the representation of the real number x in the computer, then applies:

$$a_x = x - x', \quad r_x = (x - x')/x. \tag{18.2}$$

Of particular interest is of course the relative error, which sets the absolute error in relation to the number under consideration: A large absolute error is much less important for a very large number than for a small number.

The floating-point representation (18.1) now has the advantage that it limits the relative error for numbers of all orders of magnitude: For the following, let us consider a beginner's computer with a mantissa length of 3 that works in the decimal system. We want to represent the numbers $x = 1\,234\,567$ and $y = 0.0007654321$ in it. In the floating-point representation applies:

$$x = 1\,234\,567 = 1.234567 \cdot 10^6 \mapsto 1.23 \cdot 10^6$$
$$y = 0.0007654321 = 7.654321 \cdot 10^{-4} \mapsto 7.65 \cdot 10^{-4}$$

and thus

$$a_x = 4567 \qquad r_x \approx 0.0037$$
$$a_y = 0.0000004321 \quad r_y \approx 0.00056.$$

How large is the maximum relative error? The worst case occurs when we enter the number $x = a_1.a_2a_35 \cdot 10^m$, it is rounded to $a_1.a_2(a_3 + 1) \cdot 10^m$. This makes the relative error

$$\frac{-0.005 \cdot 10^m}{a_1.a_2a_35 \cdot 10^m} = \frac{-0.005}{a_1.a_2a_35}.$$

This fraction is largest when the denominator is smallest, that is, when $x = 1.005 \cdot 10^m$, and we get a relative error of about $4.98 \cdot 10^{-3}$, regardless of the size of the exponent.

You can now easily derive the rule that for a mantissa length of n the maximum relative error is:

$$|r_{max}| \le 5 \cdot 10^{-n}.$$

Similarly, in the binary system, with a mantissa length of n, the maximum relative error is $|r_{max}| \le 2^{-n}$.

Propagation of Errors

If you carry out mathematical operations with the already rounded numbers, the errors will affect the result, and it can also happen that error-free numbers can no longer be represented precisely after an operation and therefore have to be rounded.

Let's think for a moment about what happens to the relative error when adding and multiplying: We use $r_x x = x - x'$ and $x' = x(1 - r_x)$. This follows directly from (18.2). For simplicity, we assume the sum and product of x' and y' are again numbers that can be represented in the computer. Then we get:

$$r_{x+y} = \frac{(x + y) - (x' + y')}{x + y} = \frac{(x - x') + (y - y')}{x + y}$$

$$= \frac{r_x x + r_y y}{x + y} = r_x \frac{x}{x + y} + r_y \frac{y}{x + y}, \qquad (18.3)$$

$$r_{xy} = \frac{xy - x'y'}{xy} = \frac{xy - x(1 - r_x)y(1 - r_y)}{xy}$$

$$= 1 - (1 - r_x)(1 - r_y) = r_x + r_y - r_x r_y. \qquad (18.4)$$

Since the relative errors are small numbers, $r_x r_y$ is very small compared to r_x or r_y and therefore $r_{xy} \approx r_x + r_y$.

In multiplication the relative error behaves quite wel, the two relative errors add up. In addition nothing particularly bad happens at first, when x and y have the same sign: Then the fractions $x/(x + y)$ and $y/(x + y)$ are each less than 1 and here too the worst that can happen is an addition of the relative errors.

A catastrophe can occur, however, if x and y are about the same size and have different signs: Then the values $x/(x + y)$ and $y/(x + y)$ are very large because the denominators are small, and the relative error can explode! This behavior is called *catastrophic cancellation*.

Never try to check the difference of two real numbers against 0 in a logical condition, it can have fatal consequences!

Calculation Errors in Systems of Linear Equations

The numerical examples that I presented to you in Chap. 8 for the Gaussian algorithm consisted of integers and were designed so the transformations usually resulted in integers again; I chose them so that the methods are easy to follow. Unfortunately, reality is different. In one and the same system of equations, very large, very small and very crooked non-integers can occur. Let's take a look at what can happen to systems of linear equations due to the described errors. We solve a few systems of equations with our computer, which works with 3-digit mantissas. If more than three significant digits occur in an operation, we round.

First, I want to calculate the solution of the following system of equations $Ax = b$:

$$(A, b) = \begin{pmatrix} 1 & 1 & 2 \\ 1 & 1.001 & 2.001 \end{pmatrix}.$$

The rank of A and that of (A, b) is 2, so there is a unique solution that you can easily determine to $(1, 1)$.

Unfortunately, when entering this system, the numbers have to be rounded, because 1.001 and 2.001 have 4 significant digits. So in our computer only the following system of equations arrives:

$$(A, b) = \begin{pmatrix} 1 & 1 & 2 \\ 1 & 1 & 2 \end{pmatrix}.$$

The rank of this matrix is 1, the solution space is the set $\{(1, 1) + \lambda(1, -1) \mid \lambda \in \mathbb{R}\}$. Already at the input, a system of linear equations can thus get completely different properties due to rounding errors.

Now let's look at a system of equations that does not undergo any rounding on entry:

$$(A, b) = \begin{pmatrix} 203 & 202 & 406\,000 \\ 1 & 1 & 2010 \end{pmatrix}. \tag{18.5}$$

The rank of the matrix is 2, so there is a unique solution again. Calculating by hand gives $x_1 = -20$, $x_2 = 2030$.

Now let's solve the equations with the computer using the Gaussian algorithm, see Sect. 8.1. When performing, I always round at the third digit different from 0 and still write "=" cheeky to it, just like our computer does. First we have to subtract the $1/203 = 0.00493$-fold of row 1 from row 2 and get:

$$
\begin{pmatrix}
 & 203 & 202 & 406\,000 \\
1 - \underbrace{0.00493 \cdot 203}_{1.00} & 1 - \underbrace{0.00493 \cdot 202}_{0.996} & 2010 - \underbrace{0.00493 \cdot 406\,000}_{2000}
\end{pmatrix}
$$

$$
= \begin{pmatrix}
203 & 202 & 406\,000 \\
0 & 0.00400 & 10
\end{pmatrix}.
$$

Now we determine x_2 from row 2: $0.004 \cdot x_2 = 10$, so $x_2 = 2500$. We set this in row 1, and get $203 \cdot x_1 + 202 \cdot 2500 = 406\,000$, so $203 \cdot x_1 = 406\,000 - 505\,000 = -99\,000$, and finally $x_1 = -488$, so a solution that is completely off. You shouldn't sell that to anyone!

The disaster is completed by the test:

$$203 \cdot (-488) + 202 \cdot 2500 = -99\,100 + 505\,000 = 406\,000,$$

$$1 \cdot (-488) + 1 \cdot 2500 = 2010,$$

apparently everything is right.

There is a simple trick with which one can often increase the computational accuracy when solving systems of linear equations numerically:

Look back again into the Gaussian algorithm. In the key operation, the (a_{kj}/a_{ij})-fold of row k is subtracted from row i (cf. (8.6) after Theorem 8.3):

$$a_{kl} - (a_{kj}/a_{ij}) \cdot a_{il}, \quad l = 1,\dots,n \tag{18.6}$$

These are operations in which cancellation can occur. We do not have the values of the matrix elements in our hands, but we can control the size of the quotient $q = a_{kj}/a_{ij}$ by choosing the pivot element a_{ij}. According to (18.3) and (18.4), the relative error of an operation $a - q \cdot b$ as in (18.6) results in:

$$\frac{a}{a - qb} r_a + \frac{qb}{a - qb}(r_q + r_b).$$

In any case, it has a positive effect on the error size if q is as small as possible, because then the influence of the second term of the sum on the error becomes smaller.

In implementations of the Gaussian algorithm a *pivoting* is therefore carried out: If we want to make all elements below a_{ij} (the pivot element) to zero, we not only check whether the element $a_{ij} = 0$ (we know by now that this is anyway problematic), but we also search for the row below a_{ij} whose j-th element is maximum. Then this row is swapped with the i-th row. The quotient (a_{ki}/a_{ij}), with which row i is subsequently multiplied, is therefore always less than or equal to 1 for all $k > i$.

In addition to the described partial pivoting, a complete pivoting can also be carried out, that is, columns can be exchanged to make the pivot element as large as possible. However, one must be careful here: When swapping columns, the indices of the unknowns must also be exchanged in order to obtain the correct results at the end.

Unfortunately, pivoting is not a panacea, as the equation system (18.5) shows. Here the pivot element is already maximum. Developing good strategies for solving systems of linear equations is not trivial!

In the examples you have seen: Small rounding errors in the input and in the calculations can have a huge impact on the results. This can go so far that the results are completely worthless. Numerical problems with this property are called *ill-conditioned*. A major task of numerical analysis is to formulate problems in such a way that they are well-conditioned.

What else can you do to get problems with computational errors in your computer programs under control? Some people think we don't need much mathematics here, since we solve our problems with the computer. But this is wrong, I would like to emphasize the importance of mathematics: The further you treat a problem analytically and the later you start doing numerical calculations, the fewer errors you will make.

It is also important to choose good algorithms: If you need fewer operations, you usually not only increase the speed, but also the accuracy of the calculation.

In any case, you must be aware of the problem that rounding errors can occur and not rely on exact calculations.

18.2   Nonlinear Equations

To solve systems of linear equations, we used the Gaussian algorithm in the last section. Nonlinear functions are often not solvable analytically. Other numerical methods must be found for this. I would like to focus here on the case of an equation with one unknown. Let's start with an example. The solutions of the equation

$$e^{-x} + 1 = x.$$

are sought. This equation can also be written in another form:

$$e^{-x} + 1 - x = 0.$$

The first way of writing is of the form $F(x) = x$. Here we call x a *fixed point* of the function F. The second way of writing has the form $G(x) = 0$, and x is a *zero point* of G. I would like to introduce you to methods for determining fixed points and determining zero points. You can see from the example equations can often be brought into one form or the other as required. Unfortunately, not all methods always lead to the goal and the convergence speed of the methods is also different. First a definition that helps with the assessment of convergence:

▶ **Definition 18.1** Let $(x_n)_{n \in \mathbb{N}}$ be a convergent sequence of real numbers with limit x. Then x_n is called *linearly convergent* with *rate of convergence c*, if there is a number $0 < c < 1$ such that

$$|x_{n+1} - x| \leq c \cdot |x_n - x|.$$

x_n is called *convergent with order of convergence q*, if there is a $0 < c$ with

$$|x_{n+1} - x| \leq c \cdot |x_n - x|^q.$$

In the case $q = 2$, x_n is called *quadratically convergent*.

Note that in the case of order of convergence $q > 1$, the factor c does not have to be < 1 anymore.

Calculation of Fixed Points

Let us first try to determine the fixed points. One approach could be to begin with a starting value x_0 and to determine $x_1 = F(x_0)$. Then we evaluate $F(x_1)$: $x_2 = F(x_1)$ and so on:

$$x_n = F(x_{n-1}), \quad n = 1, 2, 3, \ldots \tag{18.7}$$

Can we hope the sequence x_n converges? Maybe even to a fixed point? If $F : \mathbb{R} \to \mathbb{R}$ is a function of one variable, then a fixed point $\hat{x}$ with $F(\hat{x}) = \hat{x}$ is an intersection of the graph of F with the main diagonal, and we can visualize the sequence x_n.

As you can see, the sequence converges to the fixed point in Fig. 18.1, in Fig. 18.2 it does not. Draw a few graphs yourself and try to find out when convergence occurs and when it does not. Do you have an idea? It seems to depend on the slope of the function: If it is too steep, it does not work anymore. If it is flat, whether rising or falling, then the sequence x_n approaches the fixed point more and more closely. In fact, it can be shown: If the function values of F are always closer together than the arguments, then convergence occurs. Behind this is a famous theorem that not only applies to real functions,

Fig. 18.1 Convergence to the fixed point

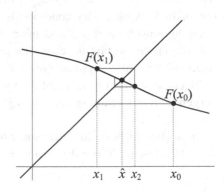

Fig. 18.2 No convergence to
the fixed point

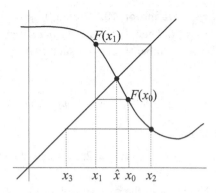

but to all functions in "complete metric spaces". I formulate it for functions in $\mathbb{R}^n$. It is named after the Polish mathematician Stefan Banach, who proved it in 1922. It is therefore one of the newer mathematical results in this book.

▶   **Definition 18.2** Let $D \subset \mathbb{R}^n$ and $f: D \to \mathbb{R}^m$ be a mapping. f is called a *contraction*, if there is a $c \in \mathbb{R}$, $c < 1$ with the property

$$\|f(x) - f(y)\| < c\|x - y\| \quad \text{for all } x, y \in D.$$

▶   **Theorem 18.3: The Banach fixed point theorem** Let $D \subset \mathbb{R}^n$ be a closed subset and $f: D \to D$ a contraction mapping of D in itself. Then holds:

1. f has in D exactly one fixed point $\hat{x}$ with $F(\hat{x}) = \hat{x}$.
2. The sequence $x_n := f(x_{n-1})$, $n = 1, 2, 3, \ldots$ converges to $\hat{x}$ for every $x_0 \in D$.
3. It is $\|x_n - \hat{x}\| \leq \dfrac{c^n}{1 - c}\|x_1 - x_0\|, n = 2, 3, \ldots$

Imagine you have a city map of the town you live in front of you on the floor. The mapping of the city onto the map is obviously a contraction mapping. The fixed point theorem now says there is exactly one point on the map that lies precisely at the point it represents. If you put a city map of Munich on the floor in Hamburg, however, you are out of luck: A necessary condition is the property of *self-mapping*: The image of the mapping must be part of the domain. But then the theorem is even constructive, which is especially important for computer scientists. It provides a method of how to find the fixed point and also says something about the rate of convergence in 3.

Because of $f(x_n) = x_{n+1}$ and $f(\hat{x}) = \hat{x}$, we can immediately read off the order of convergence from Definition 18.2: It is $\|x_{n+1} - \hat{x}\| \leq c \cdot \|x_n - \hat{x}\|$ and therefore linear convergence is given.

It is often difficult to determine whether a given function is a contraction mapping. If $f: [a, b] \to [a, b]$ is a differentiable function of one variable, it is sufficient to check whether $\text{Max}\{|f'(x)||x \in [a,b]\} = c < 1$, because from the mean value theorem of differential calculus (Theorem 15.12 in Section 15.1) it follows that for all $x, y \in [a,b]$ holds

$$\frac{|f(x) - f(y)|}{|x - y|} = |f'(x_0)| \le c \quad \text{for a } x_0 \in [x,y],$$

and thus the contraction property is fulfilled. This rule also agrees with our intuitive considerations in Figs. 18.1 and 18.2.

If you want to apply the fixed point theorem, it is sufficient to check the conditions in the vicinity of the sought-after fixed point, they do not have to be fulfilled globally for the whole function.

Calculation of Zeros

We will now examine methods for determining zeros of functions of one variable. We have already learned one algorithm: The proof of Bolzano's theorem, Theorem 14.23 in Section 14.3, was constructive. With the help of bisection, we can approximate a zero arbitrarily accurately. Unfortunately, this method converges very slowly. I would like to introduce two more methods that are often better suited to find zeros.

The *regula falsi* is based on the same assumptions as bisection: The function $f : [a,b] \to \mathbb{R}$ is continuous, $f(a) < 0$ and $f(b) > 0$ (or vice versa). Now we hope to get close to the zero between a and b more quickly if we do not simply halve the interval between them, but determine the intersection point c of the line g from $(a, f(a))$ to $(b, f(b))$ with the x-axis as the new approximation value (Figure 18.3).

The equation of g is:

$$y = g(x) = \frac{f(b) - f(a)}{b - a}(x - a) + f(a).$$

By inserting a and b into this linear equation, we immediately find $g(a) = f(a)$ and $g(b) = f(b)$, so this is the line connecting $(a, f(a))$ and $(b, f(b))$. We get the intersection c of g with the x-axis, if we insert the value c for x, set, $y = 0$ and solve for c:

Fig. 18.3 The regula falsi

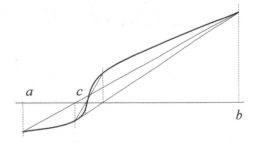

$$0 = \frac{f(b) - f(a)}{b - a}(c - a) + f(a) \quad \Rightarrow \quad c = a - \frac{f(a)(b - a)}{f(b) - f(a)}.$$

c is the first approximation for the zero. If $f(c) < 0$, we set for the next step $a = c$, otherwise $b = c$.

With this method the zero is always in the interval $[a, b]$. The process often converges much faster than the bisection method, but unfortunately not always. Under additional conditions $(f'(x) \neq 0 \neq f''(x)$ in the interval $[a, b])$ it can be shown that the regula falsi is linearly convergent.

The second algorithm is *Newton's method*. In addition, it is assumed here that the function $f : I \rightarrow \mathbb{R}$ is differentiable and $f'(x) \neq 0$ in the interval I. Remember that the tangent to the function at the point x_0 represents an approximation to the function. We therefore start with an initial value x_0, which is not too far from the zero, and take as the first approximation the intersection of the tangent in x_0 with the x-axis (Fig. 18.4).

This intersection point is the next initial value. The tangent in x_0 has the equation (see the explanations after Theorem 15.2):

$$y = g(x) = f'(x_0)(x - x_0) + f(x_0).$$

To calculate the intersection point with the x-axis, we set $y = 0$ again and solve for x:

$$0 = f'(x_0)(x - x_0) + f(x_0) \quad \Rightarrow \quad x = x_0 - \frac{f(x_0)}{f'(x_0)},$$

so we get the following recursive formula for the sequence x_n of approximations:

$$x_n = x_{n-1} - \frac{f(x_{n-1})}{f'(x_{n-1})}.$$

If the sequence x_n converges with $y := \lim_{n \to \infty} x_n$, then $f(y) = 0$.

The sequence often converges very quickly, depending on the initial value, but sometimes not at all, as you can see in Fig. 18.5. In textbooks on numerical analysis you will find theorems that describe necessary conditions for the function. For example:

Fig. 18.4 Newton's method

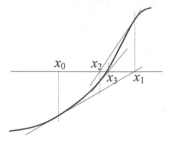

Fig. 18.5 No convergence

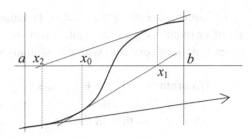

▶ **Theorem 18.4** If $f: [a, b] \to \mathbb{R}$ is twice differentiable, and if for all $x \in [a, b]$ it is $f'(x) \neq 0$ and $|f''(x)| < M$, then for a zero $\hat{x}$ of f holds:

$$|x_{n+1} - \hat{x}| \leq \frac{M}{f'(x_n)} |x_n - \hat{x}|^2.$$

In this case, we have even quadratic convergence. With each iteration, the number of correct decimals approximately doubles.

I applied the three methods bisection, regula falsi and Newton's method to the function $f: [1, 4] \to \mathbb{R}, x \mapsto x^2 - 4$. You can clearly see the different rates of convergence:

Bisection	Regula falsi	Newton (initial value 1)
2.5	1.60000000000000000888178419700	2.5
1.75	1.85714285714285720629845855500	2.04999999999999998223643160599
2.125	1.95121951219512190789373562.43	2.00060975609756086512902584.21
1.9375	1.98360655737704916212749139.95	2.00000009292229474766600105.79
2.03125	1.99452054794520550196068597.87	2.00000000000000022204460492503
1.984375	1.99817184643510059594007088.89	2
2.0078125	1.99939042974702840815837134.86	2
1.99609375	1.99979678927047355152524232.83	2

18.3 Splines

Let's return to the problem I posed at the beginning of Chap. 15.15: How can one draw a beautiful, smooth curve through n given data points? If a series of points is given that should lie on the graph of a function, then one interpolates between the points. The resulting curve is called *interpolation curve* or *spline*.

Spline means curve ruler in English. When interpolation curves could not yet be calculated numerically, one drew curves with the help of a spline, a flexible ruler that could be exactly laid out at the given points.

Of course, the curve we want to generate now must be easy to calculate. Polynomials offer themselves as candidates. You will remember that we were able to approximate given functions very well with polynomials several times. It applies:

▷   **Theorem 18.5** Let $n + 1$ points (x_i, y_i), $i = 0, \ldots, n$ in $\mathbb{R}^2$ be given and $x_i \neq x_j$ for all $i \neq j$. Then there is exactly one polynomial $f(x) \in \mathbb{R}[X]$ of degree less than or equal to n, so that all these points lie on the graph of the polynomial.

This polynomial can be specified explicitly:

▷   **Definition 18.6** Let $n + 1$ points (x_i, y_i), $i = 0, \ldots, n$ in $\mathbb{R}^2$ be given and $x_i \neq x_j$ for all $i \neq j$. Let

$$L_i(x) := \frac{(x - x_0)(x - x_1) \cdots (x - x_i) \cdots (x - x_n)}{(x_i - x_0)(x_i - x_1) \cdots (x_i - x_i) \cdots (x_i - x_n)}, \quad i = 0, \ldots, n$$

and

$$L(x) = L_0(x)y_1 + L_1(x)y_2 + \cdots + L_n(x)y_n.$$

$L(x)$ is called n-th *Lagrange interpolating polynomial* to (x_i, y_i), $i = 0, \ldots, n$.

Obviously, $L_i(x)$ is a polynomial of degree n with the property

$$L_i(x_j) = \begin{cases} 0 & \text{if } i \neq j \\ 1 & \text{if } i = j, \end{cases}$$

from which it follows that L is a polynomial of degree less than or equal to n with $L(x_i) = y_i$, so just a polynomial through the given points.

Mostly $\deg L = n$. $\deg L < n$ is possible if through the addition of the individual L_i coefficients cancel out.

Why can there not be several polynomials of degree $\leq n$ through the $n + 1$ points? If $f(x)$ and $g(x)$ were such polynomials, then $f(x) - g(x)$ would be a polynomial of degree $\leq n$ with $n + 1$ zeros. But that cannot be according to Theorem 5.24. Thus Theorem 18.5 is completely proved.   □

This type of interpolation is unfortunately mostly not well suited to solve our task of the "beautiful" connection of points. The problem is that a polynomial of degree n can have up to $n - 1$ extrema. In the worst case, the connection of 5 points can look like in Fig. 18.6. This curve does not have any corners, but it certainly did not turn out the way we wanted it to.

Fig. 18.6 A smooth
connection of points

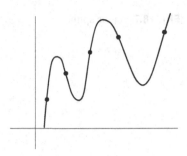

The Lagrange polynomials have rather theoretical meaning. For us, the following error formula will be important in Sect. 18.4. It provides information about how much the Lagrange polynomial can differ at most from a good and smooth function that runs through the points (x_i, y_i):

▶ **Theorem 18.7** Let $x_0 < x_1 < \cdots < x_n$, $f : [x_0, x_n] \to \mathbb{R}$ be a $(n+1)$-times continuously differentiable function with $f(x_i) = y_i$. If $L(x)$ is the Lagrange interpolating polynomial to the points (x_i, y_i), then for all $x \in [x_0, x_n]$:

$$f(x) - L(x) = \frac{f^{(n+1)}(\vartheta)}{(n+1)!}(x - x_0)(x - x_1) \cdots (x - x_n) \quad \text{for a } \vartheta(x) \in \,]x_0, x_n[.$$

This error formula is quite similar to the n-th remainder of the Taylor polynomial of a function, see Theorem 15.22 in Section 15.3, and also the proof is of similar quality. I would like to forego this here.

Cubic Splines

I would like to present you a practical solution of the problem of the smooth connection of points. There are many approaches to this. In general, the interpolant function is defined piecewise and it is ensured that the curve parts fit together well at the seams. I would like to introduce you to such a method. With the widely used *cubic splines* the curve segments are generated by third degree polynomials.

So we are looking for a "nice" curve $s : [x_0, x_n] \to \mathbb{R}$ through (x_i, y_i), where it should be $x_0 < x_1 < \cdots < x_n$ (Fig. 18.7). Between these points we interpolate by polynomials, that means we define s piecewise: Between x_i and x_{i+1} the following should apply

$$s\big|_{[x_i, x_{i+1}]} = s_i : [x_i, x_{i+1}] \to \mathbb{R},$$
$$x \mapsto a_i(x - x_i)^3 + b_i(x - x_i)^2 + c_i(x - x_i) + d_i. \tag{18.8}$$

Fig. 18.7 cubic splines

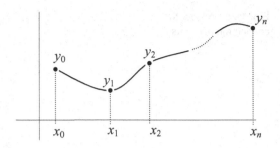

The first and second derivatives of s are:

$$s_i'(x) = 3a_i(x - x_i)^2 + 2b_i(x - x_i) + c_i, \tag{18.9}$$

$$s_i''(x) = 6a_i(x - x_i) + 2b_i. \tag{18.10}$$

If our curve consists of $n + 1$ points, we have n curve segments s_0 to s_{n-1} and thus $4 \cdot n$ unknowns a_i, b_i, c_i, d_i.

Now we have to formulate the conditions that describe the good fit: First, the s_i should of course run through the interpolation points on the left and right edges, so:

$$s_i(x_i) = y_i, \qquad i = 0, \dots, n - 1,$$
$$s_i(x_{i+1}) = y_{i+1}, \qquad i = 0, \dots, n - 1. \tag{18.11}$$

I introduce a shortcut: Let $\Delta x_i := (x_{i+1} - x_i), i = 0, \dots, n - 1$. Then, according to (18.8), by inserting x_i and x_{i+1} respectively:

$$d_i = y_i,$$
$$a_i \Delta x_i^3 + b_i \Delta x_i^2 + c_i \Delta x_i + d_i = y_{i+1}, \tag{18.12}$$

these are $2 \cdot n$ linear equations for the unknowns.

Here you can see why we wrote down the polynomials in the initially somewhat strange form (18.8): If we insert x_i, the evaluation of the polynomial and its derivatives is very easy.

If (18.11) is fulfilled, the graph has no gaps, it is continuous. In addition, the curve should be smooth at the seams. We therefore require that the first derivatives should agree at the interpolation points:

$$s_i'(x_{i+1}) = s_{i+1}'(x_{i+1}), \qquad i = 0, \dots, n - 2.$$

For the points (x_0, y_0) and (x_n, y_n) at the edges, this condition does not make sense, because here no two curve sections meet, so we get from it by insertion into (18.9) only $n - 1$ new linear equations:

$$3a_i \Delta x_i^2 + 2b_i \Delta x_i + c_i = c_{i+1}, \qquad i = 0, \dots, n - 2. \tag{18.13}$$

Now we have $3n - 1$ linear equations for our $4n$ unknowns. So we can set up more conditions.

Let us simply require that the curve should not only be smooth, but especially smooth: The first derivative should also have no kinks, that is, it should be differentiable. This means that the second derivatives should also agree at the inner points. Thus s is twice differentiable. This gives us $n - 1$ more equations:

$$s_i''(x_{i+1}) = s_{i+1}''(x_{i+1}), \quad i = 0, \ldots, n - 2$$

or after insertion into (18.10):

$$6a_i \Delta x_i + 2b_i = 2b_{i+1}, \quad i = 0, \ldots, n - 2, \tag{18.14}$$

so that we now have a system of linear equations with $4n - 2$ equations and $4n$ unknowns. The solution space of this system has dimension at least 2.

There are different approaches to choosing a suitable function. Often, the requirement

$$s_0''(x_0) = 0, \quad s_{n-1}''(x_n) = 0, \tag{18.15}$$

is made at the edge points, which means the function is not curved at the edges, one could continue it linearly to the left and right. In this way, the *natural splines* are obtained. The *clamped splines* are obtained if the slope is given at the edges:

$$s_0'(x_0) = m_0, \quad s_{n-1}'(x_n) = m_1.$$

Finally, the *periodic splines* play an important role, which assume that the function is periodic. Here one demands

$$s_0(x_0) = s_{n-1}(x_n), \quad s_0'(x_0) = s_{n-1}'(x_n), \quad s_0''(x_0) = s_{n-1}''(x_n).$$

From all these boundary conditions, two more linear equations for the coefficients can be derived, so that finally the interpolation curve is uniquely determined. For the natural splines, we get from (18.10):

$$2b_0 = 0, \quad 6a_{n-1}(x_n - x_{n-1}) + 2b_{n-1} = 0. \tag{18.16}$$

How do you calculate the coefficients a_i, b_i, c_i, d_i concretely? If you move the equations back and forth a bit more, you can reduce the problem of the $4n$ equations to the solution of n other linear equations with other unknowns, which also have a very simple form. I would like to sketch the procedure, you can follow the calculations on paper. I restrict myself to the natural splines.

The new unknowns are exactly the second derivatives at the interpolation points, which I denote with $z_i := s_i''(x_i)$, $i = 0, \ldots, n - 1$. The following formulas become somewhat simpler if I formally introduce the $(n + 1)$-th unknown $z_n = 0$.

For $i = 0, \ldots, n - 1$ we get:

- from (18.12): $d_i = y_i$,
- from (18.10): $b_i = z_i/2$,

and thus for $i = 0, \ldots, n - 2$ by insertion in (18.14) respectively for $i = n - 1$ in (18.16) ($z_n = 0$!):

$$a_i = \frac{z_{i+1} - z_i}{6 \Delta x_i}.$$

If we finally insert the just calculated d_i, b_i, a_i in (18.12) and solve for c_i, we get:

$$c_i = \frac{y_{i+1} - y_i}{\Delta x_i} - \frac{1}{6} \Delta x_i (z_{i+1} + 2z_i), \quad i = 0, \ldots, n - 1.$$

x_i, y_i are known in these four determination equations for a_i, b_i, c_i, d_i, so that only the unknowns z_i have to be determined.

As complicated as the equations now look: If we insert the calculated coefficients a_i, b_i, c_i, d_i in (18.13), much will dissolve into nothing. For $i = 0, \ldots, n - 2$ we get equations of the form

$$\alpha_i z_i + \beta_i z_{i+1} + \gamma_i z_{i+2} = \delta_i,$$

where it holds

$$\alpha_i = \Delta x_i,$$
$$\beta_i = 2(\Delta x_i + \Delta x_{i+1}),$$
$$\gamma_i = \Delta x_{i+1},$$
$$\delta_i = 6 \left(\frac{y_{i+2} - y_{i+1}}{\Delta x_{i+1}} - \frac{y_{i+1} - y_i}{\Delta x_i} \right).$$

In the natural splines, according to (18.15), further $z_0 = 0$ and the complete system of linear equations looks as follows:

$$
\begin{pmatrix}
1 & 0 & 0 & 0 & 0 & \cdots & 0 \\
\alpha_0 & \beta_0 & \gamma_0 & 0 & 0 & \cdots & 0 \\
0 & \alpha_1 & \beta_1 & \gamma_1 & 0 & \cdots & 0 \\
0 & 0 & \alpha_2 & \beta_2 & \gamma_2 & \ddots & \vdots \\
0 & 0 & 0 & \ddots & \ddots & \ddots & 0 \\
\vdots & \vdots & \vdots & \ddots & \alpha_{n-2} & \beta_{n-2} & \gamma_{n-2} \\
0 & 0 & 0 & 0 & 0 & 0 & 1
\end{pmatrix}
\begin{pmatrix}
z_0 \\
z_1 \\
z_2 \\
z_3 \\
\vdots \\
z_{n-1} \\
z_n
\end{pmatrix}
=
\begin{pmatrix}
0 \\
\delta_0 \\
\delta_1 \\
\delta_2 \\
\vdots \\
\delta_{n-2} \\
0
\end{pmatrix}.
$$

The Gaussian algorithm is in general computationally expensive, the runtime is of order n^3. However, this tridiagonal matrix can be brought into row-echelon form in linear time. You were asked to calculate this in the exercises of Chap. 8. Check on the basis of this exercise that the solution is unique.

Parametric Splines

Often, points cannot be connected by the graph of a function, the given points do not have to follow each other monotonically. If you want to connect a series of points (x_i, y_i) in the plane by a curve smoothly, you have to look for a parameter representation of the curve. See Definition 16.15 in Sect. 16.2. On the searched curve $(x(t), y(t))$ for certain parameter values t_i it always should be $x(t_i) = x_i$ and $y(t_i) = y_i$. Now we have to solve two interpolation problems: The points (t_i, x_i) and (t_i, y_i) are each connected by cubic splines. The result are the smooth parameter functions $x(t), y(t)$.

How to choose the parameter points t_i? Of course, we could simply take the values $0, 1, 2, 3, \ldots$ and interpolate the points (i, x_i) and (i, y_i). However, it is advisable to include the distance between the points (x_i, y_i) in the parameterization. For example, you can use:

$$t_0 = 0, \quad t_i = t_{i-1} + \sqrt{(x_i - x_{i-1})^2 + (y_i - y_{i-1})^2}, \quad i = 1, \ldots, n.$$

In addition to the cubic splines calculated here, there are many other methods for curve interpolation: polynomials of different degrees can be taken, the boundary conditions are not unique, and completely different approaches are possible, such as interpolation with Bezier curves. Every CAD system offers many different types of splines. This flexibility can cause gray hair to a constructor: For example, in CAD, a car body part is defined by only a finite number of points, the shape between these points is interpolated. If the constructor's and manufacturer's CAD systems are not absolutely identically configured, one might wonder about a dent in the door.

18.4 Numerical Integration

Integral calculation is a tricky business and requires imagination. In the libraries you find large integral tables with integrals calculated by someone at some time. With the help of computer algebra systems, many integrals can be solved analytically. Nevertheless, there are very simple functions that have been proven not to have an elementary function as antiderivative. For example, the function $e^{-x^2/2}$, which describes the bell curve. Statisticians help themselves with tables. Also the so-called *elliptic integral* $\int \sqrt{1 - k^2 \sin^2 t}\, dt$, $0 < k < 1$, which occurs in the calculation of the circumference of an ellipse, is not computable. Numerical integration therefore plays an important role in applied mathematics.

We divide the interval $[a, b]$ into n equal parts of length $h_n = (b - a)/n$ with the interpolation points $a = x_0, x_1, \ldots, x_n = b$. As a first attempt, we can simply use the Riemann sums, our definition of the integral in Theorem 16.3 was constructive after all: Then we get for the area

$$\int_a^b f(x)dx \approx \sum_{i=1}^n f(x_i)h_n.$$

Fig. 18.8 Approximation by
rectangles

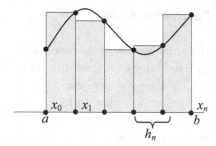

Fig. 18.9 The trapezoidal rule

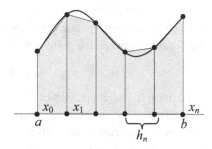

Unfortunately, this method converges very badly (Fig. 18.8). We get a significant
improvement by the *trapezoidal rule*: Between two interpolation points, not a bar, but a
trapezoid is placed, whose upper endpoints represent the function values of the interpolation points (Fig. 18.9).

The area of the trapezoid between x_i and x_{i+1} is $h_n \cdot \dfrac{f(x_i) + f(x_{i+1})}{2}$, and thus we get
for the total area:

$$
\begin{aligned}
F_n &= h_n \left(\frac{f(x_0) + f(x_1)}{2} + \frac{f(x_1) + f(x_2)}{2} + \cdots + \frac{f(x_{n-1}) + f(x_n)}{2} \right) \\
&= h_n \left(\frac{f(x_0)}{2} + f(x_1) + f(x_2) + \cdots + f(x_{n-1}) + \frac{f(x_n)}{2} \right).
\end{aligned}
$$

What is the error we make with this approximation? Now the Lagrange interpolating polynomials from Definition 18.6 come into play: We assume that f is twice continuously
differentiable and first examine the section between x_i and x_{i+1}. The Lagrange polynomial $L(x)$, which connects the two points $(x_i, f(x_i))$ and $(x_{i+1}, f(x_{i+1}))$, has degree 1 and
therefore represents the line between these two points and thus the boundary line of the
trapezoid. If we also use the error calculation from Theorem 18.7, we get the error ΔF_i
for the area calculation of the i-th partial section as

$$
\Delta F_i = \int_{x_i}^{x_{i+1}} f(x) - L(x)\,dx = \int_{x_i}^{x_{i+1}} \frac{f''(\vartheta)}{2!}(x - x_i)(x - x_{i+1})\,dx.
$$

Here ϑ is a value between x_i and x_{i+1}.

If you've gotten this far, then integrating a quadratic polynomial is no longer art. I will give you the result:

$$\Delta F_i = \frac{f''(\vartheta)}{12}(x_i - x_{i+1})^3.$$

Now we just have to sum up all the ΔF_i. Let it be $M = \text{Max}\{|f''(x)| \mid x \in [a, b]\}$, then

$$\left|\sum_{i=0}^{n-1} \Delta F_i\right| \leq \sum_{i=0}^{n-1} \frac{M}{12} h_n{}^3 = \frac{nh_n}{12} h_n{}^2 M = \frac{b-a}{12} h_n{}^2 M.$$

So in total

$$\left|\int_a^b f(x)dx - F_n\right| \leq \frac{(b-a)}{12} h_n{}^2 \cdot M.$$

The error therefore goes to zero with the square of the step size.

Even better approximations can be achieved by connecting the function values of the support points with spline curves: In *Simpson's rule,* quadratic splines are used: The three points a, b and $(a+b)/2$ are connected by the uniquely determined parabola section that runs through these three points.

If $g(x) = cx^2 + dx + e$ is the parabola through $f(a)$, $f((a+b)/2)$ and $f(b)$ (Fig. 18.10), then the area F_2 under this parabola is equal to

$$F_2 = \int_a^b g(x)dt = \frac{c}{3}(b^3 - a^3) + \frac{d}{2}(b^2 - a^2) + e(b - a).$$

You can easily check the following transformations:

$$F_2 = \frac{b-a}{6}[2c(b^2 + ab + a^2) + 3d(b + a) + 6e]$$

$$= \frac{b-a}{6}\left[(cb^2 + db + e) + 4\left(c\left(\frac{a+b}{2}\right)^2 + d\left(\frac{a+b}{2}\right) + e\right) + (ca^2 + da + e)\right]$$

$$= \frac{b-a}{6}\left[g(b) + 4g\left(\frac{a+b}{2}\right) + g(a)\right].$$

Fig. 18.10 Simpson's rule

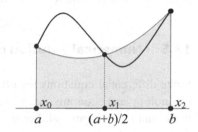

$$x_0 \qquad x_1 \qquad x_2$$
$$a \qquad (a+b)/2 \qquad b$$

Fig. 18.11 The Simpson rule

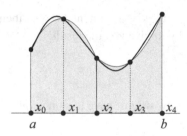

$b, (a+b)/2$ and a are precisely the points at which f and g match. So we get

$$F_2 = \frac{b-a}{6}\left[f(b) + 4f\left(\frac{a+b}{2}\right) + f(a)\right],\qquad(18.17)$$

a formula that is easy to evaluate. The parabola parameters c, d, e have disappeared as if by magic. The parabola doesn't have to be calculated explicitly.

Now we subdivide the interval $[a, b]$ again into n equally large parts h_n, where n must be even this time, and apply Simpson's rule to each of the sub-intervals (shown in Fig. 18.11 for $n = 4$). Then, by combining the areas, we get the *composite Simpson's rule*:

$$F_n = \frac{h_n}{3}\big[f(x_0) + 4(f(x_1) + f(x_3) + \cdots + f(x_{n-1}))$$
$$+ 2(f(x_2) + f(x_4) + \cdots + f(x_{n-2})) + f(x_n)\big].$$

Since $b - a$ in (18.17) is just $2 \cdot h_2$, the denominator becomes 3.

I would also like to give you an error estimate for this method. It can be calculated quite similarly to the trapezoidal rule from the Lagrange polynomials. If f is four times continuously differentiable and $M = \text{Max}\{|f^{(4)}(x)| \mid x \in [a, b]\}$, then

$$\left|\int_a^b f(x)dx - F_n\right| \le \frac{(b-a)}{180} \cdot h_n^4 \cdot M.$$

An interesting consequence of this estimate, I do not want to withhold from you: Since the fourth derivative of a polynomial of degree three is equal to 0 and thus $M = 0$, polynomials of degree three can be integrated exactly by using Simpson's rule.

In Chap. 22 on statistical methods, we will get to know another method of numerical integration, the Monte-Carlo integration, see the example at the end of Sect. 22.3. This plays an important role for multidimensional integrals.

18.5   Numerical Solution of Differential Equations

Since differential equations are often very difficult to solve, but play an immense role in technology and economy, numerical methods for solving differential equations are very important and widespread. For example, the weather forecast is essentially based on the development of pressure, temperature and humidity. This development is determined by differential equations in space and time. Prerequisites for solving these equations are good and dense initial values, powerful numerical methods and the corresponding computer capacities. In 2011, the German Weather Service calculated with a world model that uses initial values in a network with 30 km node distance in 60 different heights, a total of about 39 million points. The network is even denser over Europe. Computers for weather forecasts and climate models are regularly represented in the list of the fastest supercomputers.

In this section, I would like to introduce you to some basics of numerical integration of differential equations. We restrict ourselves to first order differential equations. A nth order differential equation can be reduced to n first order differential equations, so that in this case the same numerical methods can be used.

So we want to solve the initial value problem

$$y' = f(x,y), \quad y(x_0) = y_0$$

in the interval $[x_0, b]$. Of course, the result will not be an analytical formula, but a set of approximation points $(x_i, y(x_i))$ of the sought function.

We divide the interval between $[x_0, b]$ into n parts of width $h = (b - x_0)/n$ and denote the interpolation points with $x_0, x_1, \ldots, x_n = b$. The slope of the function $y(x)$ at the point x_0 is given, it is just $f(x_0, y(x_0)) = f(x_0, y_0)$. From this slope we first guess the function value $y(x_1) =: y_1$ as a linear continuation with this slope: $y_1 = y_0 + f(x_0, y_0) \cdot h$. From this we can calculate $y'(x_1) = f(x_1, y_1)$, that is, the slope at the point x_1 and continue with this slope linearly until x_2. We obtain the recursive formula for Euler's method (Fig. 18.12):

$$y_{k+1} = y(x_{k+1}) := y_k + f(x_k, y_k) \cdot h.$$

A simple improvement is obtained in *Heun's method* (Fig. 18.13): When transitioning to x_1, we do not continue with the slope at the point x_0, but determine an average slope between x_0 and x_1: First, we calculate the point $z_1 := y_0 + f(x_0, y_0) \cdot h$ and the slope at this point $f(x_1, z_1)$ as above, but then we start again from (x_0, y_0) with the average of the slopes $(f(x_0, y_0) + f(x_1, z_1))/2$. We get

Fig. 18.12 Euler's method

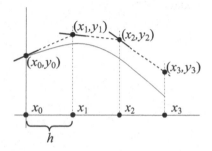

Fig. 18.13 Heun's method

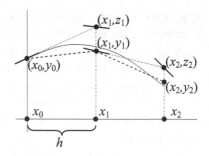

$$y_1 = y_0 + \left(\frac{f(x_0, y_0) + f(x_1, \overbrace{y_0 + f(x_0, y_0) \cdot h}^{z_1})}{2} \right) \cdot h,$$

or as the k-th step:

$$y_{k+1} = y_k + \left(\frac{f(x_k, y_k) + f(x_{k+1}, y_k + f(x_k, y_k) \cdot h)}{2} \right) \cdot h.$$

Finally, I would like to introduce you to the *Runge-Kutta method*. Like the previous methods, it is based on the idea of guessing the correct slope to the next curve point (x_{k+1}, y_{k+1}) from the point (x_k, y_k). In Euler's method, the initial slope was used, in Heun's method the average of two slopes was formed. In the Runge-Kutta method, we shoot four arrows towards the next value (x_{k+1}, y_{k+1}) and form an weighted average of these four slopes. In Fig. 18.14 you can see the procedure outlined.

Starting from the point $P = (x_k, y_k)$, we first proceed as in Euler's method, but only up to half of the interval to be bridged. The slope in the point P is $m_1 = f(x_k, y_k)$. We get the point $(x_k + h/2, y_k + m_1 \cdot h/2)$ and there the slope $m_2 = f(x_k + h/2, y_k + m_1 \cdot h/2)$. Now we start from P with the new slope m_2 to the middle of the interval and get there the point $(x_k + h/2, y_k + m_2 \cdot h/2)$ with the slope $m_3 = f(x_k + h/2, y_k + m_2 \cdot h/2)$. With this third slope, we finally go from P to the right edge of the interval and reach the point

Fig. 18.14 Runge-Kutta method

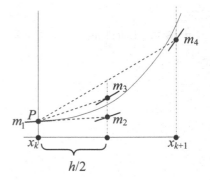

$(x_k + h, y_k + m_3 \cdot h)$ with the slope $m_4 = f(x_k + h, y_k + m_3 \cdot h)$. From these four slopes we form an average, whereby the slopes m_2 and m_3, which were formed in the middle of the interval, are double-weighted. This average then gives us the final direction with which we calculate the point (x_{k+1}, y_{k+1}). In summary, the following steps result:

$$
\left.
\begin{aligned}
m_1 &= f(x_k, y_k) \\
m_2 &= f(x_k + h/2, y_k + m_1 \cdot h/2) \\
m_3 &= f(x_k + h/2, y_k + m_2 \cdot h/2) \\
m_4 &= f(x_k + h, y_k + m_3 \cdot h)
\end{aligned}
\right\} \text{ calculation of 4 slopes}
$$

$$
\left.
m = \frac{1}{6}(m_1 + 2m_2 + 2m_3 + m_4)
\right\} \text{calculation of weighted average}
$$

$$
\left.
y_{k+1} = y_k + m \cdot h
\right\} \text{calculation of next function value}
$$

This approach looks somewhat arbitrary, but it is not: If you choose a differential equation $y' = f(x)$ that does not depend explicitly on y, then you get for the mean slope m:

$$
m = \frac{1}{6}(f(x_k) + 4f(x_k + h/2) + f(x_k + h))
$$

and thus

$$
y(x_{k+1}) = y(x_k) + \frac{h}{6}(f(x_k) + 4f(x_k + h/2) + f(x_k + h)). \tag{18.18}
$$

We can also integrate $f(x)$ directly, which gives us:

$$
\int_{x_k}^{x_{k+1}} f(x)dx = y(x_{k+1}) - y_k(x_k).
$$

If you evaluate this integral numerically using Simpson's rule (see Sec. 18.4), you will get the result (18.18) we obtained by solving the differential equation using the Runge-Kutta method.

Try it yourself to see that for differential equations of the form $y' = f(x)$, Heun's method corresponds to the trapezoidal rule and Euler's method to integration using Riemann sums.

In the solution methods for differential equations, one speaks of methods of order n if the difference between the calculated slope and the actual slope of the function $y(x)$ is of order $O(h^n)$, where h represents the step size of the method. The order of Euler's method is 1, the order of Heun's method is 2, and that of the Runge-Kutta method even 4: The error has order $O(h^4)$.

Theoretically, the approximation to the true solution will be arbitrarily accurate if only the step size is made small enough. In practice, however, this reduction in step size is limited by the fact that the number of operations and the calculation errors increase. So the task of finding the optimal step size is not trivial.

18.6   Comprehension Questions and Exercises

Comprehension questions

1. If a and b are numbers with numerical errors, in which operations can the error cause greater problems: addition or multiplication?
2. You want to transform an equation for computing zeros into a fixed point equation. Does that always work?
3. If the Banach fixed point theorem cannot be applied to a nonlinear equation because it is not a contraction mapping, can the equation still have a fixed point?
4. You know that a function has a zero. Which method always returns the zero?
5. In cubic splines you have the freedom to specify the first or second derivative at the edges. Can such a specification also have an effect on the inner curve sections of a spline?
6. Simpson's rule can be used to integrate polynomials of degree 3 exactly. Can you make a similar statement for integration with the trapezoidal rule? Take a look at the given error estimates again.
7. In the numerical solution of differential equations, the solution can be made arbitrarily accurate in theory if only the step size is made small enough. What are the limits of this accuracy in practice?

Exercises

1. Implement the Gaussian algorithm. It should be implemented recursively and row pivoting should be performed. Count the number of multiplications and divisions you need to solve the system.
2. Implement the Gaussian algorithm for tridiagonal matrices. Count the number of multiplications and divisions here as well.
3. Implement the algorithms for solving nonlinear equations. Use them to calculate the solutions of $e^{-x} + 1 = x$. How accurately can you calculate the solutions? How fast do the different methods converge?
4. Given the function $f: \mathbb{R} \to \mathbb{R}, x \mapsto x^2 + 3x + 1$. Calculate the zeros of the function using Newton's method. Determine the four approximations x_1, x_2, x_3, x_4 for the initial values $x_0 = 0, -1, -2$ and compare the results with the zeros you get by solving the quadratic equation.
5. Numerically calculate all zeros of the polynomial $x^4 - x^3 - 6x^2 + x + 6$.

6. Calculate the Lagrange polynomial and the natural spline functions for n given points in $\mathbb{R}^2$. Use a graphics program to plot your results.

7. Implement the trapezoidal rule and Simpson's composite rule for numerical integration. Use this to calculate the next decimal for the table in the appendix. Estimate beforehand how fine you need to choose the subintervals in order to be as efficient as possible. Use $\int_{-\infty}^{0} \frac{1}{\sqrt{2\pi}} e^{\frac{-t^2}{2}} dt = 0.5$

8. Show that for differential equations of the form $y' = f(x)$, Euler's method is equivalent to integrarion with Riemann sums and Heun's method is equivalent to the trapezoidal rule.

9. Implement Euler's method, Heun's method and the Runge-Kutta method for the numerical solution of first-order differential equations. Use this to determine the rabbit curve

$$y'(t) = y(t)(1 - 1/1000 \cdot y(t)), \quad y(0) = 10$$

in the interval $[0, 10]$ numerically. Compare the values with the exact solution.

Part III
Probability and Statistics

Probability Spaces

19

Abstract

Probability theory is the basis for statistics. This is what we deal with in this chapter. At the end of it

- you will know and understand important issues in statistics,
- you will know what a probability space is, and you will know the basic rules for calculating probabilities,
- you will be able to deal with conditional probabilities and independent events,
- you will know what a uniform probability space, what a Bernoulli process and what an urn problem is, and
- you will be able to calculate probabilities in Bernoulli processes and urn experiments.

It may initially look like a tightrope walk to describe processes with the exact language of mathematics whose results are unpredictable. Probability has something to do with uncertainty, but mathematics only knows true and false. Nevertheless, precise statements can also be made about probabilities. However, these can sometimes sound a bit weird, and you have to look at them closely to see what's really behind them.

In public, statistics are occasionally referred to as a refined form of lies. You know the saying which is attributed to Churchill: "The only statistics you can trust are those you falsified yourself." Maybe this reputation is also due to the fact that statistical statements are often presented in a shortened form in the media and thus the precision is lost. However,

Supplementary Information The online version contains supplementary material available at https://doi.org/10.1007/978-3-658-40423-9_19.

the exact statements are often difficult to digest for the uninitiated reader and – I'm afraid – sometimes also for the publicist.

Statistics permeate our lives, and computers are of course increasingly the tools with which they are generated. The mathematics behind it is complex. There are many text-books for statistics users, which are essentially extensive formula collections with many examples of applications. I am taking a different approach: In the following chapters I will introduce the mathematical basics that are necessary in order to work sensibly with statistical tools and to interpret the results. I will describe a few statistical methods by way of example. This should help you to be able to use other methods if necessary, with-out having to fully understand their mathematical background.

19.1   Problems in Statistics and Probability Theory

I would like to introduce you to some typical problems that statisticians deal with. Based on these questions, we can see which toolbox we need to be able to give the right answers.

The Election Poll

In the elections to the German Bundestag, about 65 million citizens are entitled to vote. For the forecast on election night, an exit poll is carried out, that is, after leaving the polling station, a certain number of voters are asked about their vote. Let's assume that all respondents are happy to tell the truth openly.

For example, 1000 voters could be asked. Of these, 400 voted for party A, 350 for party B and 60 for party C. From this, the polling company creates the following forecast:

$$\text{party A: } 40\,\% \quad \text{party B: } 35\,\% \quad \text{party C: } 6\,\%$$

This is announced on television. Most of the time, the reporter then says something like "There can still be fluctuations of plus/minus x percent."

Where does this number x come from? It is clear that the estimate will be the bet-ter the more voters are asked. If only 100 voters are asked, no reliable prediction can be expected. The statistician will therefore determine a number x which depends on the sample size and which characterizes the possible deviations.

But does this give us a precise statement? Even if 10 million citizens are asked, there is, albeit naturally vanishingly small, the possibility that only voters of party A are caught and that the forecast is therefore completely wrong.

You know that the first election forecasts are usually very good nowadays, but some-times they do change. These are then the more exciting election nights, but afterwards the mistake is sought among the mathematicians.

However, the clever statistician, to hedge his bets, has formulated his statement some-thing like this:

"With a probability of 95%, my forecast deviates from the actual final result by less than 1%."

The 95% is left out by the newsreader, he speaks of a maximum of 1% deviation and is usually right. The polling company delivers the figures as ordered: It can also guarantee a deviation of $\pm 1\%$ with a probability of 99%, but then it wants more money from the client because it has to question more voters.

Quality Checking

In industrial production of a part, for example a ball bearing, often not all balls are checked because it is too expensive. Samples are taken at regular intervals and the number of defective parts in this sample is recorded in a control chart. If this number exceeds a certain threshold, production must be intervened. The problem to be solved is the same as with the election forecast. One can make statements like "with a probability of 99%, no more than 0.01% of the balls are defective". The 99% and the 0.01% are threshold values set by the manufacturer. Based on these values, the sample size and the allowed number of defective parts in the sample must be determined.

Determining Estimates

Carp ponds for fish farming are very common in Germany. How many carp swim in the pond? If it is not possible to fish the pond completely, you can catch and mark 100 fish for example. After a few days you catch 100 fish again. If 10 marked fish are contained in it, you can guess that a total of about 1000 carp live in the pond. Again the question arises: What does "about" mean? With the help of the first two examples you can already find out what form the statement of the statistician will have.

> We also encounter a problem here that lies outside of mathematics and which we cannot consider further: How good is the method with which the sample is obtained? Are there systematic procedural errors that distort the results? Could it be, for example, that the fish form shoals and that there is no mixing at all in the pond? Then the experiment is completely wrongly designed. In the case of election polls too, a large part of the know-how of the companies lies not in the mathematical evaluation, but in the clever construction of the selected sample. But that is another matter.

In the examples given so far, an object is selected randomly from a population whose elements have certain characteristics and the characteristics of this object are checked. The mathematicians have constructed a prototype for such tasks, the urn model: In an urn there are n balls with different colors, from which m balls are selected randomly. We will carry out this experiment several times and always assume that there are no procedural errors: The balls are well mixed and the selection is really random.

Testing a Hypothesis

A few years ago, a British physicist proved Murphy's law: He computed when a slice of toast falls, it falls on the buttered side more often than on the unbuttered side. This happens for physical reasons that are related to our gravitational constant, table height, air pressure, specific weight of butter and toast, and other factors. We want to approach the question with statistical methods. I don't believe in Murphy and I put forward the hypothesis: The probability for "butter on the floor" is exactly 50%

Now we drop the toast 100 times: 54 times it falls on the butter side, 46 times on the other. What can we conclude from this? Probably only: "The result does not contradict the hypothesis."

If the toast falls on the buttered side 80 times out of 100 attempts, we will most likely say: "The deviation is significant, the hypothesis must be rejected, Murphy is right".

What does "significant" mean in this context? Somewhere there is a threshold, at which our confidence in the hypothesis is broken. What would you say? 65, 60, 70 slices?

Even with the rejection of the hypothesis in the second case, we can not be quite sure: Maybe the toast falls on the side without butter the next 100 times. With what probability do we make a mistake with the rejection? In test theory, statements of this kind are made.

If the hypothesis can not be rejected, can it be accepted? No, because even if the toast falls on the buttered side exactly 50 times, the actual probability could be, for example, 52%. In this case, no statement is possible! Here, mistakes are often made, be careful.

More serious applications of the hypothesis test can be found, for example, in the approval process for drugs or chemicals.

Random Events Over Time

Queueing theory plays an important role in computer science: Requests arrive at a web server at random intervals, the average number per day is known. Each request needs a certain amount of time, which also varies randomly, until it has been served. How must the server's capacity be designed to avoid long waiting times?

Probabilistic Algorithms

There are many problems in computer science that can not be solved exactly in a reasonable time, for example the np-complete problems. For many such tasks (for example, the problem of the traveling salesman), there are probabilistic approaches: we can find solutions that are correct only with a certain probability, or that are very likely to deviate from the optimal solution by at most a certain percentage.

We have developed such a probabilistic algorithm in Sect. 5.7: To find large prime numbers needed in cryptography, one performs prime number tests. The result is then only "probably a prime number", but for practical applications that is sufficient.

Monte Carlo Methods

Can you determine the number π by dropping a needle on a piece of paper? Perform the following experiment (*Buffon's needle problem*): Draw a series of parallel lines on a piece of paper, whose distance is just as large as the length of a needle. Focus firmly on the number π, drop the needle often on the paper, and check each time whether the needle hits one of the lines or not. Count the number N of attempts as well as the number T of hits. Then the fraction N/T will get closer and closer to $\pi/2$.

Before you believe that I am drifting into esotericism, some mathematics to this experiment: For simplicity, let's assume that the needle is 2 cm long, so the distance between the lines is also 2 cm. The needle can only hit the closest line, by which I mean the line that has the shortest distance from the center of the needle. I call this distance d ($0 \leq d \leq 1$) (Fig. 19.1). α is the acute angle that the needle forms with this line ($0 \leq \alpha \leq \pi$). The needle hits the line exactly when $d < \sin \alpha$. So for each throw there is a value pair (α, d) in the rectangle $[0, \pi] \times [0, 1]$. Then $d < \sin \alpha$ if and only if (α, d) lies below the graph of $\sin x$, in Fig. 19.2 in the grey area. If you carry out the experiment randomly, the points (α, d) are evenly distributed in the rectangle, it applies:

$$\frac{\text{number of hits}}{\text{number of attempts}} = \frac{T}{N} \approx \frac{\text{grey area}}{\text{total area}}.$$

The grey area is $F = \int_0^\pi \sin x\, dx = 2$, the total area $G = \pi$, so the $T/N \approx 2/\pi$, from which $N/T \approx \pi/2$ results.

What we have done here is nothing other than a numerical integration: The integral to be calculated is enclosed in a rectangle and from this rectangle a large number of points are randomly selected. Then the ratio "integral to rectangle" is equal to the ratio "points in the integral area to total number of points".

Fig. 19.1 Buffon's needle problem 1

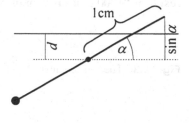

Fig. 19.2 Buffon's needle problem 2

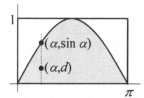

Procedures in which random numbers or random results of experiments are used in some form are called *Monte Carlo methods*.

You probably know another Monte Carlo method from computer science: The *quicksort algorithm* works recursively according to the principle:

Sort the list *L* with quicksort:
Search for a random element *z* in the list.
Sort the sub-list containing all elements smaller than *z* with quicksort
Sort the sub-list containing all elements greater than *z* with quicksort

Quicksort has an average runtime of $O(n \cdot \log n)$, so it is a very good sorting algorithm.

Distributions

In Fig. 19.3 you can see in a very abbreviated form the first chapters of this book. I compressed them and then read them in as a sequence of 32-bit integers. For each of the first 100 000 integers, I counted the number of ones in the binary representation.

All values, with the exception of 0 times one and 32 times one, occurred. The most frequent was 16 times one: exactly 13 346 times. I entered the individual results between 0 and 32 in the diagram in Fig. 19.3. You can see that the distribution of frequencies fits very well with the dashed curve. This is the *bell curve,* which occurs again and again in the analysis of many data sets and plays an important role in statistical applications.

Implement such a "one counter" yourself and apply it to text files, binary files, or sequences of random numbers. What do your results look like?

In the German board game "Man, don't get angry", you can start with your peg only after rolling a 6. In a dice experiment, I tested how long one has to wait for the six. The result of "1000 times rolling until you get a 6" can be seen in Fig. 19.4. Twice the 6 only

Fig. 19.3 The "one counter"

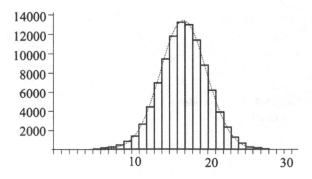

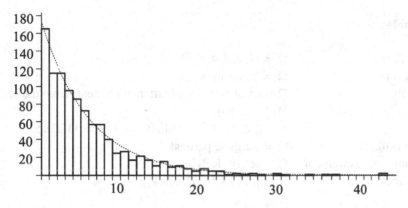

Fig. 19.4 Waiting for the six

came on the 43rd throw! Also here we can put a curve over the bars: It is an exponential function.

The probabilities of events can be described using such distributions. Some of these distributions occur again and again in statistics. One task of the next chapters will be to characterise important such distributions and to determine for which experiments or for which data sets these distributions are applicable.

19.2   The Concept of Probability

The term probability plays a central role in all the examples of statistical problems presented here. The foundation of all statistics is probability theory: Probabilities are assigned to random events, which can then be combined and interpreted in certain ways. We will examine such random events and their probabilities below.

Random Events

The statistician conducts experiments. With these experiments, the set of all possible outcomes is usually known, but it cannot be predicted which specific result will occur. Such an experiment is called a *random experiment.*

We have already encountered some random experiments: rolling dice, an election poll, dropping a slice of toast, randomly selecting a point in a rectangle. Other examples include playing the lottery, flipping a coin, or measuring a temperature.

The set of all possible outcomes of an experiment is called the *outcome space,* and is often denoted by the Greek letter Ω.

Rolling dice:	$\Omega = \{1, 2, 3, 4, 5, 6\}$.
Flipping a coin:	$\Omega = \{\text{head, tail}\}$.
Powerball:	$\Omega =$ set of 5-tuples of different elements from the set
	$\{1, 2, \ldots, 69\}$.
	It is $\Omega \subset \{1, 2, \ldots, 69\}^5$, $|\Omega| = \binom{69}{5} = 11\,238\,513$.
Election poll:	$\Omega = \{\text{eligible parties}\}$.
Temperature measurement:	$\Omega = \mathbb{R}^+$ (in Kelvin).
Point in rectangle:	$\Omega = [a, b] \times [c, d]$, $a, b, c, d \in \mathbb{R}$. ◀

Certain subsets of the outcome space Ω we call events. If $A \subset \Omega$ is an event, then we say *"the event A occurs"* if the outcome of the experiment is an element of A. For example, the event "the dice results in an even number" corresponds to the set $\{2, 4, 6\} \subset \{1, 2, 3, 4, 5, 6\}$, $\{1\}$ is the event "1 was rolled", Frost is the event $[0, 273] \subset \mathbb{R}^+$.

Ω itself is called the *certain event*; in any case, the experiment has an element from Ω as outcome, Ω thus always occurs. The empty set $\emptyset$ is called the *impossible event*.

We can apply our usual set operations to events:

$A \cap B$	is the event "A and B occur ",
$A \cup B$	is the event "A or B occurs",
$\overline{A}$	is the event "A does not occur".

The events A and B are called *incompatible*, if $A \cap B = \emptyset$. One-element events are also called *elementary events*.

Below, we will assign probabilities to events. It is important to distinguish between the outcome of an experiment ω and the event $\{\omega\}$ which states that ω has occurred.

Which subsets of a n outcome space should be events? Of course, it makes sense to use the sets $\emptyset$ and Ω as events, with A also the set $\overline{A}$, and with several sets also their union and their intersection. A system of sets that satisfies these properties is called an *algebra of events*:

▶   **Definition 19.1** Let Ω be a set. A system $\mathcal{A}$ of subsets of Ω is called an *algebra of events* in Ω, if the following is true:

(A1) $\Omega \in \mathcal{A}$.

(A2)   If $A \in \mathcal{A}$, then $\overline{A} \in \mathcal{A}$ ($\overline{A}$ is the complement of A).

(A3)   If $A_n \in \mathcal{A}$ for $n \in \mathbb{N}$, then $\bigcup_{n=1}^{\infty} A_n \in \mathcal{A}$.

The subsets that belong to the algebra of events $\mathcal{A}$ are called *events*.

Because of the possibility of building the complement, ∅ is included in $\mathcal{A}$ and all intersections of sets from $\mathcal{A}$ also belong to $\mathcal{A}$.

> The concept of algebra of events plays – except in the next definition – no more practical
> role in our further considerations. For all finite and countably infinite sets Ω you can confidently think of events as elements of the power set. Every subset of Ω is then an event. With
> uncountable sets, such as subsets of real numbers, however, one can construct wild subsets
> for which no reasonable probability can be assigned anymore. The restriction to certain
> algebras of events therefore has mathematical reasons, all "good" subsets A of Ω can also be
> referred to as events.

The probability of events or the possibility of an event occurring at all is not yet stated.
For example, when measuring temperature, there is a physical upper limit somewhere,
where the real numbers do not stop for a long time. But it does not matter if we include
these large numbers in the outcome space. For example, in the lottery we could choose
$\Omega = \{1, 2, \ldots, 69\}^5$ as the outcome space. Some of the results would then have probability 0.

Probability Spaces

How can we assign probabilities to specific events? The key here is the concept of
probability space. As we have done many times in this book, we choose an axiomatic
approach. Some basic and intuitive properties of probabilities are collected and written down as axioms. The theorems of probability theory are derived from these axioms.
If the axiom system is good, then one can prove many theorems which show good agreement with the real-world results.

The following axiom system fulfills these requirements, even though it is almost
unbelievable how simple it is. It was formulated in the 1930s by the Russian mathematician Kolmogoroff:

▶ **Definition 19.2** Let Ω be a set. A function p, which is defined on an algebra of
events $\mathcal{A}$ in Ω and which fulfills the following axioms, is called a *probability
distribution* or a *probability measure* on Ω, (Ω, p) is called *probability space*.

(W1) $0 \leq p(A) \leq 1$ for all events $A \in \mathcal{A}$.
(W2) $p(\Omega) = 1$,
(W3) If $A_i \in \mathcal{A}$ for $i \in \mathbb{N}$ are pairwise disjoint events, then

$$p\left(\bigcup_{i=1}^{\infty} A_i\right) = \sum_{i=1}^{\infty} p(A_i).$$

The axioms (W1) and (W2) are immediately clear: We measure probability with numbers between 0 and 1, where 1 means the certain event. About (W3) one can stumble at first. But for finite spaces Ω we can replace (W3) by the rule:

$$p(A \cup B) = p(A) + p(B), \quad \text{if } A \cap B = \emptyset. \tag{19.1}$$

In an example, this means if the probability of rolling a 1 is p and the probability of rolling a 2 is q, then the probability of rolling 1 or 2 is just $p + q$. This corresponds to common sense.

From (19.1) follows by mathematical induction:

$$p(A_1 \cup A_2 \cup \cdots \cup A_n) = p(A_1) + p(A_2) + \cdots + p(A_n), \quad \text{if } A_i \cap A_j = \emptyset \text{ for } i \neq j$$

This is almost the axiom (W3), for infinite spaces it has been shown that one must also allow countable unions of sets in the axiom.

▶   **Theorem 19.3: First conclusions from the axioms** Let (Ω, p) be a probability space and $A, B, A_1, A_2, \ldots A_n$ events. Then holds:

a) $p(\overline{A}) = 1 - p(A)$.
b) $p(\emptyset) = 0$.
c) From $A \subset B$ it follows that $p(A) \leq p(B)$.
d) If $A_i \cap A_j = \emptyset$ for all $i \neq j$, then:

$$p(A_1 \cup A_2 \cup \cdots \cup A_n) = p(A_1) + p(A_2) + \cdots + p(A_n).$$

e) $p(A \cup B) = p(A) + p(B) - p(A \cap B)$.

As an example, I would like to derive property e). The points a) to d) can be calculated in a very similar way:

I remind you of the distributive laws for the set operations. From these we first get

$$A \cup B = (A \cup \overline{A}) \cap (A \cup B) = A \cup (\overline{A} \cap B),$$
$$B = (A \cup \overline{A}) \cap B = (A \cap B) \cup (\overline{A} \cap B),$$

where on the right side there is always the union of two disjoint events. This results in

$$p(A \cup B) = p(A \cup (\overline{A} \cap B)) = p(A) + p(\overline{A} \cap B), \tag{19.2}$$

$$p(B) = p((A \cap B) \cup (\overline{A} \cap B)) = p(A \cap B) + p(\overline{A} \cap B). \tag{19.3}$$

From (19.3) we get $p(\overline{A} \cap B) = p(B) - p(A \cap B)$, we set it in (19.2), so the assertion follows.   □

The following is a first example for a probability space:

Example

In coding theory, a *source of information* $Q = (A, p)$, consists of a source alphabet A with the characters $x_1, x_2, \ldots, x_n$, these are assigned occurrence probabilities $p_1, p_2, \ldots, p_n$. In a text of the German language, for example, the most frequent character e has an occurence probability of 0.174, followed by n with 0.098. The rarest character is the q with an occurence probability of 0.002. If you assign in the set $A = \{x_1, x_2, \ldots, x_n\}$ to each subset the sum of the probabilities of the individual elements, then $Q = (A, p)$ is a probability space.

A message N of length k consists of a sequence of k characters of the alphabet, $N = (a_1, a_2, \ldots, a_k) \in A^k$. If each such message is again assigned a probability, then A^k is a probability space. ◄

We now want to construct further concrete models for the abstract concept of the probability space:

Uniform Probability on Finite Spaces

Let us assume in the following that an experiment can only have finitely many results, and that all these results have the same probability. Examples for this are

the coin toss: $p(\{\text{head}\}) = p(\{\text{tail}\})$,
the dice: $p(\{1\}) = p(\{2\}) = \cdots = p(\{6\})$,
the lottery: $p(\{(1, 2, 3, 4, 5)\}) = \cdots = p(\{(7, 23, 34, 36, 45)\}) = \ldots$

This does not include, for example, an opinion poll, a temperature measurement or the tossing a biased coin.

> The mathematician speaks of a *fair coin* or a *fair dice*, if the probabilities are exactly equally distributed. Of course, this does not exist in reality. But the lottery companies make great efforts to make their experiments as fair as possible.

▶ **Definition 19.4: Uniform probability space** A finite probability space $\Omega = \{\omega_1, \omega_2, \ldots, \omega_n\}$, in which applies

$$p(\{\omega_1\}) = p(\{\omega_2\}) = \cdots = p(\{\omega_n\}),$$

is called *(finite) uniform probability space* .

> This notation implies of course that all $\{\omega_i\}$ (and thus, by Definition 19.1, all subsets of Ω) are events. I will no longer mention this explicitly.

With the help of the axiom (W3) we can now determine the probability for each $\{\omega\} \subset \Omega$ and then also for each subset A of Ω: It is namely

$$1 = p(\Omega) = p(\{\omega_1\}) + p(\{\omega_2\}) + \cdots + p(\{\omega_n\}) = n \cdot p(\{\omega_i\}) \quad \Rightarrow \quad p(\{\omega_i\}) = 1/n$$

and if A is a subset with k elements, then $p(A) = \sum_{\omega_i \in A} p(\{\omega_i\}) = k/n$, so for every $A \subset \Omega$ applies:

$$p(A) = \frac{\text{number of elements of } A}{\text{number of elements of } \Omega}.$$

The elements of A are called the *favorable outcomes* for the event, so that we can also say

$$p(A) = \frac{\text{number of favorable outcomes}}{\text{number of possible outcomes}}.$$

Many real experiments can be mapped to uniform probability spaces, even if the probabilities are not initially evenly distributed.

Examples

1. In the calculation of probabilities in the lottery game, it makes sense to work with the set of 5-tuples of *different* elements from $\{1, 2, \ldots, 69\}$ and not with the set of *all* 5-tuples. There are 11 238 513 different 5-tuples and therefore the probability for a five is $1/11\,238\,513$.

2. We roll two dice and want to calculate the probabilities with which the sums of eyes $2, 3, 4, \ldots, 12$ are thrown. The space $\Omega = \{2, 3, \ldots, 12\}$ is unsuitable, because for example $p(\{2\}) \neq p(\{3\})$. We construct another space in which the results are the possible pairs of throws:

$$\Omega := \{(1, 1), (1, 2), (1, 3), (1, 4), (1, 5), (1, 6),$$
$$(2, 1), (2, 2), (2, 3), (2, 4), (2, 5), (2, 6),$$
$$\vdots$$
$$(6, 1), (6, 2), (6, 3), (6, 4), (6, 5), (6, 6)\}.$$

Each of these 36 elementary events has the same probability 1/36. Now we can simply count and, for example, get:

$$p(\text{sum of eyes} = 2) = p(\{(1, 1)\}) = 1/36,$$
$$p(\text{sum of eyes} = 3) = p(\{(1, 2), (2, 1)\}) = 2/36,$$
$$p(\text{sum of eyes} = 7) = p(\{(1, 6), (2, 5), \ldots, (6, 1)\}) = 6/36,$$
$$p(6 \leq \text{sum of eyes} \leq 8) = \underbrace{5/36}_{\text{sum} = 6} + \underbrace{6/36}_{\text{sum} = 7} + \underbrace{5/36}_{\text{sum} = 8} = 4/9.$$

3. You may be familiar with the birthday problem: What is the probability that k people all have birthdays on different days?
   Let's construct a uniform probability space Ω. We neglect leap years and any seasonal birthday clustering. The elements of Ω (the possible outcomes) should be all possible birthday distributions, the event A_k (the favorable outcomes) should be all the distributions in which the birthdays fall on k different days.

How many elements does Ω have? If we number the days from 1 to 365, the possible distributions correspond to all k-tuples of the set $\{1, 2, \ldots, 365\}$, where at the i-th position of such a k-tuple is the birthday of person i. So is $\Omega = \{1, 2, \ldots, 365\}^k$. This set has 365^k elements, which we all assume to be equally likely.

How many elements does A_k have? For this, we need to answer the question: In how many ways can k different days be selected, namely the k different birthdays, out of 365 days? With this selection, the order must be taken into account: For example, two people can be born on the days i and j or on the days j and i, these are two cases. The answer to the number gives us Theorem 4.7 in Sect. 4.1: There are $365 \cdot 364 \cdot \ldots \cdot (365 - k + 1)$ possibilities. Now we can calculate the quotient of the favorable by the possible outcomes:

$$p(A_k) = \frac{365 \cdot 364 \cdot \ldots \cdot (365 - k + 1)}{365^k}.$$

Even more interesting is the question with which probability at least two of the k persons have their birthday on the same day. This is the event $\overline{A}_k$ with probability $p(\overline{A}_k) = 1 - p(A_k)$. Here are some (rounded) values for it:

$$p(\overline{A}_{10}) = 0.11 \quad p(\overline{A}_{20}) = 0.41 \quad p(\overline{A}_{23}) = 0.507$$
$$p(\overline{A}_{30}) = 0.70 \quad p(\overline{A}_{40}) = 0.89 \quad p(\overline{A}_{50}) = 0.97$$

A very surprising result: Already from $k = 23$, the probability for a double birthday is greater than 0.5. If I am standing in front of 40 students in a lecture, I can bet on it quite well.

The birthday problem also occurs in computer science: In a hash table, the probability of a collision between two data sets is much greater than initially assumed. Now you can calculate this. ◄

Geometric Probabilities

Remember Buffon's needle problem in Sect. 19.1: In the numerical integration carried out there, we assumed that a randomly selected point in the selected rectangle lies at any position with the same probability.

▶ **Definition 19.5** If the set Ω consists of curves, surfaces or volumes and the probability for an event (a subset of the curve, the surface or the volume) is proportional to the size of this subset, then Ω is called space with *geometric probability*.

The rectangle $\Omega = [0, \pi] \times [0, 1]$ in the needle experiment contains an infinite number of points that can all be selected with the same probability. So the probability of a single point is strictly speaking 0, Ω is not a finite uniform probability space. We can only specify positive probabilities for subsets of Ω that have an area different from 0. Since the probability $p(\Omega) = 1$ and the total area is π, we get for each subarea A the probability $p(A) = A/\pi$. In this way we were able to determine the integral of the sine.

> Reality is somewhat different: If we work with a random number generator that, for example, can generate 2^{32} different pairs of points, we still have a finite uniform space in which the probability of each point is $1/2^{32}$. $(e, \pi/4)$ is, for example, not an element of Ω. Of course, it makes sense to still calculate with the geometric probability in this case.

A similar example is the position of the hand of a clock. It makes little sense to ask with what probability the minute hand is at 1.4, but one can give the probability with which the hand is between 2 and 3.

19.3   Conditional Probability and Independent Events

Different events of an experiment are often not independent of each other, the probability of an event can be different depending on whether another event has occurred or not.

Let's take the distribution of cards in the German card game Skat as an example: Alex, Bob and Charly each get 10 cards, 2 cards remain, these form the so-called Skat. With what probability is the ace of spades with Alex? Of course with the probability 10/32. With probability 2/32 the Ace of Spades is in the Skat. But if Bob already knows that he did not receive the Ace of Spades and is asking this question, then the probability that Alex has the card is now 10/22, the probability that the card is in the Skat becomes 2/22.

Are A and Y events, then we call the probability of A under the condition that already Y has occurred $p(A|Y)$. If A is the event "Alex has the Ace of Spades", and Y is the event "Bob does not have the Ace of Spades", then $p(A) = 10/32$ and $p(A|Y) = 10/22$.

Let us try to find the conditional probability of events in a uniform probability space Ω. For a set M I denote with $|M|$ the number of elements. Let A and Y be events in Ω, we want to calculate $p(A|Y)$. Y has thus occurred, which means that the result of the experiment is an element $\omega_0 \in Y$. First, let us determine the probability $p(\{\omega\}|Y)$ for all $\omega \in \Omega$. For $\omega \in Y$ the probability that $\omega = \omega_0$ is equal to $1/|Y|$, because there are $|Y|$ equally likely possibilities for ω. The event $\{\omega\}$ occurs precisely when $\omega = \omega_0$. In the case $\omega \notin Y$ is in any case $\omega \neq \omega_0$, thus $\{\omega\}$ cannot occur. We therefore have:

$$p(\{\omega\}|Y) = \begin{cases} 1/|Y| & \text{if } \omega \in Y \\ 0 & \text{if } \omega \notin Y. \end{cases}$$

With the addition rule from Theorem 19.3 we find $p(A|Y)$ is the sum of the probabilities of all elements of A. Only the elements of $A \cap Y$ contribute to this, each with $1/|Y|$, so it is

$$p(A|Y) = \frac{|A \cap Y|}{|Y|}.$$

We can convert this a bit more: For each subset M of Ω is $p(M) = |M|/|\Omega|$, and we obtain for the conditional probability in the uniform probability space:

$$p(A|Y) = \frac{|A \cap Y|}{|Y|} = \frac{|A \cap Y|/|\Omega|}{|Y|/|\Omega|} = \frac{p(A \cap Y)}{p(Y)}.$$

This calculation serves as motivation for the following definition of conditional probability, which has also proven to be the right concept for non-uniform spaces:

▷   **Definition 19.6** Let (Ω, p) be a probability space, let A and B be events and $p(B) > 0$. Then

$$p(A|B) := \frac{p(A \cap B)}{p(B)}$$

is called the *conditional probability* of A under the condition B.

The following two theorems state important calculation rules for conditional probabilities. The probability space Ω is partitioned into disjoint subsets B_i, $i = 1, \ldots, n$ (Fig. 19.5).

▷   **Theorem 19.7: The theorem of total probability** Let (Ω, p) be a probability space, let $B_1, B_2, \ldots, B_n$ be pairwise disjoint events with $p(B_i) > 0$ for all i and $\bigcup_{i=1}^{n} B_i = \Omega$. Then for all events $A \subset \Omega$:

$$p(A) = \sum_{i=1}^{n} p(B_i)p(A|B_i).$$

▷   **Theorem 19.8: Bayes' law** As in the last theorem, let $B_1, B_2, \ldots, B_n$ be a disjoint partition of the probability space Ω with $p(B_i) > 0$ for all i. Let A be an event with $p(A) > 0$. Then for all $k = 1, \ldots, n$:

$$p(B_k|A) = \frac{p(B_k)p(A|B_k)}{\sum_{i=1}^{n} p(B_i)p(A|B_i)} = \frac{p(B_k)p(A|B_k)}{p(A)}.$$

Fig. 19.5 The total probability

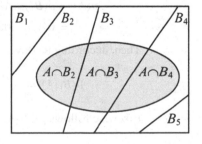

To prove Theorem 19.7: A can be written as

$$A = (A \cap B_1) \cup (A \cap B_2) \cup \cdots \cup (A \cap B_n),$$

where the sets $A \cap B_i$ are pairwise disjoint.

From Definition 19.6 follows $p(A \cap B_i) = p(B_i)p(A|B_i)$ for all i. According to Theorem 19.3d) the assertion follows:

$$p(A) = \sum_{i=1}^{n} p(A \cap B_i) = \sum_{i=1}^{n} p(B_i)p(A|B_i). \qquad \square$$

Theorem 19.8 follows immediately from this:

$$p(B_k|A) = \frac{p(A \cap B_k)}{p(A)} = \frac{p(B_k)p(A|B_k)}{p(A)} = \frac{p(B_k)p(A|B_k)}{\sum_{i=1}^{n} p(B_i)p(A|B_i)}. \qquad \square$$

With the help of Bayes' law, the calculation of the probability of B_i under the condition A can be traced back to the calculation of the inverse probabilities of A under the conditions B_i.

Examples

1. An automobile manufacturer gets a part for its production delivered by three different subcontractors, in different proportions and in different quality:

	Supplier 1	Supplier 2	Supplier 3
Share	45%	35%	20%
Scrap	2%	3%	1%

- What is the probability that a delivered part is defective?
- What is the probability that a defective part comes from supplier 1, 2 or 3?
  Let B_i be the event "part comes from supplier i", A be the event "part is defective". Then $p(A|B_1) = 0.02$, $p(A|B_2) = 0.03$, $p(A|B_3) = 0.01$. Let's first calculate the total probability $p(A)$:

$$p(A) = p(B_1)p(A|B_1) + p(B_2)p(A|B_2) + p(B_3)p(A|B_3)$$
$$= 0.45 \cdot 0.02 + 0.35 \cdot 0.03 + 0.2 \cdot 0.01 = 0.0215.$$

Then, according to Bayes' law:

$$p(B_1|A) = \frac{p(B_1) \cdot p(A|B_1)}{p(A)} = \frac{0.45 \cdot 0.02}{0.0215} \approx 0.42.$$

Likewise follows $p(B_2|A) \approx 0.49$, $p(B_3|A) \approx 0.09$.

2. In Sect. 5.7 we dealt with cryptography. To generate keys for the RSA algorithm, large prime numbers are needed. These can be found with the help of prime number tests.

If the number q is prime, then every prime number test is successful. If q is not prime, then the probability that n prime number tests are successful is less than $1/4^n$. In the checked range of numbers of 512 bits in length, the probability that a randomly chosen number is prime is approximately 0.0028 (see Exercise 9 in Sect. 14.4). What is the probability that the randomly chosen number q is not prime, although it has survived n prime number tests?

Let B_1 be the event "q is prime," B_2 be the event "q is not prime," and A be the event "n prime tests were successful." We obtain $p(A|B_1) = 1$ and $p(A|B_2) < 1/4^n$. This gives the desired probability

$$p(B_2|A) = \frac{p(B_2)p(A|B_2)}{p(B_1)p(A|B_1) + p(B_2)p(A|B_2)} < \frac{0.9972 \cdot 1/4^n}{0.0028 \cdot 1 + 0.9972 \cdot 1/4^n} \tag{19.4}$$

$$= \frac{0.9972}{4^n \cdot 0.0028 + 0.9972}.$$

For $n = 25$ the result is, for example, $p(B_2|A) < 3 \cdot 10^{-13}$, for $n = 30$ is $p(B_2|A) < 3 \cdot 10^{-16}$.

In (19.4) you must still think for a moment about the "<"-sign: Check that for $a > 0$ and for $x < y$ always applies $\frac{x}{a+x} < \frac{y}{a+y}$. Or calculate with $p(A|B_2) = 1/4^n$, that would be the worst case. ◀

Independent Events

Two events A, B are said to be independent of each other if the probability of A does not depend on the occurrence of B, that is, if $p(A|B) = p(A)$.

Rolling two dice apparently results in two independent events: If A is the event "First dice results in 6" and B is the event "Second dice results in 6", then $p(A) = 1/6 = p(A|B)$. The two dice have nothing to do with each other.

From Definition 19.6 it follows for independent events, if $p(B) > 0$, after multiplication with the denominator: $p(A)p(B) = p(A \cap B)$. Conversely, this relation, after division by $p(B)$ implies $p(A|B) = p(A)$. The following definition is therefore appropriate:

▶ **Definition 19.9: Independent events** Two events A, B of the probability space Ω are called *independent* if:

$$p(A \cap B) = p(A) \cdot p(B).$$

This definition is symmetrical in A and B, which makes sense, and it should also be valid for events with probability 0.

Examples

1. In the example after Theorem 19.3 I defined the concept of an source of informa-tion. If $Q = (A, p)$ is a source of information with the alphabet $A = \{x_1, x_2, \ldots, x_n\}$ and associated probabilities of occurrence $p_1, p_2, \ldots, p_n$, then Q is called a *memo-ryless source of information* if in a message the probability of the occurrence of the character x_i is always independent of the already sent characters p_i. In such a source of information, the probability that in a message the characters x_i and x_j fol-low each other is equal to $p_i \cdot p_j$.

   Write down which probability space and which events describe this assertion.

   A German text is apparently not a memoryless source of information: For example, the character q is followed by the character u with probability nearly 1. However, in coding theory, memoryless sources are often assumed.

2. Let's roll again with 2 dice and look at the events

   A: First die shows even number,
   B: Second die shows odd number,
   C: Sum of eyes is even.

   It is $p(A) = p(B) = p(C) = 1/2$.

   For A and B this is clear, for C you can easily calculate it, flip back to example 2 after Definition 19.4. Furthermore, we can calculate:

   $$p(A \cap B) = 1/4, \quad p(B \cap C) = 1/4, \quad p(A \cap C) = 1/4.$$

   A, B are independent of each other, just as B, C and A, C. Do A, B, C have anything to do with each other at all? It is

   $$p((A \cap B) \cap C) = 0 \neq p(A \cap B)p(C) = p(A)p(B)p(C) \blacktriangleleft$$

You can see that while A, B, C are pairwise independent, C is however dependent on $A \cap B$. For this reason, we extend the definition of independence for more than two sets as follows:

▶ **Definition 19.10: The mutual independence of events** Three events A, B, C are called *mutually independent*, if any two of the events are independent and if

$$p(A \cap B \cap C) = p(A)p(B)p(C)$$

n events $A_1, A_2, \ldots, A_n$ are called *mutually independent*, if each $n - 1$ of these events are mutually independent and if

$$p(A_1 \cap A_2 \cap \cdots \cap A_n) = p(A_1)p(A_2) \cdots p(A_n).$$

Again a recursive definition. Expressed a little less formally: With mutually independ-ent events, the probability of intersections of the events can always be calculated by the product of the probabilities, no matter how many of the events are involved.

If the events are A, B independent, then

$$p(A \cap B) = p(A)p(B) = p(A)(1 - p(\overline{B})) = p(A) - p(A)p(\overline{B})$$
$$\Rightarrow \quad p(A) = p(A \cap B) + p(A)p(\overline{B}), \tag{19.5}$$

and further using the distributive law and the addition rule:

$$p(A) = p(A \cap (B \cup \overline{B})) = p((A \cap B) \cup (A \cap \overline{B})) = p(A \cap B) + p(A \cap \overline{B}). \tag{19.6}$$

By comparison of (19.5) and (19.6) we find that $p(A)p(\overline{B}) = p(A \cap \overline{B})$, so A and $\overline{B}$ are also independent from each other.

You can show this quite analogously for more than 2 events:

▶   **Theorem 19.11** If the events $A_1, A_2, \ldots, A_n$ are mutually independent, then the events remain mutually independent if you replace any number of them with their complement.

Examples of such independent events are those that arise from repeating an experiment under the same initial conditions, such as dice, lottery or roulette.

The random drawing of numbered or colored balls from an urn is an important prototype for idealized experiments for the mathematician. The drawing of several balls from an urn with the replacement of the drawn ball before the next draw, represents independent events. If the drawn balls are not replaced, however, the result of the second draw depends on the outcome of the first, no independent events are to be expected here.

It can be shown that the events "prime number test", which I described in Sect. 5.7 in the part on key generation, are mutually independent events for different numbers a . If the probability that a composite number q passes a prime number test is $< 1/4$, so the probability that q passes all n prime number tests is $< 1/4^n$.

19.4   Bernoulli Processes and Urn Problems

▶   **Definition 19.12: The Bernoulli process** In a random experiment, the event A occurs with the probability p. The experiment is repeated n times. If the events "A occurs on the ith trial" are mutually independent, the experiment is called *Bernoulli process*.

In a Bernoulli process of length n, we denote with $h_n(A)$ the number of trials in which A occurs, that is, the *absolute frequency* of the occurrence of A, and with $r_n(A)$ the quotient $h_n(A)/n$, the *relative frequency* of the occurrence of A.

If we carry out this experiment often, we expect the relative frequency $r_n(A)$ finally approaches the probability p. For rolling dice in the long run, the ratio "number of sixes/

number of throws" will be close to $1/6$, while for coin tossing the ratio "heads/tosses" will be close to $1/2$. We are tempted to make a statement like "$\lim_{n\to\infty} r_n(A) = p$".

Strictly speaking, the limit of the sequence $r_n(A)$ however does not exist in the form in which we defined it in Definition 13.2 in Sect. 13.1: For any given ε-neighborhood of p, the sequence $r_n(A)$ every now and then jumps out. This is the nature of chance. However, the following law of large numbers can be derived from the axioms of probability, which I would like to cite here without proof. In words, the somewhat cryptic formula states that at least the probability for such jumps from $r_n(A)$ becomes arbitrarily small: For each $\varepsilon > 0$ the probability that $|r_n(A) - p| \le \varepsilon$ tends to 1.

▶   **Theorem 19.13: Bernoulli's law of large numbers** In a Bernoulli process, the event A occurs with the probability p. Let $r_n(A)$ denote the relative frequency of the occurence of A when the experiment is performed n times. Then holds for $\varepsilon > 0$:

$$\lim_{n\to\infty} p(\{\omega \in \Omega \mid |r_n(A) - p| \le \varepsilon\}) = 1$$

Let's look at some probabilities in a Bernoulli process of length n. What does the probability space look like in which our experiment takes place? One possible result of the Bernoulli process has, for example, the form:

$$\underbrace{(A, \overline{A}, \overline{A}, A, A, A, \overline{A}, A, \ldots, \overline{A})}_{n \text{ elements}}.$$

The outcome space consists of all such n-tuples of elements from $\{A, \overline{A}\}$, so $\Omega = \{A, \overline{A}\}^n$. The elements $\omega \in \Omega$ have the form $\omega = (\omega_1, \omega_2, \ldots, \omega_n)$ with $\omega_i \in \{A, \overline{A}\}$.

We calculate the probability for such a ω. Since the events in the individual trials of the process are independent of each other, for example in a Bernoulli process of length 2:

$$p(\{(A, A)\}) = p^2, \quad p(\{(A, \overline{A})\}) = p(\{(\overline{A}, A)\}) = p(1 - p), \quad p(\{(\overline{A}, \overline{A})\}) = (1 - p)^2$$

and correspondingly for a process of length n:

$$p(\{\omega\}) = p^{\text{number of } A \text{ in } \omega} \cdot (1 - p)^{\text{number of } \overline{A} \text{ in } \omega} = p^{h_n(A)} \cdot (1 - p)^{n - h_n(A)}. \quad (19.7)$$

What is the probability that in a Bernoulli process of length n the event A occurs exactly k times? For this we need to know how many $\omega \in \Omega$ contain exactly k times the event A. That is just the question of in how many ways one can choose k elements from n elements. We know the answer: There are $\binom{n}{k}$ elementary events $\{\omega\}$, in which exactly k times A is contained. What is the probability for such an event? Of course, just $p^k \cdot (1 - p^{n-k})$. The probability for the set of all these events is obtained as the sum, and thus the important theorem that we will need frequently in the next chapters follows:

▶ **Theorem and Definition 19.14: The binomial distribution** In a Bernoulli process
of length n let $p(A) = p$. Then the probability that A occurs exactly k times is

$$\binom{n}{k} p^k (1-p)^{n-k} =: b_{n,p}(k).$$

The function $b_{n,p}(k)$ is called *binomial distribution.*

In the following table I give the values rounded to three decimal places for $b_{10,1/6}(k)$ and
$b_{10,1/2}(k)$. For example, it is the probability of k six when rolling the dice ten times or the
probability of k times head when flipping a coin ten times. Fig. 19.6 shows bar charts for
this.

k	0	1	2	3	4	5	6	7	8	9	10
$b_{10,1/6}(k)$	0.162	0.323	0.291	0.155	0.054	0.013	0.002	0.000	0.000	0.000	0.000
$b_{10,1/2}(k)$	0.001	0.010	0.044	0.117	0.205	0.246	0.205	0.117	0.044	0.010	0.001

In Sect. 19.1 I carried out the rolling experiment "waiting for the 6", a typical example of
a Bernoulli process, under the heading "distributions". Now we can calculate the probability that an event occurs for the first time on the kth trial:

▶ **Theorem 19.15** In a Bernoulli process, let $p(A) = p$. Then the probability that A
occurs for the first time on the kth trial is equal to

$$p(1-p)^{k-1}.$$

For this we only have to carry out the experiment k times and determine the probability
of the event $\{(\overline{A}, \overline{A}, \overline{A}, \overline{A}, \ldots, \overline{A}, A)\}$, where A is at the kth position. According to (19.7)
this is just $p(1-p)^{k-1}$.   □

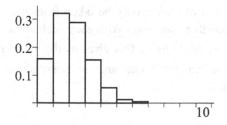 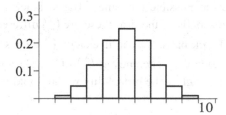

Abb. 19.6 The binomial distribution

19.5 Urn Problems

I have already mentioned the urn experiments as an ideal thought experiment of the mathematician. With their help we will derive important results for later applications.

▷   **Satz 19.16: The urn problem "drawing with replacement"** There are N balls in an urn, S black and W white, where $S + W = N$ From the urn, n balls are drawn, after each draw the ball is put back. There are n_s black and n_w white balls drawn. Then the probability of drawing exactly n_s black balls is equal to

$$p(\text{number of black balls} = n_s) = \binom{n}{n_s} \cdot \left(\frac{S}{N}\right)^{n_s} \cdot \left(\frac{W}{N}\right)^{n_w}. \tag{19.8}$$

Since the balls are replaced, there is the same starting situation at each trial, so it is aBernoulli process of length n. Are S black balls in the urn, the probability p for the event A, to draw a black ball, is just S/N. Thus the probability for the event A to occur exactly n_s times, is given by Theorem 19.14. It is equal to

$$\binom{n}{n_s} \cdot p^{n_s} \cdot (1 - p)^{n - n_s} = \binom{n}{n_s} \cdot \left(\frac{S}{N}\right)^{n_s} \cdot \left(\frac{W}{N}\right)^{n_w}. \qquad \square$$

▷   **Theorem 19.17: The urn problem "drawing without replacement."** There are N balls in an urn, S black and W white, where $S + W = N$. From the urn, n balls are drawn in succession, of which n_s balls are black and n_w balls are white. Then the probability of drawing exactly n_s black balls is equal to

$$p(\text{number of black balls} = n_s) = \binom{S}{n_s} \cdot \binom{W}{n_w} \bigg/ \binom{N}{n}. \tag{19.9}$$

Here the different trials of the process are not independent of each other, the number of balls and the ratio of black to white balls change after each draw. So we don't have a Bernoulli process. Instead, we examine a uniform probability space that contains all possible draws of n balls. In total, there are $\binom{N}{n}$ possibilities to draw the n balls, these are the possible outcomes. The n_s black balls can of course only be taken from the S black balls of the urn, these are $\binom{S}{n_s}$ different possible outcomes. With each such choice of black balls, we can in exactly $\binom{W}{n_w}$ ways add n_w white balls. This gives us the number of favorable outcomes as $\binom{S}{n_s} \cdot \binom{W}{n_w}$. The quotient "favorable outcomes by possible outcomes" gives the probability we are looking for. $\square$

1. In a box there are 20 chocolates, 8 of which are filled with marzipan. You take 5 chocolates from the box. What is the probability of getting 0, 1, 2, 3, 4 or 5 marzipan chocolates?

   Here we have the experiment "drawing without replacement". The marzipan chocolates are the black balls, and so we get with $N = 20$, $S = 8$, $W = 12$, $n_s = 0, 1, 2, 3, 4, 5$, $n_w = 5, 4, 3, 2, 1, 0$:

$$p(n_S = 0) = \binom{8}{0}\binom{12}{5}\bigg/\binom{20}{5} \approx 0.051,$$

$$p(n_s = 1) = \binom{8}{1}\binom{12}{4}\bigg/\binom{20}{5} \approx 0.255,$$

   and so on. In the following table I have written down all the results:

Marzipan pralines	0	1	2	3	4	5
p	0.051	0.255	0.397	0.238	0.054	0.004

2. Now "drawing with replacement": To allow for a comparison with the experiment "drawing without replacement", we take the same numerical values as above: The urn contains 20 balls (chocolates are not put back), 8 of them are black, 5 balls are drawn. What is the probability of getting 0, 1, 2, 3, 4 or 5 black balls?

$$p(n_s = 0) = \binom{5}{0} \cdot 0.4^0 \cdot 0.6^5 \approx 0.078,$$

$$p(n_s = 1) = \binom{5}{1} \cdot 0.4^1 \cdot 0.6^4 \approx 0.259$$

   and so on. This gives:

Black balls	0	1	2	3	4	5
p	0.078	0.259	0.346	0.230	0.077	0.01

◄

19.6   Comprehension Questions and Exercises

1. What are the characteristic properties of a finite uniform probability space?
2. Under what conditions is $p(A \cap B) = p(A)p(B)$. Under what conditions is $p(A \cup B) = p(A) + p(B)$?
3. If A, B and A, C are pairwise independent events, are B and C also independent events?
4. Rats can be sent through a maze with two exits. Behind one of the exits is food to be found. Is the event "rat finds the food" a Bernoulli process under the following conditions?
   a) One (smart) rat is sent through the maze 50 times.
   b) 50 rats are sent through the maze one after the other.
5. In a very large urn with many balls, the experiments "drawing with replacement" and "drawing without replacement" are very similar. Why?
6. An election poll can also be seen as an urn experiment. Is this an experiment with or without replacement?

1. Let your computer perform Buffon's needle experiment and determine the number π to 4 decimals. How many trials do you need? Perform the experiment several times and compare the number of trials each time.
2. In a box with 100 lottery tickets there are 40 duds. You buy 6 tickets. What is the probability that you will get at least 3 wins?
3. Assume the hypothesis to be true that a slice of toast falls on the buttered side just as often as on the unbuttered side when it is dropped. What is the probability then that out of 100 trials, the toast falls on the buttered side more than 52 times?
4. A disease occurs in 0.5% of the population. A test finds 99% of the sick people (test positive) but also responds in 2% of the healthy population. What is the probability a tested personis sick if the test is positive?
5. A multiple-choice test is conducted in an exam. For one question, there are n possible answers, exactly one is correct. Well-prepared students circle the correct answer, poorly prepared students circle randomly. $(p \cdot 100)\%$ of the participants are well prepared. With which (conditional) probability does a correct result come from a well-prepared student?
6. In 4000 draws of the German 6 out of 49 lottery, the number sequence 15, 25, 27, 30, 42, 48 was drawn twice: on 12-20-1986 and on 6-21-1995. This caused quite a stir among lottery players. Calculate whether this event was really unlikely.

7. Calculate the probability of getting a three, four or five in the 6 out of 49 lottery.

8. The following game was carried out in a similar form on an American TV quiz show: The contestant stands in front of three doors. Behind one of them is a car, behind the other two are sheep. The contestant is allowed to choose one of the doors, but not to open it yet. The quizmaster opens one of the other two doors, namely one behind which a sheep is standing. Now the contestant may choose between the two closed doors again. He receives what is behind this door. Calculate the probabilities for a (car) win with the following procedures for the second decision:

   a) The contestant tosses a coin.
   b) The contestant always sticks to his original decision.
   c) The contestant always changes his original choice.

   For a while, a bizarre discussion went on in the American media, in which serious people (including respected mathematicians) tore each other to pieces in writing, trying to solve the question of which was the optimal strategy for the contestant. After the quizmaster has opened the door, the prize is behind one of the other two doors with a probability of 0.5. Can the contestant increase the probability of winning?

Random Variables

20

Abstract

After finishing this chapter

- you will know what discrete and continuous random variables are,
- you will know the connection between random variables, probability distributions and cumulative distribution functions,
- you will have understood the meaning of the terms expected value and variance of random variables, can calculate and deal with them,
- and you will have gained a brief insight into information theory.

20.1 Random Variables and Probabilty Distributions

Take a look at example 2 after Definition 19.4 again, in which we calculated the probabilities for the sum of eyes when rolling the dice twice. In it we constructed a uniform probability space and the sum of eyes was assigned to the elements of this space. The elements of the space Ω itself were actually unimportant. This is often the case: In statistical surveys, it is not the surveyed persons who are of interest, but perhaps the weight, age or voting behavior. In a Bernoulli process, it is not the sequence of results that is in the foreground, but for example the number of times an event has occurred.

In all these cases, a characteristic is assigned to the elements of Ω, we have a mapping from Ω into the set of characteristics. As the codomain of the mapping, we only allow the

Supplementary Information The online version contains supplementary material available at https://doi.org/10.1007/978-3-658-40423-9_20.

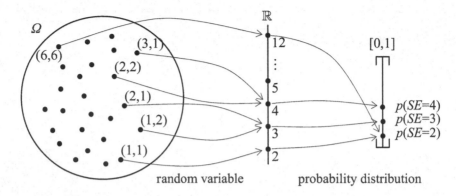

Fig. 20.1 Random variable and probability distribution

set of real numbers, because we know real mappings well. Often this is possible directly (sum of eyes, age, weight), and if not, one can often map the characteristics to numbers, for example the parties in an election can be assigned numbers.

In the case of the dice experiment, we therefore get a function "sum of eyes": $SE \colon \Omega \to \mathbb{R}, (a, b) \mapsto a + b$. Such a function is called a *random variable*.

In Fig. 20.1 we can still recognize another function: Each function value n of the function SE a probability is assigned. The function

$$V \colon \{2, 3, 4, \ldots, 12\} \to [0, 1], \quad x \mapsto p(SE = x)$$

is called *probability distribution* of the random variable.

I have written down the results of the experiment in the form $p(SE = n)$. This is a somewhat sloppy way of writing, because p is defined on the elements of Ω. More precisely, $p(SE = n) := p(\{\omega \in \Omega \mid \text{sum of eyes in } \omega = n\})$.

It will turn out that working with random variables makes it much easier to deal with probabilities and their properties. The often very complicated probability spaces recede more and more into the background and we can work with real functions that are now very familiar to us.

Discrete Random Variable

▶   **Definition 20.1** Let (Ω, p) be a probability space. A function $X \colon \Omega \to \mathbb{R}$ with a countable image is called a *discrete random variable* on Ω.

The fact that the image M is finite or has at most as many elements as $\mathbb{N}$, means that it can be written in the form $M = \{m_i \mid i \in \mathbb{N}\}$. This is sometimes useful for calculations.

Let us conduct an experiment with the result ω, the function value $X(\omega)$ belongs to this result. This function value is also called *realization* of the random variable X. The sum of eyes 12 in a specific throw with two dice is a realization of the random variable "sum of eyes".

▶ **Definition 20.2** If X is a discrete random variable on the probability space (Ω, p) and $W = X(\Omega)$ the image of X, the function

$$V: W \to [0, 1], \quad x \mapsto p(\{\omega \in \Omega \,|\, X(\omega) = x\})$$

is called *probability distribution* of the random variable X.

As with the sum of eyes of the dice, we introduce abbreviated notations for events. For example

$$X = x := \{\omega \in \Omega | X(\omega) = x\},$$
$$X \leq x := \{\omega \in \Omega | X(\omega) \leq x\},$$
$$a < X < b := \{\omega \in \Omega | a < X(\omega) < b\},$$

and so on. So it is $V(x) = p(X = x)$.

The binomial distribution from Theorem 19.14 is a probability distribution in this sense: The Bernoulli process is described by the probability space $\Omega = \{A, \overline{A}\}^n$, and

$$X_A: \Omega \to \mathbb{R}, \quad \omega \mapsto \text{number of } A \text{ in } \omega$$

is a random variable and the corresponding probability distribution

$$V: \{0, 1, 2, \ldots n\} \to [0, 1], \quad k \mapsto p(X_A = k) = b_{n,p}(k)$$

is just the binomial distribution.

Just as we have here summarized results for events that are mapped to the same characteristic, it is often interesting to not have a single result of the random variable X to investigate, but a set of results. In a series of studies, for example, one does not ask for the probability of weight 56.9 kg, but for the probability of a weight between 56 and 57 kg. In particular, for infinitely large Ω often only probabilities for subsets of $\mathbb{R}$ can be given, not for individual real numbers: Strictly speaking, the probability that the hand of the clock is exactly at 12 is 0, the probability that the hand is between 2 and 3 is $1/12$. This is the reason why we introduce a third concept after the random variable and the probability distribution, the cumulative distribution function:

▶ **Definition 20.3** If X is a random variable on (Ω, p), so the function

$$F: \mathbb{R} \to [0, 1], \quad x \mapsto p(X \leq x)$$

is called the *cumulative distribution function* of X.

Probability distribution and cumulative distribution function of the random variable X "Result of rolling a die" are shown in Fig. 20.2.

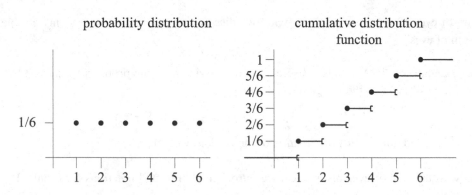

Fig. 20.2 Probability distribution and cumulative distribution function

For all $x < 1$ is $F(x) = 0$, for $1 \le x < 2$ is $F(x) = 1/6$ and so on. The cumulative distribution function of a discrete random variable is always a monotonically increasing step function that starts somewhere to the left at 0 and ends somewhere to the right at the function value 1. Jumps of the function can only occur at the function values of the random variable X. Without proof, I summarize this in the obvious theorem:

▶ **Theorem 20.4** If X is a discrete random variable with the image $W = \{x_i \mid i \in \mathbb{N}\} \subset \mathbb{R}$, so it applies to the associated cumulative distribution function F:

a) For all $x \in \mathbb{R}$ is $F(x) = p(X \le x) = \sum\limits_{x_i \le x} p(X = x_i)$.

b) If $x < y$, then $F(x) \le F(y)$.

c) $\lim\limits_{x \to -\infty} F(x) = 0$, $\lim\limits_{x \to \infty} F(x) = 1$.

In order for the sum in a) to exist at all, we use for the first time the fact that the random variable is discrete, that is, W has only finitely or at most countably infinitely many elements. If W is infinite, we have to form an infinite sum here. Remember the analysis, Theorem 13.12 in Sect. 13.2: The sequence of partial sums is monotonic and bounded, so the series is convergent.

The following theorem shows how we can easily calculate the probability of intervals using the cumulative distribution function:

▶ **Theorem 20.5** If F is the cumulative distribution function of the discrete random variable X, then for all real numbers $a < b$ holds:

a) $p(a < X \le b) = F(b) - F(a)$

b) $p(a < X) = 1 - F(a)$

Here a) follows from the relation:

$$F(b) = p(X \leq b)$$
$$= p(\{X \leq a\} \cup \{a < X \leq b\})$$
$$= p(X \leq a) + p(a < X \leq b)$$
$$= F(a) + p(a < X \leq b)$$

and b) from:

$$1 = p(\Omega)$$
$$= p(\{a < X\} \cup \{X \leq a\})$$
$$= p(a < X) + p(X \leq a)$$
$$= p(a < X) + F(a) \qquad \square$$

The graph of the probability distribution of a discrete random variable consists of a discrete number of points, as you have seen in Fig. 20.2. Usually, the form of histograms is chosen for such a graphical representation: a column is drawn over each argument. This is particularly practical if the arguments, as in the case of dice, all have the same distance from each other. Then we draw the columns all the same width, so that they just touch each other. We calculate the height of the columns so that the area of the column represents the probability of the event occurring. You can see the histogram for the probability distribution of rolling a die in Fig. 20.3.

Let's look at coin tossing as another example. The random variable X should count the number of heads in 20 throws. According to Theorem 19.14 is $p(X = k) = b_{20,1/2}(k)$. The area of the column over the value k in Fig. 20.4 gives the probability with which, in 20 trials, head occured k times.

A very nice property of these area-proportional histograms is that you can determine the probability of a range, for example $p(10 \leq X \leq 14)$ exactly as the area of the columns over the values from 10 to 14.

Do you see where I am going with this? We will approximate some common probability distributions by continuous functions; then we can calculate the probability that X lies between a and b as the integral of the distribution from a to b. To carry out this program, we first need to deal with continuous random variables.

Fig. 20.3 Histogram of the dice probability distribution

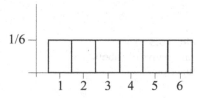

Fig. 20.4 Histogram of the
binomial distribution

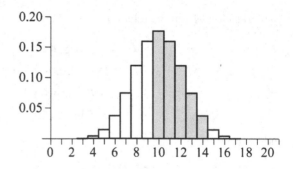

Continuous Random Variables

Let's look at the random variable "position of the hour hand of the clock". Ω consists
of the set of possible hand positions, a value between 0 and 12 is assigned to a hand
position. The random variable $X: \Omega \rightarrow \,]0, 12]$ no longer has a discrete image. We can't
give probabilities for individual elements ω anymore, but we can for example determine:
$p(x \leq 12) = 1$, $p(X \leq 6) = 1/2$, $p(2 < X \leq 3) = 1/12$.

Similarly to the case of discrete random variables (see Definition 20.3), we now look
for the cumulative distribution function F with the property $F(x) = p(X \leq x)$. Of course,
$F(x) = 0$ for $x \leq 0$ and $F(x) = 1$ for $x \geq 12$. In between: $F(x) = x/12$. In Fig. 20.5 you
can see the graph of the function F.

Just like in Theorem 20.5, for example $p(4 < X \leq 7) = F(7) - F(4)$.

Now we are still missing the analogue to the probability distribution of a discrete random variable. Let us recall: In the probability distribution V the area between a and b is
the probability that the result lies between a and b which is just $F(b) - F(a)$. We already
know this property from integral calculus:

The function F is the antiderivative of the probability distribution V:

$$\int_a^b V(t)dt = F(b) - F(a)$$

thus V the derivative of F. In our example we get for V the function (Fig. 20.6):

$$V: \rightarrow [0, 1], \quad x \mapsto \begin{cases} 0 & x \leq 0 \text{ and } x > 12 \\ 1/12 & 0 < x \leq 12. \end{cases}$$

Fig. 20.5 The
cumulative distribution
function F

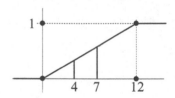

Fig. 20.6 The density V

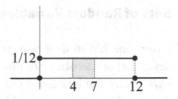

We now have a cumulative distribution function and a kind of probability distribution for the given non-discrete random variable. Mathematically, we're putting the cart before the horse: We call a function a continuous random variable if it has a nice cumulative distribution function:

▶ **Definition 20.6** Let (Ω, p) a probability space. A function $X \colon \Omega \to \mathbb{R}$ is called a *continuous random variable*, if there is an integrable, non-negative real function w with the property

$$p(X \le x) = \int_{-\infty}^{x} w(t)dt.$$

The function $F \colon \to [0, 1],\ x \mapsto p(X \le x) = \int_{-\infty}^{x} w(t)dt$ is called the *cumulative distribution function* of X, the function w is called the *probability density function* or just the *density* of the random variable X.

The position of the hour hand of a clock is a continuous random variable in this sense.

I state the following two theorems without proof, which correspond exactly to Theorems 20.4 and 20.5 for discrete random variables:

▶ **Theorem 20.7** If X a continuous random variable with cumulative distribution function F and density w, then:

a) For all $x \in \mathbb{R}$ is $F(x) = p(X \le x) = \int_{-\infty}^{x} w(t)dt$.

b) If $x < y$ then $F(x) \le F(y)$.

c) $\lim\limits_{x \to -\infty} F(x) = 0,\ \lim\limits_{x \to \infty} F(x) = 1$.

Thus, in the transition from the discrete to the continuous random variable, only the sum $\sum\limits_{x_i \le x} p(X = x_i)$ is replaced by the integral $\int_{-\infty}^{x} w(t)dt$.

▶ **Theorem 20.8** If F is the cumulative distribution function of the continuous random variable X with density w, then for all real numbers $a < b$:

a) $p(a < X \le b) = F(b) - F(a) = \int_{a}^{b} w(t)dt$.

b) $p(a < X) = 1 - F(a) = \int_{a}^{\infty} w(t)dt$.

Sets of Random Variables

Often one has to deal with more than one random variable: In a probability space Ω several characteristics are observed at the same time, for example in opinion polls. An experiment can be carried out several times in succession and each single experiment is described by a random variable.

A finite sequence of random variables $(X_1, X_2, \ldots, X_n)$ is called a *random vector* or a *multivariate random variable*. A random vector of length n is a function from Ω to $\mathbb{R}^n$.

Examples

1. The random selection of a point from a rectangle $[a, b] \times [c, d]$ can be described by two random variables: X is the random variable "choose a point from $[a, b]$", Y is the random variable "choose a point from $[c, d]$". A realization of the random vector (X, Y) consists of a specific point (x, y) of the rectangle.
2. The two random variables H and W for heigth and weight (in m and kg) form thus the random vector (H, W). The result of an experiment is a tuple, for example $(186, 82)$.
3. If you roll several dice at the same time, for example 5 dice, you get the random variables $(X_1, X_2, X_3, X_4, X_5)$, where X_i is the random variable "rolling the i-th die". The result of an experiment is an element $(x_1, x_2, x_3, x_4, x_5)$ of $\mathbb{R}^5$, where x_i is the number of eyes of the i-th die. We also say $(x_1, x_2, x_3, x_4, x_5)$ is a realization of the random vector $(X_1, X_2, X_3, X_4, X_5)$.
4. The drawing of n balls from an urn with black and white balls can be described by the random vector $(X_1, X_2, \ldots, X_n)$, X_i is the i-th draw, which has the possible outcomes black or white, respectively, the results 0 or 1, if we work with real random variables. The result of the experiment, the realization, is then a vector of the form $(1, 0, 0, 1, 0, \ldots, 0) \in \mathbb{R}^n$, where the 1 means a white and the 0 a black ball. ◀

Also, infinitely many random variables can occur. Sequences of random variables, such as $(X_n)_{n \in \mathbb{N}}$ or $(X_t)_{t \in \mathbb{R}}$ are called *stochastic processes*. We will deal with such processes in Sect. 21.3.

Different real random variables can be linked together with mathematical operations, just as we already know from real functions.

Examples

5. Currently, the body mass index is modern as a measure for a reasonable ratio between height and weight: It is "weight in kg divided by the square of the height in meters" and should be in the range of 20 to 25. With the random variables H for the height and W for the weight this gives a random variable W/H^2, where $W/H^2(\omega) := W(\omega)/H^2(\omega)$.
6. When rolling 5 dice, the random variable $X = X_1 + X_2 + X_3 + X_4 + X_5$ is the sum of eyes of the 5 dice.

7. When drawing n balls from an urn, the random variable $X = X_1 + X_2 + \cdots + X_n$ counts the number of drawn white balls, if the X_i are the random variables from Example 4 from above. ◀

Independent Random Variables

Just as with different events, you can also ask about values of random variables whether they have something to do with each other or are independent of each other. Weight and height of a person are certainly not independent of each other, but maybe height and income? With the random vector (X, Y), which describes the random selection of a point from a rectangle, the results of X and Y should be independent of each other. If we draw balls from an urn, the 2nd draw is independent of the first, provided we carry out the experiment "drawing with replacement". If we draw without replacement, the later draws are dependent on the earlier ones! When rolling dice, the result of the first die certainly has nothing to do with the result of the second die, the random variables $X_i =$ "result of the i-th throw" are in this case independent.

We mathematically describe this situation by reducing it to the already known concept of independent events: We call two discrete random variables X, Y independent of each other if the events $X = x$ and $Y = y$ are independent of each other for all possible values of x and y:

▶ **Definition 20.9: Independent discrete random variables** Two discrete random variables X and Y with the images $W(X)$, $W(Y)$ are called *independent*, if for all possible value pairs $x \in W(X)$, $y \in W(Y)$ holds:

$$p(X = x, Y = y) = p(X = x) \cdot p(Y = y) \tag{20.1}$$

The n discrete random variables $X_1, X_2, \ldots, X_n$ are called *independent*, if for all possible n-tuples $(x_1, x_2, \ldots, x_n)$ from $W(X_1) \times W(X_2) \times \cdots \times W(X_n)$ holds:

$$p(X_1 = x_1, X_2 = x_2, \ldots, X_n = x_n) = p(X_1 = x_1) \cdot p(X_2 = x_2) \cdots \cdot p(X_n = x_n).$$

I use the intuitively clear notation for probabilities of several random variables, such as

$$p(X = x, Y = y) := p(\{X = x\} \cap \{Y = y\}),$$
$$p(X_1 \leq a, X_2 \leq b) := p(\{X_1 \leq a\} \cap \{X_2 \leq b\}).$$

In the Definition 20.9 I have taken this in advance.

The concept of independence can also be formulated for continuous random variables. However, since the probability of an element $p(X = x)$ is not a meaningful quantity, we have to require independence of events belonging to intervals of $\mathbb{R}$. I write down the definition for two variables:

▶   **Definition 20.10: Independent continuous random variables** Two continuous
random variables X and Y are called *independent* if for all $x, y \in \mathbb{R}$ it holds

$$p(X \le x, Y \le y) = p(X \le x) \cdot p(Y \le y). \tag{20.2}$$

From (20.1) and (20.2) it can be derived that for both discrete and continuous independent random variables for all $a, b, c, d \in \mathbb{R}$ holds:

$$p(a < X \le b, c < Y \le d) = p(a < X \le b) \cdot p(c < Y \le d).$$

We will often have to deal with Bernoulli processes, that is, experiments carried out over and over again independently. The random variables X_i, which describe the i-th trial of a Bernoulli process, are in any case all independent of each other.

20.2   Expected Value and Variance of Random Variables

Let's take a look at the game of roulette. The ball can land in one of the pockets from 0 to 36 and we assume that $\Omega = \{0, 1, 2, \ldots, 36\}$ is a uniform probability space, so we can give the probability for each subset of Ω.

First, let's bet 1 € on a single number, for example 7. If the ball lands in the pocket with the number 7, you will receive a 36 € pay out (from which the stake is deducted). You usually have to wait for this, because the probability of winning is $1/37$. To shorten this waiting time, we could also bet 1 € on the odd numbers at the same time. The pay out for this is 2 €. Finally, we bet on the street (1,2,3), for which we still have a chance to win 12 €. Now we are interested in what the chances of winning look like if we play this combination for a longer period of time. Can we calculate how big the average payout is in one game?

First, let's write down the payout for the possible results.

Numbers	$5, 9, 11, 13, 15, \ldots, 35$ (15 values)	1, 3	2	7	the rest (18 values)
Payout	2	$12 + 2$	12	$36 + 2$	0

If X is the random variable describing the pay out in such a game, then the image $W(X) = \{0, 2, 12, 14, 38\}$

Now let's play n times. Let f_k be the number of cases in which the pay out k occurred. Then our total payout P:

$$P = 2 \cdot f_2 + 14 \cdot f_{14} + 12 \cdot f_{12} + 38 \cdot f_{38} + 0 \cdot f_0$$

the average profit per game is therefore:

$$\overline{P} = \frac{P}{n} = \frac{2 \cdot f_2 + 14 \cdot f_{14} + 12 \cdot f_{12} + 38 \cdot f_{38} + 0 \cdot f_0}{n}$$

$$= 2 \cdot r_2 + 14 \cdot r_{14} + 12 \cdot r_{12} + 38 \cdot r_{38} + 0 \cdot r_0$$

if $r_k = f_k/n$ is the relative frequency of the occurrence of the pay out k. Very likely, for large n, the relative frequency of the payout k will be of the order of the probability of k occuring:

$$\overline{P} \approx 2 \cdot p(X = 2) + 14 \cdot p(X = 14) + 12 \cdot p(X = 12)$$
$$+ 38 \cdot p(X = 38) + 0 \cdot p(X = 0)$$

Surprisingly, the right side of this equation no longer depends on the specific course of the game, it is independent of the experiment and only determined by the random variable X and its probability distribution. We call this right side of the equation the expected value $E(X)$ of the random variable X. The value $E(X)$ will usually differ from the average pay out, but the longer we play, the more it will approach it. Let's calculate the expected value in the concrete example:

$$E(X) = 2 \cdot \frac{15}{37} + 14 \cdot \frac{2}{37} + 12 \cdot \frac{1}{37} + 38 \cdot \frac{1}{37} = \frac{108}{37}.$$

That looks good at first, but unfortunately the bank deducts the stake of 3 € in each game, so that in total an average loss of $108/37 - 3 = -3/37$ € remains, that is about 2.7%.

Is there a better strategy? No, you will find that with all possible combinations of bets, on average, for every 1 euro bet, 1/37 euros go to the bank. The reason for this is something we will learn about in Theorem 20.18.

A game is called fair if the expected value for the winning is 0, if in other words, winnings and losses cancel each other out. In this sense, roulette is not fair; but if you consider that from this 1/37 all of the casino's expenses are covered, you can't really complain.

▶ **Definition 20.11: The expected value of a discrete random variable** Let X be a discrete random variable. If the sum

$$E(X) = \sum_{x \in W(X)} x \cdot p(X = x)$$

in which the sum is taken over all x from the image $W(X)$ of X, uniquely exists, it is called the *expected value* of X. The expected value $E(X)$ of a random variable X is also often denoted by μ.

A trap lies in the inconspicuous remark, "if the value exists uniquely". If the image of X is finite, then $E(X)$ always exists. However, if it is infinite, the series does not have to converge, and even if it converges, the limit may depend on the order of summation. Then there is no expected value. If I write $E(X)$ in the future, I always assume that $E(X)$ really exists.

When forming the expected value, all possible products "function value times probability of function value" must be added up. If the random variable X describes the numbers rolled with a die, then:

$$E(X) = 1 \cdot \frac{1}{6} + 2 \cdot \frac{1}{6} + 3 \cdot \frac{1}{6} + 4 \cdot \frac{1}{6} + 5 \cdot \frac{1}{6} + 6 \cdot \frac{1}{6} = 3.5. \qquad (20.3)$$

The expected value does not necessarily have to be in the image of the random variable: On average, you just roll 3.5.

I give a similar definition for continuous random variables. This can be motivated, for example, by the fact that the continuous random variable is approximated by a sequence of discrete random variables with existing expected values. I don't want to do this, but I just want to point out the analogy that has already emerged in the comparison of the Theorems 20.4 and 20.7. The sum over the probabilities of the function values of the discrete random variable X just corresponds to the integral of the density of the continuous random variable:

▷   **Definition 20.12: The expected value of a continuous random variable** Let X be a continuous random variable with density w. If the integral

$$E(X) = \int_{-\infty}^{\infty} x \cdot w(x)dx$$

exists, it is called the *expected value* of the random variable X.

Let's look back at the roulette example: If I multiply my bet by five, all winnings and losses are also multiplied by five. We get a new random variable $Y = 5 \cdot X$, which describes the winnings. What is the expected value of Y? If I am interested in the random variable "net profit", I have to construct the new random variable $Z = X - 3$ from X. What is the expected value of Z?

The following result answers these questions, it should not be very surprising:

▷   **Theorem 20.13** Let X be a discrete or continuous random variable with expected value $E(X)$. For $a, b \in \mathbb{R}$ the expected value of the random variable $aX + b$ is:

$$E(aX + b) = aE(X) + b$$

In particular, then $E(aX) = aE(X)$ and $E(X + b) = E(X) + b$.

I would like to prove this statement for discrete random variables on behalf of a whole range of subsequent results on expected values and later also on variances (the Theorems 20.15, 20.16, 20.18 and 20.19). The proof technique is similar in the other theorems, although the proofs are sometimes slightly more complex, especially when it comes to continuous random variables.

So let X be a discrete random variable and $Y = aX + b$. Then for all $\omega \in \Omega$

$$X(\omega) = x \quad \Leftrightarrow \quad Y(\omega) = aX(\omega) + b = ax + b, \qquad (20.4)$$

thus $\{\omega \mid X(\omega) = x\} = \{\omega \mid Y(\omega) = ax + b\}$ and it follows $p(X = x) = p(Y = ax + b)$. Because the image $W(Y)$ is just the set of all elements $ax + b$ with $x \in W(X)$, we get for the expected value:

$$
E(Y) = \sum_{x \in W(X)} (ax + b)p(Y = ax + b)
$$

$$
= \sum_{x \in W(X)} ax \cdot p(X = x) + \sum_{x \in W(X)} b \cdot p(X = x)
$$

$$
= a \cdot \underbrace{\sum_{x \in W(X)} x \cdot p(X = x)}_{=E(X)} + b \cdot \underbrace{\sum_{x \in W(X)} p(X = x)}_{=1} = aE(X) + b.
$$

I used that the sum over all possible probabilities $p(X = x)$ just results in 1.

In the case of continuous random variables, one can first calculate the density function of the new random variable Y and then the expected value from this. □

In the game of roulette, let X_A be the random variable that describes the "win when betting 1 € on 7", X_B the random variable "win when betting 1 € on the odd numbers" and X_C is finally "win when betting 1 € on 1,2,3". The expected values of all three random variables are equal:

$$
E(X_A) = 36 \cdot p(X_A = 7) + 0 \cdot p(X_A \neq 7) = 36 \cdot (1/37) = 36/37,
$$

$$
E(X_B) = 2 \cdot p(X_B \text{ is odd}) + 0 \cdot p(X_B \text{ is even}) = 2 \cdot (18/37) = 36/37,
$$

$$
E(X_C) = 12 \cdot p(X_C = 1, 2, 3) + 0 \cdot p(X_C \neq 1, 2, 3) = 12 \cdot (3/37) = 36/37.
$$

The thrill is much greater when betting on a number than, say, betting on odd. How do these two strategies differ? In the first case, the possible win is much greater, but so is the risk of loss. Betting on 1, 2, 3 is somewhere in between in terms of risk.

It would be interesting to know how the individual results are scattered around the expected value for different strategies, that is, how far possible wins and losses deviate from the expected value on average. Let's apply what we've already learned about random variables and expected values: If $E(X) = \mu$, then the random variable $Y = X - \mu$ describes the deviation of the win from the expected value μ. So we would need to obtain the average deviation from the expected value of the random variable Y. Let's calculate it: It is

$$
E(Y) = E(X - \mu) = E(x) - \mu = 0.
$$

Too bad: The positive and negative deviations cancel each other out. But on closer inspection this is also reasonable, the expected value is in the middle of the values that occur.

To avoid this cancellation effect, we have to count all the deviations positive, so we could examine $Y = |X - \mu|$, for example. If you have gotten this far in mathematics, you also know that a mathematician only calculates with absolute values when there is no avoiding it: There are always such ugly case distinctions. For this reason we take the random variable $Y = (X - \mu)^2$ here; this describes the squares of the deviations, and they are of course all positive.

▷ **Definition 20.14: The variance** If $E(x) = \mu$ is the expected value of the random variable X, so the value

$$\sigma^2 := \mathrm{Var}(X) := E((X - \mu)^2),$$

if it exists, is called the *variance* of X, and $\sigma = +\sqrt{\mathrm{Var}(X)}$ is called the *standard deviation* or *dispersion* of X.

▷ **Theorem 20.15: The calculation of variance** For a discrete random variable X with expected value μ it holds

$$\mathrm{Var}(X) = \sum_{x \in W(X)} (x - \mu)^2 \cdot p(X = x)$$

$$= \left(\sum_{x \in W(X)} x^2 \cdot p(X = x) \right) - \mu^2$$

$$= E(X^2) - \mu^2$$

The proof is elementary, similar to that of Theorem 20.13. The relation $\mathrm{Var}(X) = E(X^2) - \mu^2$ also applies to continuous random variables. Note the small but important difference between $E(X^2)$, the expected value of the random variable X^2 and $\mu^2 = E(X)^2$, the square of the expected value of X.

Let us calculate the variances of the random variables X_A, X_B, X_C from the roulette game:

$$\mathrm{Var}(X_A) = \left(\sum_{x \in W(X)} x^2 \cdot p(X = x) \right) - \mu^2 = 36^2 \cdot \frac{1}{37} - \left(\frac{36}{37} \right)^2 \approx 34.08,$$

$$\mathrm{Var}(X_B) = 2^2 \cdot \frac{18}{37} - \left(\frac{36}{37} \right)^2 \approx 0.999,$$

$$\mathrm{Var}(X_C) = 12^2 \cdot \frac{3}{37} - \left(\frac{36}{37} \right)^2 \approx 10.73.$$

If we go from the random variable X to $Y = aX + b$, the variance also changes:

▷ **Theorem 20.16** If the variance of the random variable X exists, then for $a, b \in \mathbb{R}$:

$$\mathrm{Var}(aX + b) = a^2 \mathrm{Var}(X).$$

The multiplicative factor a thus goes quadratically into the variance, a simple shift of the values of the random variable does not change the variance. That is obvious, then with such a shift, all deviations from the expected value remain unchanged.

▶ **Definition 20.17: Standardization of a random variable** If X is a random variable with expected value μ and variance σ^2, then the random variable

$$X^* := \frac{X - \mu}{\sigma}$$

is called the *standardization* of X.

For the standardization X^* of X holds because of $X^* = \frac{1}{\sigma}X - \frac{\mu}{\sigma}$:

$$E(X^*) = \frac{1}{\sigma}E(X) - \frac{\mu}{\sigma} = \frac{\mu}{\sigma} - \frac{\mu}{\sigma} = 0,$$

$$\mathrm{Var}(X^*) = \frac{1}{\sigma^2}\mathrm{Var}(X) = \frac{\sigma^2}{\sigma^2} = 1.$$

This conversion is especially useful when you want to compare random variables with certain given probability distributions. These given distributions are often only tabulated in standardized form. Other distributions can then be easily converted into the standardized form.

Sums and Products of Random Variables

In the game of roulette, the expected value of X_B "Bet on odd" and of X_C "Bet on 1, 2, 3" was 36/37 respectively. What happens if you combine both options in one game?

Let $S = X_B + X_C$ the game in which we bet on odd and on 1, 2, 3. If an odd number comes, you win 2 € at X_B. If 1, 2 or 3 comes, the win in X_C is 12 €. The possible function values of S are therefore 2 (for the 16 odd numbers without 1 and 3), 12 (for 2), 14 (for 1 and 3) and 0 for the rest. What is the expected value of S?

$$E(S) = 2 \cdot \frac{16}{37} + 12 \cdot \frac{1}{37} + 14 \cdot \frac{2}{37} = \frac{72}{37}$$

thus just the sum of the two individual expected values.

At the beginning of this section, we calculated the expected value of $X = X_A + X_B + X_C$: It was $108/37 = 3 \cdot (36/37)$, that is also the sum of the individual expected values. This always applies:

▶ **Theorem 20.18** If X, Y are random variables with the expected values $E(X)$ and $E(Y)$, then

$$E(X + Y) = E(X) + E(Y).$$

This statement can also be transferred to n random variables by mathematical induction.

The rules of the game in roulette are such that for every simple bet the expected value for winning is $36/37$ of the stake. Theorem 20.18 says you cannot increase this expected value by combining different types of bets.

Now let's look at the product P of the two random variables X_B and X_C. This product does not correspond to any real game, but we can calculate P and $E(P)$. $P(\omega)$ is only then unequal to 0, if $X_B(\omega)$ and $X_C(\omega)$ both are different from 0, so only in the results 1 and 3. There, the product is $X_B(\omega) \cdot X_C(\omega) = 2 \cdot 12 = 24$. This gives us for the expected value $E(P) = 24 \cdot (2/37) = 48/37$. This is different from the product of the expected values of X_B and X_C.

Things look different in a dice experiment: If we roll twice in a row and X_i is the result of the i-th throw, one can calculate with some effort that the expected value of the product $X_1 \cdot X_2 = 12.25 = 3.5^2$ is exactly the product of the expected values. How does dice differ from roulette? The random variables X_1 and X_2 of the throws carried out one after the other are independent of each other, in contrast to the random variables X_B, X_C that describe the roulette game. In fact, it applies:

▶ **Theorem 20.19** If X, Y are independent random variables with the expected values $E(X)$, $E(Y)$, then

$$E(X \cdot Y) = E(X) \cdot E(Y)$$

This result can also be transferred to n independent random variables.

If we assume independence, variances can also be added:

▶ **Theorem 20.20** If X, Y are independent random variables with the variances $\mathrm{Var}(X)$ and $\mathrm{Var}(Y)$, then

$$\mathrm{Var}(X + Y) = \mathrm{Var}(X) + \mathrm{Var}(Y).$$

If the variances of the n random variables $X_1, X_2, \ldots, X_n$ exist and these variables are pairwise independent, then:

$$\mathrm{Var}(X_1 + X_2 + \cdots + X_n) = \mathrm{Var}(X_1) + \mathrm{Var}(X_2) + \cdots + \mathrm{Var}(X_n).$$

I prove this statement for two random variables diligently using the rules of sums and products of expected values:

$$\begin{aligned}
\mathrm{Var}(X + Y) &= E((X + Y)^2) - E(X + Y)^2 \\
&= E(X^2 + 2XY + Y^2) - (E(X) + E(Y))^2 \\
&= E(X^2) + 2 \cdot E(XY) + E(Y^2) - E(X)^2 - 2E(X)E(Y) - E(Y)^2 \\
&= \underbrace{E(X^2) - E(X)^2}_{=\mathrm{Var}(X)} + \underbrace{E(Y^2) - E(Y)^2}_{=\mathrm{Var}(Y)} + 2 \cdot \underbrace{(E(XY) - E(X)E(Y))}_{=0,\text{ if } X \text{ and } Y \text{ are independent.}}. \qquad \square
\end{aligned}$$

▶ **Definition 20.21: The covariance** If X and Y are random variables with the expected values $E(X) = \mu_X$, $E(Y) = \mu_Y$, then the number

$$\text{Cov}(X, Y) := E((X - \mu_X)(Y - \mu_Y)),$$

if it exists, is called the *covariance* of X and Y. The random variables X and Y are called uncorrelated if $\text{Cov}(X, Y) = 0$.

The calculation rules for the expected value show that:

$$\text{Cov}(X, Y) = E(XY - \mu_X Y - \mu_Y X + \mu_X \mu_Y)$$
$$= E(XY) - \mu_X E(Y) - \mu_Y E(X) + \mu_X \mu_Y$$
$$= E(XY) - \mu_X \mu_Y$$

This expression is 0 if X and Y are independent. Independent random variables are therefore uncorrelated. The reverse is not true.

A Short Excursion into Information Theory

The founder of modern information theory, Claude E. Shannon, assigned an *information content* to the signs of an information source. He was guided by the following observations:

1. The information content of a sign does not depend on the sign itself, but only on the probability of the occurrence of this sign: different signs with the same probability of occurrence carry the same amount of information. If two probabilities differ only slightly, the information content should not be very different.
2. Signs with a small probability of occurrence carry more information than signs with a large probability of occurrence.

   In a political speech, the important information is not contained in the recurring clichés, but hidden in between at large intervals. An information source that always tells the same thing has no real information content, the content is in the rare and therefore surprising signs. Shannon called the information content of a message also the "surprisal" of the message.

3. The information content behaves additively when transmitting several signs.

   Two bytes of equal probability of occurrence can therefore, for example, transmit twice as much information as one byte.

These observations lead to an axiomatic definition of the concept of information:

Let $Q = (A, p)$ be a memoryless information source (see Example 1 after Definition 19.9) with the alphabet $A = \{x_1, x_2, \ldots, x_n\}$ and associated probabilities $p_1, p_2, \ldots, p_n$.

Information content that is associated with the character x_i must have the following properties:

1. The information content of a character is a continuous function of the probability of the character: $I(x_i) = f(p(x_i)) = f(p_i)$.
2. For $p_i < p_j$ is $I(x_i) < I(x_j)$
3. If $I(x_i x_j)$ is the information content of two consecutive characters, then $I(x_i x_j) = I(x_i) + I(x_j)$.

Since the source is memoryless, the two events x_i and x_j are independent of each other, the probability of $x_i x_j$ is therefore equal to $p_i p_j$ and therefore follows from 3. in particular

$$f(p_i p_j) = f(p_i) + f(p_j). \tag{20.5}$$

We know from analysis a function that satisfies this condition, it is the logarithm. In fact, one can show that the only continuous functions that fulfill (20.5) are the functions $a \cdot \log x$, where $a \in \mathbb{R}$ and the logarithm can be formed to any base. Often one chooses the logarithm to the base 2.

> After Theorem 14.31 in Section 14.3 we have seen that logarithms to different bases only differ by a constant factor. This means that a function that fulfills (20.5) always has the form $a \log_2 x$ for a $a \in \mathbb{R}$.

Since the function f is supposed to be monotonically decreasing, the factor a must be negative. First, we arbitrarily normalize $a = -1$. Finally, we use the fact $-\log p = \log(1/p)$, so we can define an information content that satisfies the three required properties:

▷   **Definition 20.22** Let $Q = (A, p)$ be a memoryless information source with the alphabet $A = \{x_1, x_2, \ldots, x_n\}$ and associated probabilities $p_1, p_2, \ldots, p_n$. Then

$$I(x_i) := \log_2(1/p_i)$$

is called the *information content* of the character x_i.

The information content is a random variable on the probability space Q. The expected value of this random variable, that is, the average amount of information transmitted, is the entropy of the source:

▷   **Definition 20.23: The entropy** Let $Q = (A, p)$ be a memoryless information source with the alphabet $A = \{x_1, x_2, \ldots, x_n\}$ and associated probabilities of occurrence $p_1, p_2, \ldots, p_n$. Then

$$H(Q) := \sum_{i=1}^{n} p_i \cdot \log_2(1/p_i)$$

is called *entropy* of the information source Q.

If in the alphabet all characters x_1 to x_n have the same probability of occurrence $1/n$, then the entropy

$$H(Q) = \sum_{i=1}^{n} 1/n \cdot \log_2(n) = \log_2(n). \qquad (20.6)$$

It can be shown that this value is the maximum entropy of an information source with n characters. As soon as the probabilities of individual characters deviate from the value $1/n$ the entropy decreases.

Do not confuse the information content thus defined with the meaning of a message interpreted by humans. Otherwise you would prefer a book filled with randomly distributed letters to the present book. That would be a shame. But you can also look at it this way: The book you are reading now has a size of about 13 MB as an electronic document. If I compress it with zip, the size decreases to about 4 MB. The amount of information is the same in both documents, so the information content per character must be greater in the second document. What is behind it? When compressing with zip, the source text alphabet is encoded in another alphabet in which, in contrast to the English clear text, each character occurs approximately equally often. The entropy of the compressed text is therefore greater than that of the uncompressed text. The better the compression, the more evenly distributed the individual characters are and the greater the entropy.

In data processing, a source is usually binary coded, that is, translated into a code word of 0 and 1. We have learned such a coding procedure, for example, with the Huffman code (at the end of Sect. 11.2). In the source $Q = (A, p)$ the character x_i is encoded in a code word of length m bits. The function $l(x_i) =$ length of the code word of the character x_i also represents a random variable on the probability space Q, and the expected value of this random variable

$$L(Q) = \sum_{i=1}^{n} p_i \cdot l(x_i)$$

is the *average codeword length* in the transmission. Of course, one would like to keep the average codeword length as short as possible in an encoding of the source alphabet. Is there a lower limit for the codeword length? First two

Examples

1. Let's take the ASCII code first and assume that all characters are equally likely. Then $p_i = 1/256$ and $\log_2(1/p_i) = 8$. According to (20.6) this is the entropy. It corresponds exactly to the constant word length of 8 bits. In this case, entropy and average word length match. This is the reason for the initially arbitrary normalization of the information content.

2. When introducing the Huffman code at the end of Sect. 11.2 we defined an alphabet with the characters $\{a,b,c,d,e,f\}$ and associated probabilities $\{0.04, 0.15, 0.1, 0.15, 0.36, 0.2\}$. The entropy of this code is:

$$0.04 \cdot \log_2(25) + 0.15 \cdot \log_2(6.66) + 0.1 \cdot \log_2(10)$$
$$+ 0.15 \cdot \log_2(6.66) + 0.36 \cdot \log_2(2.78) + 0.2 \cdot \log_2(5) \approx 2.33.$$

If we encode the 6 characters without compression in binary, we need 3 bits. The average word length is therefore 3 in this case. The Huffman code we have set up for this alphabet has different word lengths from one to four bits. The average word length of the Huffman code is shorter, it is:

$$0.04 \cdot 4 + 0.15 \cdot 3 + 0.1 \cdot 4 + 0.15 \cdot 3 + 0.36 \cdot 1 + 0.2 \cdot 3 = 2.42.$$

You can see that this value is only slightly above the entropy. ◀

The first source coding theorem by Shannon establishes an important connection between the average word length L and the entropy H here. I would like to quote it without proof:

▶   **Theorem 20.24** Let $Q = (A, p)$ be a memoryless information source with entropy $H(Q)$. Then there exists a binary prefix code for this source with average code word length L such that:

$$H(Q) \leq L \leq H(Q) + 1$$

The theorem is usually stated more generally, without the restriction to a binary coding.

We can infer two statements from this theorem: The entropy is the lower limit for the average word length in a prefix code, it cannot be shorter. And further, one can always find a prefix code whose average word length differs from the entropy by at most one bit. This is an upper bound, in the above example we were even closer.

The Huffman code was very close to the entropy in the example. This is no coincidence: It can be shown that the Huffman code is the best prefix code. There is no other prefix code with shorter average word length.

20.3   Comprehension Questions and Exercises

Comprehension Questions

1. Explain what a (discrete) random variable, a probability distribution, and a cumulative distribution function is.
2. Why can there be at most countably many function values in a discrete random variable?

3. When rolling 2 dice, the random variable X_i should be the result of the i-th dice. Are the random variables X_1 and X_2 independent of each other? Are the random variables X_1 and $X_1 + X_2$ independent of each other?

4. Does each discrete random variable have an expected value?

5. What type of deviation from the expected value does the variance of a random variable describe?

6. Let X and Y independent random variables. Is then $\text{Var}(X \cdot Y) = \text{Var}(X) \cdot \text{Var}(Y)$?

Exercises

1. Calculate the expected value, the variance and the standard deviation of the sum of eyes when rolling 5 dice.

2. The random variable P describes the product of eyes when rolling dice twice. Draw the probability distribution of P. Calculate the expected value and the variance of P.

3. The random variable X describes the number of jacks in skat after the deal (32 cards with four jacks, 3 players each receive 10 cards, 2 cards are in the skat). Determine the probability distribution and the expected value of X,
   a) without information about the card distribution among the 3 players,
   b) with the knowledge that the first player did not receive any jacks.

4. In the game of roulette, a player bets 10 Euros on red. If he wins (18 of 37 fields are red), he will receive a pay out of 20 Euros (net profit $= 10$ Euros). If he loses, he will double his bet until he wins or until the bank's betting limit of 1000 Euros is reached. Describe the random variable "net profit" in this strategy and calculate the expected value and variance of this game.

5. Use a spreadsheet program or a computer algebra system to plot histograms of the function n and p for different values of $b_{n,p}(k)$.

6. The following ad appeared in the Süddeutsche Zeitung a few years ago:[1]

---

[1] A reader sent me this ad.

Let us assume that the scientific prediction is a coin toss. The probability of a girl's birth is 0.465. Calculate the approximate annual income of the institute if about 100 requests per month arrive. How can the institute improve its income (at the same rate)? (For this purpose, set up a random variable "profit" and calculate the expected value.)

7. Calculate the entropy and the average word length of the Huffman code from Exercise 6 in Chap. 11.

Important Distributions, Stochastic Processes

Abstract

The probabilities of events can be described using probability distributions. At the end of this chapter you will know

- the most important discrete probability distributions: binomial distribution, geometric and hypergeometric distribution as well as the Poisson distribution as an approximation to the binomial distribution
- the standard normal distribution and the general normal distribution, and you will know why these can be used as an approximation to the binomial distribution,
- the central limit theorem,
- the exponential distribution and the chi-square distribution.

You will also learn about some stochastic processes:

- the Poisson process for describing events over time,
- Markov chains and transition matrices with their stationary states,
- As an application, you will calculate the evolution of queues depending on the arriving and processing of requests.

When a statistician conducts experiments or analyses data sets, a number of probability distributions repeatedly occur. If you know the right distribution underlying an

Supplementary Information The online version contains supplementary material available at https://doi.org/10.1007/978-3-658-40423-9_21.

experiment, you can use it to determine probabilities, expected values, variances, and other parameters.

We will now take a closer look at some of these distributions. The data sets we are dealing with are always finite, but sometimes it is more clever to work with continuous distributions and thus approximate the discrete distributions. We will also get to know such distributions.

21.1   Discrete Probability Distributions

Some discrete distributions have already been encountered in Sect. 19.3, from these and from other distributions we will now calculate the corresponding expected values and variances in particular.

The Discrete Uniform Distribution

The simplest probability distribution is the uniform distribution. If Ω is a uniform probability space with n elements and X the random variable that assigns the value $x_i \in \mathbb{R}$ to the $\omega_i \in \Omega$, then each function value is equally likely. If $x_1, x_2, \ldots, x_n$ are the possible values of the random variable X, then for all i the probability $p(X = x_i) = 1/n$. Then, according to Definition 20.11 for the expected value holds:

$$E(X) = \sum_{i=1}^{n} x_i \cdot \frac{1}{n} = \frac{1}{n} \cdot \sum_{i=1}^{n} x_i,$$

that is the arithmetic mean of the function values. The variance is, according to Theorem 20.15

$$\mathrm{Var}(X) = \frac{1}{n} \sum_{i=1}^{n} x_i^2 - E(X)^2.$$

For the uniformly distributed random variable "rolling a dice", we have calculated the expected value in (20.3) after Definition 20.11, it results in 3.5. The variance when rolling the dice is

$$\mathrm{Var}(X) = \sigma^2 = \frac{1}{6}(1^2 + 2^2 + 3^2 + 4^2 + 5^2 + 6^2) - 3.5^2 = 2.917,$$

the standard deviation σ thus approximately 1.7.

The Binomial Distribution

The binomial distribution describes in a Bernoulli process the probability that in n trials the event A occurs k times. The prototype for this is the urn model "drawing with replacement": If there are N balls in the urn, of which B are black, then the probability for the event A to draw a black ball is just $p = B/N$.

For the random variable X, which is given by

$$X = k \quad \Leftrightarrow \quad A \text{ occurs exactly } k \text{ times in } n \text{ trials,}$$

applies according to Theorem 19.15: $p(X = k) = \binom{n}{k} p^k (1 - p)^{n-k}$.

▶   **Definition and Theorem 21.1** A discrete random variable X, which in a Bernoulli process of size n specifies the number of trials in which the event A with $p(A) = p$ occurs, is called *binomially distributed* with parameters n and p or shortly $b_{n,p}$-distributed. It is

$$b_{n,p}(k) := p(X = k) = \binom{n}{k} p^k (1 - p)^{n-k}. \tag{21.1}$$

The random variable "number of black balls in n draws with replacement" is $b_{n,p}$-distributed.

Let's calculate the expected value of X:

$$E(X) = \sum_{k=0}^{n} k \cdot p(X = k) = \sum_{k=0}^{n} k \binom{n}{k} p^k q^{n-k} = ?$$

How can we compute this sum? A little trick helps us with that. I formulate it as a lemma because we will use it often, especially in the next chapter.

A binomially distributed random variable X can be composed additively from n individual random variables X_i:

▶   **Lemma 21.2** In a Bernoulli process with a size of n let be $p(A) = p$. For $k = 1, \ldots, n$ let the random variable X_i be defined by

$$X_i(\omega) = \begin{cases} 1 & \text{if } A \text{ occurs at the } i\text{-th trial} \\ 0 & \text{else} \end{cases}$$

Then $X = X_1 + X_2 + \cdots + X_n$ is the $b_{n,p}$-distributed random variable, indicating the number of trials in which A occurs.

For it is $X = k$ if and only if A has occured k times.   □

The X_i from the lemma are independent and therefore, with the calculation rules for expected values and variances, which we derived in Theorems 20.18 and 20.20:

$$E(X) = \sum_{k=1}^{n} E(X_i), \quad \text{Var}(X) = \sum_{k=1}^{n} \text{Var}(X_i).$$

The expected value and variance of the X_i are easy to calculate, because X_i almost always has the value 0. First the expected value:

$$E(X_i) = 1 \cdot p(X_i = 1) + 0 \cdot p(X_i = 0) = 1 \cdot p,$$

and thus $E(X) = n \cdot p$.

For the variances applies:

$$\text{Var}(X_i) = 1^2 \cdot p(X_i = 1) - p^2 = p - p^2 = p(1 - p),$$

so $\text{Var}(X) = np(1 - p)$.

▶   **Theorem 21.3** For a $b_{n,p}$-distributed random variable X holds $E(X) = np$, $\text{Var}(X) = np(1 - p)$.

Example of using the binomial distribution

The production of electronic components requires particularly clean rooms. A standard specifies how many particles of which size may still be present in such a clean room in order to meet the requirements of a certain cleanliness class. For example, in a room of class 4, a maximum of 352 particles up to size 0.5 μm may be present per cubic meter, in a room of class 5, up to 3520 such particles.

> The crooked numbers are probably due to the fact that the number of particles per cubic foot was still specified in the last standard.

What is the probability of finding 0, 1, 2 or 3 particles in a liter of air in a room of class 5? We assume the worst case, that 3520 particles are in a cubic meter.

To calculate, we choose a fixed liter from the cubic meter for observation. The particles are independent of each other and so we can regard the problem as a Bernoulli process: The i-th trial of the experiment is the location of the i-th particle.

In this example, you can see that the individual trials of a Bernoulli process do not always have to be carried out one after the other, as when drawing balls from an urn, but they take place here parallel to each other.

The observed event $A =$ "particle i is located in the selected liter of air" has the probability $p = 1/1000$. Then the random variable Z, which indicates the number of particles in the liter of air, is binomially distributed with the parameters $n = 3520$, $p = 1/1000$ and we get:

$$p(Z = k) = b_{3520,1/1000}(k) = \binom{3520}{k} \left(\frac{1}{1000}\right)^k \cdot \left(\frac{999}{1000}\right)^{3520-k}.$$

Here are the results for the first values of k:

k	0	1	2	3	4	5
$p(Z = k)$	0.02955	0.10411	0.18337	0.21524	0.18944	0.13335

$$\tag{21.2}$$

◀

The binomial distribution $b_{n,p}(k)$ can be calculated recursively from $b_{n,p}(k-1)$. We will need this formula again later:

▶ **Theorem 21.4** For all $k = 0, \ldots, n-1$ holds

$$b_{n,p}(k+1) = \frac{n-k}{k+1} \cdot \frac{p}{1-p} \cdot b_{n,p}(k), \quad b_{n,p}(0) = (1-p)^n.$$

If we use $\binom{n}{k+1} = \frac{n \cdot (n-1) \cdot \ldots \cdot (n-k)}{1 \cdot 2 \cdot 3 \cdot \ldots \cdot (k+1)} = \binom{n}{k} \cdot \frac{n-k}{k+1}$, we get with $q = 1 - p$:

$$b_{n,p}(k+1) = \binom{n}{k+1} p^{k+1} q^{n-k-1} = \binom{n}{k} \frac{n-k}{k+1} \cdot p^k p \cdot q^{n-k} \cdot q^{-1}$$

$$= \frac{n-k}{k+1} \cdot \frac{p}{q} \cdot b_{n,p}(k). \qquad \square$$

The Hypergeometric Distribution

The distribution belonging to the urn model "drawing without replacement" is called *hypergeometric distribution*: We derived this distribution in Theorem 19.17 :

▶ **Definition and Theorem 21.5** An urn contains N balls, B of which are black. A discrete random variable Y, which after n draws without replacement from the urn gives the number k of the black balls drawn, is called *hypergeometrically distributed* with parameters N, B and n. It is

$$h_{N,B,n}(k) := p(Y = k) = \binom{B}{k}\binom{N-B}{n-k} \bigg/ \binom{N}{n}. \tag{21.3}$$

Without proof, I give the expected value and variance for the hypergeometric distribution. They can be calculated similarly to the binomial distribution:

▶ **Theorem 21.6** Let Y be hypergeometrically distributed with parameters N, B and n, let $p := B/N$. Then:

$$E(Y) = n \cdot p, \quad \text{Var}(Y) = n \cdot p \cdot (1-p) \cdot \frac{N-n}{N-1}.$$

Compare the binomial distribution with the hypergeometric distribution:

The expected values are the same in both distributions, the variances differ the closer n is to N. If $n = N$ balls are drawn without replacement, that is, all balls, then in the limit case $E(Y) = N \cdot (S/N) = S$ and $\text{Var}(Y) = 0$, because then of course you exactly catch all S black balls.

Example

In a raffle, 10 sellers sell tickets. Each has 50 tickets in his box, of which 20 are winners. You want to buy 10 tickets. Is your chance of winning greater if you buy all the tickets from one seller, or if you buy only one ticket from each seller and thus always draw from the full?

The purchase of tickets from one seller is hypergeometrically distributed, from several sellers it is binomially distributed, since the same initial situation prevails again with each purchase. Calculate the probabilities for $0, 1, 2, \ldots, 10$ wins in the two cases. You can see that these always differ slightly, sometimes one is larger, sometimes the other. But the profit expectation, that is, the expected value, is the same in both cases, it is $n \cdot p = 10 \cdot 0.4 = 4$. The variance differs slightly: For one seller it is about 1.96, for several sellers 2.4. Where should you buy if you are risk-averse? ◄

How do binomial distribution and hypergeometric distribution differ in larger populations N? We also have an

---

Example

There are 1000 ball bearing balls in a box, 100 of them faulty. You take 10 balls, once without putting them back, the second time with putting them back.

Let's calculate the probability of k faulty balls among the 10 balls taken out first with the help of the hypergeometric distribution and then with the help of the binomial distribution. The results are entered in the following table:

k	0	1	2	3	4	5
$h_{N,B,n}(k)$	0.3469	0.3894	0.1945	0.0569	0.0108	0.00139
$b_{n,p}(k)$	0.3487	0.3874	0.1937	0.0574	0.0112	0.00149

k	6	7	8	9	10	
$h_{N,B,n}(k)$	$1.2 \cdot 10^{-4}$	$7.4 \cdot 10^{-6}$	$2.9 \cdot 10^{-7}$	$6.5 \cdot 10^{-9}$	$6.6 \cdot 10^{-11}$	
$b_{n,p}(k)$	$1.4 \cdot 10^{-4}$	$8.7 \cdot 10^{-6}$	$3.6 \cdot 10^{-7}$	$9.0 \cdot 10^{-9}$	$1.0 \cdot 10^{-10}$	

The expected value is in both cases $10 \cdot 0.1 = 1$, the variance in the first case is about 0.89, in the second case 0.9. ◄

You can see that the results of the formulas in this case are almost indistinguishable. For a large population N and a relatively small number n of selected elements, the probabilities of drawing without replacement and drawing with replacement are almost the same.

We can write this result in a theorem that I would like to cite without proof:

▶   **Theorem 21.7** If $B < N$, $n < N$ and n and $p = B/N$ are constant, then for $k = 0, 1, \ldots, n$:

$$\lim_{N \to \infty} h_{N,B,n}(k) = b_{n,p}(k).$$

If you compare the formulas (21.1) and (21.3) for calculating the distributions, you will find that for large populations, the binomial distribution (21.1) is much easier to evaluate than the hypergeometric distribution (21.3): with large numbers N and B it is easy to get sweaty when calculating the binomial coefficients. Later we will see that the binomial distribution can also often be approximated by other, simpler distributions.

This property is often used: If it is possible, statisticians work with the model "drawing with replacement". In an election poll, for example, the sample consists of a selection of people, each of whom is only interviewed once, so strictly speaking it is a drawing without replacement. However, with 65 million eligible voters and a few thousand respondents, the probability of catching a voter twice when he is "put back", is practically zero, so it is quite reasonable to assume that the drawing is with replacement.

The Geometric Distribution

The probability that the event A occurs for the first time in the k-th trial in a Bernoulli process is described by the *geometric distribution*. We know it from Theorem 19.15:

▶   **Definition and Theorem 21.8** A discrete random variable X, which in a Bernoulli process indicates at which trial the event A with $p(A) = p$ occurs for the first time, is called *geometrically* distributed with parameter p. It is

$$p(X = k) = p \cdot (1 - p)^{k-1}.$$

Let us also calculate the expected value and variance. There is an essential difference to the distributions examined so far: The number n of experiments carried out is not limited any more, we have to carry out the experiment as often as necessary. For the first time, a random variable with an infinite image occurs here. For the expected value, we have:

$$E(X) = \sum_{k=1}^{\infty} k \cdot p(X = k) = \sum_{k=1}^{\infty} k \cdot p(1 - p)^{k-1} = p \cdot \sum_{k=1}^{\infty} k(1 - p)^{k-1}, \qquad (21.4)$$

and therefore we have to calculate the limit of an infinite series. Our knowledge of series, which we acquired in the second part of the book, helps us further: First of all, we know that for real numbers x with $|x| < 1$ applies:

$$\sum_{k=0}^{\infty} x^k = \frac{1}{1-x} \tag{21.5}$$

This is the geometric series, see Example 2 in Sect. 13.2. In Theorem 15.21 we learned that the function determined by this series can be differentiated piecewise. If we differentiate (21.5) to the right and left, piecewise to the left and to the right according to the chain rule, we obtain for all x with $|x| < 1$ the identity:

$$\sum_{k=1}^{\infty} k \cdot x^{k-1} = \frac{1}{(1-x)^2}$$

If we set x to the number $1 - p$ we get:

$$\sum_{k=1}^{\infty} k \cdot (1-p)^{k-1} = \frac{1}{(1-(1-p))^2} = \frac{1}{p^2},$$

and thus according to (21.4)

$$E(X) = \frac{1}{p}.$$

With a similar trick you can calculate the variance. The result is:

$$\text{Var}(X) = \frac{1-p}{p^2}.$$

When rolling dice, the expected value for waiting for a number is 6, the variance is 30. The coin toss has an expected value of 2 for head, the variance is also 2 in this example.

Fig. 21.1 shows the geometric distribution in the histogram for the dice experiment, i.e. $p = 1/6$. Please compare the diagram with the one in Fig. 19.4. There the real results of the experiment "Waiting for the 6" were recorded.

The geometric distribution has an interesting property called memorylessness: the probability of rolling a 6 for the first time after k trials is independent of whether I have just started rolling, or whether I have already s futile attempts behind me. Expressed as a formula, this property means:

$$p(X = s + k | X > s) = p(X = k) \tag{21.6}$$

Fig. 21.1 The geometric distribution

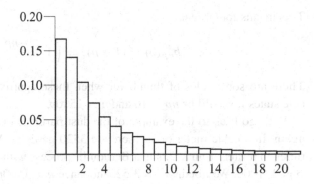

Let's check (21.6): The probability $p(X > s)$ for s futile attempts is $(1-p)^s$. With the help of the Definition 19.6 of the conditional probability we get:

$$p(X = s+k|X > s) = \frac{p(\{X = s+k\} \cap \{X > s\})}{p(X > s)} = \frac{p(X = s+k)}{p(X > s)}$$

$$= \frac{p(1-p)^{s+k-1}}{(1-p)^s} = p(1-p)^{k-1} = p(X = k)$$

Because of the memorylessness of the geometric distribution, it is pointless to bet, for example, in roulette on a number that has not been drawn for a long time. The probability of drawing a number is always $1/37$, even though it is long overdue.

The Poisson Distribution

For a large number of trials in a Bernoulli process, the binomial distribution quickly becomes unwieldy. If the probability for the observed event is small, the Poisson distribution is a good and easy to calculate approximation to the binomial distribution.

If in a Bernoulli process n is large compared to k and p is small, then $1 - p \approx 1$ and $n - k \approx n$. Now we use the recursive formula from Theorem 21.4 to calculate the probabilities $b_{n,p}(k)$. We can simplify this formula a bit by replacing $n - k$ by n and $1 - p$ by 1:

$$b_{n,p}(k+1) = \frac{(n-k)p}{(k+1)(1-p)} \cdot b_{n,p}(k) \approx \frac{np}{(k+1)} \cdot b_{n,p}(k). \qquad (21.7)$$

To calculate $b_{n,p}(0) = (1-p)^n$ we use the limit of a sequence, which is known from analysis. For all real numbers x holds:

$$\lim_{n\to\infty} \left(1 + \frac{x}{n}\right)^n = e^x.$$

The number e is often defined as the limit of the sequence $(1 + \frac{1}{n})^n$.

This means for large n:

$$b_{n,p}(0) = (1 - p)^n = \left(1 + \frac{-np}{n}\right)^n \approx e^{-np}.$$

There are some rules of thumb for when these approximations can be used. One such rule states it should be $np < 10$ and $n > 1500p$.

Let's go back to the example of the dust particles in a clean room after Theorem 21.3 again. In a cubic meter of air there are 3520 particles. What is the probability for k particles in a liter of air? This is a Bernoulli process with 3520 trials (the particles), where $p = 1/1000$. We obtain $np = 3.52$, and since $n > 1500p$ is true, we can use the approximation. For the first values of k we get:

$$b_{3520,1/1000}(0) \approx e^{-3.52} = 0.02960$$

$$b_{3520,1/1000}(1) \approx \frac{3.52}{1} e^{-3.52} \approx 0.10419,$$

$$b_{3520,1/1000}(2) \approx \frac{3.52}{2} \cdot \left(\frac{3.52}{1} e^{-3.52}\right) \approx 0.18337,$$

$$b_{3520,1/1000}(3) \approx \frac{3.52}{3} \left(\frac{3.52}{2} \frac{3.52}{1} e^{-3.52}\right) \approx 0.21516,$$

$$b_{3520,1/1000}(4) \approx \frac{3.52}{4} \left(\frac{3.52}{3} \frac{3.52}{2} \frac{3.52}{1} e^{-3.52}\right) \approx 0.18934,$$

$$\vdots$$

$$b_{3520,1/1000}(k) \approx \frac{3.52^k}{k!} e^{-3.52}.$$

Compared to Table (21.2) in the clean room example, which we calculated with the binomial distribution, you can see that the results match well.

The approximation gets better and better the larger n and the smaller p is. It is common to refer to the product $n \cdot p$ with the letter λ.

▶   **Theorem 21.9** If $n \cdot p = \lambda$ is constant, then $\lim\limits_{n \to \infty} b_{n,p}(k) = \dfrac{\lambda^k}{k!} e^{-\lambda}$.

I want to sketch the proof, it uses the known rules of limits. For each fixed number k holds:

$$\lim_{n \to \infty} b_{n,p}(k) = \lim_{n \to \infty} \underbrace{\frac{n \cdot (n-1) \cdots (n-k+1)}{1 \cdot 2 \cdot 3 \cdot 4 \cdots \cdot k}}_{\binom{n}{k}} \underbrace{\left(\frac{\lambda}{n}\right)^k}_{p} \underbrace{\left(1 - \frac{\lambda}{n}\right)^{n-k}}_{1-p}$$

$$= \frac{\lambda^k}{k!} \lim_{n \to \infty} \underbrace{\frac{n \cdot (n-1) \cdots (n-k+1)}{n^k}}_{\to 1} \cdot \underbrace{\left(1 - \frac{\lambda}{n}\right)^{n-k}}_{\to e^{-\lambda}} = \frac{\lambda^k}{k!} e^{-\lambda}. \qquad \Box$$

▶ **Definition and Theorem 21.10** A discrete random variable X, for which

$$p(X = k) = \frac{\lambda^k}{k!} e^{-\lambda}$$

is called *Poisson distributed* with parameter λ. A Poisson distributed random variable has the expected value $E(X) = \lambda$ and the variance $\text{Var}(X) = \lambda$.

I do not want to derive the expected value and variance of the Poisson distribution, but in comparison with the binomial distribution you can see that for $\lambda = n \cdot p$ the expected values match. The variance of the binomial distribution is $n \cdot p \cdot (1 - p)$, also this value is for small p close to λ.

As an immediate consequence of Theorem 21.9 we obtain the

▶ **Rule of thumb 21.11** For large n and small p the binomial distribution can be replaced by the Poisson distribution with parameter $\lambda = n \cdot p$. This replacement can be made for $\lambda \leq 10$ and for $n \geq 1500 \cdot p$.

Example

Let's calculate the probability for k particles in one cubic centimeter for a Class 4 cleanroom. Here, a maximum of 352 particles per cubic meter is allowed. In the worst case, n is equal to 352, and p is now $1/10^6$. For the parameter λ holds: $\lambda = n \cdot p = 0.000352 < 10$ and $n > 1500 \cdot p = 0.0015$.

The conditions of rule 21.11 are therefore met, so the probability for k particles in one liter is

$$\frac{0.000352^k}{k!} e^{-0.000352}$$

In the following table I have compiled the first results:

k	0	1	2
Class 4	0.9996	0.00035	$6.2 \cdot 10^{-8}$

Try to calculate these probabilities on the calculator using the binomial distribution and you will see why the approximation by the Poisson distribution is so useful here. ◀

21.2   Continuous Probability Distributions, The Normal Distribution

The Continuos Uniform Distribution

The simplest continuous distribution is, as in the discrete case, the uniform distribution. Take a look again at the example before Definition 20.6 in Sect. 20.1. If the possible results of an experiment are evenly distributed in an interval $[a, b] \subset \mathbb{R}$, we can no longer give the probability for a single number to occur, but only for subsets of $[a, b]$.

The density of the uniform distribution in the interval $[a, b] \subset \mathbb{R}$ has the form (Fig. 21.2):

$$w : \mathbb{R} \to \mathbb{R}, \quad x \mapsto \begin{cases} 1/(b-a) & a \leq x \leq b \\ 0 & \text{else} \end{cases}$$

The corresponding cumulative distribution function is $F(x) = \int_{-\infty}^{x} w(t)dt$ (Fig. 21.3):

$$F : \mathbb{R} \to [0, 1], \quad x \mapsto \begin{cases} 0 & x < a \\ (x-a)/(b-a) & a \leq x \leq b \\ 1 & x > b \end{cases}$$

The expected value of a uniformly distributed random variable is, by Definition 20.12, equal to

$$E(X) = \int_{-\infty}^{+\infty} x \cdot w(x)dx = \int_{a}^{b} \frac{x}{b-a}dx = \frac{a+b}{2},$$

which is exactly the mean of a and b.

The results of a random number generator that generates numbers between 0 and 1 are ideally uniformly distributed. The expected value is 0.5 and, for example, $p(0.1 < X < 0.2) = F(0.2) - F(0.1) = 0.2 - 0.1 = 0.1$. Is your random number generator working properly? You can generate as many numbers as you want and analyze the

Fig. 21.2 The density of the uniform distribution

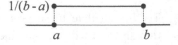

Fig. 21.3 The cumulative distribution function of the uniform distribution

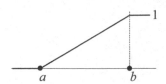

resulting data set. The task of statistics is to check whether this data set has the correct probability distribution. More on this in the next chapter.

The Standard Normal Distribution

The uniform distribution serves me mainly for mental warming up at the beginning of a difficult task: We come to the most important probability distribution of all, the normal distribution. With this we have to deal extensively. I hope I can give you an understanding in the following section how this distribution arises and why it is so important.

As in the derivation of the Poisson distribution, we start from the binomial distribution. We had seen there that for large n and small p the binomial distribution can be replaced by the Poisson distribution: If $\lambda = np$ is constant, then

$$\lim_{n \to \infty} b_{n,p}(k) = \frac{\lambda^k}{k!} e^{-\lambda}.$$

If n gets bigger, p must get smaller. But what happens if p remains constant and n gets bigger and bigger? Think of a Bernoulli process that is carried out very often. In Fig. 21.4 I have plotted for $p = 0.5$ the binomial distributions for different values of n as histograms.

You can see that the distribution becomes flatter and flatter for larger n, the maximum moves further and further to the right. This is not very surprising, because we know that the expected value is $E(b_{n,p}) = n \cdot p$, so it moves to the right with increasing n. For the variance applies $\text{Var}(b_{n,p}) = np(1 - p)$, so the dispersion of the values around the expected value becomes greater and greater, that is, the histograms always become wider and wider. This sequence of distributions therefore certainly does not converge to a reasonable limit function, it dissipates with increasing n.

In order to find a limit distribution nevertheless, we carry out a trick. Do you remember that you can normalize the expected value and variance of a random variable? To the

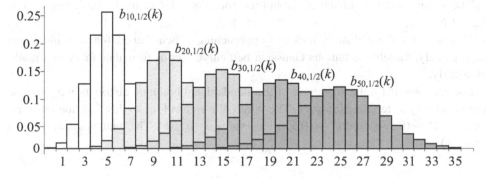

Fig. 21.4 The binomial distribution with increasing n

random variable X with expected value μ and variance σ^2 we have formed in Definition 20.17 the standardized variable $X^* = (X - \mu)/\sigma$. This has the expected value 0 and the variance 1. We now carry this out for the binomial distributions. The effect of this standardization is that all distributions are shifted with their peak, the expected value, to 0, and that they are squeezed together until they all have the same variance 1.

Let us now examine the histograms of the standardizations: If we set $k^* := (k - \mu)/\sigma$ we get:

$$X = k \quad \Leftrightarrow \quad \underbrace{(X - \mu)/\sigma}_{X^*} = \underbrace{(k - \mu)/\sigma}_{k^*} \quad \Leftrightarrow \quad X^* = k^*.$$

The values of k^*, for $k = 0, 1, \ldots, n$ are the function values of the random variable X^*, that is, the points at which the bars are to be drawn.

In the case of a binomially distributed random variable is $k^* = (k - np)/\sqrt{np(1 - p)}$. Let's take $b_{16,1/2}$: It is $\mu = np = 8$ and $\sigma^2 = np(1 - p) = 4$, so $\sigma = 2$. This results in the values:

k	0	1	2	3	4	5	6	7	8	9	10	11	12	13	14	15	16
k^*	-4	-3.5	-3	-2.5	-2	-1.5	-1	-0.5	0	0.5	1	1.5	2	2.5	3	3.5	4

We now want to draw area-proportional histograms. This means that we plot over k^* the bar with the area $p(X^* = k^*)$. The bars should all be the same width and abut each other. The distance between two values of k^* in the example is just 0.5, and in the general case the bar width for X^* is:

$$(k + 1)^* - k^* = \frac{(k + 1) - \mu}{\sigma} - \frac{k - \mu}{\sigma} = \frac{1}{\sigma}.$$

In the random variable X the width of the bar was exactly 1, and since $p(X^* = k^*) = p(X = k)$, the bar over k^* must be stretched by the factor σ in order to obtain the same area. In Fig. 21.5a you can see the histogram of $b_{16,1/2}$, in Fig. 21.5b the histogram to $b^*_{16,1/2}$.

The standardizations of the distributions from Fig. 21.4 result in the histograms in Fig. 21.6.

You see that these distributions are approaching a bell-shaped function more and more closely, namely the famous Gaussian bell curve. Let us formulate this result mathematically:

For each number n the histogram of the standardized binomial distribution $b^*_{n,p}$ can be represented by a step function $\varphi_n(x)$ The width of a step is $1/\sigma$, so this function has for x between $k^* - (0.5/\sigma)$ and $k^* + (0.5/\sigma)$ the value $\sigma \cdot b_{n,p}(k)$. The mathematical formula for $\varphi_n(x)$ is:

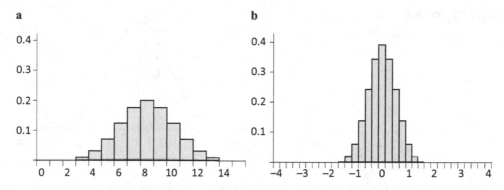

Fig. 21.5 $b_{16,1/2}(k)$ **(a)** and $b^*_{16,1/2}(k)$ **(b)**

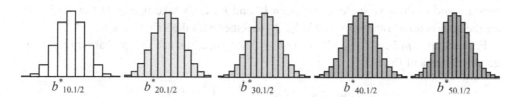

Fig. 21.6   Standardized binomial distributions

$$\varphi_n(x) = \begin{cases} \sigma b_{n,p}(k) & \text{for } \frac{k-\mu-0.5}{\sigma} \le x < \frac{k-\mu+0.5}{\sigma} \\ 0 & \text{else,} \end{cases} \qquad \mu = np, \ \sigma = \sqrt{np(1-p)}.$$

$$(21.8)$$

The following, very difficult theorem, a version of the *Moivre-Laplace theorem*, states that this sequence of functions actually converges:

▷   **Theorem 21.12: The Moivre-Laplace theorem**   For all p between 0 and 1 holds:

$$\lim_{n \to \infty} \varphi_n(x) = \frac{1}{\sqrt{2\pi}} \cdot e^{-\frac{x^2}{2}}.$$

Fig. 21.7 shows the bell curve, which is defined by the rule $\varphi(x) = \frac{1}{\sqrt{2\pi}} \cdot e^{-\frac{x^2}{2}}$. The graph is not drawn to scale, the maximum is approximately at 0.4, so the curve is very flat. Interesting is that the convergence does not only take place for $p = 0.5$ but also for all other probabilities p: Then the binomial distributions are not symmetrical from the beginning, the convergence takes a little longer, but eventually the same form is reached.

Fig. 21.7 The bell curve

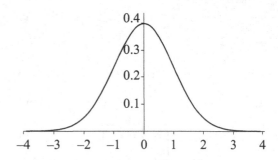

What do we get from this limit curve? If X is binomially distributed and n is large enough, we can use it to calculate the probability $p(k_1 \leq X \leq k_2)$: Since $p(k_1 \leq X \leq k_2) = p(k_1^* \leq X^* \leq k_2^*)$, we only have to look at the area of the bars of the standardized random variable X^* between k_1^* and k_2^*. Let's take $b_{64,1/2}$. In Fig. 21.8 you see the standardized random variable $b_{64,1/2}^*$ together with the values of k and k^*.

For example, $p(28 \leq X \leq 32) = p(-1 \leq X^* \leq 0)$, and since this probability just represents the area of the bars,

$$p(28 \leq X \leq 32) \approx \frac{1}{\sqrt{2\pi}} \int_{-1}^{0} e^{-\frac{x^2}{2}} dx.$$

In general, for a binomially distributed random variable X

$$p(X \leq k) = p(X^* \leq k^*) \approx \frac{1}{\sqrt{2\pi}} \int_{-\infty}^{k^*} e^{-\frac{x^2}{2}} dx.$$

The function $\frac{1}{\sqrt{2\pi}} \cdot e^{-\frac{x^2}{2}}$ is integrable, even if the integral is not analytically solvable. It is always greater than 0 and the area under the entire function is 1. This gives a continuous random variable (see Definition 20.6):

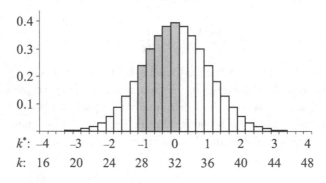

Fig. 21.8 Calculation of probabilities with the bell curve

▶   **Theorem 21.13** The function $\varphi(x) = \frac{1}{\sqrt{2\pi}} \cdot e^{-\frac{x^2}{2}}$ is the density of a continuous random variable N with expected value 0 and variance 1. N has the cumulative distribution function

$$\Phi(x) = p(N \le x) = \frac{1}{\sqrt{2\pi}} \int_{-\infty}^{x} e^{-\frac{t^2}{2}} dt.$$

▶   **Definition 21.14** The random variable N from Theorem 21.13 is called *standard normally distributed* or $N(0, 1)$-distributed.

The cumulative distribution function Φ can only be evaluated numerically. In the exercises for integral calculus, you could calculate a table, the result of this numerical integration can be found in the appendix. Since $\varphi(x)$ is symmetrical, a simple consideration shows that for $x > 0$ always holds $\Phi(-x) = 1 - \Phi(x)$, therefore usually only the positive values are tabulated.

The standardized binomial distributions are approximately standard normally distributed for large n, therefore

$$p(X \le k) \approx p(X^* \le k^*) = \Phi(k^*) = \Phi\left(\frac{k - \mu}{\sigma}\right),$$

$$p(k_1 < X \le k_2) \approx \Phi(k_2^*) - \Phi(k_1^*) = \Phi\left(\frac{k_2 - \mu}{\sigma}\right) - \Phi\left(\frac{k_1 - \mu}{\sigma}\right). \tag{21.9}$$

The convergence in Theorem 21.12 works best if p is close to 0.5. The more p differs from 0.5, the larger n must be chosen. There are several rules of thumb for this in the literature. The approximation of the binomial distribution by the normal distribution is already very good for $np > 5$ and $n(1 - p) > 5$. Another rule requires that it should be $np(1 - p) > 9$. In both rules you can see that the further p deviates from 0.5, the larger n must be chosen. I summarize our considerations in the following rule of thumb:

▶   **Rule of thumb 21.15** Let X be a $b_{n,p}$-distributed random variable. Then, for $np > 5$ and $n(1 - p) > 5$ the following calculation is possible:

a) $p(k_1 \le X \le k_2) = \Phi\left(\dfrac{k_2 - np + 0.5}{\sqrt{np(1-p)}}\right) - \Phi\left(\dfrac{k_1 - np - 0.5}{\sqrt{np(1-p)}}\right),$

$0 \le k_1 \le k_2 \le n,$

b) $p(X \le k) = \Phi\left(\dfrac{k - np + 0.5}{\sqrt{np(1-p)}}\right),\ p(X < k) = \Phi\left(\dfrac{k - np - 0.5}{\sqrt{np(1-p)}}\right),$

$0 \le k \le n.$

The ± 0.5 in the two formulas arise because in the binomial distribution one always has to go to the left or to the right edge of the bar over k^*, see (21.8). This correction only plays a role for small values of n.

Example 1

In an election, 10% of voters voted for party A. In the election poll, 1000 voters are surveyed and a election forecast is created. How big is the probability that this forecast deviates by no more than $\pm 2\%$ from the actual election result?

The survey is a Bernoulli process, the random variable $X = $ "number of voters for party A" is binomially distributed with the parameters $n = 1000$ and $p = 0.1$. The probability sought is $p(80 \leq X \leq 120)$, because then the prediction lies between 8% and 12%. It is $np = 100, n(1 - p) = 900$, so we can apply rule 21.15a):

$$p(80 \leq X \leq 120) = \Phi\left(\frac{120 - 100 + 0.5}{\sqrt{90}}\right) - \Phi\left(\frac{80 - 100 - 0.5}{\sqrt{90}}\right)$$

$$= \Phi(2.16) - \Phi(-2.16) = 2 \cdot \Phi(2.16) - 1$$

$$= 2 \cdot 0.9846 - 1 = 0.9692.$$

The probability we are looking for is therefore about 96.9%. Calculate for yourself what the result is if we leave out the ± 0.5 in the formula: we get about 96.5%.

Of course, we could also solve this task directly with the binomial distribution: But then the probabilities $p(X = 80), p(X = 81)$ and so on until $p(X = 120)$ all have to be calculated individually, a rather cumbersome procedure. ◀

From the example you can see that we are slowly approaching the statistical questions that I introduced in Sect. 19.1. In reality, the actual result, 10%, is unfortunately not given. We have to guess a percentage from the survey and give a so-called confidence interval to it.

Example 2

Assume the hypothesis to be true that a slice of toast falls on the buttered side as often as on the unbuttered side when it falls down. What is the probability the toast will fall on the buttered side more than 52 times out of 100 trials?

The associated random variable X is $b_{100,1/2}$-distributed, we can use rule 21.15b):

$$p(X > 52) = 1 - p(X \leq 52) = 1 - \Phi\left(\frac{52 - 50 + 0.5}{\sqrt{25}}\right)$$

$$= 1 - \Phi(0.5) = 1 - 0.6915 = 0.3085.$$

In the exercises to Chap. 19 you were asked to solve the problem using the binomial distribution. The result of this calculation was (rounded) 0.3087. ◀

The General Normal Distribution

Just as we were able to normalize the binomial distribution by shifting and pressing it into a distribution with expected value 0 and variance 1, we can transform the standard normal distribution into a similar distribution with expected value μ and variance σ^2: If X is a $N(0,1)$-distributed random variable, then according to Theorems 20.13 and 20.16 the random variable $Y = \sigma X + \mu$ has expected value μ and variance σ^2. Then $X = (Y - \mu)/\sigma$.

▶ **Definition 21.16** A random variable X with expected value μ and variance σ^2 is called $N(\mu,\sigma^2)$-distributed or $N(\mu,\sigma^2)$-normally distributed, if the standardization $X^* = (X - \mu)/\sigma$ is standard normally distributed.

For such a random variable holds $X = \sigma \cdot N(0,1) + \mu$.

▶ **Theorem 21.17** Let X be a $N(\mu,\sigma^2)$-distributed random variable. Then for the associated cumulative distribution function F holds:

$$F(x) = p(X \le x) = \Phi\left(\frac{x - \mu}{\sigma}\right)$$

$$F(y) - F(x) = p(x < X \le y) = \Phi\left(\frac{y - \mu}{\sigma}\right) - \Phi\left(\frac{x - \mu}{\sigma}\right). \tag{21.10}$$

X has the density function

$$\phi(x) = \frac{1}{\sqrt{2\pi\sigma^2}} \cdot e^{-\frac{(x-\mu)^2}{2\sigma^2}}.$$

The density function is given for completeness only, we do not need it further; it can be derived from the density function of the standard normal distribution. For the cumulative distribution function $F(x)$ of the general normal distribution holds because of $X \le x \Leftrightarrow (X - \mu)/\sigma \le (x - \mu)/\sigma$:

$$F(x) = p(X \le x) = p\left(\frac{X - \mu}{\sigma} \le \frac{x - \mu}{\sigma}\right) = p\left(X^* \le \frac{x - \mu}{\sigma}\right) = \Phi\left(\frac{x - \mu}{\sigma}\right).$$

Analogously follows the second line of the formula (21.10). □

If X is a binomially distributed random variable with expected value μ and variance σ^2, we now know that $X^* = (X - \mu)/\sigma$ is approximately standard normally distributed. But that means X is approximately $N(\mu,\sigma^2)$-distributed. Therefore:

▶ **Rule of thumb 21.18** For $np > 5$ and $n(1 - p) > 5$ a $b_{n,p}$-distributed random variable is approximately $N(np, np(1 - p))$-distributed.

It is therefore not surprising that the formulas (21.9) and (21.10) agree with each other except for the "$\approx$".

Example

It is known that the diameter of ball bearing balls from a special production is $N(45, 0.01^2)$ -distributed. A ball is unusable if it deviates from the nominal 45 mm by more than 0.03 mm. What is the probability of such a deviation?

$$p(\text{ball usable}) = p(45 - 0.03 \leq X \leq 45 + 0.03)$$
$$= F(45 + 0.03) - F(45 - 0.03)$$
$$= \Phi\left(\frac{45 + 0.03 - 45}{0.01}\right) - \Phi\left(\frac{45 - 0.03 - 45}{0.01}\right)$$
$$= \Phi(3) - \Phi(-3) = 2 \cdot 0.99865 - 1 = 0.9973.$$

So 0.27% of the balls are unusable. ◀

▶ **Theorem 21.19** If X is a $N(\mu, \sigma^2)$-distributed random variable, the deviations from the expected value are:

$$p(|X - \mu| \leq \sigma) \approx 0.6826,$$
$$p(|X - \mu| \leq 2\sigma) \approx 0.9546,$$
$$p(|X - \mu| \leq 3\sigma) \approx 0.9973.$$

For the proof: It is

$$p(|X - \mu| \leq k\sigma) = p(\mu - k\sigma \leq X \leq \mu + k\sigma)$$
$$= F(\mu + k\sigma) - F(\mu - k\sigma) =$$
$$= \Phi\left(\frac{\mu + k\sigma - \mu}{\sigma}\right) - \Phi\left(\frac{\mu - k\sigma - \mu}{\sigma}\right)$$
$$= \Phi(k) - \Phi(-k) = 2\Phi(k) - 1.$$

The given values for $k = 1, 2, 3$ can be determined using the table in the appendix.   □

The interesting thing about this statement is that these numerical values apply regardless of μ and σ (Fig. 21.9): 68.3% of the results of X differ from the expected value by less than σ, 95.4% by less than 2σ and 99.7% less than 3σ. These multiples of the standard deviation are therefore often used as simple indicators to determine how "normal" a particular result is.

In the example of the ball bearings, the permissible deviation of 0.03 mm happened to be exactly three times the standard deviation. We could therefore have directly taken the result from Theorem 21.19.

Fig. 21.9 The standard deviation

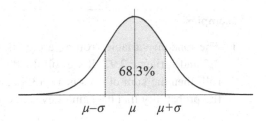

The normal distribution occurs again and again with many random variables in practice. Why is it so common? We know it as the limit distribution of the binomial distribution, but there are other ways to get the normal distribution. A very important one results from the central limit theorem of probability theory, which I would like to quote in a special form:

▶   **Theorem 21.20: The central limit theorem** Let $X_1, X_2, \ldots, X_n$ independent random variables that all have the expected value μ and the variance σ^2, and all have the same distribution. Let S_n be the random variable $S_n = X_1 + X_2 + \cdots + X_n$ and S_n^* the corresponding standardization. Then S_n^* converges to the standard normal distribution, that is

$$\lim_{n \to \infty} p(S_n^* \leq x) = \Phi(x).$$

A random variable that is composed additively of many individual influences, where the individual influences are equally distributed random variables, is thus normally distributed. The conditions on the variables X_i in this limit theorem can still be significantly weakened, but then the theorem becomes very unreadable. The proof is very involved. Many random variables meet the conditions of the theorem. This is where the great importance of the normal distribution in statistics lies.

The convergence takes place quite quickly, already for $n \geq 30$ we can apply the rule:

▶   **Rule of thumb 21.21** Let $X_1, X_2, \ldots, X_n$ be independent random variables with expected value μ and variance σ^2, let all X_i have the same distribution. Then the random variable $S_n = X_1 + X_2 + \cdots + X_n$ has expected value $n\mu$ and variance $n\sigma^2$ and is approximately $N(n\mu, n\sigma^2)$-distributed. For $n \geq 30$ holds

$$p(S_n \leq x) \approx \Phi\left(\frac{x - n\mu}{\sqrt{n\sigma^2}}\right).$$

With the help of the Theorems 20.18 and 20.20 we determine the expected value and variance of S_n as the sum of the individual expected values and the individual variances. According to the limit theorem, then S_n is $N(n\mu, n\sigma^2)$-distributed and with the help of Theorem 21.17 we can calculate the distribution function as specified.   □

1. The random variable "rolling a die" is uniformly distributed with expected value 3.5 and variance 2.92. If we roll the die 1000 times and X_i is the result of the i-th roll, then the sum of eyes is a $N(3500, 2920)$-distributed random variable. What is the probability that the sum of eyes deviates by more than 100 from 3500?

$$p(3400 \leq S_{1000} \leq 3600) = \Phi\left(\frac{3600 - 3500}{54}\right) - \Phi\left(\frac{3400 - 3500}{54}\right)$$
$$= \Phi(1.85) - \Phi(-1.85) = 2 \cdot \Phi(1.85) - 1$$
$$= 2 \cdot 0.9678 - 1 = 0.9356.$$

The probability is therefore about 6.4%.

2. In a random number of 32 bits in length, let X_i be the bit value of the i-th position in the binary representation. All X_i are uniformly distributed, and so the random variable $X = X_1 + X_2 + \cdots + X_{32}$, the sum of the ones of the integer, is approximately normally distributed. Flip back to Sect. 19.1, to the first example of a probability distribution. Now we can understand how the normal distribution comes about in Fig. 19.3: The integers that come from a compressed file are not random numbers, but a good compression method is characterized by the fact that existing structures in the numbers are destroyed, otherwise further compression could be carried out. Fig. 19.3 is an indication that the compression carried out is efficient. ◀

The Exponential Distribution

The exponential distribution is the continuous variant of the geometric distribution. It describes the waiting time until an event occurs. The discrete number of experiments in the geometric distribution is thus replaced by the continuously running time.

The characteristic property of the exponential distribution is the memorylessness, which I mentioned already in the geometric distribution: the probability that I have to wait at a street corner for another t minutes until a taxi arrives is unfortunately independent of whether I have just arrived or whether I have been waiting for an hour. If the random variable X describes my waiting time, then:

$$p(X \geq s + t | X \geq s) = p(X \geq t). \tag{21.11}$$

The function

$$w: \mathbb{R} \to \mathbb{R}, \quad t \mapsto \begin{cases} \lambda e^{-\lambda t} & t \geq 0 \\ 0 & t < 0, \end{cases} \quad \lambda > 0,$$

is the density of a continuous distribution that has this property, as we will see shortly: The antiderivative of $\lambda e^{-\lambda x}$ is $-e^{-\lambda x}$ and so we can calculate the cumulative distribution function F:

$$F : \mathbb{R} \to [0, 1], \quad t \mapsto p(X \leq t) = \int_{-\infty}^{t} w(x)dx = \int_{0}^{t} \lambda e^{-\lambda x}dx = 1 - e^{-\lambda t}.$$

For $\lambda > 0$ is always $1 - e^{-\lambda t}$ between 0 and 1.

Now we can check (21.11). Because of $p(X \geq t) = 1 - F(t) = e^{-\lambda t}$ it applies:

$$p(X \geq s + t | X \geq s) = \frac{p(\{X \geq s + t\} \cap \{X \geq s\})}{p(X \geq s)} = \frac{p(X \geq s + t)}{p(X \geq s)}$$

$$= \frac{1 - F(s + t)}{1 - F(s)} = \frac{e^{-\lambda(s+t)}}{e^{-\lambda s}} = e^{-\lambda t} = 1 - F(t) = p(X \geq t).$$

$$(21.12)$$

Fig. 21.10 shows the graph of the density of the exponential distribution. If you compare this image with Fig. 21.1 of the geometric distribution, you will see the similarity of the two distributions.

In addition to the exponential distribution, there is no other memoryless continuous distribution. After all, given such a distribution, just as in (21.12), must hold:

$$\frac{1 - G(t + s)}{1 - G(s)} = 1 - G(t).$$

If we denote with $\tilde{G}(t)$ the function $1 - G(t)$, then $\tilde{G}(t + s) = \tilde{G}(t)\tilde{G}(s)$. In Theorem 14.31 we saw that then for $a = \tilde{G}(1)$ applies:

$$\tilde{G}(t) = a^t = e^{\ln a \cdot t}.$$

Here, $0 < a < 1$. But this just means that $G(t) = 1 - \tilde{G}(t) = 1 - e^{\ln a \cdot t}$ is the distribution function of the exponential distribution to the parameter $\lambda = -\ln a$.

In the first example of partial integration after Theorem 16.12 in Sect. 16.1 we calculated the integral $\int xe^x dx = (x - 1)e^x$. Just as there, you can calculate the expected value of the exponential distribution:

$$E(X) = \int_{0}^{\infty} x\lambda e^{-\lambda x}dx = -\frac{1 + \lambda \cdot x}{\lambda}e^{-\lambda x}\Big|_{0}^{\infty} = \frac{1}{\lambda}.$$

Fig. 21.10 The exponential distribution

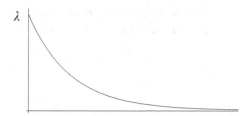

A similar calculation for the variance results in the value $1/\lambda^2$. I summarize the results:

▷   **Definition and Theorem 21.22: The exponential distribution** The function

$$
w : \mathbb{R} \to \mathbb{R}, \quad t \mapsto
\begin{cases}
\lambda e^{-\lambda t} & t \geq 0 \\
0 & t < 0,
\end{cases}
\quad \lambda > 0,
$$

is the density of a continuous random variable X with expected value $1/\lambda$ and variance $1/\lambda^2$. The random variable X is called exponentially distributed with expected value $1/\lambda$. It has the distribution function $F(t) = p(X \leq t) = 1 - e^{-\lambda t}$.

In practice, the exponential distribution is often used as a distribution for lifetime calculations. Here, the memorylessness states that the expected lifetime of a component is independent of how long it has been in operation. Such components are called fatigue-free. The expected value in the exponential distribution then corresponds to the expected lifetime of the component.

Example

The manufacturer of a hard disk specifies a value of 70 years as the mean time to failure *(MTTF)*. What is the probability that the hard disk in a server will fail next year or in the next two years? The *MTTF* is the expected value $1/\lambda$. So we are looking for $p(X \leq 1)$ or $p(X \leq 2)$ with $\lambda = 0.014286$. It is

$$
p(X \leq 1) = F(1) = 1 - e^{-0.014286} = 1 - 0.9858 = 0.0142
$$
$$
p(X \leq 2) = F(2) = 1 - e^{-2 \cdot 0.014286} = 1 - 0.9718 = 0.0282
$$

What is the probability that all hard disks running in 50 computers in continuous operation will survive one year or two years without failure?

It is $p(X > 1) = 0.9858$ and $p(X > 2) = 0.9718$. Since the hard disk failures are independent of each other, it applies:

$$
p(\text{all disks last 1 year}) = (0.9858)^{50} = 0.489
$$
$$
p(\text{all disks last 2 years}) = (0.9718)^{50} = 0.239
$$

In fatigue-free systems, the failure rate does not change over the course of its life. Reality looks different. None of the hard disks in operation today will still be running in 70 years, even if one were to try. So why can the *MTTF* still be a reasonable measure?

The failure rate of a component often follows a bathtub curve: at the beginning of its use, there is a higher number of failures that can be caused by production or material errors. Such failures are hopefully caught by the warranty. This is followed by a longer period of regular operation, during which the probability of failure remains almost unchanged. At some point, then, the aging processes set in, and the probability of failure becomes greater again. The specified failure rate, 70 years in the example, only applies at the bottom of the bathtub, where the exponential distribution can be applied well. The number "70 years" given in the example must under no circumstances be interpreted as the mean service life of the hard disk. ◀

The Chi-Square Distribution

As the last continuous distribution, I would like to introduce you to the chi-square distribution, which plays an important role in test theory. In the next chapter we will need this for a statistical test.

The χ^2-distribution consists of several independent standard normal distributions:

▶ **Definition 21.23** If the random variables $X_1, X_2, \ldots, X_n$ are independent and $N(0, 1)$-distributed, the probability distribution of the random variable

$$\chi_n^2 = X_1^2 + X_2^2 + \ldots + X_n^2$$

is called χ^2-distribution with n degrees of freedom.

The χ^2-distribution is a probability distribution of a continuous random variable. The density function $g_n(x)$ belonging to degree n is dependent on n and can be computed from the density of the standard normal distribution by means of mathematical induction. I give you the density functions for $n = 1$ to 5. Since the function values of χ^2 are all positive, it follows for $x \leq 0$ is also $g_n(x) = 0$. For $x > 0$ holds:

$$g_1(x) = \frac{1}{\sqrt{2\pi x}} \cdot e^{-\frac{x}{2}}, \quad g_2(x) = \frac{1}{2} \cdot e^{-\frac{x}{2}},$$

$$g_3(x) = \sqrt{\frac{x}{2\pi}} \cdot e^{-\frac{x}{2}}, \quad g_4(x) = \frac{x}{4} \cdot e^{-\frac{x}{2}},$$

$$g_5(x) = \sqrt{\frac{x^3}{18\pi}} \cdot e^{-\frac{x}{2}}.$$

In Fig. 21.11 you see the graphs of the first five chi-square density functions.

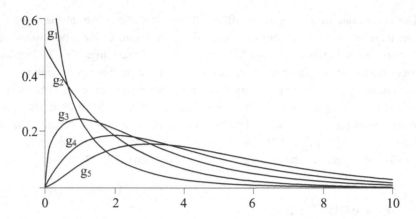

Fig. 21.11 The χ^2-distribution

From g_3 on the densities have a maximum, the expected value of χ_n^2 is n, the variance is $2n$. For large n is χ_n^2 approximately $N(n, 2n)$-distributed. This can also be concluded from the central limit theorem.

21.3   Stochastic Processes

Sequences of random variables, such as $(X_n)_{n \in \mathbb{N}}$ or $(X_t)_{t \in \mathbb{R}}$ are called *stochastic processes*. The index can take continuous values ($t \in \mathbb{R}$) or take discrete values ($n \in \mathbb{N}$). Such processes often describe the temporal development of a random variable. The index t then denotes the current observation time of the variable.

1. The *Brownian motion* of a particle is a stochastic process: The random variable X_t has in this case the value set $\mathbb{R}^3$, it gives the position of a particle in space at time t.
2. The random occurrence of an event over time is a stochastic process: X_t is the number of events that have occurred from time 0 to time t. If the events occur randomly and independently of each other, we obtain the *Poisson process*.
3. Do you remember the long queues at enrollment? The number of students who have arrived at the enrollment office by the time t, the number of students who have been enrolled by the time t and the number of students in the queue at the time t are stochastic processes that are studied in queueing theory.

Carrying out an experiment means observing the values of X_t for all t, i.e. for example, observing a specific queue between t_0 and t_1. In this interval, each X_t is assigned the observed value x_t, that is, the number of students waiting at time t. The result of the experiment, the realization, thus represents a function here: $[t_0, t_1] \to \mathbb{R}, t \mapsto x_t$. This function is called the *trajectory* of the stochastic process. The trajectory is the result of the specific experiment.

The Poisson Process

We first want to investigate events that occur randomly and independently of each other over time. Examples of this are:

- Nuclear decay,
- The arrival of service requests at a server,
- Calls in a call center,
- The occurrence of software errors in a program system.

Let's start with a classic random event: the nuclear decay of a particle, for example plutonium. The half-life is known from physics, so for a given amount of plutonium one knows the decay rate, that is, one knows how many decays are to be expected in a given time interval. Unknown, however, is when a single particle decays and how the decays are distributed in this time interval. Since the decay is a random process, only a probability can be given that in an interval T exactly k particles decay. We now want to calculate this probability.

Assume that in a sample of material there are about 7200 decays per hour. This decay rate can be calculated from the known half-life. What is the probability that in a period of 1 second 0, 1, 2 or more decays will occur?

To do this, we choose a fixed second T from the hour of observation. Physics tells us that the particles decay independently of each other and so we can regard the process as a Bernoulli process: The i-th trial of the process is the fate of the i-th decaying particle. The observed event A = "Particle decays in T" has the probability $p = 1/3600$. Then the random variable Z, which indicates the number of particles that decay in second T, is binomially distributed with the parameters $n = 7200$, $p = 1/3600$. We can apply rule 21.11 and obtain with $\lambda = 7200 \cdot 1/3600 = 2$:

$$p(Z = k) = \frac{\lambda^k}{k!}e^{-\lambda} = \frac{2^k}{k!}e^{-2} \tag{21.13}$$

λ plays the role of an expected value for the examined time interval: We expect an average of 7200 decays per hour, which corresponds to 2 decays per second.

In the calculation of the number of decays in one second, I made a mistake: I assumed that exactly 7200 particles decay, which was the number n of trials. However, the number 7200 is only an average, there will always be deviations from it. Now let's look at a longer period of time, say 1000 hours, then the relative deviation from the average and thus the calculation error is certainly much smaller. In 1000 hours we expect 7 200 000 decays. Then $p = 1/3\,600\,000$ and the same calculation as in (21.13) results with $\lambda = 7\,200\,000 \cdot 1/3\,600\,000 = 2$ surprisingly again in:

$$p(Z = k) = \frac{2^k}{k!}e^{-2}.$$

The result is therefore independent of the length of the observed time period. If we con-sider Theorem 21.9, we can make another interesting observation: I chose the longer time period of 1000 hours to reduce the error caused by the deviation from the mean decay rate. But at the same time, the longer the period under investigation, the better the approximation of the binomial distribution by the Poisson distribution. In fact, the Pois-son distribution not only approximately describes the distribution of nuclear decays over time, but exactly, provided that the decay rate is known exactly.

Let us now consider the stochastic process Z_t which describes the probability for k decays in t seconds. For $n = 7200$ the probability p has the value $t/3600$ and it is $\lambda = 7200 \cdot t/3600 = 2t$. Of course we get the same value for λ with $n = 7\,200\,000$ and $p = t/3\,600\,000$. It results in:

$$p(Z_t = k) = \frac{(2t)^k}{k!} e^{-2t}.$$

The random variable Z_t is therefore Poisson distributed for each t with parameter $2t$. Nuclear decay is a prototype for a class of stochastic processes that count independent and random events. These are called *Poisson processes* and they all have the following properties:

▶   **Definition 21.24: The Poisson process** For $t \in \mathbb{R}^+$ let X_t be the random variable that describes the number of occurrences of an event in the period from 0 to t. For these events it should apply:

   a) The probability for the occurrence of k events in an interval I of length t only depends on k and on t not on the position of the interval I on the time axis, that is $p(X_{s+t} - X_s = k) = p(X_t = k)$.
   b) The numbers of events in disjoint time intervals are independent random variables.
   c) No two events ever occur at the same time.

   Then $(X_t)_{t \in \mathbb{R}^+}$ is called *Poisson process*.

The conclusions we have drawn for the atomic decay process can be generalized:

▶   **Theorem 21.25** In a Poisson process $(X_t)_{t \in \mathbb{R}^+}$ the random variable X_t is Poisson distributed for all t with parameter λt, that is,

$$p(X_t = k) = \frac{(\lambda t)^k}{k!} e^{-\lambda t}.$$

In this theorem, λt takes on the role of λ from the Poisson distribution: λt is the expected value for the number of events in the period from 0 to t and λ is something like the "expected value per unit of time". λ is called *rate* of the events. In the example of the nuclear decay, on average, 7200 decays per hour are expected, which corresponds to 2 decays per second, that is $\lambda = 2/\text{sec}$. The rate for the events in a Poisson process is often known.

Example

A call center receives an average of 120 requests per hour. It is overloaded if 5 or more requests arrive in one minute. In what percentage of intervals of one minute does an overload occur?

The expected value λ is $120/h$, for $T = 1 \min = 1/60\,h$ is $\lambda \cdot t = 2$. We have to calculate

$$p(X_{1\min} \geq 5) = 1 - p(X_{1\min} < 5).$$

It is

$$p(X_1 = 0) = \frac{1}{1}e^{-2} = 0.135, \quad p(X_1 = 1) = \frac{2}{1}e^{-2} = 0.271,$$

$$p(X_1 = 2) = \frac{4}{2}e^{-2} = 0.271, \quad p(X_1 = 3) = \frac{8}{6}e^{-2} = 0.180,$$

$$p(X_1 = 4) = \frac{16}{24}e^{-2} = 0.090,$$

and thus the sum $p(X_1 < 5) \approx 0.947$. This results in an overload in about 5.3 % of the minute intervals. ◀

If X_t is a Poisson process, then the waiting times between the events of the process are exponentially distributed:

▶ **Theorem 21.26** Let $(X_t)_{t>0}$ be a Poisson process and X_t be Poisson distributed with parameter λt. Let W_i be the time between the i-th and the $i+1$-th occurrence of the event. Then W_i is exponentially distributed with expected value $1/\lambda$.

According to Theorem 21.25, $p(X_t = 0) = \frac{(\lambda t)^0}{0!}e^{-\lambda t} = e^{-\lambda t}$. The probability that the waiting time W_0 for the first event is greater than t is equal to the probability that $X_t = 0$. We thus obtain for W_0:

$$p(W_0 > t) = p(X_t = 0) = e^{-\lambda t}$$

and therefore

$$p(W_0 \leq t) = 1 - e^{-\lambda t}.$$

W_0 therefore has the distribution function of the exponential distribution and is thus exponentially distributed. For $i > 0$ we can conclude similarly. If s is the time at which the event occurs for the i-th time, then by Theorem 21.24a) for W_i holds:

$$p(W_i > t) = p(X_{s+t} - X_s = 0) = p(X_t = 0) = e^{-\lambda t},$$

so because of $p(W_i \le t) = 1 - e^{-\lambda t}$ also W_i is exponentially distributed.   □

Markov Chains

We will now investigate stochastic processes with a discrete index set $\mathbb{N}_0$. Again, the index can be interpreted as a time parameter, but in contrast to the Poisson processes, we only look at the value of the random variable X_t at certain points in time, for example every minute or every hour. Such a process $(X_t)_{t \in \mathbb{N}_0}$ is called a *Markov process* or *Markov chain*, if the probability that $X_t = k$ depends only on the distribution of the random variable X_{t-1}, not on other random variables of the process. The image of the random variables should be a subset of $\mathbb{N}$ or $\mathbb{N}_0$. The elements of the image of X_t are called states.

▶   **Definition 21.27** Let $(X_t)_{t \in \mathbb{N}_0}$ for all $t \in \mathbb{N}_0$ be a discrete random variable with the same image $W \subset \mathbb{N}$. The stochastic process $(X_t)_{t \in \mathbb{N}_0}$ is called a *Markov chain*, if $p(X_{t+1} = k)$ only depends on X_t. Let

$$p_{ik}(t) := p(X_{t+1} = k | X_t = i).$$

$p_{ik}(t)$ is called the *transition probability* from i to k. If $p_{ik}(t) =: p_{ik}$ is constant for all $t \in \mathbb{N}_0$, then $(X_t)_{t \in \mathbb{N}_0}$ is called a *homogeneous Markov chain*. The vector or sequence $(a_0, a_1, a_2, \ldots)$ with $a_i = p(X_0 = i)$ is called *initial distribution* of the chain. If the image W is finite, the matrix $P = (p_{ij})$ is called *transition matrix* of the Markov chain.

We will only deal with homogeneous Markov chains. With probability p_{ik} the system transitions from state i to state k in an observed time period.

Such a Markov chain can be represented in the form of a directed network (compare Definition 11.25). For this purpose, an

---

Example

An admittedly somewhat crude description of the weather could consist of the three states: sunny all day, cloudy but dry, rain. I number the states with 0, 1, 2. The times t should be the days. If the weather tomorrow only depends on the weather today, then there are transition probabilities. I have entered these in the graph in Fig. 21.12.

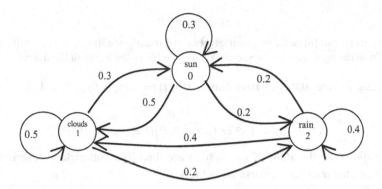

Fig. 21.12 State transition probabilities

Let's look at the adjacency matrix of this graph:

$$P = \begin{pmatrix} 0.3 & 0.5 & 0.2 \\ 0.3 & 0.5 & 0.2 \\ 0.2 & 0.4 & 0.4 \end{pmatrix}$$

The elements of the matrix are just the transition probabilities p_{ik}. At row i and column j is the probability that state i is followed by state j. The adjacency matrix is therefore just the transition matrix of the process. ◄

Starting from the state $X_t = i$ always exactly one of the states $X_{t+1} = k$, $k = 0, \ldots, n$ occurs with the probability p_{ik}. Therefore, the sum must be $\sum_{k=0}^{n} p_{ik} = 1$. This is just the sum of the elements of the i-th row of the matrix. In a transition matrix, therefore, all row sums are equal to 1.

How can we calculate the distribution at a later time from a given initial distribution $a(0) := (a_0, a_1, \ldots, a_n)$? Let us put in the law of total probability (Theorem 19.7) $B_i = \{X_t = i\}$ and $A = \{X_{t+1} = k\}$, so we get

$$p(X_{t+1} = k) = \sum_{i=0}^{n} p(X_t = i) \cdot p(X_{t+1} = k | X_t = i),$$

which means for $t = 0$:

$$p(X_1 = k) = a_0 p_{0k} + a_1 p_{1k} + \ldots + a_n p_{nk}$$

Look back at linear algebra in Sect. 7.2: That's just the k-th element in the matrix multiplication of the vector $a(0)$ with the matrix P. In total, one obtains the vector $a(1) := (p(X_1 = 0), p(X_1 = 1), \ldots, p(X_1 = n))$ as the product of the row vector $a(0)$ from the left with the transition matrix:

$$a(1) = a(0)P.$$

You have to be careful here: in linear algebra, we usually multiply matrices with column vectors from the right, here a row vector is multiplied from the left with the matrix.

It is the same for the states at later times. If $a(t) := (p(X_t = 0), p(X_t = 1), \ldots, p(X_t = n))$, then:

$$a(t) = a(t-1)P = (a(t-2)P)P = \ldots = a(0)P^t$$

The probabilities for the state of the system are therefore completely described by the initial state and the transition matrix for all times.

Let's look again at the above

Example

If the initial distribution for day 0 is given, for example $(a_0, a_1, a_2) = (0.2, 0.5, 0.3)$ for sun, clouds or rain, then the probabilities of the three states are now calculable for all times. Let's start with day 1:

$$a(1) = a(0)P = (0.2, 0.5, 0.3) \cdot \begin{pmatrix} 0.3 & 0.5 & 0.2 \\ 0.3 & 0.4 & 0.3 \\ 0.2 & 0.4 & 0.4 \end{pmatrix} = (0.27, 0.42, 0.31)$$

It follows

$$a(2) = a(1)P = (0.269, 0.427, 0.304),$$
$$a(3) = (0.2696, 0.4269, 0.3035).$$

Do you see the sum of the vector elements is always 1 again? This must be so, because there are only the three states sun, clouds or rain.

What is after 100 days? With a computer algebra system, you can immediately calculate that the values of $a(100)$ only differ from $a(2)$ after the fourth decimal place. We will see in a moment that this is no coincidence. Let's look at the (rounded) powers of the matrix P:

$$P^2 = \begin{pmatrix} 0.28 & 0.43 & 0.29 \\ 0.27 & 0.43 & 0.30 \\ 0.26 & 0.42 & 0.32 \end{pmatrix},$$

$$P^3 = \begin{pmatrix} 0.271 & 0.428 & 0.301 \\ 0.270 & 0.427 & 0.303 \\ 0.268 & 0.426 & 0.306 \end{pmatrix},$$

$$\vdots$$

$$P^{10} = \begin{pmatrix} 0.270 & 0.427 & 0.303 \\ 0.270 & 0.427 & 0.303 \\ 0.270 & 0.427 & 0.303 \end{pmatrix}.$$

For higher powers, almost nothing happens anymore. The powers converge to a matrix P^∞ and surprisingly, in the limit all lines are equal. This has an interesting consequence: For each initial distribution $a = (a_1, a_2, a_3)$ we have $aP^\infty = (0.270, 0.427, 0.303)$. Please try this yourself. This means that the state probabilities of the Markov chain become stationary. Already after a few days we get a probability of 0.270 for sun, 0.427 for clouds and 0.303 for rain. ◀

▶ **Theorem 21.28** If $P = (p_{ij})$ is the transition matrix of a homogeneous Markov chain with a finite number of states, and if there is a $m \in \mathbb{N}$, so that P^m only has positive elements, then

a) There exists the limit $P^\infty = \lim_{n \to \infty} P^n$. All rows of the matrix P^∞ are equal. The sum of the elements of this row is 1.

For the row $p = (p_0, p_1, \ldots, p_n)$ of the matrix P^∞ holds:

b) For each initial distribution $a = (a_0, a_1, \ldots, a_n)$ is $aP^\infty = p$
c) p is the only initial distribution with the property $pP = p$.

The row vector p is also called the *stationary distribution* of the Markov chain.

Remember that for the transposition of matrices holds $(AB)^T = B^T A^T$. Then $pP = p$ is the same as $P^T p^T = p^T$. In the language of eigenvalue theory, Theorem 21.28c) says: p^T is the only eigenvector of the matrix P^T to the eigenvalue 1.

The proof of part a) of the theorem requires only elementary mathematics, but is somewhat tricky. I will not carry it out here.

To b): Since the row sum is $\sum_{i=0}^{n} a_i = 1$, we get $aP^\infty = p$ simply by substitution:

$$(a_0, a_1, \ldots, a_n) \begin{pmatrix} p_0 & p_1 & \cdots & p_n \\ p_0 & p_1 & \cdots & p_n \\ \vdots & \vdots & & \vdots \\ p_0 & p_1 & \cdots & p_n \end{pmatrix}$$

$$= \left(\left(\sum_{i=0}^{n} a_i \right) p_0, \left(\sum_{i=0}^{n} a_i \right) p_1, \ldots, \left(\sum_{i=0}^{n} a_i \right) p_n \right) = (p_0, p_1, \ldots, p_n).$$

To c): Because of $P^{n+1} = P^n P$ it follows in the limit $P^\infty = P^\infty P$. If we multiply this equation from the left by p, we get $pP^\infty = pP^\infty P$. Since $pP^\infty = p$, we get $pP = p$.

Is there another stationary distribution q, which is different from p? From $qP = q$ would then for all n follow $qP^n = q$, thus also $qP^\infty = q$. But we already know that $qP^\infty = p$ is true. Therefore $p = q$. ☐

Theorem 21.28c) provides us with an elegant way to determine the stationary distribution of a homogeneous Markov chain. We do not have to laboriously calculate the limit of a matrix power, it is enough to solve the system of linear equations $pP = p$ for the unknowns $p = (p_0, p_1, \ldots, p_n)$.

> Don't be confused by the notation here. In the language of Chap. 8 we had to solve the system of equations $P^T x = x$ or $(P^T - E)x = 0$.

The solution space of this equation system is one-dimensional according to Theorem 21.28c) and the solution vector, the components of which sum up to 1, is our stationary state p.

We will use this method to calculate stationary states of queues.

Queues

Queues occur again and again in many areas of everyday life: at the checkout in the supermarket, in the doctor's waiting room, when calling a call center, when requesting a server. In queueing theory, one usually speaks of customers who arrive and are served at a service station. In a common type of queue, arrival and service can each be described by Poisson processes. In between is a waiting room that can usually accommodate a limited number of customors waiting. We will now investigate such queues.

The customers are to arrive in the queue with an arrival rate of λ. At the head of the queue they are served, this is to be done with a service rate of μ. λ and μ are the rates of the corresponding Poisson processes, they give the expected number of events in a certain period of time. The maximum length of the queue is n. Furthermore, let $\rho = \lambda/\mu$. The number ρ is called the utilization of the system. We want to calculate how the queue develops as a function of λ and μ. The queue can be described by a stochastic process $(X_t)_{t \in \mathbb{R}}$, where the random variable X_t is the number of customers in the system at time t, including the customer currently being served.

If we consider the queue at fixed times $t \in \mathbb{N}_0$, for example every minute, we get a stochastic process $(X_t)_{t \in \mathbb{N}_0}$, and since both entry and exit from the queue are random and independent, this process is a homogeneous Markov chain: the number of customers in the system at time $t + 1$ depends only on the number at time t, and at different times the transition probability from t to $t + 1$ is always the same. For this Markov chain, we now want to calculate the transition matrix and the stationary distribution.

For simplicity, we choose for the period of time an interval T, which should be so small that in this interval at most one customer arrives or is served. The probability that two or more of these events will occur in the interval T should be negligibly small.

Example

A doctor can treat 10 patients per hour in his practice, about 9 patients come per hour. The system should have 11 seats, 10 in the waiting room, one in the treatment room. If all seats are occupied and another patient arrives, he leaves again. The arrival rate in this example is $\lambda = 9$, the service rate $\mu = 10$. The period of time of one hour is too large for our calculations, so we choose, for example, the interval $T = 1$ minute $= 1/60$ hour. Then we calculate with the arrival rate $\lambda_T = 9/60$ and with the service rate $\mu_T = 10/60$ per minute. We now assume that never more than one patient enters or leaves the waiting room per minute.

> How large is the error we make with this approximation? According to Theorem 21.25, we get $p(X_1 \leq 1) = e^{-\lambda_T} + (\lambda_T)e^{-\lambda_T} = 0.9898$. If that is not accurate enough for you, calculate per second, then $p(X_1 \leq 1) = 0.999997$. ◀

λ_T and μ_T are the arrival and service rates in the small interval T. We will later use that the quotients are equal: $\lambda_T/\mu_T = \lambda/\mu = \rho$. The rate λ_T is the expected value for the arrival of a customer in this interval, and since we assume that only 0 or 1 customer can come, λ_T is also the probability for the arrival of a customer in T. The same is true for μ_T. The smaller T is, the smaller become λ_T and μ_T. For the following calculation, this has the pleasant side effect that the product $\lambda_T \cdot \mu_T$ is tiny and can also be neglected.

Now for the calculation of the transition matrix. It is $p_{ik} = p(X_{t+1} = k|X_t = i)$. The probability that between t and $t + 1$ a customer arrives is λ_T, the probability that a customer has been served and leaves the system is μ_T. Then the probabilities that between t and $t + 1$ no customer arrives respectively no customer leaves the system are $1 - \lambda_T$ and $1 - \mu_T$. Between t and $t + 1$ the following cases can now occur:

Event	Probability
One customer comes, one goes	$\lambda_T \cdot \mu_T$
Nothing happens	$(1 - \lambda_T) \cdot (1 - \mu_T)$
One customer comes, no one goes	$\lambda_T \cdot (1 - \mu_T)$
No customer comes, one goes	$(1 - \lambda_T) \cdot \mu_T$

Then, for $0 < i < n$

$$p_{ii} = \lambda_T \cdot \mu_T + (1 - \lambda_T) \cdot (1 - \mu_T) = 1 - \lambda_T - \mu_T + 2\lambda_T\mu_T$$
$$p_{i,i+1} = \lambda_T \cdot (1 - \mu_T) = \lambda_T - \lambda_T\mu_T$$
$$p_{i,i-1} = \mu_T \cdot (1 - \lambda_T) = \mu_T - \lambda_T\mu_T$$

All other values in the row i are 0, the number of customers in the queue can only change by 1. If we leave out the small products $\lambda_T \cdot \mu_T$, we get for row i of the transition matrix just

$$(0, 0, \ldots 0, \underset{\underset{i-1}{\uparrow}}{\mu_T}, \underset{\underset{i}{\uparrow}}{1 - \lambda_T - \mu_T}, \underset{\underset{i+1}{\uparrow}}{\lambda_T}, 0, \ldots, 0)$$

If $i = 0$, then $p_{01} = \lambda_T - \lambda_T\mu_T$ as above. Furthermore, $p_{0k} = 0$ for $k > 1$. If there is no customer in the system, then the probability that no one comes is $1 - \lambda_T$. Nobody can be served in this case, so we get $p_{00} = 1 - \lambda_T$. Similarly, for the last line $p_{nk} = 0$ for

$k < n - 1$, $p_{n,k-1} = \mu_T - \lambda_T\mu_T$ and $p_{nn} = 1 - \mu_T$. If we also eliminate the products $\lambda_T \cdot \mu_T$ in the first and last line, we get as the transition matrix:

$$P = \begin{pmatrix} 1 - \lambda_T & \lambda_T & 0 & 0 & \cdots & 0 \\ \mu_T & 1 - \lambda_T - \mu_T & \lambda_T & 0 & \cdots & 0 \\ 0 & \mu_T & 1 - \lambda_T - \mu_T & \lambda_T & \cdots & 0 \\ \vdots & & & \ddots & & 0 \\ 0 & 0 & \cdots & \mu_T & 1 - \lambda_T - \mu_T & \lambda_T \\ 0 & 0 & \cdots & 0 & \mu_T & 1 - \mu_T \end{pmatrix}$$

Although we have done some rounding, in the matrix all line sums are equal to 1, so it is a correct transition matrix of a Markov chain. And the smaller the interval T was chosen, the better this matrix describes our queue.

Does this process have a stationary state? If you calculate a few powers of the matrix, you will notice that more and more zeros disappear, eventually all elements are positive. Then we can apply Theorem 21.28: there is a stationary state, which we can calculate with the help of Theorem 21.28c). For this, the system of equations $pP = p$ has to be solved for $p = (p_0, p_1, \ldots, p_n)$.

The n equations of the system are:

$$p_0(1 - \lambda_T) + p_1\mu_T = p_0$$
$$\Rightarrow \quad -p_0\lambda_T + p_1\mu_T = 0$$
$$p_0\lambda_T + p_1(1 - \lambda_T - \mu_T) + p_2\mu_T = p_1$$
$$\Rightarrow \quad p_0\lambda_T - p_1(\lambda_T + \mu_T) + p_2\mu_T = 0$$
$$\vdots$$
$$p_{i-1}\lambda_T + p_i(1 - \lambda_T - \mu_T) + p_{i+1}\mu_T = p_i$$
$$\Rightarrow \quad p_{i-1}\lambda_T - p_i(\lambda_T + \mu_T) + p_{i+1}\mu_T = 0$$
$$\vdots$$
$$p_{n-1}\lambda_T + p_n(1 - \mu_T) = p_n$$
$$\Rightarrow \quad -p_{n-1}\lambda_T + p_n\mu_T = 0$$

From the first equation we get $p_0 = (\mu_T/\lambda_T)p_1$. Substituted into equation 2 we get

$$p_1\mu_T - p_1(\lambda_T + \mu_T) + p_2\mu_T = -p_1\lambda_T + p_2\mu_T = 0 \quad \Rightarrow \quad p_1 = (\mu_T/\lambda_T)p_2,$$

and so on. For all i, also for $i = n - 1$ is $p_i = (\mu_T/\lambda_T)p_{i+1}$, or $p_{i+1} = (\lambda_T/\mu_T)p_i$. If we now set $\rho = \lambda_T/\mu_T$ we get for the result vector:

$$p = p_0(1, \rho, \rho^2, \rho^3, \ldots, \rho^n)$$

The solution of the system of equations is a one-dimensional vector space. We now have to determine p_0 so that the sum of the vector elements is 1:

$$p_0(1 + \rho + \rho^2 + \rho^3 + \ldots + \rho^n) = 1$$

According to Theorem 3.3 is $(1 + \rho + \rho^2 + \rho^3 + \ldots + \rho^n) = (1 - \rho^{n+1})/(1 - \rho)$ and therefore $p_0 = (1 - \rho)/(1 - \rho^{n+1})$. The stationary state of the queue is therefore

$$p = (1 - \rho)/(1 - \rho^{n+1})(1, \rho, \rho^2, \rho^3, \ldots, \rho^n) \qquad (21.14)$$

How long does a customer spend on average in the system? According to Theorem 21.25, the average service time is $1/\mu$. If he arrives when i people are in the system, he has to wait on average $(i + 1) \cdot 1/\mu$ time units until he has been served. If E is the expected value for the number of people in the system, the average waiting time is therefore $(E + 1)/\mu$.

Example

In the waiting room of the doctor with $\lambda_T = 9/60$ and $\mu_T = 10/60$ is $\rho = 0.9$. That is also just λ/μ with the original rates $\lambda = 9$ and $\mu = 10$. If the system has 11 places, the sum $(1 - \rho)/(1 - \rho^{n+1})$ has the value 0.14 and the stationary state is:

$$p = (0.14, 0.13, 0.11, 0.10, 0.091, 0.082, 0.074, 0.067, 0.060, 0.054, 0.049, 0.044)$$

Position i indicates the probability that i people are in the system. The probability of an empty system (position 0) is therefore 0.14. In this case, the doctor has nothing to do. With a probability of 0.13, he is currently treating a patient and the waiting room is empty. With a probability of 0.11, there is 1 patient in the waiting room and so on.

The expected value for the number of people in the practice, calculated according to the formula in Definition 20.11, is 4.28. The average length of stay in the practice is 5.28/10 hours, that is about 32 minutes. Calculate for yourself how the queue develops if the waiting room gets bigger or if the arrival rate is 10 or 11 patients per hour. ◀

Often it is assumed in queueing theory that the length of the queue is unlimited, that n can therefore be arbitrarily large. The transition matrix then becomes arbitrarily large and we finally obtain for the stationary state p the "infinite" vector $p = p_0(1, \rho, \rho^2, \rho^3, \ldots)$. If $\rho < 1$ then for the geometric series holds $\sum_{i=0}^{\infty} \rho^i = 1/(1 - \rho)$, so $p_0 = (1 - \rho)$. Then formula (21.14) becomes a little bit easier:

$$p = (1 - \rho)(1, \rho, \rho^2, \rho^3, \ldots).$$

The random variable X with $p(X = k) = (1 - \rho)\rho^k$ describes the number of customers in the system. We can also give a simple formula for the calculation of the expected value of X: For the random variable $Y = X/\rho$ we have $p(Y = k) = (1 - \rho)\rho^{k-1}$, so Y

according to Theorem 21.8 is geometrically distributed with parameter $(1 - \rho)$ and has the expected value $1/(1 - \rho)$. Then $X = \rho Y$, according to Theorem 20.13, has the expected value $\rho/(1 - \rho)$. Now we replace ρ by $\lambda_T/\mu_T = \lambda/\mu$, and we get:

$$\frac{\rho}{1 - \rho} = \frac{\lambda/\mu}{1 - \lambda/\mu} = \frac{\lambda}{\mu - \lambda}.$$

We can now calculate the expected value of the length of the queue X_S: At position i of the vector p, the queue has the length $i - 1$ and therefore the expected value is:

$$E(X_S) = \sum_{i=1}^{\infty}(i - 1)p_i = \sum_{i=0}^{\infty} ip_i - \sum_{i=1}^{\infty} p_i$$

$$= E(X) - (1 - p_0) = \frac{\rho}{1 - \rho} - \rho = \frac{\lambda}{\mu - \lambda} - \frac{\lambda}{\mu}.$$

The mean duration of stay is as before

$$(E(X) + 1)/\mu = \left(\frac{\lambda}{\mu - \lambda} + 1\right) \cdot \frac{1}{\mu} = \frac{1}{\mu - \lambda}$$

and the average waiting time in the queue:

$$\frac{1}{\mu - \lambda} - \frac{1}{\mu}$$

We summarize the results for unlimited queues:

▶   **Theorem 21.29** If the arrival and service of customers in a queue follow Poisson processes with arrival rate λ and service rate μ, if the length of the queue is unlimited and $\lambda < \mu$, so the random variable X with

$$p(X = k) = \left(1 - \frac{\lambda}{\mu}\right)\left(\frac{\lambda}{\mu}\right)^k, \quad k \in \mathbb{N}_0$$

describes the number of customers in the system in the stationary state. Further, it is:

$\dfrac{\lambda}{\mu - \lambda}$:   The expected value for the number of customers in the system,

$\dfrac{\lambda}{\mu - \lambda} - \dfrac{\lambda}{\mu}$:   The expected value for the length of the queue,

$\dfrac{1}{\mu - \lambda}$:   The average length of stay in the system,

$\dfrac{1}{\mu - \lambda} - \dfrac{1}{\mu}$:   The average waiting time in the queue.

Let's look again in the waiting room of the doctor and assume it is unlimited. For the arrival rate 9 and the service rate 10 we then get

$$p = (0.10, 0.090, 0.081, 0.073, 0.066, 0.059,$$
$$0.053, 0.048, 0.043, 0.039, 0.035, 0.031, \ldots).$$

The expected value for the number of patients is $9/(10 - 9) = 9$ and the average waiting time until leaving the practice is $1/(10 - 9) = 1$ hour. The expected value for the length of the queue is 8.1, the waiting time in the waiting room on average $1 - 1/10$ hours, that is 54 minutes.

Look at the difference to the waiting room with 10 seats. The situation was much more relaxed there. However, about 5% of the patients were also sent home without having received medical attention, because the waiting room was full. ◄

While in the case of a finite waiting room solutions exist for all values of λ and μ, in the infinite case must necessarily be $\lambda < \mu$. It is obvious that the arrival rate cannot be greater than the service rate in the long run, otherwise the queue will always grow, there can be no stationary state. But even if $\lambda = \mu$, no stationary state is achieved. You can see very well what happens here if you play around with the example of the waiting room: If $\lambda = \mu$ then finally all lengths are equally likely. Of course, this is no longer possible if the waiting room has an infinite number of seats.

In queueing theory, many other models are considered. For example, arrival and service can follow other distributions, customers can be scheduled at certain times, instead of one service station there can be several stations with possibly different distributions, customers can be served according to different priorities. Also combinations of different queues can be examined.

21.4   Comprehension Questions and Exercises

1. Under what circumstances can the binomial distribution be approximated by the Poisson distribution?
2. Under what circumstances can the binomial distribution be approximated by the normal distribution?
3. Why does the normal distribution occur in many applications?
4. What are the similarities between geometric distribution and exponential distribution?
5. The waiting time for a randomly passing taxi is exponentially distributed. What about the waiting time for a bus?

6. How are binomial distribution and hypergeometric distribution related?
7. The normal distribution is the limiting distribution of the binomial distribution if n becomes larger and larger. How does this convergence depend on the parameter p of the binomial distribution?

1. In a light drizzle, 100 000 water droplets fall on a grid on the ground. The grid is 1 m^2 large, the holes in the grid 1 cm^2.
   Assume that the numbers of drops falling on different areas of the grid are independent of each other.
   For each sub-task, describe in detail which "experiment" is carried out and for which events probabilities are calculated. Describe the random variables that occur and justify which distribution you use for which sub-task.
   a) Determine the probability that exactly 10 raindrops fall into a certain grid hole and the probability that at least 1 drop falls into each hole.
   b) The grid is divided into 100 parts. Calculate the probability that between 900 and 1100 drops fall into such a part, as well as the probability that there is a part on which more than 1050 or more than 1100 drops fall.
   c) On 10 adjacent holes, 100 drops fall randomly. Calculate the probability that exactly 10 drops fall into the first of these 10 holes.
2. A baker bakes 100 olive breads. He throws 400 olives into the dough. Determine the probability that there are two to six olives in a bread and that you catch a bread without olives.
3. An airline knows that on average 10% of the booked flight seats are canceled and overbooks by 5%. For an airplane with 100 seats, it sells 105 tickets. What is the probability that more passengers will come to the departure than there are seats available? Solve this task with the exact distribution as well as with a possible approximation to the exact distribution.

   You will find that the rules of thumb are sometimes to be taken with a grain of salt.

4. At an highway tollbooth, 1000 cars arrive randomly distributed between 7:00 and 17:00. If more than 5 arrive in one minute, the tollbooth is overloaded. In how many percent of the intervals is this the case?
5. A company receives an average of 500 orders between 8 a.m and 1 a.m. in the morning, 800 orders between 2 p.m. and 6 p.m. in the afternoon, evenly distributed over the 5 or 4 hours. Waiting times occur when more than 3 orders arrive in one minute. Calculate the probability of this occuring in a one-minute interval for the morning and the afternoon.

6. A machine produces chips with a (randomly distributed) defect rate of 9%.
   a) What distribution determines the number of inedible chips in a number n of produced chips. Explain this.
   b) Calculate the probability of 110 or more defective chips in a production of 1000 chips using an appropriate approximation. Why are you allowed to use the approximation?

7. A cargo ship has two diesel engines that usually both run. On one day of travel, there is a 2% chance that one of the two engines will fail. This is not a problem, the ship can continue with one engine. However, if one engine fails, there is a 2% chance that the second engine will also fail the next day. The engineers can repair a failed engine with a 30% chance by the next day, but if both engines are failed, the repair probability is reduced to 15% per engine. We will neglect the fact that both engines could fail on one day.
   a) Define states and state transitions for this process. Draw a state diagram.
   b) When the ship sets sail, both engines are running. The journey from China to Europe takes 48 days. What is the probability that on the last day one engine or both engines have failed? Use sage or another tool for calculation.

   Note: If both engines are failed, there is still an emergency engine that should keep the ship maneuverable and not stranded in the Elbe off Hamburg.

Statistical Methods

22

Abstract

Now you can reap the fruits of the last chapters. If you have worked through this chapter

- then you know what a sample is, and can calculate the sample mean, the mean squared error and the sample variance,
- you can examine data using principal component analysis,
- you are familiar with the concept of the estimator and have criteria to determine whether an estimator is unbiased or consistent,
- you know estimators for the probability in a Bernoulli process and for the expected value and variance of a random variable,
- you understand the concepts of confidence interval and confidence level and can calculate confidence intervals for the probability in a Bernoulli process and determine necessary sample sizes,
- you can carry out one-sided and two-sided hypothesis tests for Bernoulli processes and know the errors of type I and type II in hypothesis tests,
- you have carried out Pearson's chi-squared test,
- and can put the finished book aside. Congratulations!

An important task of the statistician is to make inferences about properties of large data sets from a small set of data – a sample. This is usually done by making a random

---

Supplementary Information The online version contains supplementary material available at https://doi.org/10.1007/978-3-658-40423-9_22.

selection of data – a sample – from the population and analyzing it. In this chapter we will learn some such analysis methods.

22.1   Parameter Estimation

In the example of the election forecast, the task arose of guessing the election result from a sample. This election result is an unknown parameter. We now want to develop methods with which one can derive estimates for such unknown parameters, the estimands, from a sample.

First I would like to clarify the concept of the sample:

Samples

▶   **Definition 22.1** Given n observation values $x_1, x_2, \ldots, x_n$. The n-tuple is called *sample* of size n, the individual values x_i are called *sample values*. If a sample is obtained by a random experiment, it is called *random sample*.

I will only investigate random samples in what follows, I simply call them samples.

If you receive a sample value by a random experiment, it is the result of a random variable X. If the experiment is repeated n times independently, we get n sample values $x_1, x_2, \ldots, x_n$. This n-tuple can also be regarded as the result of a random vector $(X_1, X_2, \ldots, X_n)$, where X_i describes the result of the i-th experiment. A sample of size n is then a realization of this random vector.

Examples

1. Let's roll a die 10 times. The random variable X_i denotes the result of the i-th roll. A sample is a function value of the random vector $(X_1, X_2, \ldots, X_{10})$, different function values result in different samples. So, for example, $s_1 = (2, 5, 2, 6, 2, 5, 5, 6, 4, 3)$ and $s_2 = (4, 6, 2, 2, 6, 1, 6, 4, 4, 1)$ are possible samples of this random experiment.

2. There are 10 000 screws in a box, some of which are defective. For a sample, 100 screws are taken. The random variable X_i describes the condition of the i-th screw taken:

$$X_i(\omega) = \begin{cases} 0 & \text{if } i\text{-th screw is ok} \\ 1 & \text{if } i\text{-th screw is defective.} \end{cases}$$

In the large population, there is no need to distinguish between "drawing without replacement" and "drawing with replacement", we can assume that the random variables X_i are independent of each other.

A concrete sample is a realization of the random vector $(X_1, X_2, \ldots, X_{100})$ and, for example, has the value $(0, 1, 0, 0, 0, 1, 1, 0, 0, \ldots, 1)$.

New random variables can be obtained from random variables by combination. For example, one could consider:

$$H_{100}(\omega) = X_1(\omega) + X_2(\omega) + \ldots + X_{100}(\omega),$$

$$R_{100}(\omega) = \frac{1}{100}(X_1(\omega) + X_2(\omega) + \ldots + X_{100}(\omega)).$$

H_{100} describes the number of defective screws in the sample, R_{100} the relative frequency of defective screws. These are also random variables that can take different values in each sample. We already know from Theorem 21.2 that H_{100} is binomially distributed, but the parameter p of the binomial distribution is still unknown.

3. In an election poll, the sample consists of n randomly selected voters. The random experiment X_i describes the voting behavior of the i-th person surveyed. The poll is a realization of the random vector $(X_1, X_2, \ldots, X_n)$. The opinion research company conducts a sample and estimates the election result from it. Another institute receives a different sample and is therefore very likely to have a different election forecast. ◀

Just as we have assigned parameters to random variables, such as expected value and variance, we now want to define characteristics for samples:

▶ **Definition 22.2** If $(x_1, x_2, \ldots, x_n)$ is a sample, then

$$\bar{x} := \frac{1}{n}(x_1 + x_2 + \ldots + x_n)$$

is the *sample mean*. The number

$$m := \frac{1}{n} \sum_{i=1}^{n} (x_i - \bar{x})^2$$

is called the *mean squared error*, it represents the mean of the squared deviations of $\bar{x}$.

$$s^2 := \frac{1}{n-1} \sum_{i=1}^{n} (x_i - \bar{x})^2$$

is called the *sample variance* and $s = \sqrt{s^2}$ *sample standard deviation*. If $(y_1, y_2, \ldots, y_n)$ is another sample with mean $\bar{y}$, then

$$s_{xy}^2 := \frac{1}{n-1} \sum_{i=1}^{n} (x_i - \bar{x})(y_i - \bar{y})$$

is the *sample covariance*.

The expected value of a random variable is something like a mean function value, and the variance of a random variable is the mean squared deviation from it. You see the analogy to the concepts of sample mean and mean squared error of a sample. The reason why the sample variance is still introduced, in which instead of dividing by n, we divide by $n-1$, is not yet clear, we will come to it later. We will use these numbers as estimates for corresponding parameters of a random variable.

▶   **Theorem 22.3: The calculation of the sample variance** It holds

$$s^2 = \frac{1}{n-1}\left(\left(\sum_{i=1}^{n} x_i^2\right) - n\bar{x}^2\right).$$

Proof:

$$\sum_{i=1}^{n}(x_i - \bar{x})^2 = \sum_{i=1}^{n}(x_i^2 - 2x_i\bar{x} + \bar{x}^2)$$

$$= \sum_{i=1}^{n} x_i^2 - 2\bar{x}\underbrace{\sum_{i=1}^{n} x_i}_{=n\bar{x}} + n\bar{x}^2 = \sum_{i=1}^{n} x_i^2 - n\bar{x}^2 \qquad \square$$

Let's look at the dice example again. The expected value when rolling dice is 3.5, the variance 2.917. If we calculate the sample mean and sample variance of the two samples $s_1 = (2, 5, 2, 6, 2, 5, 5, 6, 4, 3)$ and $s_2 = (4, 6, 2, 2, 6, 1, 6, 4, 4, 1)$, we get:
Sample mean of s_1:

$$\bar{s}_1 = \frac{1}{10}(2+5+2+6+2+5+5+6+4+3) = \frac{40}{10} = 4$$

Sample variance of s_1:

$$s^2 = \frac{1}{9}(2^2 + 5^2 + 2^2 + 6^2 + 2^2 + 5^2 + 5^2 + 6^2 + 4^2 + 3^2 - 10 \cdot 4^2) = \frac{24}{9} \approx 2.67$$

similarly for the sample mean of s_2: $\bar{s}_2 = 3.6$, and the sample variance of s_2: $s^2 = 4.04$.

We must carefully distinguish: The expected value, variance and standard deviation of a random variable are fixed numbers that are independent of a specific experiment. The sample mean, mean square error, sample variance and sample standard deviation of a sample, on the other hand, are themselves realizations of a random variable, that is, random, trial-dependent values. If $(x_1, x_2, \ldots, x_n)$ arises as the realization of the random vector $(X_1, X_2, \ldots, X_n)$, then the associated sample mean is the realization of the random variable $\bar{X} := \frac{1}{n}(X_1 + X_2 + \cdots + X_n)$, the sample variance is the realization of $S^2 := \frac{1}{n-1}((\sum_{i=1}^{n} X_i^2) - n\bar{X}^2)$.

Estimators

▶ **Definition 22.4** Let the random variable X describe the outcome of a random experiment. This experiment is to be repeated n times. Let X_i be the random variable that describes the outcome of the i-th experiment. If p is a parameter that is associated with the random variable X and f is a function that can be used to determine an estimate $\tilde{p} = f(x_1, x_2, \ldots, x_n)$ for the parameter p from a sample $(x_1, x_2, \ldots, x_n)$, then the random variable

$$P = f(X_1, X_2, \ldots, X_n)$$

formed from the random variables X_i is called *estimator* for the parameter p. A realization of the estimator

$$P(\omega) = f(X_1(\omega), X_2(\omega), \ldots, X_n(\omega)) = f(x_1, x_2, \ldots, x_n)$$

is called an *estimate* for the parameter p.

You will have to read this definition several times to understand what is behind it. But it becomes clearer when we look at the examples from the beginning of the chapter again:

Examples

1. The random variable X describes the result of rolling a die. If we do not know the expected value of X yet, we can guess after 10 rolls with the result $(x_1, x_2, \ldots, x_{10})$:

$$E(X) \approx \frac{1}{10}(x_1 + x_2 + \cdots + x_{10}).$$

In this case, for all $n \in \mathbb{N}$, the random variable $\overline{X} = \frac{1}{n}(X_1 + X_2 + \ldots + X_n)$ is an estimator for the unknown parameter $E(X)$.

How can we estimate the variance of the random variable X? It will turn out that $S^2 := \frac{1}{n-1}((\sum_{i=1}^{n} X_i^2) - n\overline{X}^2)$, the function whose result is the sample variance of the sample, is the right estimator for the parameter $\mathrm{Var}(X)$.

2. The random experiment consists of taking a screw out of the box with 10 000 screws. X describes the state of the screw with 0 (okay) or 1 (not okay). The experiment is repeated 100 times, X_i represents the state of the i-th screw. The unknown parameter of X is the probability p with which $X = 1$. In a sample $(x_1, x_2, \ldots, x_{100})$, the number of defective screws is just the sum of the x_i. It is therefore natural to guess:

$$p(X = 1) \approx \frac{x_1 + x_2 + \ldots + x_{100}}{100}.$$

This is a realization of the estimator $R_{100} = \frac{1}{100}(X_1 + X_2 + \ldots + X_{100})$. ◀

In these examples we have guessed the estimators for a parameter with common sense, we do not yet know any criteria for when an estimator is good or bad. For example, with the screws in the box, the estimate for p can take any value between 0 and 1 depending on the sample. Is R_{100} really a good estimator?

We describe the quality of the estimators with the means of probability theory that we have already worked out:

▶   **Definition 22.5** Let $S = f(X_1, X_2, \ldots, X_n)$ be an estimator for the parameter p. Then S is called an *unbiased estimator* for p if $E(S) = p$ holds.

The function values of the estimator can of course not always exactly match the parameter p, but the estimates should be scattered around this parameter, the expected value of the estimator should be p.

For a good estimator, this is not enough: The values of the estimator must not jump too wildly around p:

▶   **Definition 22.6** Let $S_n = f(X_1, X_2, \ldots, X_n)$ be an estimator for the parameter $p.S_n$ is called a *consistent estimator*, if for all $\varepsilon > 0$ holds: $\lim_{n \to \infty} P(|S_n - p| \geq \varepsilon) = 0$.

I only included the definition in this form for the sake of mathematical correctness. For us, the following, much more intuitive theorem, which can be derived with some effort from Definition 22.6 is sufficient:

▶   **Theorem 22.7** An estimator $S_n = f(X_1, X_2, \ldots, X_n)$ for the parameter p is consistent if $\lim_{n \to \infty} \text{Var}(S_n) = 0$ holds.

This means that with increasing n, the variance of the estimator becomes smaller and smaller.

Estimator for the Probability in a Bernoulli Process

In the screw example, we chose the relative frequency as an estimator for the probability of the event "screw defective". The following theorem shows that this was a good choice:

▶   **Theorem 22.8** In a Bernoulli process, let p be the probability of the event A. Let $X_i = 1$ if A occurs on the i-th trial, otherwise let $X_i = 0$. Then

$$R_n := \frac{1}{n} \sum_{i=1}^{n} X_i$$

is an unbiased and consistent estimator for p. For large n, R_n is approximately $N(p, p(1-p)/n)$-distributed.

Let us check Definition 22.5 and Theorem 22.7. We know from Theorem 21.2 that $H_n = \sum_{i=1}^{n} X_i$ is binomially distributed. Therefore, $E(H_n) = np$ and $\mathrm{Var}(H_n) = np(1-p)$. Because of $R_n = \frac{1}{n}H_n$, just like H_n, R_n is normally distributed for large n and it is

$$E(R_n) = \frac{1}{n}E(H_n) = p, \quad \mathrm{Var}(R_n) = \frac{1}{n^2}\mathrm{Var}(H_n) = \frac{p(1-p)}{n}.$$

So R_n is unbiased and because of $\lim_{n\to\infty} \mathrm{Var}(R_n) = 0$ also consistent. $\square$

Estimators for the Expected Value and Variance of a Random Variable

Let us finally examine the expected value and variance of a random variable:

▶   **Theorem 22.9** Let the random variable X describe an experiment that is repeated n times independently, let X_i be the outcome of the i-th experiment. Let X have the expected value μ and the variance σ^2. Then

$$\overline{X} := \frac{1}{n}(X_1 + X_2 + \ldots + X_n)$$

is an unbiased and consistent estimator for the expected value of X and

$$S^2 := \frac{1}{n-1}\left(\left(\sum_{i=1}^{n} X_i^2\right) - n\overline{X}^2\right)$$

is an unbiased estimator for the variance of X.

We first show that $E(\overline{X}) = \mu$ and $\mathrm{Var}(\overline{X}) = \frac{\sigma^2}{n}$. The random variables X_i all have the same expected value and variance as X, so it follows with the help of the rules for expected values and variances (Theorem 20.18 and 20.20):

$$E(\overline{X}) = E\left(\frac{1}{n}\sum_{i=1}^{n} X_i\right) = \frac{1}{n}\sum_{i=1}^{n} E(X_i) = \frac{1}{n} \cdot n \cdot \mu = \mu$$

$$\mathrm{Var}(\overline{X}) = \mathrm{Var}\left(\frac{1}{n}\sum_{i=1}^{n} X_i\right) = \frac{1}{n^2}\sum_{i=1}^{n} \mathrm{Var}(X_i) = \frac{1}{n^2} \cdot n \cdot \sigma^2 = \frac{\sigma^2}{n}.$$

So $\overline{X}$ is unbiased and consistent.

We still have to show $E(S^2) = \sigma^2$: Since according to Theorem 20.15 for each random variable Y it holds $E(Y^2) = \mathrm{Var}(Y) + E(Y)^2$, we obtain from the assumptions, respectively from the first part of the proof:

$$E(X_i^2) = \sigma^2 + \mu^2, \quad E(\overline{X}^2) = \frac{\sigma^2}{n} + \mu^2$$

and thus:

$$E(S^2) = E\left(\frac{1}{n-1}\left(\left(\sum_{i=1}^{n} X_i^2\right) - n\overline{X}^2\right)\right) = \frac{1}{n-1}\left(\sum_{i=1}^{n} E(X_i^2) - n \cdot E(\overline{X}^2)\right)$$

$$= \frac{1}{n-1}\left(n(\sigma^2 + \mu^2) - n\left(\frac{\sigma^2}{n} + \mu^2\right)\right) = \frac{1}{n-1}((n-1)\sigma^2) = \sigma^2. \qquad \square$$

In this calculation is hidden the reason why in the definition of the sample variance we divided by $n-1$ and not by n: Otherwise, $E(S^2) = \sigma^2(n-1)/n$ and S^2 would not be an unbiased estimator.

22.2   Principal Component Analysis

In big data analysis, large, often unstructured data sets are examined. They can hardly be processed manually anymore, mathematical methods are used to evaluate them. One of these methods is the principal component analysis (PCA). In this section we will do linear algebra again. Let's start with an example: A high school graduate takes an online test to find out recommendations for a suitable field of study or vocational training. The test collects 75 features from the candidate, including school grades, personal preferences and hobbies, social and political interest and much more. Let's assume that each feature can be represented by a real number, then the test result is a point in $\mathbb{R}^{75}$. Of the features, some are certainly more important than others, some are more or less redundant, others are of great importance for the purpose of the evaluation. How can a suggestion for a study direction be read out of this?

If the test is not taken by one, but by 1000 subjects, we get a point cloud in $\mathbb{R}^{75}$. In the language of statistics, a sample of size 1000 is taken for each of the 75 features. Based on these samples, the relevance of features and the relation between different features should now be worked out. The following idea is behind it: First calculate the covariances of sample pairs. The smaller these covariances are, the less the corresponding features are correlated. Now try to generate new features through a change of basis in $\mathbb{R}^{75}$, with the goal of minimizing the covariances between the different features. With the principal component analysis we will even be able to make them 0! The new features are then linear combinations of the original data points. Now you have generated 75 largely uncorrelated features. Often you will then find that many of these new features have a very small variance. The same value results for all subjects here. This is a strong indication that these features do not contain much information that can be used to interpret the result. Such features are ignored. If you restrict yourself to the features with the largest variance, the so-called principal components, you get a point cloud in a low-dimensional space, for example, with two or three principal components in $\mathbb{R}^2$ or $\mathbb{R}^3$. At this point, human expertise is required. The two- or three-dimensional cloud can be visualized, and if the questions of the online test were well formulated, you can see areas in which data accumulate. You will now try to identify areas in $\mathbb{R}^2$ or $\mathbb{R}^3$ in which, for example, respondents with mathematical and scientific interests, economic, social, craft or other interests can be found.

Once this first manual interpretation of the reduced data has been done, the 1001st participant in the survey can be automatically assigned to one of these areas and receives a corresponding recommendation.

The algorithmic core of this program consists in finding the suitable change of basis to make the correlation of the features zero. Subsequently, the principal components are analyzed. I would now like to introduce this process. I start with a concrete example with three features: From a set of $n = 13$ people, the height, weight and body mass index (BMI) are recorded, so we have three samples:

Sub-ject	1	2	3	4	5	6	7	8	9	10	11	12	13
Height	161	181	174	170	165	191	184	162	183	171	159	166	193
Weight	56	73	57	70	59	71	74	58	72	64	54	60	80
BMI	21.60	22.28	18.50	24.91	20.94	19.46	21.86	23.62	20.90	21.89	21.36	24.68	21.48

You may remember that BMI is just determined by the person's height and weight. So it's a redundant feature in this data collection. From the raw data perspective, you can't see this immediately. But if our data analysis is any good, it should be able to tell us that. We will see.

The following calculations are hardly possible on paper anymore, you need computer support for that. I recommend that you follow the examples using a computer algebra system.

To analyze the data, it makes sense to first normalize the samples. For this purpose, we subtract the mean in the three samples S_1, S_2, S_3. If the sample variances are very different, it also makes sense to divide by the sample standard deviation to make the features quantitatively comparable. I did this in the present example. Now all samples scatter around zero and have a sample variance of 1. We enter these normalized sample values in a matrix A. The columns are assigned to the subjects and the rows to the features. The matrix of normalized features is:

$$\begin{pmatrix} -1.12 & 0.62 & 0.013 & -0.33 & -0.77 & 1.50 & 0.88 & -1.03 & 0.80 & -0.25 & -1.29 & -0.68 & 1.67 \\ -1.09 & 0.92 & -0.97 & 0.57 & -0.74 & 0.68 & 1.04 & -0.86 & 0.80 & -0.15 & -1.33 & 0.62 & 1.75 \\ -0.05 & 0.57 & -2.10 & 2.07 & 0.10 & -1.61 & 0.25 & 0.43 & -0.32 & 0.27 & -0.14 & 0.18 & -0.05 \end{pmatrix}.$$

The matrix A therefore contains at the position a_{ij} the i-th normalized feature of the j-th subject. Let us multiply the matrix A from the right by its transpose A^T, let $B = AA^T$. At the position b_{ij} of the result matrix we now have:

$$b_{ij} = \sum_{k=1}^{n} a_{ik}a_{jk}.$$

See (7.5) and note that because of the transpose in the second factor of the product, the indices are just swapped.

Since the means of the normalized features are 0, this is exactly the sample covariance of the samples i and j, except for the factor $1/(n-1)$, see Definition 22.2. We thus obtain a matrix containing all covariances:

$$C = \begin{pmatrix} \mathrm{Cov}(S_1, S_1) & \mathrm{Cov}(S_1, S_2) & \mathrm{Cov}(S_1, S_3) \\ \mathrm{Cov}(S_2, S_1) & \mathrm{Cov}(S_2, S_2) & \mathrm{Cov}(S_2, S_3) \\ \mathrm{Cov}(S_3, S_1) & \mathrm{Cov}(S_3, S_2) & \mathrm{Cov}(S_3, S_3) \end{pmatrix} = \frac{1}{n-1} A A^{\mathrm{T}}$$

In the concrete example,

$$C = \begin{pmatrix} 1.00 & 0.89 & -0.27 \\ 0.89 & 1.00 & 0.19 \\ -0.27 & 0.19 & 1.00 \end{pmatrix}.$$

In the diagonal are the sample variances themselves, after normalization these are of course 1. Now we want to perform a change of basis in the $\mathbb{R}^3$ which makes the covariances as small as possible. Theorem 10.14 from eigenvalue theory helps us here: C is a real symmetric matrix, so it has an orthonormal basis of eigenvectors. The eigenvalues of C are 1.90, 1.19 and 0.001. I sorted them by size. The transition matrix T, which changes the standard basis into this basis of eigenvectors, has in its columns exactly the corresponding eigenvectors, see Sect. 9.3. It is:

$$T = \begin{pmatrix} -0.71 & -0.18 & 0.68 \\ -0.70 & 0.27 & -0.66 \\ 0.06 & 0.95 & 0.31 \end{pmatrix}$$

I normalized the eigenvectors to length 1, T is therefore an orthogonal matrix. We now carry out this change of basis in our feature space. The new coordinates of the samples are obtained according to Theorem 9.21, if we multiply the old coordinates from the left with T^{-1}. We have written all the samples in the rows of the matrix A, and so the matrix of new features results as $A' = T^{-1}A$. Because of the orthogonality of T, the inverse matrix is according to Theorem 10.11 just the transposed matrix, so it holds that $A' = T^{\mathrm{T}}A$. In the example:

$$A' = \begin{pmatrix} 1.5 & -1.0 & 0.56 & -0.02 & 1.1 & -1.6 & -1.4 & 1.4 & -1.1 & 0.28 & 1.9 & 0.94 & -2.5 \\ -0.05 & 0.69 & -2.2 & 2.2 & 0.04 & -1.6 & 0.34 & 0.38 & 0.04 & 0.25 & -0.25 & 0.12 & 0.12 \\ -0.015 & -0.011 & 0.002 & 0.043 & 0.001 & 0.055 & -0.016 & 0.006 & -0.003 & 0.013 & -0.035 & 0.006 & -0.051 \end{pmatrix}.$$

What are the covariances of the new features now? Just like before, we have to multiply the new feature matrix from the right with its transpose:

$$C' = \frac{1}{n-1} A' A'^{\mathrm{T}} = \frac{1}{n-1}(T^{-1}A)(T^{\mathrm{T}}A)^{\mathrm{T}} = \frac{1}{n-1} T^{-1} A A^{\mathrm{T}} T = T^{-1} C T$$

$T^{-1}CT$ is just the matrix of the linear mapping C in the new basis of eigenvectors. With respect to this basis, C has according to Theorem 9.25 the diagonal matrix D, which contains the eigenvalues in the diagonal:

$$C' = \begin{pmatrix} \lambda_1 & 0 & 0 \\ 0 & \lambda_2 & 0 \\ 0 & 0 & \lambda_3 \end{pmatrix}$$

Great magic: The covariances are all 0 now, in the diagonal are the variances of the new features, and these are just the calculated eigenvalues of C. By this coordinate transformation, we have therefore achieved our goal! In the concrete example,

$$C' = \begin{pmatrix} 1.90 & 0 & 0 \\ 0 & 1.10 & 0 \\ 0 & 0 & 0.001 \end{pmatrix}.$$

You can see that λ_3 is almost 0. The variance of the third feature is very small, and therefore the values of the third row in the transformed matrix A' are also very small. This is an irrelevant feature that we can ignore, We can restrict ourselves to the first two principal components.

So the method has recognized that in the raw data the BMI is dependent on the other two features.

If we ignore the third row in A', we can draw a two-dimensional image of the new features. Compared to the original data, this is just a rotation of the normalized first two components of the raw data, see Fig. 22.1.

How do we interpret the two components? The first principal component (the x-axis) shows to the left the rather large and heavy subjects, to the right the small and light ones. Size and weight are not independent, but overweight is not correlated with size. In the second principal component, one actually finds deviations from the normal weight. The overweight people are above, the underweight ones are below.

Fig. 22.1 The principal components in Example 1

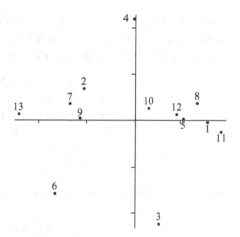

I would like to summarize the procedure for the principal component analysis:

1. Set up a matrix M of the features: The columns are assigned to the objects to be examined, the rows contain the different features of the objects.
2. Normalize the features: The mean is subtracted from each row of the matrix M, the samples. If the variances of the rows differ greatly from each other, each row is divided by its sample standard deviation. The matrix A results.
3. Calculate the matrix $C = 1/(n-1)AA^{\mathrm{T}}$. This contains all sample covariances of the examined features.
4. Calculate eigenvalues and eigenvectors of the covariance matrix. The eigenvalues are sorted by size and the eigenvectors are normalized to length 1. The orthogonal transition matrix T is the matrix that contains the normalized and sorted eigenvectors in its columns.
5. Perform the transition T with the matrix A. The matrix of new features $A' = T^{\mathrm{T}}A$ results.
6. Based on the size of the eigenvalues, it is decided how many principal components are to be examined. The principal components are the first rows of A'.
7. Now the principal components have to be interpreted.

I would like to present another example with several features, not quite as detailed. The raw data consists of six final grades of ten students (the best grade is 1):

	Alex	Bianca	Claudia	Daniel	Eva	Felix	Georg	Hanna	Ina	Kai
German	1	3	1	2	2	4	2	2	2	3
History	2	1	2	1	2	3	1	1	3	3
English	2	3	1	1	1	3	2	4	4	3
French	2	4	1	2	2	3	1	3	4	4
Math	4	2	3	1	3	1	4	1	2	4
Physics	4	2	2	1	4	1	3	2	1	3

First, the six samples are normalized again by subtracting the sample means. The sample variances do not differ greatly, so this time I will forego the division by the sample standard deviation. We obtain the feature matrix:

$$
A = \begin{pmatrix}
-1.2 & 0.8 & -1.2 & -0.2 & -0.2 & 1.8 & -0.2 & -0.2 & -0.2 & 0.8 \\
0.1 & -0.9 & 0.1 & -0.9 & 0.1 & 1.1 & -0.9 & -0.9 & 1.1 & 1.1 \\
-0.4 & 0.6 & -1.4 & -1.4 & -1.4 & 0.6 & -0.4 & 1.6 & 1.6 & 0.6 \\
-0.6 & 1.4 & -1.6 & -0.6 & -0.6 & 0.4 & -1.6 & 0.4 & 1.4 & 1.4 \\
1.5 & -0.5 & 0.5 & -1.5 & 0.5 & -1.5 & 1.5 & -1.5 & -0.5 & 1.5 \\
1.7 & -0.3 & -0.3 & -1.3 & 1.7 & -1.3 & 0.7 & -0.3 & -1.3 & 0.7
\end{pmatrix}
$$

From this we can calculate the sample covariance matrix:

$$
C = (\mathrm{Cov}(S_i, S_j)) = \frac{1}{n-1}AA^{\mathrm{T}}
$$

In the concrete example this results in:

$$C = \begin{pmatrix} 0.84 & 0.24 & 0.47 & 0.65 & -0.44 & -0.40 \\ 0.24 & 0.77 & 0.27 & 0.40 & 0.17 & -0.078 \\ 0.47 & 0.27 & 1.38 & 1.07 & -0.44 & -0.47 \\ 0.65 & 0.40 & 1.07 & 1.38 & -0.44 & -0.42 \\ -0.44 & 0.17 & -0.44 & -0.44 & 1.61 & 1.17 \\ -0.40 & -0.078 & -0.47 & -0.42 & 1.17 & 1.34 \end{pmatrix}$$

The sorted eigenvalues of this matrix are

$$3.80, \quad 1.84, \quad 0.70, \quad 0.51, \quad 0.27, \quad 0.21.$$

The eigenvalues are the variances of the new features. You can see that more than 75% of the variances fall on the first two components, in general one will work with two principal components here.

If T is the basis transition matrix, that is, the matrix of normalized eigenvectors, then we obtain the matrix of transformed features $A' = T^T A$. In the example:

$$A' = \begin{pmatrix} -2.2 & 1.5 & -1.9 & 0.23 & -2.0 & 2.5 & -2.2 & 1.6 & 2.2 & 0.28 \\ 1.0 & 0.23 & -1.2 & -2.5 & 0.20 & -0.38 & -0.06 & -0.50 & 0.75 & 2.5 \\ -0.44 & -0.75 & 0.62 & 0.23 & 0.38 & 1.4 & -0.44 & -1.6 & 0.08 & 0.62 \\ -0.41 & 0.94 & -0.94 & 0.23 & 0.75 & 0.41 & 0.17 & -0.14 & -1.4 & 0.25 \\ 0.34 & 0.12 & -0.05 & 0.38 & 0.69 & -0.47 & -1.1 & -0.01 & 0.19 & -0.01 \\ 0.25 & -0.62 & -0.17 & -0.47 & 0.47 & 0.62 & -0.11 & 0.62 & -0.16 & -0.38 \end{pmatrix}.$$

We can now plot the two principal components, that is, the first two rows of the matrix, in $\mathbb{R}^2$. In Fig. 22.2 we assign the points to the ten people again.

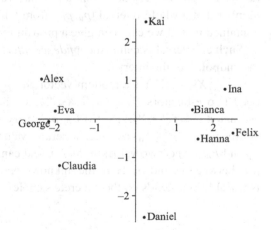

Fig. 22.2 The principal components in Example 2

For the interpretation of the principal components: The further to the right a point lies, the greater the inclination for mathematical and natural science subjects, on the left are the people with a tendency for languages. The further down a point is located, the greater the performance of the graduate, at the top are placed the rather weaker students.

One could derive this interpretation more or less well from the view of the raw data. The real performance of the principal component analysis only shows itself with data sets with very many features. So it is used, for example, in pattern recognition, in cancer diagnosis, in material or food analysis and in many other data sets with many features for analysis.

> Mathematical algorithms in themselves are neither good nor evil. But there are also very controversial applications in the use of methods. For example, facial recognition is used in China for comprehensive population surveillance. In Germany, Schufa checks the creditworthiness of applicants on the basis of secret criteria. There are well-founded suspicions that this assessment also uses the name and address. In the USA, software is used to determine from over 100 features whether a prisoner can be released on parole or not. Colored people are systematically disadvantaged here. If, based on a similar grade analysis, the decision is made about admission to a degree program, as I have carried out above, this is also highly questionable. Be aware of your responsibility as an IT specialist! The mathematician Hannah Fry has written an interesting book on this topic: Hello World: Being Human in the Age of Algorithms.

22.3   Confidence Intervals

With the help of parameter estimation, it is possible, for example, to give an estimate for the proportion p of voters of a party in an opinion poll. This estimate is the relative frequency of voters in the sample. We now know that the expected value of this estimate results in the exact right probability and that the variance of the estimate decreases with the size of the sample. But in practice we will never hit p exactly, and we cannot say how far we are off. To get out of this dilemma, we use a trick: We do not estimate a single number, but a whole interval $[p_u, p_o]$, from which we assume that the true parameter is contained in it. If we can then give a probability that p is in $[p_u, p_o]$, we are satisfied.

Such an interval is called a *confidence interval*. We find it by evaluating estimators for the endpoints of the interval:

If $(X_1, X_2, \ldots, X_n)$ is a random vector and p is the parameter to be estimated, we now need two estimators $F_u = f_u(X_1, X_2, \ldots, X_n)$ and $F_o = f_o(X_1, X_2, \ldots, X_n)$ instead of the estimator $F = f(X_1, X_2, \ldots, X_n)$ for p. We try to determine F_u and F_o so that the probability $p(F_u \leq p \leq F_o)$ assumes a certain value γ. When realizing the random variables F_u and F_o, we get two values p_u and p_o and can then say that with probability γ the value p is between p_u and p_o. Here the unknown parameter p is fixed. The confidence interval is variable, it depends on the concrete sample.

▶ **Definition 22.10** If p is a parameter to be estimated and F_u and F_o are random variables (estimators) with the property

$$p(F_u \leq p \leq F_o) = \gamma,$$

so a realization (p_u, p_o) of (F_u, F_o) is called a *confidence interval* for p, the number γ is called the *confidence level*.

If it is possible to determine such random variables, statements are possible such as "with a probability of 95%, between 9% and 11% of voters voted for Party A" or "with 99% certainty, the proportion of defective screws in the box is less than 1%".

As a rule, the confidence level γ is given and the estimators are determined in dependence on γ. The greater the desired certainty, the greater γ is chosen. Different samples then result in different intervals. Depending on the size of γ, the parameter p will be captured more or less often in the estimated interval.

Confidence intervals can be determined for different distributions and different parameters. I would like to carry out the procedure by way of example for an important case: We determine a confidence interval for the unknown probability p of the event A in a Bernoulli process.

We know the random variable X, which indicates the number of occurrences of A, is binomially distributed with parameters n and p. For large n, we can approximate the binomial distribution by the normal distribution. We assume n to be large enough anyway. Further, let the confidence level γ be given.

Now we can determine the estimators. We will see that these only depend on the $b_{n,p}$-distributed random variable X. To calculate, we carry out the following program:

1. Let $X^* = \frac{X - np}{\sqrt{np(1-p)}}$ be the standardization of X. We determine a number c with the property $p(-c \leq X^* \leq c) = \gamma$.

2. If p is the unknown parameter to be estimated, we search for functions $f_u(X)$, $f_o(X)$ with the property

$$f_u(X) \leq p \leq f_o(X) \quad \Leftrightarrow \quad -c \leq X^* \leq c.$$

3. If this is successful, then $p(f_u(X) \leq p \leq f_o(X)) = p(-c \leq X^* \leq c) = \gamma$ and thus $F_u = f_u(X)$ and $F_o = f_o(X)$ are the sought estimators.

To point 1: Because of $p(-c \leq X^* \leq c) = \Phi(c) - \Phi(-c) = 2 \cdot \Phi(c) - 1 = \gamma$, we obtain from the table of the distribution function Φ the element c with $\Phi(c) = \frac{1+\gamma}{2}$.

Point 2: It is

$$-c \leq X^* \leq c \quad \Leftrightarrow \quad -c \leq \frac{X - np}{\sqrt{np(1-p)}} \leq c$$

$$\Leftrightarrow \quad \frac{(X - np)^2}{np(1-p)} \leq c^2$$

$$\Leftrightarrow \quad (X - np)^2 \leq c^2 np(1-p)$$

$$\Leftrightarrow \quad (nc^2 + n^2)p^2 - (2nX + nc^2)p + X^2 \leq 0.$$

Let us investigate when the random variable

$$(nc^2 + n^2)p^2 - (2nX + nc^2)p + X^2 \tag{22.1}$$

assumes a value less than or equal to 0: If we carry out a sample, then X is given a fixed value $X(\omega) = k \in \{0, 1, 2, \ldots, n\}$. We now take for a moment (22.1) as a function depending on p. For each value $X(\omega)$,

$$f(p) = (nc^2 + n^2)p^2 - (2nX(\omega) + nc^2)p + X(\omega)^2$$

then represents the equation of a parabola opening to the top, because the coefficient of p^2 is greater than 0. $f(p) \leq 0$ can therefore only be valid between the two roots of $f(p)$, if there are any. We calculate the roots of $f(p)$ using the well-known formula for solving quadratic equations and reformulate the result a little. The roots $p_{1/2}$ depend on the result ω, they are:

$$p_{1/2}(\omega) = \frac{1}{c^2 + n}\left(X(\omega) + \frac{c^2}{2} \pm c\sqrt{\frac{X(\omega)(n - X(\omega))}{n} + \frac{c^2}{4}}\right).$$

Now we know: $p_1(\omega) \leq p \leq p_2(\omega)$ if and only if $f(p) \leq 0$ and thus $-c \leq X^* \leq c$. This solves point 2 of the program, p_1 and p_2 are the estimators we are looking for:

$$f_u(X) := p_1 = \frac{1}{c^2 + n}\left(X + \frac{c^2}{2} - c\sqrt{\frac{X(n - X)}{n} + \frac{c^2}{4}}\right),$$

$$f_o(X) := p_2 = \frac{1}{c^2 + n}\left(X + \frac{c^2}{2} + c\sqrt{\frac{X(n - X)}{n} + \frac{c^2}{4}}\right).$$

If we take a sample, we receive a concrete value k for X and obtain a confidence interval as a realization of $f_u(X)$ and $f_o(X)$:

$$p_{u/o} = \frac{1}{c^2 + n}\left(k + \frac{c^2}{2} \pm c\sqrt{\frac{k(n - k)}{n} + \frac{c^2}{4}}\right), \quad \text{with } c = \Phi^{-1}\left(\frac{1+\gamma}{2}\right). \tag{22.2}$$

Now we can finally deal with the example of the election forecast from Section 19.1:

We consider the poll as an experiment "drawing with replacement", so that we conduct a Bernoulli process and the number of voters of party A is binomially distributed. We set a confidence level γ of 0.95. First, we determine the number c with $\Phi(c) = (1+\gamma)/2 = 0.975$ from the table of the distribution function and get $c = 1.96$. If the survey is conducted with 1000 voters, of which 400 vote for party A, we get:

$$p_{u/o} = \frac{1}{1000 + 1.96^2}\left(400 + \frac{1.96^2}{2} \pm 1.96 \cdot \sqrt{\frac{400(1000 - 400)}{1000} + \frac{1.96^2}{4}}\right).$$

$$(22.3)$$

This results in $p_u = 0.370$ and $p_o = 0.431$. With 95% probability, the result of party A lies between 37.0% and 43.1%.

If we increase the confidence level to 99%, we get for c the value 2.57 and further $p_u = 0.361$, $p_o = 0.440$.

If we calculate with a sample of 10 000 and 5 500 voters of party A, then at the confidence level of 95% we get $p_u = 0.390$, $p_o = 0.410$. ◀

If you look at these results, you can assess the quality of the so-called "election day question", which in Germany is answered monthly in the media and from which we are to infer how the federal election would turn out, if elections were held today. As a rule, a survey of around 1000 voters is conducted for this purpose, which must also represent a representative cross-section of the population, i.e. age groups, education, regions and so on. You can see that with a confidence level of 95%, deviations of ±3% are quite possible for each party. Judge for yourself the informative value of this statistic. Larger samples are usually only conducted on election day itself, in the exit poll or by evaluating partial results of the election.

If in (22.2) the numbers k, n and $n - k$ are large, then in comparison the summands containing the number c^2 are negligibly small. We let them fall under the table and after this really hard work we get the very simple to use rule:

▶ **Calculation rule 22.11** A Bernoulli process with an unknown probability $p = p(A)$ is carried out n times, the event A occurs k times. Let k and $n - k$ be greater than 30. For the confidence level γ, the confidence interval $[p_u, p_o]$ for the unknown probability p of A is obtained by

$$p_{u/o} = \frac{k}{n} \pm \frac{c}{n}\sqrt{\frac{k(n-k)}{n}} \quad \text{with } c = \Phi^{-1}\left(\frac{1+\gamma}{2}\right).$$

Let's take a closer look at this rule: k/n is the relative frequency of the event A. This value is the estimate for p and is exactly in the middle of the confidence interval. This interval has the width

$$B = \frac{2c}{n}\sqrt{\frac{k(n-k)}{n}}.$$

The width of the interval shrinks with increasing n, the prediction will therefore always be better if the sample is larger. If you increase the confidence level γ at a fixed sample size, c will also increase, making the interval wider. The higher certainty can only be bought with larger allowed deviations.

According to rule 21.18, the approximation of the binomial distribution by the normal distribution requires $np > 5$ and $n(1 - p) > 5$. We cannot check this condition because p is unknown. But it should at least be fulfilled for the estimate of p. This is the case: If k and $n - k$ are greater than 30, then $n \cdot (k/n) = k > 30 > 5$ and $n(1 - k/n) = n - k > 30 > 5$.

What error do we make by canceling the summands with c^2? Let's calculate the example of the election poll again with rule 22.11: From

$$p_{u/o} = \frac{400}{1000} \pm \frac{1.96}{1000} \cdot \sqrt{\frac{400(1000 - 400)}{1000}}$$

we get $[p_u, p_o] = [0.3696, 0.4304]$, with the precise calculation (22.3) we get $[p_u, p_o] = [0.3701, 0.4307]$, an error we can live with.

The Sample Size

We have just seen the connection between sample size, confidence level and width of the confidence interval. Of course, one always wants high confidence levels and small confidence intervals. But surveys cost money. If the client gives the research company the task of making a statement with confidence level γ and confidence interval width B, the company must know how large the sample must be to meet these requirements. It prepares a cost estimate. If this is too high for the client, he must either go down with the confidence level or allow a larger confidence interval.

How can you determine the necessary sample size for given γ and given interval width B? From γ we can determine c. The task is to find the number n so that for all possible values of k applies:

$$\frac{2c}{n} \sqrt{\frac{k(n - k)}{n}} \leq B. \tag{22.4}$$

There is still one unknown too many in here: Let's try to eliminate k first. We are looking for the k for which $f(k) = k(n - k)$ is maximized. Then the expression on the left side of (22.4) is also maximized, that is the worst case. The maximum is obtained as the zero of $f'(k)$, it is at $k = n/2$. So (22.4) is fulfilled if it holds:

$$\frac{2c}{n} \sqrt{\frac{(n/2)(n - (n/2))}{n}} = \frac{2c}{n} \sqrt{\frac{n}{4}} \leq B$$

Because of

$$\frac{2c}{n} \sqrt{\frac{n}{4}} \leq B \quad \Leftrightarrow \quad \frac{4c^2}{n^2} \frac{n}{4} \leq B^2 \quad \Leftrightarrow \quad \frac{c^2}{n} \leq B^2 \quad \Leftrightarrow \quad n \geq \frac{c^2}{B^2}$$

we get the rule:

▶ **Calculation rule 22.12** In a Bernoulli process, let the confidence level γ be given. The confidence interval around the unknown parameter $p = p(A)$ has a maximum width of B if the sample size is:

$$n \geq \frac{c^2}{B^2}, \quad \text{with } c = \Phi^{-1}\left(\frac{1+\gamma}{2}\right).$$

Example

Let's try to a numerical integration with a Monte-Carlo method. For this, please also take another look at Buffon's needle problem section 19.1 on Monte-Carlo methods. Now I want to calculate $\Phi(1)$. It is

$$\Phi(1) = 0.5 + \int_0^1 \varphi(x)dx \quad \text{with } \varphi(x) = \frac{1}{\sqrt{2\pi}}e^{-\frac{x^2}{2}}.$$

It is therefore sufficient to determine the integral over $\varphi(x)$ from 0 to 1. In numerical integration, the function $\varphi(x)$ is enclosed in a rectangle, here in the rectangle $[0,1] \times [0, 1/2]$ with the area $R = 0.5$(Fig. 22.3). The Bernoulli experiment that we carry out is "choose a random point (x, y) in the rectangle". The event A with the unknown probability p is: "$y < \varphi(x)$", because exactly then the point lies in the integral area. If F is the integral, then $p = F/R$.

We carry out the experiment n times. If the event occurs k times, then $\tilde{p} = k/n$ is an estimate for p and thus $R \cdot (k/n)$ is an estimate for F.

How often must the experiment be carried out in order to determine with 99.9% confidence the integral with an error of less than 10^{-3}?

For $\gamma = 0.999$ we get $c = 3.29$. The area R is 0.5. If $\tilde{F}$ is the estimate for F, then it should be $|\tilde{F} - F| = |0.5 \cdot \tilde{p} - 0.5 \cdot p| < 10^{-3}$, that is $|\tilde{p} - p| < 2 \cdot 10^{-3}$. The deviation from p is allowed in both directions, so we can specify the width of the confidence interval with $B = 2 \cdot 2 \cdot 10^{-3}$. The required sample size is then $c^2/B^2 = 676\,506$. The estimate for F with this number of trials gave me 0.341235, that is $\Phi(1) = 0.841235$. From the table in the appendix we can read the result 0.8413. ◀

You can see that this method requires very many trials to achieve reasonably good results, it shows a very poor convergence behavior. With this you can hardly generate

Fig. 22.3 Monte Carlo Integration

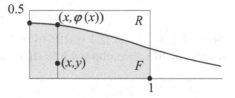

the table of the normal distribution from the appendix. For simple integrals, the method is practically not usable. But it shows its strengths in multi-dimensional integrals, which in comparison to the one-dimensional case become much more complex for the usual numerical methods, while the effort of the Monte Carlo integration remains almost unchanged.

22.4   Hypothesis Testing

In hypothesis testing, it is also about examining parameters of random variables. However, here certain assumptions are made about the parameters. For example, if you want to check whether a player's coin is fair, you can set up the hypothesis $p(\text{head}) = 1/2$ or $p(\text{head}) \neq 1/2$. If you want to compare the healing probability p of a medication with that of another, which is p_0, you will assume $p \geq p_0$ or $p \leq p_0$. For many data sets that are to be analyzed, we have assumptions about their distribution or about expected values and variances, then we test whether these really apply. The advantage of hypothesis testing over parameter estimation is that you can use the hypothesis in the calculation, so you have more in hand.

Parameter Testing

Let's test the coin of a player. We observe the event $A = $ "coin shows head". $p_0 = p(A)$ is unknown, but we take the hypothesis $p_0 = 1/2$. Now we carry out the experiment 100 times. Let X_i be the random variable that describes the outcome of the i-th trial, and $X = X_1 + X_2 + \cdots + X_{100}$. We know that $R = X/100$ is an estimator for the parameter p_0. How far may the estimate p, obtained by a sampling, deviate from $p_0 = 1/2$, without shaking our confidence in the hypothesis?

First let's try an intuitive approach: We want to allow a deviation δ of ± 0.1 of p_0 from $p_0 = 1/2$, otherwise we reject the hypothesis. Of course, a sample can also deviate more even though the hypothesis is correct. Then we commit an error by rejecting the hypothesis. How often does that happen?

R is a $N(p_0, p_0(1 - p_0)/n) = N(0.5, 0.0025)$-distributed estimator for p_0 according to Theorem 22.8. We are looking for the probability $p(0.4 < R < 0.6)$:

$$p(0.4 \leq R \leq 0.6) = \Phi\left(\frac{0.6 - 0.5}{\sqrt{0.0025}}\right) - \Phi\left(\frac{0.4 - 0.5}{\sqrt{0.0025}}\right)$$

$$= \Phi(2) - \Phi(-2) = 2 \cdot \Phi(2) - 1 = 0.9546.$$

So $p(R \notin [0.4, 0.6]) = 1 - 0.9546 = 0.0454$. The probability for a larger deviation than $\delta = 0.1$ is therefore about 4.5 %. If p_0 is correct and such a deviation occurs, we reject

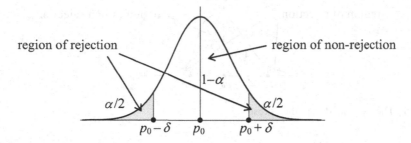

Fig. 22.4 The region of rejection

the hypothesis and commit an error. The probability of error α is 4.5 %. The area to the left and right of $p_0 \pm \delta$ under the bell curve is $\alpha/2$, see Fig. 22.4.

The problem is usually stated the other way around: Given is a *probability of error* α and an estimator T for the parameter p. In test theory, this random variable T is often called the *test statistic*. For the parameter p, we make a hypothesis. This hypothesis is called the *null hypothesis* and is denoted by H_0. The alternative to this hypothesis is called H_1. The probability of error α is usually small, for example 1% or 5%. The number $1 - \alpha$ is called the *significance level*. Depending on α, a region of rejection for H_0 is determined.

Now a sample is taken and thus an estimate value p for p_0 is determined. If p is in the region of rejection, the hypothesis H_0 is rejected. Then the alternative hypothesis H_1 is automatically accepted.

Errors can be made here:

- The *type I error* is the error that the hypothesis is rejected although it is correct.
- The *type II error* is the error that the hypothesis is not rejected although it is false.

The type I error has just the probability α.

The type II error can be very large depending on α, as we will see in examples. If the result of the sample is in the region of non-rejection, the test result should be formulated as follows: "The result does not contradict the hypothesis". The hypothesis can by no means be accepted! If one wants to make further statements about the parameter in the non-rejection case, a confidence interval for the parameter can be determined.

First, let's look at the *two-sided hypothesis test*: The null hypothesis H_0 here is $p = p_0$, the alternative H_1 is: $p \neq p_0$. The test statistic T is an estimator for p and α is the probability of error. How large does the deviation of an estimate for p from the parameter p_0 have to be in order to reject the hypothesis with this probability of error?

With the help of the known probability distribution of T, numbers δ_1 and δ_2 are computed so that

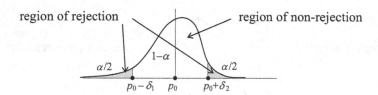

Fig. 22.5 The two-sided hypothesis test

$$p(T < p_0 - \delta_1) = \frac{\alpha}{2} \quad \text{und} \quad p(T > p_0 + \delta_2) = \frac{\alpha}{2}.$$

The distribution does not have to be symmetrical in general, but there are two equally large regions of rejection to the right and left of p_0, see Fig. 22.5.

Let's carry out this test in the specific case of a Bernoulli process. The probability p of the event A is unknown, we set up the hypothesis H_0: "$p = p_0$". The random variable R, which describes the relative frequency of the event A, is the test statistic for the parameter p. It is $N(p_0, p_0(1 - p_0)/n)$-distributed under the assumption of H_0. R is symmetrical, so here the two rejection regions are equally far from p_0, as shown in Fig. 22.4.

So we are looking for the number δ for which is $p(p_0 - \delta \leq R \leq p_0 + \delta) = 1 - \alpha$.

δ can be calculated using Theorem 21.19:

$$p(p_0 - \delta \leq R \leq p_0 + \delta)$$
$$= \Phi\left(\frac{\cancel{p_0} + \delta - \cancel{p_0}}{\sqrt{p_0(1 - p_0)/n}}\right) - \Phi\left(\frac{\cancel{p_0} - \delta - \cancel{p_0}}{\sqrt{p_0(1 - p_0)/n}}\right) \tag{22.5}$$
$$= 2\Phi\left(\frac{\delta}{\sqrt{p_0(1 - p_0)/n}}\right) - 1 = 1 - \alpha,$$

that is

$$1 - \frac{\alpha}{2} = \Phi\left(\frac{\delta}{\sqrt{p_0(1 - p_0)/n}}\right) \quad \Rightarrow \quad \delta = \Phi^{-1}\left(1 - \frac{\alpha}{2}\right)\sqrt{p_0(1 - p_0)/n}.$$

We can write the result in a calculation rule:

▶ **Calculation rule 22.13** In a Bernoulli process, the null hypothesis $p(A) = p_0$ is tested. The probability of error α is given. In n trials, the event A occurs k times. Let

$$\delta = c \cdot \sqrt{p_0(1 - p_0)/n}, \quad c := \Phi^{-1}\left(1 - \frac{\alpha}{2}\right).$$

Then the hypothesis is rejected with a probability of error of α if holds:

$$k/n \notin [p_0 - \delta, p_0 + \delta] \quad \text{resp.} \quad k \notin [np_0 - n\delta, np_0 + n\delta]$$

Let's look at the coin toss example again.

With $n = 100$ tosses and a probability of error $\alpha = 10\%$ we get:

$$\delta = \Phi^{-1}(0.95)\sqrt{0.25/100} = 1.65 \cdot 0.05 \approx 0.083.$$

With a relative frequency of the event $A =$ "head" of less than 0.417 or more than 0.583, the hypothesis is rejected with a probability of error of 10%. In 100 trials this means: for $k \notin [42, 58]$ H_0 is rejected, the region of non-rejection is $[42, 58]$.

If we increase the number of trials to 200 while keeping the same probability of error, we get $\delta \approx 0.058$, the region of rejection becomes larger.

In 100 trials with a probability of error of 1%, the calculation results in $\delta \approx 0.128$. With a smaller probability of error, the region of rejection shrinks, here the hypothesis is only rejected for $k \notin [37, 63]$. ◀

It is very important to note that if the hypothesis is not rejected, it cannot be accepted! Let's take the coin toss example again: In 100 tosses, the coin should show head 57 times. Neither with a probability of error of 1% nor with a probability of error of 10% is the result in contradiction to the hypothesis $p_0 = 1/2$. However, the hypothesis cannot be accepted either, because it would be quite possible that the parameter, for example, is 0.57, 0.58 or 0.55. We can determine a confidence interval: With rule 22.11 we get that with a confidence level of 90%, the true parameter lies between 0.49 and 0.65. Unfortunately, this statement does not help us much in our question.

You can see it again here: It is easier to be destructive than constructive. Hypotheses can be rejected more easily than accepted. If you want to substantiate a hypothesis, you should therefore formulate the test in such a way that the exact opposite is assumed as the null hypothesis.

This is usually not possible with a two-sided hypothesis test, but with a *one-sided hypothesis test* the situation looks different: The hypothesis H_0 here is $p \leq p_0$ or $p \geq p_0$, we can choose which of the two statements we want to refute, and again determine the Type I error. A complication compared to the two-sided hypothesis test is that the exact distribution of the test statistic T is not known, it depends on the true value p of the parameter, but we only know $p \leq p_0$ or $p \geq p_0$.

Let's calculate an example for the case $H_0 = "p \leq p_0"$. The alternative hypothesis H_1 is then $"p > p_0"$.

At least in continuous distributions we can assume the value 0 for the probability $p = p_0$, so that there is no difference between $p > p_0$ and $p \geq p_0$.

This is exactly the test we can use to confirm Murphy's Law (see Sect. 19.1 "Testing a Hypothesis"): We conduct a Bernoulli process and observe the event $B =$ "Toast falls on the buttered side". $p = p(B)$ is unknown, but I would like to confirm the hypothesis "$p > 1/2$". So we take the opposite as the null hypothesis: $H_0 = "p \leq 1/2"$. If we can reject H_0, Murphy is right.

Fig. 22.6 One-sided
hypothesis test 1

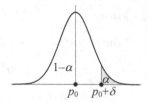

If X_i is the result of the i-th trial, then $R = 1/n \cdot (X_1 + X_2 + \ldots + X_n)$ is again the test statistic for the probability p after n trials. R is normally distributed with expected value p, but in contrast to the two-sided case, this value p is unknown and thus the correct normal distribution is unknown. How should we determine the region of rejection?

We will try to keep it simple at first with $p = p_0 = 1/2$ as in the two-sided case: Then R is distributed as before $N(p_0, p_0(1 - p_0)/n)$ and for the given probability of error α we can determine the threshold value δ with the property

$$p(R > p_0 + \delta | p_0) = \alpha \qquad (22.6)$$

(Figure 22.6). In the one-sided test we only need a one-sided region of rejection, because we have nothing against small values of p, they do not contradict the hypothesis. I write $p(A|p_0)$ here to indicate that the probability of the event A has been calculated under the assumption $p = p_0$.

But what if the true parameter p_1 is less than p_0? Then R is a $N(p_1, p_1(1 - p_1)/n)$-distributed random variable. But since $p_1 < p_0$, R has a smaller expected value compared to the first case: $R|p_1$ is shifted to the left compared to $R|p_0$, see Fig. 22.7.

You can calculate that

$$p_1 \le p_0 \quad \Rightarrow \quad \alpha' = p(R > p_0 + \delta | p_1) \le p(R > p_0 + \delta | p_0) = \alpha. \qquad (22.7)$$

But this means that the δ calculated in (22.6) covers the worst possible case, for all $p \le p_0$ is $p(R > p_0 + \delta | p) \le \alpha$ and thus in any case $p(R > p_0 + \delta) \le \alpha$.

Now if the experiment is carried out n times and B occurs k times, we can reject the hypothesis with a probability of error of at most α if for the estimated parameter holds $k/n > p_0 + \delta$.

The calculation of the value δ from (22.6) is exactly the same as in the two-sided hypothesis test, with the only difference that the region of rejection α now focuses on one side:

Fig. 22.7 One-sided
hypothesis test 2

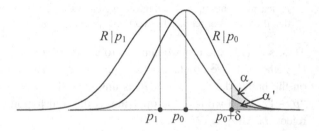

$$p(R \leq p_0 + \delta) = \Phi\left(\frac{p_0 + \delta - p_0}{\sqrt{p_0(1-p_0)/n}}\right) = \Phi\left(\frac{\delta}{\sqrt{p_0(1-p_0)/n}}\right) = 1 - \alpha$$

and thus

$$\delta = \Phi^{-1}(1-\alpha)\sqrt{p_0(1-p_0)/n}.$$

Before we carry out the example with concrete numerical values, we can again formulate a calculation rule for this test in the Bernoulli process:

▶   **Calculation rule 22.14** In a Bernoulli process, the null hypothesis $p(A) \leq p_0$ is tested. The probability of error α is given. In n trials, the event occurs k times. Let

$$\delta = c \cdot \sqrt{p_0(1-p_0)/n}, \quad c := \Phi^{-1}(1-\alpha).$$

Then the hypothesis is rejected with a probability of error of α if

$$k/n > p_0 + \delta \quad \text{resp.} \quad k > np_0 + n\delta$$

Similarly, the hypothesis $p(A) \geq p_0$ is rejected with a probability of error of α if

$$k/n < p_0 - \delta \quad \text{resp.} \quad k < np_0 - n\delta.$$

(22.7) states that the function $\varphi(p) = p(R > c|p)$ is monotonically increasing. That's why our calculation worked. $\varphi(p) = p(T > c|p)$ has this property for many test statistics T, in these cases the threshold value δ can be determined similarly in a one-sided test as in the example of the Bernoulli process.

Examples

1. We want to confirm the hypothesis that Murphy's Law applies, that is, $p(B) = p(\text{butter on the floor}) > 1/2$ is true. To do this, we have to assume the opposite as the null hypothesis: $H_0 = p(B) \leq 1/2$. Now let's sacrifice some sticks of toast and a big piece of butter and let 100 pieces of buttered toast fall on the floor. In this test, 61 times the butter side comes to lie on the floor.
First, the probability of error $\alpha = 5\%$ is given. Then $c = 1.64$ and $\delta = 1.64 \cdot \sqrt{0.25/100} = 0.082$. For $k > 50 + 8.2 = 58.2$, the hypothesis can be rejected with this probability of error. This is the case here. With a probability of error of no more than 5%, Murphy is right.
Can we increase our significance level? Let's take $\alpha = 1\%$: Now it's $c = 2.33$ and $\delta = 0.1165$, the region of rejection is $[62, 100]$. The result does not contradict the hypothesis.
Let's look at the type II error in this case: The hypothesis is not rejected, although it is false. With the normally distributed test statistic R and $p = p(B) > 1/2$, we

have to calculate the probability that R has a value in the region of non-rejection, that is

$$p(R < 1/2 + \delta|p),$$

where p can take all values greater than $1/2$. A similar problem as in the derivation of the rule 22.14: Also here, the limiting case $p = 1/2$ represents the worst case, as you can see for yourself in a small drawing. So let's calculate $p(R < 1/2 + \delta|1/2)$. Since R is continuous, we have

$$p(R < 1/2 + \delta|1/2) = 1 - p(R > 1/2 + \delta|1/2) = 1 - \alpha,$$

because $R > 1/2 + \delta$ is just the region of rejection for α! This gives us the sad result that the type II error can be arbitrarily close to $1 - \alpha = 99\%$.

Calculate the confidence interval for α yourself with the confidence level 99% (Rule 22.11): With a probability greater than 99%, p_0 is between 0.48 and 0.74.

2. The manufacturer of a biometric authentication tool specifies the following error rates in its sales prospectus: The FRR, the False Rejection Rate, is less than 0.01, the FAR, the False Acceptance Rate, is less than 0.0001. With a probability of less than 0.01, the authentication of a person is therefore wrongly rejected, with a probability of less than 0.0001 a false authentication is performed. In a test center, the product is tested with 10 000 authentication attempts. In doing so, 81 authentication attempts are wrongly rejected, twice a false authentication takes place. How are the advertising claims of the manufacturer to be assessed?

Let's look at the rejection of the authentications. The expected value is 100, with 81 actual rejections, it makes sense to check the null hypothesis H_0: "$p \geq 0.01$", in the hope of being able to reject it. It is

$$\delta = c \cdot \sqrt{0.01 \cdot 0.99/10\,000} = c \cdot 9.95 \cdot 10^{-4}, \quad c = \Phi^{-1}(1 - \alpha).$$

For $\alpha = 1\%$ $(c = 2.33)$ we get $\delta = 2.32 \cdot 10^{-3}$, so the hypothesis can be rejected for $k < 100 - 23.2 = 76.8$, the result does not contradict the hypothesis H_0.

For $\alpha = 5\%$ $(c = 1.64)$ we get $\delta = 1.63 \cdot 10^{-3}$, the region of rejection is $k < 100 - 16.3 = 83.7$. With a probability of error of 5%, H_0 can be rejected and the manufacturer's statement confirmed.

Now for the false authentications: We assume $p > 10^{-4}$, and therefore check the null hypothesis $p \leq 10^{-4}$. Now

$$\delta = c \cdot \sqrt{0.0001 \cdot 0.9999/10\,000} = c \cdot 9.9995 \cdot 10^{-5}, \quad c = \Phi^{-1}(1 - \alpha).$$

For $\alpha = 5\%$ we get $\delta = 1.64 \cdot 10^{-4}$, the region of rejection is $k > 1 + 1.64 = 2.64$, no contradiction to the hypothesis. Only with a probability of error of about 16% would result a region of rejection $k < 2$, so that the rejection of the hypothesis would be justified.

If our test had not revealed a single false authentication, we could still not be able to confirm the manufacturer's statement. ◀

In the last example you can see that it is very difficult to make statements with high significance when there are only a few cases. The probability of error is always very high here. This problem occurs in many studies, especially when the sample size is limited.

For the Japanese government, the necessary sample size is an argument for why several hundred whales have to be caught for "scientific purposes" every year. Small case numbers also lead to groups with different interests creating statistics that appear to contradict each other, but which unfortunately all have the same low predictive value. Think of such controversial issues as the increased incidence of leukemia in the vicinity of nuclear power plants or in the vicinity of strong radio transmitters.

Pearson's Chi-Squared Test

In many cases, the statistician has to examine data sets in which he already has a suspicion about the distribution present. Very often he will expect a normal distribution, for example, when many individual characteristics accumulate additively. In such cases, Pearson's chi-squared test is carried out: The hypothesis does not contain a parameter, but the assumption that a certain distribution is present. I would like to present such a test as an example of this test form.

Let a sample be given as a realization of the random variable X. We have the suspicion that X has a certain distribution and want to test this hypothesis. First, we divide the image of X into various disjoint sub-ranges $I_1, I_2, \ldots, I_m$. For small images, each value can be its own range, for larger images or for continuous random variables, values are combined or a series of intervals is formed. If A_k is the event that the value of X lies in I_k, then we can calculate the probabilities $p_k = p(A_k)$ from the hypothesis about the distribution. The hypothesis H_0, which we want to check, is then formulated as follows:

$$H_0 : p(A_1) = p_1, p(A_2) = p_2, \ldots, p(A_m) = p_m$$

Of course it is $p_1 + p_2 + \ldots + p_m = 1$.

Let X_k be the random variable that counts the number of times A_k occurs in n trials. X_k is b_{n,p_k}-distributed and therefore has expected value np_k and variance $np_k(1 - np_k)$. If we set

$$Y_k^2 = \frac{(X_k - np_k)^2}{np_k}$$

, then Y_k^2 represents the squared deviation of X_k from the expected value divided by the expected value. The sum over all Y_k^2,

$$\chi^2 := \sum_{k=1}^{m} Y_k^2 = \sum_{k=1}^{m} \frac{(X_k - np_k)^2}{np_k}$$

is the test statistic for our null hypothesis. It measures the sum of all squared deviations from the respective expected value, each in relation to the expected value. The distribution of this test statistic is known, but difficult to calculate, it is χ^2-distributed with $m - 1$ degrees of freedom. Already if for all k it is $n \cdot p_k > 5$, this is a good approximation.

> Beware of the trap: If you look at the definition of the χ^2-distribution at the end of Sect. 21.2 you might be tempted to simply take $\sum_{k=1}^{m} X_k^{*2}$ as a test statistic, the sum of the squared standardized X_k. Isn't this χ^2-distributed? No, because the X_k are not independent of each other! If, for example, X_1 occurs frequently, then X_2 to X_k can only occur less often. This is the reason for the number of degrees of freedom of the test function: $m - 1$ and not m.

Please read the properties of the chi-square distribution again at the end of Sect. 21.2. The expected value of χ^2 is $m - 1$, so if the null hypothesis is true, the sample values will be distributed around $m - 1$. What does the region of rejection look like? Is it two-sided or one-sided? The Y_k all have the expected value 0, and even if it is not very likely that all Y_k and thus all Y_k^2 have a small value at the same time, this does not contradict the hypothesis. Deviations from the expected value of the Y_k are, on the other hand, squared and summed up in the test statistic. Large deviations result in large values of χ^2. The region of rejection is therefore one-sided. We specify a probability of error α and determine the δ with

$$p(\chi^2 > \delta) = \alpha.$$

The hypothesis can be rejected with the probability of error of α if, for a sample of size n, holds $\chi^2(\omega) > \delta$.

Let's take a dice as an example. We want to test if it is fair. The ranges $I_1, I_2, \ldots, I_6$ should be the possible dice results 1 to 6 here. The hypothesis says that the dice results are evenly distributed, that is

$$H_0 : p(A_1) = 1/6, \quad p(A_2) = 1/6, \quad \ldots, \quad p(A_6) = 1/6.$$

I rolled the dice 100 times and noted the results. They were:

k	1	2	3	4	5	6
x_k	17	21	17	18	18	9

Inserted into the test function, we get

$$\sum_{k=1}^{6} \left(\frac{x_k - 100/6}{\sqrt{100/6}} \right)^2 \approx 4.88.$$

Because of $n \cdot p_k \approx 16.7 > 5$ the test function is approximately χ^2-distributed with 5 degrees of freedom.

The χ^2 distribution functions are tabulated. At χ_5^2 (see Fig. 22.8) we find from the table

$$p(\chi_5^2 \leq 9.24) = 0.9; \; p(\chi_5^2 \leq 15.09) = 0.99.$$

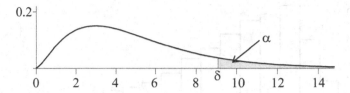

Fig. 22.8 The χ^2 adaptation test

For the probability of error of 10%, the region of rejection is therefore $]9.24, \infty[$, for $\alpha = 1\%$ the region of rejection is $]15.09, \infty[$. For the test value, we therefore get no contradiction to the hypothesis, both with an probability of error of 10% and with an probability of error of 1%.

More interesting than the dice is a random number generator. For example, this can generate random integers between 0 and $2^{32} - 1$ and these should also be evenly distributed. As a division into disjoint areas, I first chose the 1000 remainders modulo 1000 here. If A_k is the event that the value of X lies in I_k, then $p_k = 1/1000$. To achieve $n \cdot p_k > 5$, I carried out $n = 10\,000$ trials. The test function is χ^2_{999}-distributed, so approximately $N(999, 1998)$ distributed. We have to find the δ with

$$p(\chi^2_{999} \leq \delta) = \Phi\left(\frac{\delta - 999}{\sqrt{1998}}\right) = 1 - \alpha.$$

For $\alpha = 1\%$ we get $\delta = 1103$, for $\alpha = 5\%$ we get $\delta = 1072$. The test yielded the value 1037.6, so no contradiction to the hypothesis at the 95% and at the 99% significance level.

But this is only true for the special choice of I_k as remainders modulo 1000. Are small or large numbers perhaps preferred? In a second attempt, I chose 1000 equally large intervals as I_k. Here the test yielded 1017.4, also no contradiction to the hypothesis.

But beware: The fact that the hypothesis "uniform distribution" cannot be rejected does not mean that the random number generator is really good: It could still have accumulations with different decompositions. Furthermore, not only the number of values in individual number ranges must be evenly distributed, but also the order of appearance must be random. For the first edition of this book, I carried out these tests for the random number generator of a common C++ compiler, the hypothesis of uniform distribution could not be rejected. But if I only evaluated every second random number, the first test yielded 10 938. The hypothesis of uniform distribution could thus be rejected with a probability of error of 0.1%. On closer inspection, it turned out that the random number generator had the strange habit of always producing even and odd numbers alternately, a pretty disaster. Meanwhile, this error has been eliminated.

If you have to prove results experimentally in a research project and are perhaps under pressure from the client who wants to see results, you can easily get into the situation of wanting to sweep results under the table that do not fit into the desired picture. What does it matter if you repeat the test under the same conditions?

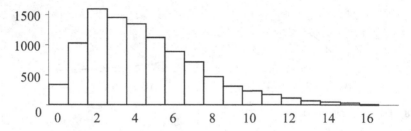

Fig. 22.9 Test results

Don't do that if you should find yourself in that situation: The results of such a test are worthless! See the last experiment for this:

My computer is patient, I have not only simulated the dice test I described earlier once, but 10 000 times and written down the test results each time. In Fig. 22.9 you can see the frequency of the individual test results as a histogram.

You can clearly see the shape of the χ_5^2 distribution. But you can also see that of course at some point very unlikely results will occur. 108 times a test result was greater than 15. If I only test often enough, almost every desired result will occur at some point and thus also every hypothesis will be rejected. Therefore, when testing, the sample size must be determined before the test is carried out and then the test is carried out and evaluated exactly once.

22.5   Comprehension Questions and Exercises

Comprehension questions

1. What is meant by a parameter estimation?
2. If a sample is won by a random experiment, is there a relationship between the expected value and the variance of the random variable and the sample mean and sample variance? What is this relationship?
3. You want to calculate a confidence interval. If the confidence level is increased, will the confidence interval be smaller or larger? If the sample size is increased, will the confidence interval be smaller or larger?
4. What is the type I and type II error in a hypothesis test?
5. How should a hypothesis be formulated if one wants to confirm a conjecture?
6. If the probability of error is reduced in a hypothesis test, will the region of rejection be smaller or larger?
7. Can one carry out a hypothesis test several times if one does not like the result of a test that has been carried out?

1. Rolling 10 times with 2 dice results in the pairs $(2, 4),(5, 6),(2, 2),(6, 2),(2, 6),$
   $(5, 1),(5, 6),(6, 4),(4, 4),(3, 1)$.
   Calculate the sample mean and sample variance for the sum of eyes of these samples. Give an estimate for the probability of a double and for the expected value of
   the sum of the eyes.
   Why can't you calculate a confidence interval for p(double) using our standard formula?
   Show that with a fair die, the probability that 0, 1, 2 or 3 doubles will occur in 10
   throws is more than 90%.

2. Try to follow the two numerical examples in the principal component analysis
   (Sec. 22.2) using a computer algebra system.

3. Determine the defect rate of a bulk article: A sample of 1000 elements yields 94
   defective parts. Determine a confidence interval for the unknown probability p of
   the defects at the confidence level 95%.

4. The probability of a girl's birth:
   In 2017, 784 901 children were born in Germany, 382 374 of them girls. Determine a confidence interval for the probability of a girl's birth at the confidence
   level 99%.

5. In a survey, the percentage of the population that does not believe in statistics is to
   be determined. For the specifications
   a) Confidence level 98%, confidence interval width 4%,
   b) Confidence level 96%, confidence interval width 4%
   the necessary sample size is to be calculated.
   The survey is conducted with 3000 people. Of these, 1386 do not believe in statistics. Determine the corresponding confidence intervals at the confidence levels
   96% and 98%.

6. In roulette, the probability of a red number is $p(\text{red}) = 18/37$. You suspect that this
   is not true in a special drum. The last 5000 results are posted in lists. This resulted
   in 2355 red numbers. Make a suitable hypothesis to check your suspicion. Use the
   probability of error 2% resp. 5% and formulate the calculated result precisely.

7. 4964 draws of the German lottery "6 out of 49" since 1955 resulted in the following draw frequencies for the numbers 1 to 49 in sequence:

606	617	628	615	602	633	607	583	630	591	608	592	551
600	576	612	608	606	602	577	585	616	600	602	642	649
635	569	581	597	640	616	628	603	609	608	600	646	610
605	629	615	651	604	564	585	607	606	638			

   Does this result, with a probability of error of 1%, contradict the hypothesis of the
   uniform distribution of the 49 numbers? Write a test program.

Appendix

23

23.1 The Greek Alphabet

Traditionally, mathematicians use many Greek symbols in their formulas. Here is the Greek alphabet with the names for the letters.

A	α	Alpha	N	ν	Ny
B	β	Beta	Ξ	ξ	Xi
Γ	γ	Gamma	O	o	Omikron
Δ	δ	Delta	Π	π	Pi
E	ε	Epsilon	P	ρ	Rho
Z	ζ	Zeta	Σ	σ	Sigma
H	η	Eta	T	τ	Tau
Θ	θ	Theta	Υ	υ	Ypsilon
I	ι	Iota	Φ	φ	Phi
K	κ	Kappa	X	χ	Chi
Λ	λ	Lambda	Ψ	ψ	Psi
M	μ	My	Ω	ω	Omega

© The Author(s), under exclusive license to Springer Fachmedien Wiesbaden GmbH, part of Springer Nature 2023
P. Hartmann, *Mathematics for Computer Scientists*,
https://doi.org/10.1007/978-3-658-40423-9_23

23.2 The Standard Normal Distribution

$$\Phi(x) = p(N(0,1) \le x) = \frac{1}{\sqrt{2\pi}} \int_{-\infty}^{x} e^{-\frac{t^2}{2}} dt, \quad \Phi(-x) = 1 - \Phi(x), \quad \Phi^{-1}(y) = -\Phi^{-1}(1-y)$$

x	$\Phi(x)$	x	$\Phi(x)$	x	$\Phi(x)$	x	$\Phi(x)$	x	$\Phi(x)$	x	$\Phi(x)$	x	$\Phi(x)$	x	$\Phi(x)$
0.00	0.5000	0.46	0.6772	0.92	0.8212	1.38	0.9162	1.84	0.9671	2.30	0.9893	2.76	0.99711	3.22	0.99936
0.01	0.5040	0.47	0.6808	0.93	0.8238	1.39	0.9177	1.85	0.9678	2.31	0.9896	2.77	0.99720	3.23	0.99938
0.02	0.5080	0.48	0.6844	0.94	0.8264	1.40	0.9192	1.86	0.9686	2.32	0.9898	2.78	0.99728	3.24	0.99940
0.03	0.5120	0.49	0.6879	0.95	0.8289	1.41	0.9207	1.87	0.9693	2.33	0.9901	2.79	0.99736	3.25	0.99942
0.04	0.5160	0.50	0.6915	0.96	0.8315	1.42	0.9222	1.88	0.9700	2.34	0.9904	2.80	0.99744	3.26	0.99944
0.05	0.5199	0.51	0.6950	0.97	0.8340	1.43	0.9236	1.89	0.9706	2.35	0.9906	2.81	0.99752	3.27	0.99946
0.06	0.5239	0.52	0.6985	0.98	0.8365	1.44	0.9251	1.90	0.9713	2.36	0.9909	2.82	0.99760	3.28	0.99948
0.07	0.5279	0.53	0.7019	0.99	0.8389	1.45	0.9265	1.91	0.9719	2.37	0.9911	2.83	0.99767	3.29	0.99950
0.08	0.5319	0.54	0.7054	1.00	0.8413	1.46	0.9279	1.92	0.9726	2.38	0.9913	2.84	0.99774	3.30	0.99952
0.09	0.5359	0.55	0.7088	1.01	0.8438	1.47	0.9292	1.93	0.9732	2.39	0.9916	2.85	0.99781	3.31	0.99953
0.10	0.5398	0.56	0.7123	1.02	0.8461	1.48	0.9306	1.94	0.9738	2.40	0.9918	2.86	0.99788	3.32	0.99955
0.11	0.5438	0.57	0.7157	1.03	0.8485	1.49	0.9319	1.95	0.9744	2.41	0.9920	2.87	0.99795	3.33	0.99957
0.12	0.5478	0.58	0.7190	1.04	0.8508	1.50	0.9332	1.96	0.9750	2.42	0.9922	2.88	0.99801	3.34	0.99958
0.13	0.5517	0.59	0.7224	1.05	0.8531	1.51	0.9345	1.97	0.9756	2.43	0.9925	2.89	0.99807	3.35	0.99960
0.14	0.5557	0.60	0.7258	1.06	0.8554	1.52	0.9357	1.98	0.9762	2.44	0.9927	2.90	0.99813	3.36	0.99961
0.15	0.5596	0.61	0.7291	1.07	0.8577	1.53	0.9370	1.99	0.9767	2.45	0.9929	2.91	0.99819	3.37	0.99962
0.16	0.5636	0.62	0.7324	1.08	0.8599	1.54	0.9382	2.00	0.9773	2.46	0.9931	2.92	0.99825	3.38	0.99964
0.17	0.5675	0.63	0.7357	1.09	0.8621	1.55	0.9394	2.01	0.9778	2.47	0.9932	2.93	0.99831	3.39	0.99965
0.18	0.5714	0.64	0.7389	1.10	0.8643	1.56	0.9406	2.02	0.9783	2.48	0.9934	2.94	0.99836	3.40	0.99966
0.19	0.5754	0.65	0.7422	1.11	0.8665	1.57	0.9418	2.03	0.9788	2.49	0.9936	2.95	0.99841	3.41	0.99968
0.20	0.5793	0.66	0.7454	1.12	0.8686	1.58	0.9430	2.04	0.9793	2.50	0.9938	2.96	0.99846	3.42	0.99969
0.21	0.5832	0.67	0.7486	1.13	0.8708	1.59	0.9441	2.05	0.9798	2.51	0.9940	2.97	0.99851	3.43	0.99970
0.22	0.5871	0.68	0.7518	1.14	0.8729	1.60	0.9452	2.06	0.9803	2.52	0.9941	2.98	0.99856	3.44	0.99971
0.23	0.5910	0.69	0.7549	1.15	0.8749	1.61	0.9463	2.07	0.9808	2.53	0.9943	2.99	0.99861	3.45	0.99972
0.24	0.5948	0.70	0.7580	1.16	0.8770	1.62	0.9474	2.08	0.9812	2.54	0.9945	3.00	0.99865	3.46	0.99973
0.25	0.5987	0.71	0.7612	1.17	0.8790	1.63	0.9485	2.09	0.9817	2.55	0.9946	3.01	0.99869	3.47	0.99974
0.26	0.6026	0.72	0.7642	1.18	0.8810	1.64	0.9495	2.10	0.9821	2.56	0.9948	3.02	0.99874	3.48	0.99975
0.27	0.6064	0.73	0.7673	1.19	0.8830	1.65	0.9505	2.11	0.9826	2.57	0.9949	3.03	0.99878	3.49	0.99976
0.28	0.6103	0.74	0.7704	1.20	0.8849	1.66	0.9515	2.12	0.9830	2.58	0.9951	3.04	0.99882	3.50	0.99977
0.29	0.6141	0.75	0.7734	1.21	0.8869	1.67	0.9525	2.13	0.9834	2.59	0.9952	3.05	0.99886	3.51	0.99978
0.30	0.6179	0.76	0.7764	1.22	0.8888	1.68	0.9535	2.14	0.9838	2.60	0.9953	3.06	0.99889	3.52	0.99978

x	$\Phi(x)$	x	$\Phi(x)$	x	$\Phi(x)$	x	$\Phi(x)$	x	$\Phi(x)$	x	$\Phi(x)$	x	$\Phi(x)$	x	$\Phi(x)$
0.31	0.6217	0.77	0.7794	1.23	0.8907	1.69	0.9545	2.15	0.9842	2.61	0.9955	3.07	0.99893	3.53	0.99979
0.32	0.6255	0.78	0.7823	1.24	0.8925	1.70	0.9554	2.16	0.9846	2.62	0.9956	3.08	0.99896	3.54	0.99980
0.33	0.6293	0.79	0.7852	1.25	0.8944	1.71	0.9564	2.17	0.9850	2.63	0.9957	3.09	0.99900	3.55	0.99981
0.34	0.6331	0.80	0.7881	1.26	0.8962	1.72	0.9573	2.18	0.9854	2.64	0.9959	3.10	0.99903	3.56	0.99981
0.35	0.6368	0.81	0.7910	1.27	0.8980	1.73	0.9582	2.19	0.9857	2.65	0.9960	3.11	0.99906	3.57	0.99982
0.36	0.6406	0.82	0.7939	1.28	0.8997	1.74	0.9591	2.20	0.9861	2.66	0.9961	3.12	0.99910	3.58	0.99983
0.37	0.6443	0.83	0.7967	1.29	0.9015	1.75	0.9599	2.21	0.9865	2.67	0.9962	3.13	0.99913	3.59	0.99983
0.38	0.6480	0.84	0.7996	1.30	0.9032	1.76	0.9608	2.22	0.9868	2.68	0.9963	3.14	0.99916	3.60	0.99984
0.39	0.6517	0.85	0.8023	1.31	0.9049	1.77	0.9616	2.23	0.9871	2.69	0.9964	3.15	0.99918	3.61	0.99985
0.40	0.6554	0.86	0.8051	1.32	0.9066	1.78	0.9625	2.24	0.9875	2.70	0.9965	3.16	0.99921	3.62	0.99985
0.41	0.6591	0.87	0.8079	1.33	0.9082	1.79	0.9633	2.25	0.9878	2.71	0.9966	3.17	0.99924	3.63	0.99986
0.42	0.6628	0.88	0.8106	1.34	0.9099	1.80	0.9641	2.26	0.9881	2.72	0.9967	3.18	0.99926	3.64	0.99986
0.43	0.6664	0.89	0.8133	1.35	0.9115	1.81	0.9649	2.27	0.9884	2.73	0.9968	3.19	0.99929	3.65	0.99987
0.44	0.6700	0.90	0.8159	1.36	0.9131	1.82	0.9656	2.28	0.9887	2.74	0.9969	3.20	0.99931	3.66	0.99987
0.45	0.6736	0.91	0.8186	1.37	0.9147	1.83	0.9664	2.29	0.9890	2.75	0.9970	3.21	0.99934	3.67	0.99988

Bibliography

Mathematics ...

[1] Aigner, M., Ziegler, G.: *Proofs from the BOOK*, Springer, Berlin, 2018.

[2] Artmann, B.: *Lineare Algebra*, Birkhäuser, Basel, 1986.

[3] Barner, M., Flohr, F.: *Analysis I*, DeGruyter, Berlin, 1974.

[4] Bauer, H.: *Wahrscheinlichkeitstheorie und Grundzüge der Maßtheorie*, DeGruyter, Berlin, 1974.

[5] Beutelspacher, A.: *Lineare Algebra*, Vieweg, Wiesbaden 1994.

[6] Beutelspacher, A., Zschiegner, M.: *Diskrete Mathematik für Einsteiger*, Vieweg+Teubner, Wiesbaden 2011.

[7] Biggs, N.L.: *Discrete Mathematics*, Oxford University Press, Oxford, 2002.

[8] Blatter, C.: *Analysis I+II*, Springer, Berlin, 1974.

[9] Bosch, K.: *Elementare Einführung in die Statistik*, Vieweg, Wiesbaden, 1994.

[10] Bosch, K.: *Elementare Einführung in die Wahrscheinlichkeitstheorie*, Vieweg+Teubner, Wiesbaden, 2011.

[11] Bosch, K.: *Statistik für Nichtstatistiker*, Oldenbourg, Munich, 1994.

[12] Brill, M.: *Mathematik für Informatiker*, Hanser, Munich, 2001.

[13] Domschke, W., Drexl, A.: *Einführung in Operations Research*, Springer, Berlin, 2011.

[14] Dörfler, W., Peschek, W.: *Mathematik für Informatiker*, Hanser, Munich, 1988.

[15] Engel, A.: *Wahrscheinlichkeitsrechnung und Statistik, Vol. 2*, Klett, Stuttgart, 1976.

[16] Forster, O.: *Algorithmische Zahlentheorie*, Vieweg, Wiesbaden, 1996.

[17] Forster, O.: *Analysis I+II*, Vieweg+Teubner, Wiesbaden, 2011.

[18] Gathen, J. von zur: *CryptoSchool*, Springer, Berlin 2015.

[19] Greiner, M., Tinhofer, G.: *Stochastik für Studienanfänger der Informatik*, Hanser, Munich, 1996.

[20] Handl, A., Kuhlenkasper, T.: *Multivariate Analysemethoden*, Springer Spektrum, Berlin, 2017.

[21] Huppert, B., Willems, W.: *Lineare Algebra*, Vieweg+Teubner, Wiesbaden 2010.

[22] Knorrenschild, M.: *Numerische Mathematik, Eine beispielorientierte Einführung*, Hanser, Munich, 2010.

[23] Rosen, Kenneth H.: *Elementary Number Theory and Its Applications*, Pearson, Boston, 2005.

[24] Rießinger, T.: *Mathematik für Ingenieure*, Springer, Berlin, 1996.

[25] Schöning, U.: *Logik für Informatiker*, Spektrum Akademischer Verlag, Heidelberg, 2000.

© The Editor(s) (if applicable) and The Author(s), under exclusive license to Springer Fachmedien Wiesbaden GmbH, part of Springer Nature 2023
P. Hartmann, *Mathematics for Computer Scientists*,
https://doi.org/10.1007/978-3-658-40423-9

[26] Teschl, G.,Teschl, S.: *Mathematik für Informatiker Band I*, Springer, Berlin 2013.
[27] Waerden, B. L. van der: *Algebra I*, Springer, Berlin, 1971.
[28] Walter, W.: *Gewöhnliche Differentialgleichungen*, Springer, Berlin, 1976.
[29] Weller, F.: *Numerische Mathematik für Ingenieure und Naturwissenschaftler*, Vieweg, Wiesbaden, 1996.

... for computer scientists

[30] Beutelspacher, A.: *Kryptologie*, Vieweg+Teubner, Wiesbaden 2009.
[31] Dillmann, R., Huck, M.: *Informationsverarbeitung in der Robotik*, Springer, Berlin, 1991.
[32] Foley, J. D., Dam A. van, Feiner, S., Hughes J.: *Computer Graphics*, Addison-Wesley, Reading, 2000.
[33] Goldschlager, L., Lister, A.: *Informatik*, Hanser, München, 1990.
[34] Langville, A. N., Meyer, C. D.: *Google's PageRank and Beyond: The Science of Search Engine Rankings*, Princeton University Press, Princeton, 2012.
[35] Nissanke, N.: *Introductory Logic and Sets for Computer Scientists*, Pearson, Edinburgh, 1999.
[36] Ottmann, T., Widmann, P.: *Algorithmen und Datenstrukturen*, BI, 1990.
[37] Salomon, D.: *Data Compression*, Springer, New York, 1998.
[38] Schneier, B.: *Applied Cryptography*, John Wiley & Sons, Inc, 1995.
[39] Schulz, R.-H.: *Codierungstheorie*, Vieweg, Wiesbaden, 2003.
[40] Sedgewick, R.: *Algorithms in C++*, Addison-Wesley, Reading, 1992.
[41] Tanenbaum, A.S.: *Computer Networks*, Pearson, 2010.
[42] Werner, M.: *Information und Codierung*, Vieweg+Teubner, Wiesbaden, 2008.

... and something else to relax in the evening

[43] Bell, E. T.: *Men of Mathematics*, Simon and Schuster, 1937.
[44] Davis, P. J., Hersh, R.: *The Mathematical Experience*, Birkhauser, Boston, 1995.
[45] Dubben, H. H., Beck-Bornholdt, H. P.: *Mit an Wahrscheinlichkeit grenzender Sicherheit*, Rowohlt Taschenbuch Verlag, Hamburg, 2005.
[46] Fry, H.: *Hello World: Being Human in the Age of Algorithms*, W. W. Norton & Company, 2018.
[47] Guedj, D.: *The Parrot's Theorem*, Weidenfeld & Nicolson, 2000.
[48] Hofstadter, D. R.: *Gödel, Escher, Bach*, Basic Books, 1979.
[49] Singh, S.: *Fermat's Last Theorem*, HarperCollins Publishers, 2002.
[50] Singh, S.: *The Code Book*, Anchor, 2000.
[51] Singh, S.: *The Simpsons and Their Mathematical Secrets*, Bloomsbury USA, 2014.
[52] Wallwitz, G. von: *Meine Herren, dies ist keine Badeanstalt*, Berenberg, Berlin, 2017.

Printed in the United States
by Baker & Taylor Publisher Services